Curves and Surfaces in Geometric Design

Curves and Surfaces in Geometric Design

edited by

Pierre-Jean Laurent
Université Joseph Fourier
Grenoble, France

Alain Le Méhauté
Ecole Nationale Supérieure
Télécommunications
de Bretagne, France

Larry L. Schumaker
Vanderbilt University
Nashville, Tennessee, USA

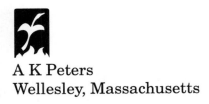

A K Peters
Wellesley, Massachusetts

Cat / Sep

Editorial, Sales, and Customer Service Office

A K Peters, Ltd.
289 Linden Street
Wellesley, MA 02181   *MATH*

Copyright © 1994 by A K Peters, Ltd.

Library of Congress Cataloging-in-Publication Data

Curves and surfaces in geometric design / [edited by] Pierre-Jean
    Laurent, Alain Le Méhauté, Larry L. Schumaker.
       p. cm.
    Papers from the Second International Conference on Curves and
    Surfaces, held in Chamonix-Mont-Blanc, France, June 10–16, 1993.
    Includes bibliographical references.
    ISBN 1-56881-039-3 :
       1. Approximation theory—Congresses. 2. Spline theory—
Congresses. 3. Surfaces—Computer simulation—Congresses.
4. Curves—Computer simulation—Congresses. I. Laurent, Pierre-
Jean, 1937 –   . II. Le Méhauté, Alain. III. Schumaker, Larry L.,
1939 –   . IV. Conference on Curves and Surfaces (2nd : 1993 :
Chamonix-Mont-Blanc, France)
QA221.C872 1994
516.3'52—dc20   94-11335
                                              CIP

Printed in the United States of America
98 97 96 95 94   10 9 8 7 6 5 4 3 2 1

CONTENTS

QA 221
C 872
1994
MATH

Preface ix

Contributors xi

Properties of Interpolating Means with Exponential-Type Weights
    G. Allasia and R. Besenghi 1

Approximating Reachable Sets by Extrapolation Methods
    R. Baier and F. Lempio 9

Unimodality of Quartic B-splines
    P. J. Barry 19

Object Oriented Spline Software
    R. Bartels 27

On Polynomial Functions Defining the Geometric Continuity
between Two (SBR) Surfaces
    J. L. Bauchat 35

A 3D Generalized Voronoi Diagram for a Set of Polyhedra
    E. Bertin and J.-M. Chassery 43

Variational Design of Rational Bézier Curves and Surfaces
    G. P. Bonneau and H. Hagen 51

An Introduction to Padé Approximations
    C. Brezinski 59

Discrete Curves and Curvature Constraints
    H. G. Burchard 67

A Necessary and Sufficent Condition for the C^k Continuity
of Triangular Rational Surfaces
    J. C. Canonne 75

Monotonicity Preserving Representations
    J. M. Carnicer and J. M. Peña 83

Splines Focales
    P. de Faget de Casteljau 91

A Surface-Surface Intersection Algorithm with a Fast
Clipping Technique
    M. Daniel and A. Nicolas 105

Zonoidal Surfaces
    O. Daoudi, B. Lacolle, N. Szafran, and P. Valentin 113

Spline Conversion: Existing Solutions and Open Problems
    T. Dokken and T. Lyche 121

A Stepwise Algorithm for Converting B-Splines
  M. Eck and J. Hadenfeld 131

Best Constrained Approximations of Planar Curves by Bézier Curves
  E. F. Eisele 139

Projective Blossoms and Derivatives
  G. Farin 147

Characterizations of the Set of Rational Parametric Curves with
Rational Offsets
  J. C. Fiorot and Th. Gensane 153

A Necessary and Sufficient Condition for Joining B-Rational Curves
with Geometric Continuity G^3
  J. C. Fiorot and P. Jeannin 161

Generalizations of Bézier Curves and Surfaces
  I. Gânscǎ, Gh. Coman, and L. Ţâmbulea 169

Corner Cutting Algorithms and Totally Positive Matrices
  M. Gasca and J. M. Peña 177

Piecewise Polynomial Approximation of Spheres
  S. Glærum 185

Non-polynomial Polar Forms
  D. Gonsor and M. Neamtu 193

Curvature of Rational Quadratics Splines
  T. N. T. Goodman 201

B-Spline Knot-Line Elimination and Bézier Continuity Conditions
  R. Gormaz 209

Applications of Constrained Polynomials to Curve and
Surface Approximation
  R. J. Goult 217

Semi-regular B-spline Surfaces: Generalized Lofting by B-splines
  J. Gravesen 225

On Best Convex Interpolation of Curves
  C. Henninger and K. Scherer 233

Approximate Conversion and Data Compression of Integral and
Rational B-spline Surfaces
  J. Hoschek and F.-J. Schneider 241

A Geometrical Approach to Interpolation on Quadric Surfaces
  B. Jüttler and R. Dietz 251

Finding Shortest Paths on Surfaces
  R. Kimmel, A. Amir and A. M. Bruckstein 259

Polygonalization of Algebraic Surfaces
R. Klein 269

A Knot Removal Strategy for Spline Curves
J. C. Koua Brou 277

Computation of Curvatures Related to Surface-Surface Blending
I. Lef 285

Least-Squares Optimization of Thread Surfaces
M. Léger, J.-M. Morvan, and M. Thibaut 293

A Multivariate Generalization of the de Boor-Fix Formula
S. Lodha and R. Goldman 301

A Metric for Parametric Approximation
T. Lyche and K. Mørken 311

Mathematical Modelling of Free-Form Curves and Surfaces
from Discrete Points with NURBS
W. Ma and J.-P. Kruth 319

Evaluating Surface Intersections in Lower Dimensions
D. Manocha, A. Varshney, and H. Weber 327

The Iterative Solution of a Nonlinear Inverse Problem from Industry:
Design of Reflectors
A. Neubauer 335

Splines with Prescribed Modified Moments
H. J. Oberle and G. Opfer 343

G^2-Continuous Cubic Algebraic Splines and Their Efficient Display
M. Paluszny and R. R. Patterson 353

Splines in a Topological Setting
S. Pesco and G. Tavares 361

A Characterization of Connecting Maps as Nonlinear Roots
of the Identity
J. Peters 369

Applications of the Dual Bézier Representation
of Rational Curves and Surfaces
H. Pottmann 377

Bifurcation Phenomenon in a Tool Path Computation
J. F. Rameau 385

Interpolation with an Arc Length Constraint
J. A. Roulier and B. Piper 393

Curves Reconstruction
J.-C. Roux 401

The Ubiquitous Ellipse
G. Sapiro and A. M. Bruckstein 409

Axial Convexity: a Well-shaped Shape Property
T. Sauer 419

Variation Diminution and Blossoming for Curves and Surfaces
G. Schmeltz 427

Approximation with Helix Splines
G. Seemann 435

Simplex Splines Support Surprisingly Strong Symmetric Structures
and Subdivision
H.-P. Seidel and A.H. Vermeulen 443

An Object-Oriented Framework for Curves and Surfaces with Applications
Ph. Slusallek, R. Klein, A. Kolb, and G. Greiner 457

Designing of a Progressive Lens
M. Tazeroualti 467

Multiplication as a General Operation for Splines
K. Ueda 475

Symmetric TB–spline Schemes
M. G. Wagner and H. Pottmann 483

PREFACE

During the week of June 10–16, 1993, an International Conference on *Curves and Surfaces* was held in Chamonix-Mont-Blanc (France). It was organized by the *Association Française d'Approximation* (A. F. A.). The organizing committee consisted of D. Apprato (Pau), M. Attéia (Toulouse), J.-M. Chassery (Grenoble), P. Chenin (Grenoble), J.-P. Dedieu (Toulouse), J.-C. Fiorot (Valenciennes), J. Gaches (Toulouse), B. Lacolle (Grenoble), Y. Lafranche (Rennes), P.-J. Laurent (Grenoble), A. Le Méhauté (Brest), M.-L. Mazure (Grenoble), C. Rabut (Toulouse), P. Sablonnière (Rennes), L. L. Schumaker (Nashville, USA), and C. Vercken (Paris).

The Conference was attended by 260 mathematicians from 22 different countries, and the program included 10 invited one-hour lectures and 140 research talks. The survey lectures dealt with several particularly active subareas of *Approximation Theory*, including *approximation of images, computer-aided geometric design, constrained approximation, finite elements, multivariate splines, Padé approximation, subdivision, surface compression,* and *wavelets*. A number of the research talks were presented in minisymposia on *Spline Conversion, Software Infrastructure for CAGD, Rational Approximation, Wavelets,* and *Constrained Approximation,* organized by J. Hoschek, R. Bartels, J.-C. Fiorot, A. Cohen, and F. Utreras, respectively.

The proceedings of this conference consist of a total of 106 papers. This volume contains 58 papers relating to Computer Aided Geometric Design. The volume *Wavelets, Images, and Surface Fitting,* also published by A K PETERS, contains 48 papers.

We would like to thank the following institutions and companies for their financial support: Universités J. Fourier, Grenoble and P. Sabatier, Toulouse, and Vanderbilt University, Nashville; Inst. Nat. Sciences Appliquées de Toulouse; Institut IMAG; Laboratoire de Modélisation et de Calcul (Grenoble); Centre National de la Recherche Scientifique (C.N.R.S.); Direction des Recherches et Etudes Doctorales (DRED); Direction des Recherches et Etudes Techniques (DRET, Contract 93-1110/A000/DRET/DS/SR); US Air Force European Office of Aerospace Research and Development (EOARD); US Army Research, Development and Standardization Group, UK (USARDSG); US Navy, Office of Naval Research, European Office; and Elf-Aquitaine.

Local organization of the Conference was carried out by the LMC-IMAG Institute (Grenoble). We would like to thank D. Iglesias and his colleagues at the printing service. In addition to the members of the organizing committee, several of our colleagues and students helped with the organization of the Conference, including O. Daoudi, M. Duc-Jacquet, and A. Eberhard. Finally, we want to express our special thanks to our wives Guizou Laurent, Michelle Le Méhauté, and Gerda Schumaker.

Nashville, Tennessee   February 28, 1994

CONTRIBUTORS

Numbers in parentheses indicate pages on which authors' contributions begin.

GIAMPIETRO ALLASIA (1), *Dpt. di Matematica, Università di Torino, Via Carlo Alberto 10, 10123 Torino, ITALY* [allasia@dm.unito.it]

ARNON AMIR (259), *Dept. of Computer Science, Technion – Israel Institute of Technology, Haifa 32000, ISRAEL* [arnon@csc.cs.technion.ac.il]

ROBERT BAIER (9), *Chair of Applied Mathematics, University of Bayreuth, D-95440 Bayreuth, GERMANY*

PHILLIP J. BARRY (19), *Computer Science Dept., Univ. of Minnesota, 4-192 EE/CSci Bldg., 200 Union St. SE, Minneapolis, MN 55455* [barry@mail.cs.umn.edu]

RICHARD H. BARTELS (27), *Computer Science Department, University of Waterloo, Waterloo, Ontario N2L 3G1, CANADA* [rhbartel@watcgl.uwaterloo.ca]

JEAN-LUC BAUCHAT (35), *LCRMAO, ENSAM, 8 Bd Louis XIV, 59046 Lille, FRANCE* [bauchat@ensam.decnet.citilille.fr]

ETIENNE BERTIN (43), *TIMC-IMAG, University Joseph Fourier, BP 53, 38041 Grenoble cedex 9, FRANCE* [Etienne.Bertin@imag.fr]

RENATA BESENGHI (1), *Dpt. di Matematica, Università di Torino, Via Carlo Alberto 10, 10123 Torino, ITALY* [besenghi@dm.unito.it]

GEORGES-PIERRE BONNEAU (51), *Universität Kaiserslautern, Fachbereich Informatik, P. O. 3049, 67653 Kaiserlautern, GERMANY* [bonneau@informatik.uni-kl.de]

CLAUDE BREZINSKI (59), *Laboratoire d'Analyse Numérique et d'Optimisation, U.F.R. I.E.E.A. - M3, Université des Sciences et Technologies de Lille, 59655–Villeneuve d'Ascq cedex, FRANCE* [brezinsk@citil.citille.fr]

ALFRED M. BRUCKSTEIN (259, 409), *Dept. of Computer Science, Technion – Israel Institute of Technology, Haifa 32000, ISRAEL* [freddy@cs.technion.ac.il]

HERMANN G. BURCHARD (67), *Dept. of Mathematics, Oklahoma State University, Stillwater, OK 74078-0613* [burchard@math.okstate.edu]

J. C. CANONNE (75), *LIMAV, I. U. T. G. E. I. I., Univ. de Valenciennes, Le Mont Houy, 59326 Valenciennes cedex, FRANCE* [canonne@univ-valenciennes.fr]

J. M. CARNICER (83), *Departamento de Matemática Aplicada, Universidad de Zaragoza, Edificio de Matemáticas, Planta 1a, 50009 Zaragoza, SPAIN* [carnicer@cc.unizar.es]

PAUL DE FAGET DE CASTELJAU (91), *4 Avenue du Commerce, 78000 Versailles, FRANCE*

J.-M. CHASSERY (43), *TIMC-IMAG, University Joseph Fourier, BP 53, F-38041 Grenoble cedex 9, FRANCE* [chassery@imag.fr]

GH. COMAN (169), *Dept. of Mathematics, Technical University, Cluj-Napoca, ROUMANIA*

MARC DANIEL (105), *Ecole Centrale de Nantes, I.R.I.N., 1 rue de la Noë, 44072 Nantes cedex 03, FRANCE* [Marc.Daniel@ec-nantes.fr]

O. DAOUDI (113), *Laboratoire LMC-IMAG, BP 53, F38041 Grenoble cedex 9, FRANCE* [daoudi@busard.imag.fr]

ROLAND DIETZ (251), *Technische Hochschule Darmstadt, Fachbereich Mathematik, AG Differentialgeometrie und Kinematik, Schlossgartenstr. 7, D – 64289 Darmstadt, GERMANY* [rdietz@mathematik.th-darmstadt.de]

TOR DOKKEN (121), *SINTEF SI, P. O. Box 124, Blindern, 0314 Oslo, NORWAY* [tor.dokken@si.sintef.no]

MATTHIAS ECK (131), *Department of Mathematics, University of Technology and Science, Schlossgartenstrasse 7, D-64289 Darmstadt, GERMANY* [eck@mathematik.th-darmstadt.de]

EBERHARD F. EISELE (139), *Mathematisches Institut B, Univ. Stuttgart, 70550 Stuttgart, GERMANY* [e.eisele@rus.uni-stuttgart.de]

GERALD FARIN (147), *Computer Science, Arizona State University, Tempe, AZ 85287-5406* [farin@asu.edu]

JEAN-CHARLES FIOROT (153, 161), *Université de Valenciennes et du Hainaut-Cambraisis, ENSIMEV, Laboratoire IMAV B.P. 311, F-59304 Valenciennes - cedex, FRANCE*

IOAN GÂNSCĂ (169), *Dept. of Mathematics, Technical University, Cluj-Napoca, ROUMANIA*

MARIANO GASCA (177), *Departamento de Matemática Aplicada, Universidad de Zaragoza, 50009 Zaragoza, SPAIN* [gasca@cc.unizar.es]

THIERRY GENSANE (153), *Université des Sciences et Technologies de Lille, Laboratoire d'Analyse Numérique et d'Optimisation, UFR IEEA - M3, 59655 Villeneuve d'Ascq - cedex, FRANCE*

SIGURD GLÆRUM (185), *Universitetet i Oslo, Institutt for Informatikk, Box 1080 Blindern, N-0316 Oslo, NORWAY* [sigurdg@ifi.uio.no]

RON GOLDMAN (301), *Department of Computer Science, Rice University, Houston, TX 77251-1892* [rng@cs.rice.edu]

DAN E. GONSOR (193), *Kent State University, Department of Mathematics and Computer Science, Kent, OH 44242* [dgonsor@mcs.kent.edu]

T. N. T. GOODMAN (201), *Dept. of Mathematics and Computer Science, The University, Dundee DD1 4HN, SCOTLAND* [tgoodman@mcs.dundee.ac.uk]

RAÚL GORMAZ (209), *Universidad de Chile, Facultad de Ciencias Físicas y Matemáticas, Departamento de Ingeniería Matemática, Casilla 170/3, Correo 3, Santiago, CHILE* [rgormaz@uchcecvm.cec.uchile.cl]

R. J. GOULT (217), *LMR Systems, 33 Filgrave, Newport Pagnell, Bucks, MK 16 9ET, ENGLAND*

J. GRAVESEN (225), *Mathematical Institute, The Technical University of Denmark, Building 303, DK-2800 Lyngby, DENMARK* [gravesen@mat.dth.dk]

G. GREINER (457), *Universität Erlangen, IMMD IX- Graphische Datenverarbeitung, Am Weichselgarten 9, D-91058 Erlangen, GERMANY* [greiner@informatik.uni-erlangen.de]

JAN HADENFELD (131), *Department of Mathematics, University of Technology and Science, Schlossgartenstrasse 7, D-64289 Darmstadt, GERMANY* [hadenfeld@mathematik.th-darmstadt.de]

HANS HAGEN (51), *Universität Kaiserslautern, Fachbereich Informatik, P. O. 3049, 67653 Kaiserlautern, GERMANY* [hagen@informatik.uni-kl.de]

CHRISTOPH HENNINGER (233), *Institut für Angewandte Mathematik, Wegelerstr. 6, D – 53115 Bonn, GERMANY*

J. HOSCHEK (241), *University of Technology, Dept. of Mathematics, Schlossgartenstr. 7, 64289 Darmstadt, GERMANY* [hoschek@mathematik.th-darmstadt.de]

PIERRE JEANNIN (161), *Département de Mathématiques Appliquées, Université du Littoral, Bâtiment Henri Poincaré, B.P. 699, 62228 Calais cedex, FRANCE*

BERT JÜTTLER (251), *Technische Hochschule Darmstadt, Fachbereich Mathematik, AG Differentialgeometrie und Kinematik, Schlossgartenstr. 7, D – 64289 Darmstadt, GERMANY* [juettler@mathematik.th-darmstadt.de]

RON KIMMEL (259), *Dept. of Electrical Engineering, Technion – Israel Institute of Technology, Haifa 32000, ISRAEL* [ron@techunix.technion.ac.il]

REINHARD KLEIN (269, 457), *Universität Tübingen, WSI/GRIS, Auf der Morgenstelle 10, D-72076 Tübingen, GERMANY* [reinhard@gris.informatik.uni-tuebingen.de]

A. KOLB (457), *Universiät Erlangen, IMMD IX- Graphische Datenverarbeitung, Am Weichselgarten 9, D-91058 Erlangen, GERMANY* [kolb@informatik.uni-erlangen.de]

JEAN CLAUDE KOUA BROU (277), *Département Mathématiques et Systèmes de Communication, Ecole Nationale des Télécommunications de Bretagne, B. P. 832, 29285 Brest cedex, FRANCE* [Koua_brou@Gti.Enst-Bretagne.fr]

J. P. KRUTH (319), *Katholieke Universiteit Leuven, Celestijnenlaan 300B, B-3001 Heverlee, BELGIUM* [kruth@mech.kuleuven.ac.be]

BERNARD LACOLLE (113), *Laboratoire LMC-IMAG, BP 53, F38041 Grenoble cedex 9, FRANCE* [lacolle@imag.fr]

ISAAC LEF (285), *Applicon, Inc., Billerica, MA 01821* [lef@billerica.applicon.slb.com]

MICHEL LÉGER (293), *Institut Français du Pétrole, 1-4 avenue de Bois Préau, 92500 Rueil Malmaison, FRANCE* [legerm@irsun1.ifp.fr]

FRANK LEMPIO (9), *Chair of Applied Mathematics, University of Bayreuth, D-95440 Bayreuth, GERMANY*

SURESH LODHA (301), *Computer and Information Sciences, University of California, Santa Cruz, CA 95064* [lodha@cse.ucsc.edu]

TOM LYCHE (121, 311), *Institutt for Informatikk, P. O. Box 1080, Blindern, 0316 Oslo, NORWAY* [tom@ifi.uio.no]

W. MA (319), *Katholieke Universiteit Leuven, Celestijnenlaan 300B, B-3001 Heverlee, BELGIUM* [weiyin@mech.kuleuven.ac.be]

DINESH MANOCHA (327), *CB #3175, Sitterson Hall, University of North Carolina at Chapel Hill, Chapel Hill, NC 27599-3175* [manocha@cs.unc.edu]

KNUDT MØRKEN (311), *Dept. of Informatics, University of Oslo, P. B. 1080, Blindern, 0316 Oslo, NORWAY* [knutm@ifi.uio.no]

JEAN-MARIE MORVAN (293), *Université Claude Bernard, Inst. Math. Info., 43 Blv. du 11 nov. 1918, 69622 Villeurbanne cedex, FRANCE*

MARIAN NEAMTU (193), *Vanderbilt University, Department of Mathematics, Nashville, TN 37240* [neamtu@athena.cas.vanderbilt.edu]

ANDREAS NEUBAUER (335), *Institut für Mathematik, Universität Linz, A–4040 Linz, AUSTRIA* [neubauer@indmath.uni-linz.ac.at]

ALAIN NICOLAS (105), *Ecole Centrale de Nantes, I.R.I.N., 1 rue de la Noë, 44072 Nantes cedex 03, FRANCE* [Alain.Nicolas@ec-nantes.fr]

HANS JOACHIM OBERLE (343), *Universität Hamburg, Institut für Angewandte Mathematik, Bundesstraße 55, D-20146 Hamburg, GERMANY* [opfer@math.uni-hamburg.de]

GERHARD OPFER (343), *Universität Hamburg, Institut für Angewandte Mathematik, Bundesstraße 55, D-20146 Hamburg, GERMANY* [opfer@math.uni-hamburg.de]

M. PALUSZNY (353), *Escuela de Física y Matemáticas, Universidad Central de Venezuela, Apartado 47809, Los Chaguaramos, Caracas 1041-A, VENEZUELA* [paluszny@dino.conicit.ve]

R. R. PATTERSON (353), *Department of Mathematics, Indiana University – Purdue University at Indianapolis, Indianapolis, IN 46202-3216* [rpatters@indyvax.iupui.edu]

J. M. PEÑA (83, 177), *Departamento de Matemática Aplicada, Universidad de Zaragoza, Edificio de Matemáticas, Planta 1a, 50009 Zaragoza, SPAIN* [jmpena@cc.unizar.es]

SINÉSIO PESCO (361), *Departamento de Matemática, Pontifícia Universidade Católica (PUC-Rio), R. Marquês de São Vicente, 225, 22453-900 Rio de Janeiro-RJ, BRASIL* [sinesio@mat.puc-rio.br]

JÖRG PETERS (369), *Department of Computer Science, Purdue University, W-Lafayette, IN 47907-1398* [jorg@cs.purdue.edu]

BRUCE PIPER (393), *Department of Mathematics, Rensselaer Polytechnic Institute, Troy, NY 12180-3590* [piperb@rpi.edu]

HELMUT POTTMANN (377, 483), *Inst. f. Geometrie, TU Wien, Wiedner Hauptstrasse 8–10, A 1040 Wien, AUSTRIA* [pottmann@EGMVS2.una.ac.at]

JEAN F. RAMEAU (385), *Dassault Systemes, 24-28, Av du Gal de Gaulle, BP - 310, 92156 Suresnes cedex, FRANCE*

JOHN A. ROULIER (393), *Department of Computer Science and Engineering U-155, University of Connecticut, Storrs, CT 06269-3155* [jrou@brc.uconn.edu]

JEAN-CHRISTOPHE ROUX (401), *Laboratoire LMC-IMAG, Université Joseph Fourier, BP 53X, 38041 GRENOBLE Cedex FRANCE* [roux@imag.fr]

GUILLERMO SAPIRO (409), *Technion-Israel Institute of Technology, Haifa, ISRAEL 32000* [guille@techunix.technion.ac.il]

THOMAS SAUER (419), *Mathematical Institute, University of Erlangen-Nuremberg, Bismarckstr. 1½, 91054 Erlangen, GERMANY* [sauer@mi.uni-erlangen.de]

KARL SCHERER (233), *Institut für Angewandte Mathematik, Wegelerstr. 6, D – 53115 Bonn, GERMANY* [unm11c@ibm.rhrz.uni-Bonn.de]

GERD SCHMELTZ (427), *Institute of Computer Science, Hebrew University, Givat Ram, Ross Building, 91904 Jerusalem, ISRAEL* [schmeltz@cs.huji.ac.il]

F.-J. SCHNEIDER (241), *University of Technology, Dept. of Mathematics, Schlossgartenstr. 7, 64289 Darmstadt, GERMANY*

GERALD SEEMANN (435), *Fachbereich Mathematik, Technische Hochschule Darmstadt, Schlossgartenstrasse 7, D–64289 Darmstadt, GERMANY* [seemann@mathematik.th-darmstadt.de]

HANS-PETER SEIDEL (443), *Universität Erlangen, Graphische Datenverarbeitung, Am Weichselgarten 9, D-91058 Erlangen, GERMANY* [seidel@informatik.uni-erlangen.de]

PHILIPP SLUSALLEK (457), *Universität Erlangen, IMMD IX- Graphische Datenverarbeitung, Am Weichselgarten 9, D-91058 Erlangen, GERMANY* [slusallek@informatik.uni-erlangen.de]

N. SZAFRAN (113), *Laboratoire LMC-IMAG, BP 53 X, F-38041 Grenoble cedex, FRANCE*

L. ŢÂMBULEA (169), *Dept. of Mathematics, Technical University, Cluj-Napoca, ROUMANIA*

GEOVAN TAVARES (361), *Departamento de Matemática, Pontifícia Universidade Católica (PUC-Rio), R. Marquês de São Vicente, 225, 22453-900 Rio de Janeiro-RJ, BRASIL* [geovan@mat.puc-rio.br]

MOHAMMED TAZEROUALTI (467), *LMC-IMAG, Université Joseph Fourier, BP 53X, 38041 Grenoble, FRANCE* [tazer@busard.imag.fr]

MURIEL THIBAUT (293), *Université de Grenoble, IRIGM, BP 68, 38402 St. Martin d'Hères, FRANCE* [thibaut@irsun1.ifp.fr]

KENJI UEDA (475), *Ricoh Company, Ltd., 1-1-17, Koishikawa, Bunkyo-ku, Tokyo, 112, JAPAN* [ueda@src.ricoh.co.jp]

PATRICK VALENTIN (113), *ELF-ANTAR-FRANCE, Centre de Recherche ELF-Solaize, BP 22, F69360 Saint Symphorien d'Ozon, FRANCE* [elfcnrs@cism.univ-lyon1.fr]

A. VARSHNEY (327), *CB #3175, Sitterson Hall, University of North Carolina at Chapel Hill, Chapel Hill, NC 27599-3175*

AL VERMEULEN (443), *Rogue Wave Software, P. O. Box 2328, Corvallis, OR 97339* [alv@roguewave.com]

MICHAEL G. WAGNER (483), *Institute of Geometry, Technical University of Vienna, AUSTRIA* [wagner@egmvs2.una.ac.at]

H. WEBER (327), *CB #3175, Sitterson Hall, University of North Carolina at Chapel Hill, Chapel Hill, NC 27599-3175*

Properties of Interpolating Means with Exponential-Type Weights

G. Allasia and R. Besenghi

Abstract. Given a set S_n of n points, irregularly distributed on an interval $[a, b] \subset \mathbb{R}$, and a function $f \in C^r[a, b]$ whose Taylor expansion at each point of S_n is known, we consider an approximant μ of f, expressed as a weighted arithmetic mean of the Taylor expansions using exponential-type weights. We give some properties of the positive linear operator μ.

§1. Introduction.

Given a set $S_n \subset [a, b] \subset \mathbb{R}$ of n irregularly-spaced points, we consider the interpolation of a function $f \in C^r[a, b]$ by the weighted arithmetic mean μ given by (2.2), that is a positive linear functional. In order to increase the degree of approximation of μ to f, we use in the operator μ, instead of the value $f(x_k)$ of the interpolated function, the truncated Taylor expansion T_k of f up to the derivative of order r at the point x_k, given by (2.1), for $k = 1, 2, \ldots, n$. The weight w_k which multiplies the Taylor expansion T_k is given by (2.4). The weights satisfy the properties (2.3), depend on the data points and are approximately exponentially decreasing with distance; they are interesting, both theoretically and practically, and include noteworthy particular cases.

In the second section we prove that the derivatives $D^m w_k$, $(m = 1, ..., q)$, of the weights vanish at the interpolation nodes. As a consequence, we find that the derivatives $D^m \mu$ of the interpolating operator and the corresponding derivatives $D^m f$ of the interpolated function are equal at the nodes.

In the third section we examine the degree of approximation of μ to f; precisely, we give upper bounds for the differences $|D\mu - Df|$ using the modulus of continuity of $f^{(r)}$. The upper bounds proposed are not the sharpest in all cases, because our aim was mainly to obtain as simple and general a representation of them as possible; moreover, we considered the problem from a viewpoint midway between the numerical calculation and the approximation theory. Previous results, in various ways related to ours, have been obtained by a few authors, see [7,8] and references therein.

Curves and Surfaces in Geometric Design
P. J. Laurent, A. Le Méhauté, and L. L. Schumaker (eds.), pp. 1–8.
Copyright © 1994 by A K PETERS, Wellesley, MA
ISBN 1-56881-039-3.

1

The remarkable problems of the extension to the multivariate case of the operator μ and its numerical behaviour are left to a forthcoming paper [3]. The convergence problem for $n \to \infty$ of the multivariate extension of the operator μ can be reduced to the convergence in one dimension [5,3]. This is why we want manageable conditions for the convergence of μ in one dimension. Here, we limit ourselves to a study of this operator.

i) The operator μ not only interpolates to positional information but allows the interpolation to specified derivatives at the scattered points. However, the method does not necessarily require higher order derivatives in practice, but we have the option of supplying more general interpolation data available to the user. For example, instead of the Taylor expansion of the function at a nodal point x_k, we can consider any other local approximation to f at x_k built up by using functional values only, that is the values of the function at the point x_k and at the nearest points. To this end, an interpolation formula or a constrained quadratic least-squares scheme can work usefully [9].

ii) The use of exponential-type weights is suggested, e.g., by McLain [10]; considering the Shepard formula, he observes that much more accurate results can be obtained using weights rapidly decreasing with distance.

iii) The use of exponential-type weights increases the central processor evaluation time but usually this drawback can be tolerated. On the other hand, the use of such weights has the practical advantage that, if n is large, say over 500, we can divide the region into subregions and, at a slight cost in programming effort assign zero weight to the data points in remote subregions [10].

iv) In evaluating the operator μ, a substantial reduction of computer time and storage requirements can be realized by means of a suitable parallel algorithm. To achieve this aim, we partition the set S_n into p subset S_{n_j} so that the jth subset, $(j = 1, 2, ..., p)$, consists of the points $x_{j_1}, x_{j_2}, \ldots, x_{j_{n_j}}$, with $n_1 + n_2 + \cdots + n_p = n$. Then the operator μ can be expressed in the form

$$(1.1) \qquad \mu(x; T, S_n) = \sum_{j=1}^{p} \mu(x; T, S_{n_j}) \frac{A_j}{\sum_{j=1}^{p} A_j},$$

where

$$(1.2) \qquad A_j = \sum_{k_j=1}^{n_j} |x - x_{j_{k_j}}|^{-q} \exp(-\beta |x - x_{j_{k_j}}|),$$

and A_j is stored during the calculation of $\mu(x; T, S_{n_j})$. This formula is particularly suited to the calculation of μ by a SIMD machine [1,2].

v) At first glance, the operator (2.2) has a serious drawback. As for the Lagrange interpolation formula, if we want to add a new point to the given set of data, then we must repeat the whole computation to find the numerical value of μ. But this difficulty can be avoided. In order to do so, we partition the given set of points S_n in two subsets only, say $S_1 = \{x_{1_1}, x_{1_2}, \ldots, x_{1_{(n-1)}}\}$ and $S_2 = \{x_n\}$. The corresponding forms of (1.1) and (1.2) are

$$\mu(x; T, S_n) = \frac{\mu(x; T, S_1)A_1 + T_n w_n}{A_1 + w_n},$$

where

$$A_1 = \sum_{j=1}^{n-1} |x - x_j|^{-q} \exp(-\beta|x - x_j|).$$

These give a recurrence relation to compute μ. In other words, we get a form of μ which has the property that an additional point may be added to the interpolation set by simply combining an extra term with the original formula.

§2. Interpolating Means with Exponential-type Weights

Let us consider a set of $n \in \mathbb{N}$ distinct points $S_n = \{x_1, x_2, .., x_n\}$ in the interval $[a, b] \subset \mathbb{R}$, and a function $f \in C^r[a, b]$ whose truncated Taylor expansion up to rth derivative at the point $x_k \in S_n, (k = 1, 2, .., n)$,

$$T_k \equiv T_k(x; f, r) = \sum_{j=0}^{r} \frac{f^{(j)}(x_k)}{j!} (x - x_k)^j \qquad (2.1)$$

is known. Then we form the following weighted arithmetic mean μ relating to the function f and to the set S_n

$$\mu \equiv \mu(x; T, S_n) = \sum_{k=1}^{n} T_k(x; f, r) \, w_k(x; S_n), \qquad (2.2)$$

where the weights satisfy the usual properties

$$w_k(x; S_n) \geq 0, \quad \sum_{k=1}^{n} w_k(x; S_n) = 1. \qquad (2.3)$$

The weights

$$w_k \equiv w_k(x, S_n) = \frac{|x - x_k|^{-q} \, exp(-\beta|x - x_k|)}{\sum_{h=1}^{n} |x - x_h|^{-q} \, exp(-\beta|x - x_h|)} \qquad (2.4)$$

(which are approximately exponentially decreasing with distance, so we refer to them as "exponential-type weights") with q and β nonnegative real numbers, are of particular interest, both theoretically and practically [10,9 p.186]. In particular, for $\beta = 0$, we get the well-known weights attributed to Shepard [9, pp. 185-186]; while for $q = 0$ we get purely exponential weights. Later on we will sometimes write $q = r + p$, where r has the same meaning as in (2.1).

The weights (2.4) may also be written in the "product form"

$$w_k(x; S_n) = \frac{\prod_{h=1, h \neq k}^{n} |x - x_h|^q \ exp(\beta|x - x_h|)}{\sum_{i=1}^{n} \prod_{l=1, l \neq i}^{n} |x - x_l|^q exp(\beta|x - x_l|)} \tag{2.5}$$

which is useful mainly for theoretical considerations. They satisfy

$$w_k(x_k; S_n) := \lim_{x \to x_k} w_k(x; S_n) = 1, \quad w_k(x_h; S_n) = \delta_{hk}, (h, k = 1, 2, ..., n),$$

where δ_{hk} is the Kronecker operator. Since

$$\mu(x_k; T, S_n) = f(x_k),$$

the mean μ is interpolating. In particular, if we set

$$f(x) = x^j, \ (j = 0, 1, ..., r),$$

then we have

$$\mu(x; x^j, S_n) = \sum_{k=1}^{n} w_k(x; S_n) \, x^j = x^j.$$

Hence the interpolating mean of a polynomial of degree at most r (which does not depend on index k) is the polynomial itself.

The following theorem is proved in the Appendix:

Theorem 2.1. *Let m be an integer such that $0 \leq m < q$. Then*

$$D^m w_k(x_s; S_n) = \begin{cases} \delta_{ks}, & m = 0, \\ 0, & 0 < m < q, \end{cases} \tag{2.6}$$

for $k, s = 1, 2, \ldots, n$.

Theorem 2.2. *For $s = 1, 2, .., n$*

$$D^i \mu(x_s; T; S_n) = \begin{cases} D^i f(x_s), & (i = 0, 1, ..r), \\ 0, & (i = r + 1, .., [r + p]), \end{cases} \tag{2.7}$$

where $[r + p]$ is the largest integer value of $r + p$.

Proof: From (2.2) by differentiation

$$D^i \mu(x; T, S_n) = \sum_{k=1}^{n} \sum_{j=0}^{i} \binom{i}{j} D^j w_k(x; S_n) D^{i-j} T_k(x; f, r).$$

Then by Theorem 2.1

$$D^i \mu(x_s; T, S_n) = \sum_{k=1}^{n} \delta_{ks} D^i f(x_s) = D^i f(x_s) \quad (i = 0, 1, \ldots, r);$$

which is the first part of (2.7). The remainder of the proof is immediate. ∎

§3. Rate of Convergence

We now give an estimate of the degree of approximation of the interpolating mean function $\mu(x; T, S_n)$ with exponential-type weights (2.4) to the interpolated function $f \in C^r[a, b]$, when n tends to infinity and x is any fixed value.

Given a set S_n of n points $a \leq x_1 < x_2 < \cdots < x_n \leq b$, put

$$D_{S_n} = \max_k(x_{k+1} - x_k), \qquad d_{S_n} = \min_k(x_{k+1} - x_k) \qquad (3.1)$$

for $k = 1, 2, .., n-1$, and

$$K_{S_n} = \frac{D_{S_n}}{d_{S_n}}. \qquad (3.2)$$

Moreover, let $x_1 - a \leq D_{S_n}/2$ and $b - x_n \leq D_{S_n}/2$. The following result holds true.

Theorem 3.1. *Let $f \in C^r[a, b]$, and suppose $\beta, p \in \mathbb{R}$ are nonnegative, and $r \in \mathbb{N}$. Then*

$$\left| f(x) - \mu(x; T, S_n) \right| < \frac{(1+c)c^r}{r!} \frac{1}{(n+1)^r} K_{S_n}^{q+\epsilon} \times$$
$$\times \left[1 + 3s \, exp\left(c\beta K_{S_n} \frac{1}{n+1} \right) \right] \omega\left(f^{(r)}; \frac{1}{n+1} \right), \qquad (3.3)$$

where $c = (b - a)/2$,

$$s = s(\beta, p, n) = \sum_{l=1}^{n-1} \frac{2}{(2l-1)^{p-1} exp\left(\beta d_{S_n} \frac{2l-1}{2} \right)} \qquad (3.4)$$

$$\epsilon = \epsilon(\beta, r, p) = \begin{cases} 0, & \text{for } r = 0 \text{ or } \beta = 0, r \geq 1, p = 1, \\ 1, & \text{otherwise.} \end{cases}$$

The proof is very complicated and is sketched in [3]. The sum s in (3.4) converges when $n \to +\infty$ for $\beta > 0$ or $\beta = 0$ and $p > 2$. It diverges in the other cases. In the case of convergence, we can rewrite (3.3) in the following more compact form.

Corollary 3.2. *If $s(\beta, p, n)$ is bounded for $n \to +\infty$ and $K_{S_n} = o(n+1)$ at least for $\beta > 0$, then*

$$\left| f(x) - \mu(x; T, S_n) \right| < const \, \frac{1}{(n+1)^r} K_{S_n}^{q+\epsilon} \, \omega\left(f^{(r)}; \frac{1}{n+1} \right), \qquad (3.5)$$

and, moreover, if $f^{(r)} \in Lip_M\alpha$

$$\left| f(x) - \mu(x; T, S_n) \right| < const \, \frac{1}{(n+1)^{r+\alpha}} K_{S_n}^{q+\epsilon}. \qquad (3.6)$$

On the other hand, when s diverges, it is also possible to write inequalities analogous to (3.5) and (3.6).

Corollary 3.3. *If $\beta = 0$ and $r \geq 0$, then*

a) *for $p = 2$, and $s(0, 2, n) = O(\log n)$,*

$$\left| f(x) - \mu(x, T, S_n) \right| < const \; \frac{1}{(n+1)^r} K_{S_n}^{q+\epsilon} \times$$

$$\times \log(n+1) \; \omega \left(f^{(r)}; \frac{1}{n+1} \right), \tag{3.7}$$

b) *for $0 \leq p < 2$, and $s(0, p, n) = O(n^{2-p})$,*

$$\left| f(x) - \mu(x, T, S_n) \right| < const \; \frac{1}{(n+1)^r} K_{S_n}^{q+\epsilon} \times$$

$$\times (n+1)^{2-p} \; \omega \left(f^{(r)}; \frac{1}{n+1} \right). \tag{3.8}$$

§4. Appendix.

In order to prove Theorem 2.1 it is convenient to state beforehand a few lemmas, following the approach of [6]. To shorten things, we will omit some details.

Lemma A.1. *If $m < q$, then*

$$D^m \left[|x - x_h|^q exp(\beta |x - x_h|) \right] = \left[sign(x - x_h) \right]^m \times$$

$$\times P_m(|x - x_h|)|x - x_h|^{q-m} exp(\beta |x - x_h|) \tag{A.1}$$

where $P_m(|x - x_h|)$ is a polynomial of degree $\leq m$ in the variable $|x - x_h|$.

Proof: First we have

$$D^m \left[|x - x_h|^q exp(\beta |x - x_h|) \right] = \sum_{j=0}^{m} \binom{m}{j} D^j |x - x_h|^q D^{m-j} exp(\beta |x - x_h|),$$

and

$$D^j |x - x_h|^q = \left[sign(x - x_h) \right]^j \binom{q}{j} j! |x - x_h|^{q-j}$$

$$D^{m-j} exp(\beta |x - x_h|) = \left[sign(x - x_h) \right]^{m-j} \beta^{m-j} exp(\beta |x - x_h|) .$$

Substituting the two derivatives in the previous one, we obtain

$$D^m[|x - x_h|^q exp(\beta |x - x_h|)] = [sign(x - x_h)]^m exp(\beta |x - x_h|)|x - x_h|^{q-m} \times$$

$$\times \sum_{j=0}^{m} \binom{m}{j} \beta^{m-j} \binom{q}{j} j! |x - x_h|^{m-j},$$

which is (A.1). ∎

We observe that Lemma A.1 implies

$$\lim_{x \to x_h} D^m[|x - x_h|^q exp(\beta|x - x_h|) = 0$$

for $0 < m < q$.

Let B be the set of all positive functions $G(x)$ defined on some domain $Q \subset \mathbb{R}$ whose derivatives of order less than q are bounded. That is, $G \in B$ means that there exist numbers $a_G > 0$ and $M_G < +\infty$ so that $G(x) \geq a_G$ and $|D^m G(x)| < M_G$ for all m such that $0 \leq m < q$.

Lemma A.2. *If $G(x) \in B$, then $G^2(x) \in B$.*

Lemma A.3. *If $G \in B$, then $D^m[G(x)]^{-1}$ is bounded.*

Lemma A.4. *Let*

$$G(x) = \sum_{i=1}^{n} \prod_{h=1, h \neq i}^{n} |x - x_h|^q exp(\beta|x - x_h|)$$

in any bounded region Q. Then $G(x) \in B$ and so $|D^m[G(x)]^{-1}| \leq M < +\infty$ for $0 \leq m < q$ and $x \in Q$.

Proof of Theorem 2.1. The case $m = 0$ is clear by simple inspection of (2.4). For $0 < m < q$, by (2.4)

$$D^m w_k(x; x_k, S_n) = \sum_{j=0}^{m} \binom{m}{j} D^j \left[\prod_{h=1, h \neq k}^{n} |x - x_h|^q exp(\beta|x - x_h|) \right] \times$$

$$\times D^{m-j} \left[\sum_{i=1}^{n} \prod_{l=1, l \neq i}^{n} |x - x_h|^q exp(\beta|x - x_h|) \right]^{-1}.$$

The general term in the sum contains the product of two factors involving derivatives. The second factor is bounded by Lemma A.4, whereas the first factor is

$$D^j \left[\prod_{h=1, h \neq k}^{n} |x - x_h|^q exp(\beta|x - x_h|) \right] =$$

$$D^j \left[|x - x_s|^q exp(\beta|x - x_s|) \prod_{\substack{h=1 \\ h \neq k, s}}^{n} |x - x_h|^q exp(\beta|x - x_h|) \right] =$$

$$\sum_{l=0}^{j} \binom{j}{l} D^l[|x - x_s|^q exp(\beta|x - x_s|)] D^{j-l} \left[\prod_{\substack{h=1 \\ h \neq k, s}}^{n} |x - x_h|^q exp(\beta|x - x_h|) \right].$$

This factor tends to zero when $x \to x_s$ for all l, $0 < l \leq j \leq m$, as we observed after Lemma A.1. Then the sum vanishes for all $1 \leq s \leq n$ and $0 < m < q$. ∎

Acknowledgements. This work has been supported by the Italian Ministry of Scientific and Technological Research (M.U.R.S.T.) and the National Research Council (C.N.R.).

References

1. Allasia, G., Some physical and mathematical properties of inverse distance weighted methods for scattered data interpolation, Calcolo **29**, 1-2 (1992), 97–109.
2. Allasia, G., Parallel and recursive computation of Shepard type formulas, Univ. Torino, Dept. Mathematics, Report 1992.
3. Allasia, G. and R. Besenghi, Multivariable interpolating means with exponential–type weights for scattered data, to appear.
4. Allasia, G., R. Besenghi and V. Demichelis, Weighted arithmetic means possessing the interpolation property, Calcolo **25**, 3 (1988), 203–217.
5. Allasia, G., R. Besenghi and V. Demichelis, Multivariable interpolation by weighted arithmetic means at arbitrary points, Calcolo **29**, 3-4 (1992), 301–311.
6. Barnhill, R. E., R. P. Dube and F. F. Little, Properties of Shepard's surfaces, Rocky Mountain J. Mathematics **13** (1983), 365–382.
7. Della Vecchia, B. and G. Mastroianni, Pointwise simultaneous approximation by rational operators, Journ. Approx. Th. **65** (1991), 140–150.
8. Farwig, R., Multivariate interpolation of scattered data by moving least squares methods, in *Algorithms for approximation*, M. C. Cox and J. C. Mason (eds.), Clarendon Press, Oxford, 1987, 193–211.
9. Franke, R., Scattered data interpolation: test of some methods, Math. Comp. **38** (1982), 161–200.
10. McLain, D. H., Drawing contours from arbitrary data points, Computer J. **17** (1974), 318–324.

Giampietro Allasia and Renata Besenghi
Dpt. di Matematica
Università di Torino
Via Carlo Alberto 10
10123 Torino, ITALY
allasia@dm.unito.it

Approximating Reachable Sets by Extrapolation Methods

R. Baier and F. Lempio

Abstract. Order of convergence results with respect to Hausdorff distance are summarized for the numerical approximation of Aumann's integral by an extrapolation method which is the set-valued analogue of Romberg's method. This method is applied to the discrete approximation of reachable sets of linear differential inclusions. For a broad class of linear control problems, it yields at least second order of convergence; for problems with additional implicit smoothness properties even higher order of convergence.

§1. Introduction

Curves, surfaces, and higher dimensional manifolds, which are implicitly defined as submanifolds of reachable sets of controlled dynamical systems, constitute a challenging object of approximation methods. In this paper, our main interest lies in extrapolation methods, especially in the visualization of order of convergence results, for the discrete approximation of reachable sets with respect to Hausdorff distance.

We concentrate on a special approach for the numerical approximation of reachable sets of linear differential inclusions which is based on the computation of Aumann's integral for set-valued mappings. It consists in exploiting ordinary quadrature formulae with nonnegative weights for the numerical approximation of the dual representation of Aumann's integral via its support functional. Theoretical roots of this approach could be traced back via [11] to [5]. The paper [4] is the first one with explicit numerical computations, exploiting mainly composite closed Newton-Cotes formulae for set-valued integrands, and including an outline of proof techniques for error estimates with respect to Hausdorff distance, which avoid the embedding of families of convex sets into abstract spaces (cf. [13,14]). All proofs are based on error estimates using weak assumptions on the regularity of single-valued integrands (see [15,7,8,4]).

Curves and Surfaces in Geometric Design   9
P. J. Laurent, A. Le Méhauté, and L. L. Schumaker (eds.), pp. 9–18.
Copyright © 1994 by A K PETERS, Wellesley, MA
ISBN 1-56881-039-3.

In Section 2 we sketch the error estimate for the discrete approximation of Aumann's integral for set-valued mappings by an adaptation of Romberg's method ([6]). Contrary to [4], we admit perturbations of the set-valued integrand and put emphasis on extrapolation methods from the very beginning. Since every column of the extrapolation tableau has to be interpreted by quadrature formulae with nonnegative weights, we restrict ourselves to equidistant grids with Romberg's sequence of stepsizes. As is familiar from integration of single-valued functions, the starting column is given by composite trapezoidal rule, the first extrapolation step by composite Simpson's rule for set-valued mappings. The following columns of the extrapolation tableau can be regarded as well as applications of quadrature formulae with nonnegative weights on an equidistant grid. Thus, every extrapolation step defines an approximation of Aumann's integral by a certain Minkowski sum of convex sets. Exploiting this interpretation of the extrapolation procedure numerically in a direct way or by the dual approach pursued in Sections 2 and 3 is a real challenge for computational geometry, especially for higher dimensional problems. Naturally, the order of convergence with respect to Hausdorff distance depends on the smoothness of the set-valued integrand in an appropriately defined sense. For a broad class of integrands, exploiting results in [9,16], at least order of convergence equal to 2 can be expected. For smooth integrands, extrapolation based on Romberg's integration scheme yields even higher order approximations, as is demonstrated by several examples in Section 3.

Most important are adaptations of these extrapolation methods to linear differential inclusions. As a result, in Section 3 we get higher order methods for the discrete approximation of reachable sets of special smooth classes of linear control problems. Contrary to [3] and [4], we present in Example 2 a control region which is not even strictly convex and in Example 3 a control region with lower dimension than state space dimension, both nevertheless admitting arbitrarily high order discrete approximations of the reachable sets by extrapolation methods. For linear control systems, especially non-autonomous ones, a fundamental solution of the according homogeneous system has to be computed numerically. This can be done by Runge-Kutta methods of appropriate orders, cf. [4], or, as in Section 3, by extrapolation methods using the hybrid method announced in [3].

In the final Section 4, we outline some open questions and possible directions of future research.

§2. Set-Valued Integration

According to [2], we use the following definition of an integral of a set-valued mapping.

Definition. Let $I = [a, b]$ with $a < b$ be a compact interval, and $F : I \Rightarrow \mathbb{R}^n$ a set-valued mapping of I into the set of all subsets of $\mathbb{R}^n$. Then the set

$$\int_I F(\tau)d\tau = \{z \in \mathbb{R}^n : \text{there exists an integrable selection}$$

$$f(\cdot) \text{ of } F(\cdot) \text{ on } I \text{ with } z = \int_I f(\tau)d\tau\}$$

is called *Aumann's integral* of $F(\cdot)$ over I.

Our objective is to approximate Aumann's integral numerically by extrapolatory quadrature formulae which are motivated by classical Romberg quadrature. Choose Romberg's sequence of stepsizes

$$h_0 = b - a, \qquad h_i = 2^{-i}h_0 \qquad (i = 1, \ldots, r)$$

corresponding to the sequence of grids

$$a = t_{i,0} < t_{i,1} < \ldots < t_{i,2^i} = b, \qquad t_{i,j} = a + jh_i \qquad (j = 0, \ldots, 2^i)$$

and compute as first column of the extrapolation tableau the corresponding weighted Minkowski sums of sets

$$T_{i0}(F) = h_i \left[\frac{1}{2}\overline{\text{co}}(F(a)) + \sum_{j=1}^{2^i-1} \overline{\text{co}}\left(F(t_{i,j})\right) + \frac{1}{2}\overline{\text{co}}(F(b)) \right]. \qquad (1)$$

Here $\overline{\text{co}}(\cdot)$ denotes the closed convex hull operation. This is just the set-valued analogue of composite trapezoidal rule. In fact, up to now, due to the computational complexity of this rule, the calculation in (1) is done for the dual representation of $T_{i0}(F)$ by means of its support functional

$$\delta^\star(l, T_{i0}(F)) = \sup_{z \in T_{i0}(F)} (l|z)$$

$$= h_i \left[\frac{1}{2}\delta^\star(l, F(a)) + \sum_{j=1}^{2^i-1} \delta^\star\left(l, F(t_{i,j})\right) + \frac{1}{2}\delta^\star(l, F(b)) \right]$$

for all $l \in \mathbb{R}^n$, where $(\cdot|\cdot)$ denotes the usual inner product in $\mathbb{R}^n$ with induced Euclidean norm $\|\cdot\|_2$.

Because of the fact that for an integrably bounded measurable set-valued mapping $F(\cdot)$ with nonempty and closed values Aumann's integral is convex and compact (cf. [1]) the following equality holds

$$\delta^\star\left(l, \int_I F(\tau)d\tau\right) = \int_I \delta^\star(l, F(\tau))\, d\tau = \delta^\star(l, T_{i0}(F)) + R_{i0}(l, F)$$

with a remainder term $R_{i0}(l, F)$ depending on $l \in \mathbb{R}^n$ and $F(\cdot)$. Motivated by classical Romberg integration, this relation suggests the following dual extrapolation scheme

$$\delta^\star (l, T_{ik}(F)) = \frac{4^k \delta^\star (l, T_{i,k-1}(F)) - \delta^\star (l, T_{i-1,k-1}(F))}{4^k - 1} \tag{2}$$

for $i = 1, \ldots, r$, $k = 1, \ldots, s$, $k \leq i$ with some $s \leq r$. It is well-known (see [12]) that the right-hand side of (2) can be written also as a quadrature formula with nonnegative weights for the integrand $\delta^\star(l, F(\cdot))$, e.g., for $k = 1$ one gets the set-valued analogue of composite Simpson's rule. Therefore, the left-hand side $\delta^\star (l, T_{ik}(F))$ is in fact a value of a support functional of a well-defined closed convex set $T_{ik}(F)$.

Moreover, due to the well-known relation between Hausdorff distance $\mathrm{haus}(\cdot, \cdot)$ with respect to Euclidean norm and support functionals, cf. e.g., [13], the representation holds

$$\mathrm{haus} \left(\int_I F(\tau) d\tau, \ T_{ik}(F) \right) = \sup_{\|l\|_2 = 1} \ | \ \delta^\star \left(l, \int_I F(\tau) d\tau \right) - \delta^\star (l, T_{ik}(F)) \ | \tag{3}$$

Hence, exploiting error estimates for classical Romberg integration under weak regularity assumptions and admitting, contrary to [4], perturbations of F of suitable order with respect to Hausdorff distance, we get the following fundamental order of convergence result.

Theorem. *Let $F : I \Rightarrow \mathbb{R}^n$ be a measurable and integrably bounded set-valued mapping with nonempty compact values. Assume that the support function $\delta^\star(l, F(\cdot))$ has an absolutely continuous $(2s)$-th derivative and that its $(2s+1)$-st derivative is of bounded variation with respect to t uniformly for all $l \in \mathbb{R}^n$ with $\|l\|_2 = 1$. Moreover, assume that $\tilde{F} : I \Rightarrow \mathbb{R}^n$ is a perturbation of F with nonempty compact convex values such that the Hausdorff distance*

$$\mathrm{haus}(\overline{\mathrm{co}}(F(t)), \ \tilde{F}(t)) \leq c_1 \cdot h_r^{2s+2}$$

with a constant c_1 which is independent of h_r. Then the estimate

$$\mathrm{haus} \left(\int_I F(\tau) d\tau, \ T_{rs}(\tilde{F}) \right) \leq c_2 \cdot h_r^{2s+2}$$

holds with a constant c_2 which is independent of h_r.

§3. Approximation of Reachable Sets

Most important is the application of quadrature formulae for set-valued integrals to the approximation of reachable sets $\mathcal{R}(b, a, Y_0)$ for linear differential inclusions consisting of all possible endpoints of absolutely continuous functions $y(\cdot)$ on I which satisfy

$$\begin{aligned} y'(t) &\in A(t)y(t) + B(t)U \qquad \text{(for almost every } t \in I := [a, b]), \\ y(a) &\in Y_0. \end{aligned} \tag{4}$$

Here, $A(\cdot)$ is an integrable $n \times n$-matrix function, $B(\cdot)$ an integrable $n \times m$-matrix function, $U \subset \mathbb{R}^m$ is a compact, nonempty control region and $Y_0 \subset \mathbb{R}^n$ a compact, convex, nonempty initial set.

Denoting with $\phi(t, \tau)$ the fundamental solution of the corresponding homogeneous differential equation with $\phi(\tau, \tau) = E_n$, the reachable set of (4) could be equivalently expressed by a set-valued integral, namely

$$\mathcal{R}(b, a, Y_0) = \phi(b, a)Y_0 + \int_a^b \phi(b, \tau)B(\tau)U \, d\tau.$$

Applying the extrapolation method of Section 2 and replacing all values $\phi(b, t_{r,j})$ in $T_{rs}(\phi(b, \cdot)B(\cdot)U)$ with approximations $\widetilde{\phi}_{rs}(b, t_{r,j})$ computed with an error of order $\mathcal{O}(h_r^{2s+2})$ (e.g., with an extrapolation of the midpoint rule for sufficiently smooth $A(\cdot)$), we could compute the set

$$\widetilde{\phi}_{rs}(b, a)Y_0 + T_{rs}(\widetilde{\phi}_{rs}(b, \cdot)B(\cdot)U)$$

which approximates the reachable set with order $\mathcal{O}(h_r^{2s+2})$ on appropriate smoothness assumptions, cf. Section 2.

To demonstrate the convergence properties of the extrapolation method for various types of control regions U, we consider the following three examples. In all tables, the Hausdorff distance in (3) is approximated in the following way: the exact integral is replaced by a very precisely computed reference set and the supremum in (3) is restricted to a discretization of the boundary of the unit ball.

Example 1. *We consider the following time-dependent linear differential inclusion on $I = [1, 2]$ with*

$$A(t) = \begin{pmatrix} 0 & 1 \\ -2/t^2 & 2/t \end{pmatrix}, \quad B(t) = \begin{pmatrix} t^2 & 0 \\ t & te^t \end{pmatrix}, \quad Y_0 = \left\{ \begin{pmatrix} 0 \\ 0 \end{pmatrix} \right\},$$

where $U = B_1(0) \subset \mathbb{R}^2$ is the closed Euclidean unit ball. In particular, U is a strictly convex control region.

This example possesses typical properties which allow higher order of convergence: the matrix function $B(\cdot)$ is invertible on I, and $A(\cdot), B(\cdot)$ are sufficiently often differentiable, so that the support function

$$\delta^*(l, \phi(2, t)B(t)) = \|B(t)^* \phi(2, t)^* l\|_2$$

is also sufficiently often differentiable with bounded derivatives uniformly for all $l \in \mathbb{R}^2$ with $\|l\|_2 = 1$. Figure 1 shows the first three approximations together with T_{22} which coincides with the reachable set within plotting accuracy.

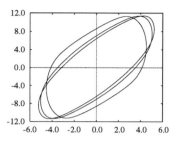 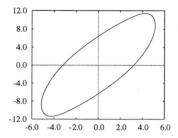

Figure 1. Approximations T_{00}, T_{10}, T_{11} resp. T_{22} for Example 1.

The corresponding convergence tables with an estimated Hausdorff distance between the approximations and the reachable set together with an estimated order of convergence are shown in Tables 1 and 2.

T_{rs}	approximation error	order
T_{00}	1.4565749402558685	——
T_{10}	0.3420734035031976	2.0902
T_{20}	0.0856358565527171	1.9980
T_{30}	0.0214188467870042	1.9993
T_{40}	0.0053554930687882	1.9998
T_{50}	0.0013389211774539	1.9999
T_{60}	0.0003347332747903	2.0000

T_{rs}	approximation error	order
T_{11}	0.1107201069639423	——
T_{21}	0.0087819059343079	3.6562
T_{31}	0.0005074987990517	4.1131
T_{41}	0.0000293793678088	4.1105
T_{51}	0.0000017870583280	4.0391
T_{61}	0.0000001111448684	4.0071

Table 1. Errors of T_{r0} and T_{r1} for Example 1.

T_{rs}	approximation error	order
T_{22}	0.0096375283496939	——
T_{32}	0.0003822727111560	4.6560
T_{42}	0.0000060351547466	5.9851
T_{52}	0.0000000724816180	6.3796
T_{62}	0.0000000010038779	6.1740

T_{rs}	approximation error	order
T_{33}	0.0004745738258154	——
T_{43}	0.0000100916525794	5.5554
T_{53}	0.0000000644178035	7.2915
T_{63}	0.0000000001809770	8.4755

Table 2. Errors of T_{r2} and T_{r3} for Example 1.

Example 2. *Consider the linear differential inclusion on $I = [0, 1]$ with*

$$A(t) = \begin{pmatrix} 1 & -1 \\ 4 & -3 \end{pmatrix}, \quad B(t) = \begin{pmatrix} 1-t & te^t \\ 3-2t & (-1+2t)e^t \end{pmatrix}, \quad Y_0 = \left\{ \begin{pmatrix} 0 \\ 0 \end{pmatrix} \right\},$$

where $U = [-1, 1]^2 \subset \mathbb{R}^2$ is the unit ball with respect to the maximum norm. Here U is a control set which has corners and is not strictly convex.

Nevertheless, all assumptions of the convergence theorem are fulfilled, since

$$\delta^*(l, \phi(1,\tau)B(\tau)U) = e^{-(1-\tau)}(|l_2| + e^\tau|l_1 + l_2|)$$

is arbitrarily often differentiable with bounded derivatives uniformly for all $l \in \mathbb{R}^2$ with $\|l\|_2 = 1$. Figure 2 shows the first three approximations together with T_{22} which again coincides with the exact reachable set within plotting precision.

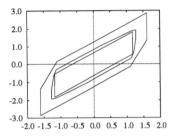 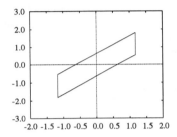

Figure 2. Approximations T_{00}, T_{10}, T_{11} resp. T_{22} for Example 2.

Convergence tables for this example can be found in Tables 3 and 4.

T_{rs}	approximation error	order
T_{00}	1.1377005895412307	——
T_{10}	0.1511442148720676	2.9121
T_{20}	0.0362760384531468	2.0588
T_{30}	0.0092155851665754	1.9769
T_{40}	0.0023132235480369	1.9942
T_{50}	0.0005788914860450	1.9985

T_{rs}	approximation error	order
T_{11}	0.0078719847420130	——
T_{21}	0.0005332856739972	3.8837
T_{31}	0.0000340540243489	3.9690
T_{41}	0.0000021400309889	3.9921
T_{51}	0.0000001339354307	3.9980

Table 3. Errors of T_{r0} and T_{r1} for Example 2.

T_{rs}	approximation error	order
T_{22}	0.0000486148810883	——
T_{32}	0.0000008339770790	5.8652
T_{42}	0.0000000133505798	5.9650
T_{52}	0.0000000002098739	5.9912

T_{rs}	approximation error	order
T_{33}	0.0000000755683551	——
T_{43}	0.0000000003248339	7.8619
T_{53}	0.0000000000012932	7.9726

Table 4. Errors of T_{r2} and T_{r3} for Example 2.

Example 3. *Modifying Example 2 only slightly, we choose*

$$B(t) = \begin{pmatrix} te^t \\ (-1 + 2t)e^t \end{pmatrix}$$

and $U = [-1, 1] \subset \mathbb{R}$ as a control region with a lower dimension than the state space dimension.

The support function

$$\delta^*(l, \phi(1, \tau)B(\tau)U) = e^{-(1-2\tau)}|l_1 + l_2|$$

still fulfills all assumptions of the convergence theorem. Due to unavoidable errors in the computation of the fundamental system, the reachable set is approximated by solid polygons which converge quickly to the straight line shown in Figure 3. One observes the expected order of convergence in Tables 5 and 6.

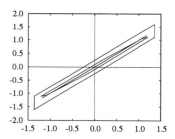 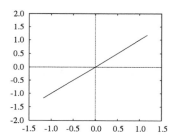

Figure 3. Approximations T_{00}, T_{10}, T_{11} resp. T_{22} for Example 3.

T_{rs}	approximation error	order
T_{00}	0.4713014578701207	——
T_{10}	0.0668209378745849	2.8183
T_{20}	0.0197452231727039	1.7588
T_{30}	0.0051496006755192	1.9390
T_{40}	0.0013012561272192	1.9846
T_{50}	0.0003261847432624	1.9961

T_{rs}	approximation error	order
T_{11}	0.0047332585545146	——
T_{21}	0.0003239110494004	3.8692
T_{31}	0.0000207428109016	3.9649
T_{41}	0.0000013044760001	3.9911
T_{51}	0.0000000816565897	3.9978

Table 5. Errors of T_{r0} and T_{r1} for Example 3.

T_{rs}	approximation error	order
T_{22}	0.0000481425650156	——
T_{32}	0.0000008264432341	5.8643
T_{42}	0.0000000132322462	5.9648
T_{52}	0.0000000002080203	5.9912

T_{rs}	approximation error	order
T_{33}	0.0000000754808080	——
T_{43}	0.0000000003244811	7.8618
T_{53}	0.0000000000012885	7.9763

Table 6. Errors of T_{r2} and T_{r3} for Example 3.

§4. Concluding Remarks

We tried to point out the intrinsic relation between set-valued numerical integration by extrapolation methods and higher order discrete approximations of reachable sets for linear control problems. In principle, each discrete approximation is a weighted Minkowski sum of closed convex sets. Especially for higher state space dimension, the direct computation of these sums or of their dual representation by support functionals is a real challenge. Admitting errors up to a certain order in the different terms of the Minkowski sum (resp. in the set-valued integrand) could ease this task.

For a remarkably broad class of linear control problems, one gets at least second order of convergence. We have shown by several examples that higher order of convergence can be achieved if the underlying problem has additional smoothness properties, even if the control region is not strictly convex or if the dimension of the control region is smaller than the state space dimension. A characterization of broader classes of such problems with additional implicit smoothness properties would be very desirable.

For nonlinear problems, reachable sets are no longer necessarily convex, and an integral representation by Aumann's integral is not available. Nevertheless, first order of convergence can be achieved by Euler's method (see [10]), and second order of convergence by modified Euler method for special problem classes ([17]). The development of higher order methods is an interesting and challenging field of ongoing research.

References

1. Aubin, J.-P., and H. Frankowska, *Set-Valued Analysis*, Birkhäuser, Boston–Basel–Berlin, 1990.
2. Aumann, R. J., Integrals of set-valued functions, J. Math. Anal. Appl. **12**, no. 1 (1965), 1–12.
3. Baier, R., Extrapolation methods for the computation of set-valued integrals and reachable sets of linear differential inclusions, ZAMM **74**, no. 6 (1993), to appear.
4. Baier, R., and F. Lempio, Computing Aumann's integral, in *Proceedings of the IIASA Workshop on Modelling Techniques for Uncertain Systems*, Sopron (Hungary), July 6–10, 1992, to appear.
5. Balaban, E. I., On the approximate evaluation of the Riemann integral of many-valued mapping, U.S.S.R. Comput. Maths. Math. Phys. **22**, no. 2 (1982), 233–238.
6. Bulirsch, R., Bemerkungen zur Romberg-Integration, Numer. Math. **6** (1964), 6–16.
7. Chartres, B. A., and R. S. Stepleman, Actual order of convergence of Runge-Kutta methods on differential equations with discontinuities, SIAM J. Numer. Anal. **11**, no. 6 (1974), 1193–1206.
8. Chartres, B. A., and R. S. Stepleman, Convergence of linear multistep methods for differential equations with discontinuities, Numer. Math. **27** (1976), 1–10.

9. Doitchinov, B. D., and V. M. Veliov, Parametrizations of integrals of set-valued mappings and applications, J. Math. Anal. Appl. **179**, 2 (1993), 483–499.

10. Dontchev, A., and F. Lempio, Difference methods for differential inclusions: a survey, SIAM Rev. **34**, no. 2 (1992), 263–294.

11. Donchev, T. D., and E. M. Farkhi, Moduli of smoothness of vector valued functions of a real variable and applications, Numer. Funct. Anal. Optim. **11**, no. 5&6 (1990), 497–509.

12. Engels, H., *Numerical Quadrature and Cubature*, Academic Press, London–New York–Toronto–Sydney–San Francisco, 1980.

13. Hörmander, P. L., Sur la fonction d'appui des ensembles convexes dans un espace localement convexe, Ark. Mat. **3**, no. 12 (1954), 181–186.

14. Rådström, H., An embedding theorem for spaces of convex sets, Proc. Amer. Math. Soc. **3** (1952), 165–169.

15. Sendov, B., and V. A. Popov, *The Averaged Moduli of Smoothness*, John Wiley & Sons, Chichester–New York–Brisbane–Toronto–Singapore, 1988.

16. Veliov, V. M., Discrete approximations of integrals of multivalued mappings, Comptes rendus de l'Académie bulgare des Sciences **42**, no. 12 (1989), 51–54.

17. Veliov, V. M., Second order discrete approximations to strongly convex differential inclusions, Systems and Control Letters **13** (1989), 263–269.

18. Veliov, V. M., Second order discrete approximation to linear differential inclusions, SIAM J. Numer. Anal. **29**, no. 2 (1992), 439–451.

Robert Baier and Frank Lempio
Chair of Applied Mathematics
University of Bayreuth
D-95440 Bayreuth, GERMANY
frank.lempio@uni-bayreuth.de

A Unimodality Property of Quartic B-splines

P. J. Barry

Abstract. This paper shows that over any knot vector, quartic B-splines evaluated at any domain value yield a unimodal sequence. That is, given any quartic B-splines $\{N_j\}$, and any domain value $\hat{t}$, and letting i be any index such that $N_i(\hat{t}) \geq N_j(\hat{t})$ for all j, then $N_j(\hat{t}) \leq N_{j+1}(\hat{t})$ for all $j < i$, and $N_j(\hat{t}) \geq N_{j+1}(\hat{t})$ for all $j \geq i$.

§1. Introduction

Consider the following problem: let $\{N_j\}$ be a sequence of basis functions, let $\hat{t}$ be any domain value, and let i be an index such that $N_i(\hat{t}) \geq N_j(\hat{t})$ for all j. Will $N_j(\hat{t}) \leq N_{j+1}(\hat{t})$ for all $j < i$, and $N_j(\hat{t}) \geq N_{j+1}(\hat{t})$ for all $j \geq i$? The motivation for this problem was the need to identify the coefficients most affecting a curve at a given point [2]. A recent work [1] showed that many bases, including arbitrary degree Bernstein bases, any B-spline basis of degree 3 or less, arbitrary degree uniform B-splines, and arbitrary degree uniform beta-splines basis functions do possess this property. However, some bases do not possess this property. In particular, [1] provided examples of a quartic beta-spline basis, and B-splines bases of all degrees ≥ 6, that do not possess this property. A few cases, most notably degree 4 and 5 B-splines, and non-uniform cubic Beta-splines, were left open. The purpose of this paper is to prove that any degree 4 B-spline basis will possess the property.

§2. Preliminary Observations

We begin by stating the result more precisely.

Theorem. *For any knot vector $\{t_j\}$ and any domain value t, the degree 4 B-splines $\{N_j\}$ have the following unimodality property: let i be such that $N_i(t) \geq N_j(t)$ for all j. Then $N_j(t) \leq N_{j+1}(t)$ for all $j < i$, and $N_j(t) \geq N_{j+1}(t)$ for all $j \geq i$.*

Note this unimodality property fixes an arbitrary t and lets the B-spline index j vary. Another well-known unimodality property fixes j and lets t vary.

Curves and Surfaces in Geometric Design
P. J. Laurent, A. Le Méhauté, and L. L. Schumaker (eds.), pp. 19–26.
Copyright © 1994 by A K PETERS, Wellesley, MA
ISBN 1-56881-039-3.

The proof below relies on explicit formulas for the B-splines, observations about B-spline behavior, and algebraic manipulation. In this section we list the observations and formulas.

Observation 1: By a relabelling and affine transformation of the knots, it suffices to show that no counterexample to the theorem exists over the knot interval $[t_4, t_5] = [0, 1]$. Therefore we restrict our attention to this interval.

Observation 2: If there exists a knot set, value $\hat{t}$ and index i such that $N_{i-1}(\hat{t}) > N_i(\hat{t}) < N_{i+1}(\hat{t})$, then for any t within a sufficiently small neighborhood of $\hat{t}$ the same inequalities hold. Thus it suffices to show no counterexample occurs for the open interval $t \in (0, 1)$.

Observation 3: Over $[t_4, t_5]$, only $N_0, \ldots, N_4$ are non-zero, so if a counterexample exists at $\hat{t}$, either $N_0(\hat{t}) > N_1(\hat{t}) < N_2(\hat{t})$, or $N_1(\hat{t}) > N_2(\hat{t}) < N_3(\hat{t})$, or $N_2(\hat{t}) > N_3(\hat{t}) < N_4(\hat{t})$. By symmetry it suffices to consider only the first 2 cases. We will consider the first case in Section 3, the second in Section 4.

Observation 4: By the variation diminishing property of B-spline curves for any j, there can be at most one value t for which $N_j(t) = N_{j+1}(t) \neq 0$. This implies that if for some $\tilde{s} \in [0, 1]$ we have $N_{j+1}(\tilde{s}) \geq N_j(\tilde{s})$, with $N_{j+1}(\tilde{s}) > 0$, then $N_{j+1}(s) \geq N_j(s)$ for all $s > \tilde{s}$. Moreover, equality holds only when both $N_{j+1}(s)$ and $N_j(s)$ are 0. Similarly, if for some $\tilde{s} \in [0, 1]$, $N_j(\tilde{s}) \geq N_{j+1}(\tilde{s})$, with $N_j(\tilde{s}) > 0$, then $N_j(s) \geq N_{j+1}(s)$ for all $s < \tilde{s}$, with equality holding only when both $N_j(s)$ and $N_{j+1}(s)$ are 0.

Observation 5: From Observation 4, if, for any j, there exists a $t \in [0, 1]$ such that $N_{j-1}(t) \leq N_j(t) \geq N_{j+1}(t)$, with $N_j(t) > 0$, then there does not exist a $\hat{t}$ such that $N_{j-1}(\hat{t}) > N_j(\hat{t}) < N_{j+1}(\hat{t})$.

Observation 6: The only knots affecting the curve over $[t_4, t_5]$ are $t_1, \ldots, t_8$. If a given set of knots does not yield a counterexample, then, by symmetry, neither will the set of knots $\{s_i\}_{i=1}^8$ with $s_i := 1 - t_{9-i}$.

Observation 7: If there exists a knot set, value $\hat{t}$ and index i such that $N_{i-1}(\hat{t}) > N_i(\hat{t}) < N_{i+1}(\hat{t})$, then modifying the knots so that $N_{i-1}(\hat{t})$ or $N_{i+1}(\hat{t})$ increases, while $N_i(\hat{t})$ does not increase, will also provide a counterexample. Similary, modifying the knots so that $N_i(\hat{t})$ decreases, while $N_{i-1}(\hat{t})$ and $N_{i+1}(\hat{t})$ do not decrease, will still provide a counterexample.

To simplify the proof, it is useful to modify some of the usual B-spline notation. Remember $t_4 = 0$ and $t_5 = 1$. Now let

$$\alpha = -t_3, \quad \beta = t_6 - 1$$
$$\gamma = -t_2, \quad \delta = t_7 - 1 \tag{1}$$
$$\eta = -t_1, \quad \nu = t_8 - 1.$$

Then $\eta \geq \gamma \geq \alpha \geq 0$ and $\nu \geq \delta \geq \beta \geq 0$. Now the B-splines $N_0, \ldots, N_4$ over

[0, 1] are

$$N_0(t) = \frac{(1-t)^4}{(1+\alpha)(1+\gamma)(1+\eta)}$$

$$N_1(t) = \frac{(1-t)^3(t+\eta)}{(1+\alpha)(1+\gamma)(1+\eta)} + \frac{(1-t)^2(t+\gamma)(1+\beta-t)}{(1+\alpha)(1+\gamma)(1+\beta+\gamma)}$$

$$+ \frac{(1-t)(t+\alpha)(1+\beta-t)^2}{(1+\alpha)(1+\alpha+\beta)(1+\beta+\gamma)} + \frac{t(1+\beta-t)^3}{(1+\beta)(1+\alpha+\beta)(1+\beta+\gamma)}$$

$$N_2(t) = \frac{(1-t)^2(t+\gamma)^2}{(1+\alpha)(1+\gamma)(1+\beta+\gamma)} + \frac{(1-t)(t+\alpha)(1+\beta-t)(t+\gamma)}{(1+\alpha)(1+\alpha+\beta)(1+\beta+\gamma)}$$

$$+ \frac{(1-t)(t+\alpha)^2(1+\delta-t)}{(1+\alpha)(1+\alpha+\beta)(1+\alpha+\delta)} + \frac{t(1+\beta-t)^2(t+\gamma)}{(1+\beta)(1+\alpha+\beta)(1+\beta+\gamma)}$$

$$+ \frac{t(1+\beta-t)(t+\alpha)(1+\delta-t)}{(1+\beta)(1+\alpha+\beta)(1+\alpha+\delta)} + \frac{t^2(1+\delta-t)^2}{(1+\beta)(1+\delta)(1+\alpha+\delta)}$$

$$N_3(t) = \frac{(1-t)(t+\alpha)^3}{(1+\alpha)(1+\alpha+\beta)(1+\alpha+\delta)} + \frac{t(1+\beta-t)(t+\alpha)^2}{(1+\beta)(1+\alpha+\beta)(1+\alpha+\delta)}$$

$$+ \frac{t^2(1+\delta-t)(t+\alpha)}{(1+\beta)(1+\delta)(1+\alpha+\delta)} + \frac{t^3(1+\nu-t)}{(1+\beta)(1+\delta)(1+\nu)}$$

$$N_4(t) = \frac{t^4}{(1+\beta)(1+\delta)(1+\nu)}.$$

$$\text{(2)}$$

§3. Case 1

In this section we prove

Lemma 1. *There does not exist a $\hat{t} \in (0,1)$ such that $N_0(\hat{t}) > N_1(\hat{t}) < N_2(\hat{t})$.*

Proof: For any $t \in (0,1)$, note N_0 and N_1 are independent of δ, and N_2 is a strictly increasing function of $\delta \geq 0$, so, referring to Observation 7, maximize N_2 with respect to δ by making δ arbitrarily large. Also, N_0 is a strictly decreasing function of η, N_1 is a strictly increasing function of η, and N_2 is independent of η, so we can simultaneously maximize N_0 and minimize N_1 with respect to η by taking η as small as possible, namely equal to γ. Finally, N_0 is independent of β, and N_1 is a strictly increasing function of β. The B-splines sum to 1, and since we have already taken δ to be arbitrarily large, N_3 and N_4 are essentially zero. This implies that N_2 is a strictly decreasing function of β. We can therefore simultaneously maximize N_2 and minimize N_1 with respect to β by taking β as small as possible — equal to 0.

Now N_0 has a zero of multiplicity 4 at 1, while N_1 has a zero of multiplicity 3. Thus a counterexample can exist only if there exists a $\tilde{t} \in (0,1)$ such that

$N_0(\tilde{t}) = N_1(\tilde{t})$, or

$$\frac{(1-\tilde{t})^4}{(1+\alpha)(1+\gamma)^2} = \frac{(1-\tilde{t})^3}{(1+\alpha)(1+\gamma)} \left(2\frac{\tilde{t}+\gamma}{1+\gamma} + \frac{\tilde{t}+\alpha}{1+\alpha} + \tilde{t}\right). \qquad (3)$$

Solving for $\tilde{t}$ yields $\tilde{t} = (1 - 2\gamma - 3\alpha\gamma)/(5 + 4\alpha + 2\gamma + \alpha\gamma)$. Since we need $\tilde{t} > 0$, this implies $\alpha \le \gamma < .5$.

Now, by Observation 5 it suffices to show $N_1(\tilde{t}) \ge N_2(\tilde{t})$. Since $N_2(\tilde{t}) = 1 - 2N_1(\tilde{t})$, we need only show that $N_1(\tilde{t}) \ge 1/3$. To show this, we examine how small $N_1(\tilde{t}) = (.8)^4(1 + \alpha)^3(1 + \gamma)^2/(1 + .8\alpha + .4\gamma + .2\alpha\gamma)^4$ can be. Differentiating with respect to α yields $N_1(\tilde{t})$ is a strictly decreasing function of α for $0 \le \alpha \le \gamma < .5$. Making α as large as possible, namely equal to γ, yields $N_1(\tilde{t}) = (.8)^4(1 + \gamma)/(1 + .2\gamma)^4$, which is $\ge .8^4 = .4096$ for $0 \le \gamma < .5$. This concludes the proof of Lemma 1.

§4. Case 2

In this section we complete the proof of the theorem by proving

Lemma 2. *There does not exist a* $\hat{t} \in (0, 1)$ *such that* $N_1(\hat{t}) > N_2(\hat{t}) < N_3(\hat{t})$.

Proof: Using Observation 7 in a manner similar to its use in Lemma 1, we note that we can take η and ν to be arbitrarily large. Moreover, we can take γ and δ as small as possible, that is, $\gamma = \alpha$ and $\delta = \beta$. We then have

$$N_1(t) = \frac{(1-t)^3}{(1+\alpha)^2} + \frac{(1-t)^2(t+\alpha)(1+\beta-t)}{(1+\alpha)^2(1+\alpha+\beta)}$$

$$+ \frac{(1-t)(t+\alpha)(1+\beta-t)^2}{(1+\alpha)(1+\alpha+\beta)^2} + \frac{t(1+\beta-t)^3}{(1+\beta)(1+\alpha+\beta)^2}$$

$$N_2(t) = \frac{(1-t)^2(t+\alpha)^2}{(1+\alpha)^2(1+\alpha+\beta)} + \frac{2(1-t)(t+\alpha)^2(1+\beta-t)}{(1+\alpha)(1+\alpha+\beta)^2}$$

$$+ \frac{2t(1+\beta-t)^2(t+\alpha)}{(1+\beta)(1+\alpha+\beta)^2} + \frac{t^2(1+\beta-t)^2}{(1+\beta)^2(1+\alpha+\beta)} \qquad (4)$$

$$N_3(t) = \frac{(1-t)(t+\alpha)^3}{(1+\alpha)(1+\alpha+\beta)^2} + \frac{t(1+\beta-t)(t+\alpha)^2}{(1+\beta)(1+\alpha+\beta)^2}$$

$$+ \frac{t^2(1+\beta-t)(t+\alpha)}{(1+\beta)^2(1+\alpha+\beta)} + \frac{t^3}{(1+\beta)^2}$$

The proof now proceeds by considering four subcases. For each subcase we will show that certain values of α and β do not allow a counterexample. The first three parts proceed by finding a point $\tilde{t} \in [0, 1]$ such that $N_1(\tilde{t}) \le N_2(\tilde{t}) \ge N_3(\tilde{t})$, and then applying Observation 5. The fourth part examines N_0 and N_1 at $t = 0$ and uses Observation 4. The regions considered by each subcase

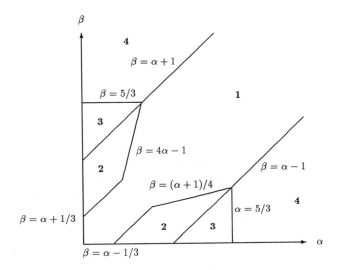

Fig. 1. The regions for the subcases in Lemma 2.

are shown in Figure 1. The bold numbers in the figure identify the subcase that considers the region.

Subcase 1: The first subcase shows no counterexample occurs for all nonnegative α, β satisfying $\beta \in \{\{[\alpha - 1/3, \alpha + 1/3] \cup [\frac{(1+\alpha)}{4}, 4\alpha - 1]\} \cap [\alpha - 1, \alpha + 1]\}$.

We prove this subcase by taking $\tilde{t} = (1 + \beta - \alpha)/2 \in [0, 1]$. Note this implies $\alpha - 1 \le \beta \le \alpha + 1$. First, for $\alpha - 1/3 \le \beta \le \alpha + 1/3$, examine $16(1 + \beta)(1 + \alpha)^2(N_2(\tilde{t}) - N_1(\tilde{t}))/(1 + \alpha - \beta)^2$. This equals

$$
\frac{(1 + \alpha + \beta)(1 + \beta)(1 + \alpha)}{(1 + \alpha - \beta)} + \frac{(1 + \beta - \alpha)(1 + \alpha + \beta)(1 + \alpha)^2}{(1 + \alpha - \beta)^2}
$$

$$
+ \frac{(1 + \beta - \alpha)^2(1 + \alpha + \beta)(1 + \alpha)^2}{(1 + \alpha - \beta)^2(1 + \beta)} - 2(1 + \alpha - \beta)(1 + \beta)
$$

$$
\ge (1 + \alpha + \beta)(1 + \beta) + (1 + \beta - \alpha)(1 + \alpha + \beta) + (1 + \beta - \alpha)^2
$$
$$
- 2(1 + \alpha - \beta)(1 + \beta) \tag{5}
$$

$$
\ge (1 + \alpha + \beta)(1 + \beta) + \frac{2}{3}(1 + \alpha + \beta) + \frac{4}{9} - 2(1 + \alpha - \beta)(1 + \beta)
$$

$$
= \frac{1}{9} - \frac{\alpha}{3} + \frac{8\beta}{3} - \alpha\beta + 3\beta^2 \ge \frac{1}{9} - \frac{\beta + 1/3}{3} + \frac{8\beta}{3} - (\beta + 1/3)\beta + 3\beta^2
$$

$$
= 2\beta + 2\beta^2 \ge 0.
$$

Next, the explicit formulas yield $N_1((1 + \beta - \alpha)/2) = N_3((1 + \alpha - \beta)/2)$ and $N_2((1 + \alpha - \beta)/2) = N_2((1 + \beta - \alpha)/2)$, so by interchanging α and β and using the same steps as above we get $N_2(\tilde{t}) - N_3(\tilde{t}) \ge 0$ for $\beta - 1/3 \le \alpha \le \beta + 1/3$, or, equivalently $\alpha - 1/3 \le \beta \le \alpha + 1/3$.

Next, examine $16(1 + \alpha)^2 (N_2(\tilde{t}) - N_1(\tilde{t}))/(1 + \alpha - \beta)$, which is

$$\geq (1 + \alpha + \beta)(1 + \alpha) - 2(1 + \alpha - \beta)^2 = -(1 + \alpha)^2 + 5(1 + \alpha)\beta - 2\beta^2. \quad (6)$$

This is nonnegative, for $\alpha - 1 \leq \beta \leq \alpha + 1$, when

$$\beta \in [\frac{5 - \sqrt{17}}{4}(1 + \alpha), \frac{5 + \sqrt{17}}{4}(1 + \alpha)] \supseteq [\frac{(1 + \alpha)}{4}, \frac{9(1 + \alpha)}{4}]. \quad (7)$$

By interchanging the roles of α and β and N_1 and N_3 we also get $N_2(\tilde{t}) - N_3(\tilde{t}) \geq 0$ when $\beta \in [\frac{4\alpha}{9} - 1, 4\alpha - 1] \cap [\alpha - 1, \alpha + 1]$. Applying Observation 5 to the region obtained by combining all these regions proves the subcase.

Subcase 2: We next consider the region of all nonnegative α, β such that $\beta \leq (1 + \alpha)/4$ and $\alpha - 1 \leq \beta \leq \alpha - 1/3$. By Observation 6, proving the result for this region will also prove it for all nonnegative α, β such that $\alpha \leq (1 + \beta)/4$ and $\beta - 1 \leq \alpha \leq \beta - 1/3$.

Again we use Observation 5. Let $\tilde{t} = -4\alpha/9 + 5(1 + \beta)/9 \in [0, 1]$ with $\alpha - 1 \leq \beta \leq \alpha - 1/3$ and $\beta \leq (1 + \alpha)/4$. Now that observe $9^4(1 + \alpha)^2 (N_2(\tilde{t}) - N_1(\tilde{t}))/(1 + \alpha + \beta)$ is equal to

$$5(4 + 4\alpha - 5\beta)^2 + 120(4 + 4\alpha - 5\beta)(1 + \alpha) + \frac{96(5 + 5\beta - 4\alpha)(1 + \alpha)^2}{(1 + \beta)}$$
$$+ \frac{16(5 + 5\beta - 4\alpha)^2(1 + \alpha)^2}{(1 + \beta)^2} - \frac{9(4 + 4\alpha - 5\beta)^3}{(1 + \alpha + \beta)}. \quad (8)$$

Since $\beta \geq \alpha - 1$ for the region under consideration, we have $24(5 + 5\beta - 4\alpha) \geq (4 + 4\alpha - 5\beta)$. Using this and $(4 + 4\alpha - 5\beta)/(1 + \alpha + \beta) \leq 4$, we get $9^4(1 + \alpha)^2(4 + 4\alpha - 5\beta)(N_2(\tilde{t}) - N_1(\tilde{t}))/(1 + \alpha + \beta)$ is greater than or equal to

$$5(4 + 4\alpha - 5\beta) + 120(1 + \alpha) + 4(1 + \alpha) - 36(4 + 4\alpha - 5\beta) = 155\beta \geq 0. \quad (9)$$

Next examine $9^4(1 + \alpha)^2(1 + \beta)^2(N_2(\tilde{t}) - N_3(\tilde{t}))/(1 + \alpha + \beta)$. This equals

$$25(4 + 4\alpha - 5\beta)^2(1 + \beta)^2 + 75(4 + 4\alpha - 5\beta)(1 + \alpha)(1 + \beta)^2$$
$$+ 60(5 + 5\beta - 4\alpha)(1 + \beta)(1 + \alpha)^2 - 4(5 + 5\beta - 4\alpha)^2(1 + \alpha)^2$$
$$- 9(5 + 5\beta - 4\alpha)^3(1 + \alpha)^2/(1 + \alpha + \beta)$$
$$\geq 25(4 + 4\alpha - 5\beta)(7 + 7\alpha - 5\beta) + 8(5 + 5\beta - 4\alpha)(1 + \alpha)^2(5 + 5\beta + 2\alpha)$$
$$- 9(5 + 5\beta - 4\alpha)^3(1 + \alpha). \quad (10)$$

Now since $\alpha \geq 5\beta/4$ for the region under consideration, we have $20 + 20\alpha - 25\beta \geq 20 + 20\beta - 16\alpha$, or $5(4 + 4\alpha - 5\beta) \geq 4(5 + 5\beta - 4\alpha)$. Thus the above quantity is greater than or equal to

$$(5 + 5\beta - 4\alpha)[20(7 + 7\alpha - 5\beta) + 8(1 + \alpha)^2(5 + 5\beta + 2\alpha) - 9(5 + 5\beta - 4\alpha)^2(1 + \alpha)]. \quad (11)$$

Since $\beta \leq \alpha - 1/3$, this is greater than or equal to

$$(5 + 5\beta - 4\alpha)[20(2\alpha + 26/3) + 8(1 + \alpha)^2(5 + 2\alpha) - 9(\alpha + 10/3)^2(1 + \alpha)]$$
$$\geq (5 + 5\beta - 4\alpha)[100 - 24\alpha + 3\alpha^2 + 7\alpha^3],$$

(12)

which is greater than or equal to 0 for α in the region under consideration.

Subcase 3: We next show no counterexample exists for all nonnegative α, β such that $\beta \leq \alpha - 1$ and $\alpha \leq 5/3$. By Observation 6, proving the result for this region will also prove it for $\alpha \leq \beta - 1$ and $\beta \leq 5/3$.

Again we use Observation 5. Let $\tilde{t} = -3\alpha/8 + 5(1 + \beta)/8 \in [0, 1]$ with $\beta \leq \alpha - 1$ and $\alpha \leq 5/3$. Thus $8^4(N_2(\tilde{t}) - N_1(\tilde{t}))$ equals

$$\frac{9(5 + 5\beta - 3\alpha)^2(1 + \alpha + \beta)}{(1 + \beta)^2} + \frac{63(5 + 5\beta - 3\alpha)(1 + \alpha + \beta)}{(1 + \beta)}$$
$$+ \frac{105(3 + 3\alpha - 5\beta)(1 + \alpha + \beta)}{(1 + \alpha)} + \frac{10(3 + 3\alpha - 5\beta)^2(1 + \alpha + \beta)}{(1 + \alpha)^2}$$
$$- \frac{8(3 + 3\alpha - 5\beta)^3}{(1 + \alpha)^2}$$

(13)

$$\geq \frac{105(1 + \alpha + \beta)(3 + 3\alpha - 5\beta)(1 + \alpha) - 8(3 + 3\alpha - 5\beta)^3}{(1 + \alpha)^2}$$
$$\geq \frac{(3 + 3\alpha - 5\beta)}{(1 + \alpha)^2}[105(1 + \alpha)^2 - 72(1 + \alpha - 5\beta/3)^2] \geq 0.$$

Also $8^4(N_2(\tilde{t}) - N_3(\tilde{t}))$ equals

$$\frac{25(3 + 3\alpha - 5\beta)^2(1 + \alpha + \beta)}{(1 + \alpha)^2} + \frac{25(3 + 3\alpha - 5\beta)(1 + \alpha + \beta)}{(1 + \alpha)}$$
$$+ \frac{15(5 + 5\beta - 3\alpha)(1 + \alpha + \beta)}{(1 + \beta)} - \frac{6(5 + 5\beta - 3\alpha)^2(1 + \alpha + \beta)}{(1 + \beta)^2}$$
$$- \frac{8(5 + 5\beta - 3\alpha)^3}{(1 + \beta)^2}$$
$$\geq \frac{25(3 + 3\alpha - 5\beta)^2(1 + \alpha + \beta)}{(1 + \alpha)^2} + \frac{3(5 + 5\beta - 3\alpha)(1 + \alpha + \beta)(6\alpha - 5 - 5\beta)}{(1 + \beta)^2}$$
$$- \frac{8(5 + 5\beta - 3\alpha)^3}{(1 + \beta)^2}.$$

(14)

Since $\alpha - 1 \geq \beta$, and since $\alpha \leq 5/3$, we have both $6\alpha - 5 - 5\beta > 0$, and

$$\frac{5(3 + 3\alpha - 5\beta)}{4(1 + \alpha)} \geq \frac{5(8 - 2\alpha)}{4(1 + \alpha)} \geq \frac{5(8 - 10/3)}{4(1 + 5/3)} \geq 2 \geq \frac{(5 + 5\beta - 3\alpha)}{(1 + \beta)}. \quad (15)$$

Therefore,

$$8^4(N_2(\tilde{t}) - N_3(\tilde{t})) \geq 64(1 + \alpha + \beta) - 32(5 + 5\beta - 3\alpha) = 32(5\alpha - 3 - 3\beta) > 0. \tag{16}$$

Subcase 4: In this subcase we show no counterexample exists for the region $\beta \in [0, \alpha - 1]$ and $\alpha \geq 5/3$. Again, by Observation 6, this will also account for the region $\alpha \in [0, \beta - 1]$ and $\beta \geq 5/3$.

We rely on Observation 4 by examining

$$\begin{aligned}
N_1(0) &= \frac{1}{(1+\alpha)^2} + \frac{\alpha(1+\beta)}{(1+\alpha)^2(1+\alpha+\beta)} + \frac{\alpha(1+\beta)^2}{(1+\alpha)(1+\alpha+\beta)^2} \\
N_2(0) &= \frac{\alpha^2}{(1+\alpha)^2(1+\alpha+\beta)} + \frac{2\alpha^2(1+\beta)}{(1+\alpha)(1+\alpha+\beta)^2}.
\end{aligned} \tag{17}$$

Note $0 \neq N_1(0) \leq N_2(0)$ when $(1+\alpha)^2\beta^2 + 2(1+3\alpha-\alpha^3)\beta + (1+4\alpha-3\alpha^3) \leq 0$, which occurs when

$$\beta \in \left[\frac{(\alpha^3 - 3\alpha - 1) - \sqrt{(\alpha^3 - 3\alpha - 1)^2 - (1+\alpha)^2(1+4\alpha-3\alpha^3)}}{(1+\alpha)^2}, \right.$$

$$\left. \frac{(\alpha^3 - 3\alpha - 1) + \sqrt{(\alpha^3 - 3\alpha - 1)^2 - (1+\alpha)^2(1+4\alpha-3\alpha^3)}}{(1+\alpha)^2} \right]. \tag{18}$$

For $\alpha \geq 5/3$ we have $1 + 4\alpha - 3\alpha^3 \leq 0$, so the left endpoint of this interval is less than or equal to 0. Next consider when the right endpoint is $\geq \alpha - 1$. Then

$$\sqrt{(\alpha^3 - 3\alpha - 1)^2 - (1+\alpha)^2(1+4\alpha-3\alpha^3)} \geq (1+\alpha)^2(\alpha-1) - (\alpha^3 - 3\alpha - 1). \tag{19}$$

Squaring both sides and simplifying yields this is true if $\alpha^4 + \alpha^3 - 4\alpha^2 \geq 0$, which is true for $\alpha \geq 5/3$.

This concludes the proof of this subcase, of Lemma 2, and of the theorem.

Acknowledgements. This work was supported in part by NSF Grant No. CCR-9113239.

References

1. Barry, P. J., J. C. Beatty, and R. N. Goldman, Unimodality properties of B-spline and Bernstein basis functions, Comput. Aided Design **24** (1992), 627–636.

2. Bartels, R. H., and J. C. Beatty, A technique for the direct manipulation of spline curves, Proceedings of the Graphics Interface '89 (1989), 33–39.

Phillip J. Barry
Computer Science Dept., Univ. of Minnesota
4-192 EE/CSci Bldg., 200 Union St. SE
Minneapolis, MN 55455
barry@ mail.cs.umn.edu

Object Oriented Spline Software

R. Bartels

Abstract. Object oriented programming provides software with a facility for imitating the mathematician's tool of theoretical abstraction. The commonality of a collection of related entities can be described in terms of a *base class*, which may be added to an object oriented language as a new data type. Each concrete entity of the collection can be described as a *derived class* by specifying whatever features distinguish it from the base, and the entity can be added as a further data type. Algorithms written for the base type can be applied to any of the derived types. This paper provides an example of these concepts applied to the design of a library of spline evaluation and refinement tools.

§1. Object Oriented Programming

The three fundamental concepts of object oriented programming are: *encapsulation, inheritance*, and *polymorphism*. Together they provide mechanisms for the programmer to extend the programming language by introducing new data types and operators. More importantly, they provide a means of writing code that is generic for collections of related data types.

Encapsulation allows one to define bundles of data and procedures, with associated operators, as new data types in the language. Variables with this type can be defined and used in programs as easily as the types provided originally. Inheritance allows one to define new data types in terms of those already defined, simply adding to the existing type the few additional items of data, procedures, and operators that distinguish the new type, and/or redefining data, procedures, and operators from the existing type. Polymorphism allows one to write code that can be executed generically on a collection of related data types.

In the next section we shall give examples using C++. The examples will be kept unrealistically simple, just containing enough components to illustrate the main points. For a general treatment of object oriented programming, see [3], and for the truth about C++, see [5].

Curves and Surfaces in Geometric Design
P. J. Laurent, A. Le Méhauté, and L. L. Schumaker (eds.), pp. 27–34.
Copyright ©1994 by A K PETERS, Wellesley, MA
ISBN 1-56881-039-3.

§2. Examples

In C++ the definition of a type is a *class*. A simple example would be a type
to represent squares, which could have the following declaration:

```
class Square
{
  public:
    Square( void )
      { _left = 0.0; _bottom = 0.0; _width = 1.0; }
    void left( double x )
      { _left = x; }
    double left( void )
      { return( _left ); }
        ...
    double area( void )
      { return( _width*_width ); }
  protected:
    double _left, _bottom, _width;
};
```

This class contains most of the conventional elements. It consists of
eight procedures and three data items. In the interests of space, four of the
procedures have been omitted in the location indicated by triple dots. The
procedures left out are two versions each of **bottom** and **width**, whose definitions are similar to those of **left**. (The capacity to have multiple definitions
for a procedure or operator, each distinguished by different argument types,
is called *overloading* and is supported in C++.) Together the procedures and
data are referred to as the *members* of the class. The data members _left,
_bottom, and _width constitute the major aspect of the class implementation;
i.e., they reflect the decision to represent a square by its width and the coordinates of its lower left corner. Any outside user of the class is prevented
from having direct access to these members by registering them as **protected**.
Indirect access is available through the interface provided by the member procedures **left**, **bottom**, and **width**, each having two definitions, one for setting
a value and one for reporting the value, and all are registered as **public**.

Indirect access is preferred, since it allows the designer of a class to
change the class implementation without disturbing the use of the class; *e.g.*,
to switch to a representation that uses the coordinates of the center of the
square, if some future consideration should reveal that as being more efficient.
The code for a member procedure such as **left** would be changed appropriately, of course, but the user of the class need never know that a change had
taken place. This capacity to separate *interface* cleanly from *implementation*
constitutes the substance of encapsulation.

The member procedure **Square**, having the same name as the class, is
a *constructor*. It specifies how a variable of type **Square** is to be initialized
upon creation.

The following would be a trivial program using the class in all its important aspects:

```
#include <iostream.h>
#include <Square.h>
void main( void )
{
  Square s;
  cout << "Initial area:  " << s.area() << endl;
  s.left(2.0); s.bottom(-3.0); s.width(6.2);
  cout << "Final area:  " << s.area() << endl;
}
```

The first **include** loads a system package of input/output routines; the second **include** provides the declaration of class **Square** (assumed to be contained in a file named **Square.h**). "**cout <<**" provides printed output, and is defined by the **iostream** package. Input is provided by the same package through "**cin >>.**" The statement "**Square s;**" creates an instance of a **Square** by invoking the constructor. **s.area()** and **s.width(6.2)** illustrate the ways in which all member procedures (except constructors) are called.

A user who might attempt to circumvent the protection of **_width**, for example, is prevented from doing so. If the statement "**s._width = 6.2;**" had been written instead of "**s.width(6.2);**", the compiler would have halted and reported an access violation.

Assume now that we wish to add rectangles to our language. In order to illustrate inheritance, we regard a rectangle as a kind of square that has a *height* in addition to a width:

```
#include <Square.h>
class Rectangle :  public Square
{
  public:
    Rectangle( void )
      { _height = 1.0; }
    void height( double h )
      { _height = h; }
    double height( void )
      { return( _height ); }
    double area( void )
      { return( _height*_width ); }
  protected:
    double _height;
};
```

Class **Rectangle**, derived from **Square**, provides a revised definition of **area** and has added **_height** together with two member procedures that report and set the value of **_height**. The code that implements **area** for the **Rectangle** class can gain access to the protected member **_width** of **Square**

by virtue of the derivation being declared public. General users of `Rectangle`
are still prevented any access to `Rectangle`'s data members `_left`, `_bottom`,
`_width`, and `_height`.

A trivial usage of rectangle would be as follows:

```
#include <iostream.h>
#include <Rectangle.h>
void main( void )
{
  Rectangle r;
  cout << "Initial area:  " << r.area() << endl;
  r.left(2.0); r.bottom(-3.0); r.width(6.2); r.height(3.7);
  cout << "Final area:  " << r.area() << endl;
}
```

Polymorphism is used to write programs that treat squares and rectangles
as abstract objects with something in common; e.g., the ability to report
their area. For example, a definition of the ">" operator that compared areas
might be useful. To do this, a *base class* must be defined that specifies what
commonality exists. For example:

```
class Shape
{
  public:
    virtual double area( void )=0;
};
```

Here the `area` procedure's information (name, argument(s), return type) are
listed, but no body of procedure code is given. Instead, the lack of an implementation for `area` is indicated by "=0." The key word `virtual` indicates
that the implementation for `area` may be supplied as appropriate from any
descendant class.

A definition for the ">" operator might look as follows:

```
int operator>( Shape& shA, Shape& shB )
{
  if( shA.area() > shB.area() )
  { return( 1 ); } else { return( 0 ); }
}
```

(The use of "&" as a modifier to the arguments is a final technicality required
by C++ in order to complete the polymorphic definition of the operator.)

Class `Square` can be made a derived class of `Shape` by changing the first
line of its declaration from "class Square" to "class Square : public
Shape" and including the file `Shape.h`, assumed to contain the declaration of
`Shape` and the code for the comparison operator. `Rectangle` thereby becomes
a descendant of `Shape`. Together they constitute an inheritance hierarchy:
Shape⇒Square⇒Rectangle.

The following trivial main program would illustrate the use of the comparison operator just defined:

```
#include <iostream.h>
#include <Rectangle.h>
void main( void )
{
  Square s;
  Rectangle r;
  s.width(5.0);
  r.width(3.0);  r.height(2.0);
  if( s>r )
    { cout << "Square is larger." << endl; }
  else
    { cout << "Rectangle is larger." << endl; }
}
```

(The inclusion of `Rectangle.h` causes a chain of inclusions for `Square` and `Shape`, making all declarations available to the program.)

§3. Spline Classes

Three inheritance hierarchies of C++ classes for use in the design and fitting of spline curves and surfaces at the University of Waterloo will be described. Each of the hierarchies was designed with polymorphism in mind to permit the implementation of general algorithms that could be used with different spline representations. The three hierarchies cooperate, each providing a collection of related services: the `FuncBasis` hierarchy, whose top classes abstract the features of basis function evaluation, the `Func` hierarchy, whose top classes abstract the assembly of basis functions into finished functions, and the `Refiner` hierarchy, whose top classes abstract the insertion of knots and, to a rudimentary extent, the conversion of spline representations.

Many of the algorithms used in the first and third hierarchies were influenced by work due to Barry and Goldman [2]. The elemental operations in that work are triads of the form

$$\mathcal{B}(\ldots, u_n, \ldots) := \alpha_{rs}(u_n, \tau_r, \tau_s)\mathcal{B}(\ldots, \tau_r, \ldots) + \beta_{rs}(u_n, \tau_r, \tau_s)\mathcal{B}(\ldots, \tau_s, \ldots)$$

involving the polar form ("blossom"), $\mathcal{B}$, of a spline. The choice of the affine (or sometimes linear) α's and β's represents the choice of the spline representation (basis), and the τ's are (or are sometimes related to) the knots.

The `FuncBasis` hierarchy uses triad schemes to produce all the nonzero basis values for any given segment of the domain. At the top of the hierarchy is the `FuncBasis` class, which has virtual public member procedures `dimension` and `evaluate`. The member `dimension` is present to report the number of basis values that the member `evaluate` will produce. None of the member procedures is implemented, of course; the `FuncBasis` class is designed for polymorphic code that would accommodate such things as trigonometric functions as well as spline basis functions.

The `BBasis` class is derived from the `FuncBasis` class and has virtual member procedures `order`, `degree`, `numKnots`, `numBreakpoints`, `knot`, `breakpoint`, and `multiplicity`. A procedure such as `knot` allows the setting and reporting of values, but is unimplemented at this level of the hierarchy to permit specific variations on triad schemes to be implemented in derived classes.

At the lowest level of the hierarchy are several classes: currently `UB-Basis`, `NUBBasis`, `BezBasis`, and `CMSBasis`, each individually derived from `BBasis`. In order they implement cardinal B-splines, general B-splines, Bézier splines, and connection matrix splines [4]. Each of these classes has member data related to knots: start and stride for the cardinal splines, knots themselves for the general B-splines, breakpoints for the Bézier splines, and finally, breakpoints with associated connection matrices for the CMS-splines. Each is responsible for implementing its own version of the `evaluate` procedure inherited from `FuncBasis` through `BBasis`.

`Func` is the top class of the `Func` hierarchy. Its main service, as far as the present discussion is concerned, is to provide an unimplemented virtual procedure `evaluate`. In contrast to the `FuncBasis` version, `Func`'s version of `evaluate` is intended to return a single value (or curve or surface point).

Derived from `Func` is `LCFunc`, to support the writing of polymorphic code for functions that are linear combinations of basis functions; *e.g.*, trigonometric sums as well as splines. This class adds member procedures to set and report coefficients (or control vertices) and implements the `evaluate` member by combining coefficients with basis function values. In order to carry out the implementation, the class contains the coefficients as member data, but in order not to be tied to specifics, the basis functions are represented as member data only by pointers to objects of type `FuncBasis`. This means that the computation of the linear combinations that implements `evaluate` is polymorphic code itself. Some further details about an earlier version of the `Func` and `FuncBasis` hierarchies are provided in [1].

At the next level of the hierarchy comes the `BFunc` class, for which the pointers to `FuncBasis` objects in `LCFunc` are now reflected in `BBasis` pointers. Finally, at the lowest level, the `Func` hierarchy has classes `UBFunc`, `NUBFunc`, `BezFunc`, and `CMSFunc`, which provide spline objects of explicit representational type. In these classes, the member data that in the `BFunc` class consisted of pointers to `BBasis` objects is, for these classes, reflected in pointers to `UBBasis`, `NUBBasis`, `BezBasis`, or `CMSBasis` objects, as appropriate. With the `Func` hierarchy it is possible to compose code that works for any function, or any function that is a linear combination of basis functions, or any function that is a linear combination of spline basis functions, or finally, functions that are specifically given in Bézier representation or the like.

The final hierarchy to be surveyed begins with the `Refiner` class. Refinement (*i.e.*, knot insertion) is implemented as a collection of classes that transform unrefined coefficients (control vertices) into refined ones based on the original knot sequence, the final knot sequence, and the spline representation. The `Refiner` class is intended to support polymorphic code that deals with refinement in general. This class has an unimplemented virtual procedure, `newCVs`, whose argument is the vector of the unrefined coefficients and whose return is the vector of the refined coefficients. The `Refiner` hierarchy is exceedingly shallow. Derived from the `Refiner` class is only a level of classes that implement the `newCVs` procedure as an appropriate instance of the refinement algorithms covered in [2]: `BezFastBreakInserter` for Barry-Goldman fast breakpoint insertion on Bézier splines, `BezRandomBreakInserter` and `BezSingleBreakInserter` for de Casteljau refinement, `NUBFastKnotInserter` for Barry-Goldman fast knot insertion on general B-splines, `NUBRandomKnotInserter` for the Oslo algorithm, and `NUBSingleKnotInserter` for Boehm's algorithm.

Finally, as a convenience to the user, the `BFunc` class and its descendants provide a member procedure, `addKnots`, that manages the entire transition from an unrefined basis with unrefined coefficients to a refined basis and refined coefficients. This procedure calls the appropriate member of the `Refiner` hierarchy to carry out its service.

The general flavor of code using these three class hierarchies is given in the examples below. In the first, a general B-spline basis is created from data read in; a B-spline function is generated; the function is evaluated and then refined. (Note that constructors can be written to require arguments.)

```
#include <iostream.h>
#include <Func/NUBFunc.h>
void main( void )
{
  unsigned order;
  double val, u;
  DoubleVec cvs;
  NumberSequence knots, addedknots;
  cin >> order >> knots >> cvs >> u >> addedknots;
  NUBBasis nb(order,knots);
  NUBFunc nf(cvs,nb);
  cout << "Original spline:   " << nf << endl;
  cout << "First derivative:  " << nf.evaluate(u,1) << endl;
  nf.addKnots(addedknots);
  cout << "Refined spline:    " << nf << endl;
}
```

The `DoubleVec` class provides resizeable arrays of doubles. The `NumberSequence` class is derived from the `DoubleVec` class by adding member procedures to keep the vector sorted. In the second example, neither basis nor function are created. The program simply uses a knot sequence and added knots to create a refinement object that is then used to transform unrefined spline coefficients into refined coefficients.

```
#include <iostream.h>
#include <Refiner/NUBRandomKnotInserter.h>
void main( void )
{
  unsigned order;
  DoubleVec cvs, newcvs;
  NumberSequence knots, addedknots;
  cin >> order >> knots >> cvs >> addedknots;
  NUBRandomKnotInserter nr(order,knots,addedknots);
  newcvs = nr.newCVs(cvs);
  cout << "Refined CV's:  " << newcvs << endl;
}
```

Acknowledgements. This work has been supported through the Canadian Government's NSERC Strategic and Operating funding programs, through the Province of Ontario's ITRC funding program, and with the assistance of General Motors.

References

1. Vermeulen, A. H., and R. H. Bartels, C++ splines classes for prototyping, *SPIE Vol. 1610: Curves and Surfaces in Computer Vision and Graphics II* (1991), 121–131.

2. Goldman, R. N., and T. Lyche (eds.), *Knot Insertion and Deletion Algorithms for B-Spline Curves and Surfaces*, SIAM, Philadelphia, PA, 1993.

3. Meyer, B., *Object-Oriented Software Construction*, Prentice Hall, New York, NY, 1988.

4. Seidel, H-P. Polar forms for geometrically continuous spline curves of arbitrary degree, ACM Trans. Graphics **12** (1993), 1–34.

5. Stroustrup, B., *The C++ Programming Language, second edition*, Addison-Wesley, Reading, MA, 1991.

Richard H. Bartels
Computer Science Department
University of Waterloo
Waterloo, Ontario N2L 3G1
CANADA
rhbartel@ watcgl.uwaterloo.ca

On Polynomial Functions Defining the Geometric Continuity Between Two (SBR) Surfaces

J. L. Bauchat

Abstract. This paper presents some new results concerning the G^2 continuity between two (SBR) surfaces. A necessary and sufficient condition on the parametrizations of the surfaces is obtained and illustrated on two examples.

§1. General Framework

According to [8,9], let us define by $\mathcal{E}$ (resp. $\mathcal{F}$) the real affine space, $\overrightarrow{\mathcal{E}}$ (resp. $\overrightarrow{\mathcal{F}}$) its associated linear vector space such that $\mathcal{E}$ (resp. $\overrightarrow{\mathcal{E}}$) is a hyperplane of $\mathcal{F}$ (resp. $\overrightarrow{\mathcal{F}}$) and $\widetilde{\mathcal{E}}$ the projective completion of $\mathcal{E}$. For our purpose, $\mathcal{E}$ is a 3-dimensional affine space and $\mathcal{F}$ is a 4-dimensional affine space. Let Ω be a point of $\mathcal{F}$ not belonging to $\mathcal{E}$. We call *massic vector space* the linear vector space $\widehat{\mathcal{E}} = (\mathcal{E} \times \mathbb{R}^*) \cup \overrightarrow{\mathcal{E}}$.

The isomorphim $\widehat{\Omega}$ between $\widehat{\mathcal{E}}$ and $\overrightarrow{\mathcal{F}}$, defined by $\widehat{\Omega}(A; \alpha) = \alpha \overrightarrow{\Omega A}$, $\widehat{\Omega}(\overrightarrow{u}) = \overrightarrow{u}$ induces an addition and an external multiplication in $\widehat{\mathcal{E}}$. The mass $\chi(a)$ of the massic vector a is defined by $\chi(A; \alpha) = \alpha$ and $\chi(\overrightarrow{u}) = 0$. We will often use the conic projection $\Pi\Omega$ of apex Ω, from $\mathcal{F} \setminus \{\Omega\}$ onto $\widetilde{\mathcal{E}}$ defined by $\Pi\Omega(\vec{v}) = m$ where m is such that $\overrightarrow{\Omega m}$ is parallel to $\vec{v}$.

Let $r = \{r_{ij}, 0 \leq i \leq n, 0 \leq j \leq p\}$ be a grid of massic vectors. $\mathcal{R}_{ij}$ is the point of $\overrightarrow{\mathcal{F}}$ defined by $\overrightarrow{\Omega \mathcal{R}_{ij}} = \widehat{\Omega}(r_{ij})$ and ρ_{ij} is the mass $\chi(r_{ij})$ of r_{ij}. The (SBR) surface, S_r, associated to this net is denoted by $SBR[r; [0,1]^2]$, and is defined as generated by the point $R(u,v) = \Pi\Omega(\mathcal{R}(u,v))$, where the function $\mathcal{R}$ is the polynomial mapping given by $\mathcal{R}(u,v) = \sum_{i=0}^{n} \sum_{j=0}^{p} B_i^n(u) B_j^p(v) \mathcal{R}_{ij}$, i.e. $\mathcal{R}(u,v) = SBP[\mathcal{R}_{ij}, 0 \leq i \leq n, 0 \leq j \leq p; [0,1]^2](u,v)$.

Similarly, let $l = \{l_{ij}, 0 \leq i \leq n1, 0 \leq j \leq p1\}$ be a rectangular grid of massic vectors. $\mathcal{L}_{ij}$ is the point of $\overrightarrow{\mathcal{F}}$ defined by $\overrightarrow{\Omega \mathcal{L}_{ij}} = \widehat{\Omega}(l_{ij})$ and λ_{ij} is

Curves and Surfaces in Geometric Design
P. J. Laurent, A. Le Méhauté, and L. L. Schumaker (eds.), pp. 35–42.
Copyright © 1994 by A K PETERS, Wellesley, MA
ISBN 1-56881-039-3.

the mass $\chi(l_{ij})$ of l_{ij}. We denote by $SBR[l; [0,1]^2]$ the surface S_l generated by the point $L(u,v) = \Pi\Omega(\mathcal{L}(u,v))$, where the function $\mathcal{L}$ is the polynomial mapping defined by $\mathcal{L}(u,v) = \sum_{i=0}^{n1} \sum_{j=0}^{p1} B_i^{n1}(u) B_j^{p1}(v) \mathcal{L}_{ij}$.

§2. G^2 Continuity between the two Surfaces S_l and S_r

Notation and definitions relevant to the general geometric framework defining the G^2 continuity between the surfaces S_l and S_r along a common edge, can be found in [1,2,3,10]. In order to avoid points at infinity on the common boundary curve, we suppose $\mathcal{L}(1,v) \neq \Omega$ and $\Pi\Omega(\mathcal{L}(1,v)) \in \mathcal{E}$, $\forall v \in [0,1]$. The same holds for S_r, i.e. $\mathcal{R}(0,v) \neq \Omega$ and $\Pi\Omega(\mathcal{R}(0,v)) \in \mathcal{E}$, $\forall v \in [0,1]$. As a consequence of these assumptions, the functions $\lambda(1,v) = \sum_{j=0}^{p1} B_j^{p1}(v) \lambda_{n1j}$ and $\rho(0,v) = \sum_{j=0}^{p} B_j^p(v) \rho_{0j}$ do not vanish on $[0,1]$.

Applied to the two (SBR) surfaces S_l and S_r, the geometric framework presented in [2] leads to the following analytical conditions [1].

Proposition 1. *A necessary and sufficient condition for G^0 continuity between S_l and S_r along $\Gamma = \{L(1,v), v \in [0,1]\}$ is*

$$\Pi\Omega(\mathcal{L}(1,v)) = \Pi\Omega(\mathcal{R}(0,v)), \forall v \in [0,1]. \tag{1}$$

Theorem 1. *When the condition (1) is satisfied, the G^1 continuity between S_l and S_r along Γ is equivalent to the existence of four continuous differentiable real functions $\{f_1, f_2, f_3, f_4\}$, defined on $[0,1]$ such that*

$$\lambda(1,v)\rho(0,v)f_2(v)f_1(v) < 0, \forall v \in [0,1]$$

$$f_1(v)\mathcal{R}_u(0,v) + f_2(v)\mathcal{L}_u(1,v) + f_3(v)\mathcal{L}_v(1,v) + f_4(v)\mathcal{L}(1,v) = 0. \tag{2}$$

In order to simplify the condition defining the G^2 continuity, we set

$$a(v) = -\frac{\lambda(1,v)f_2(v)}{\rho(0,v)f_1(v)} \quad \text{and} \quad b(v) = -\frac{\lambda(1,v)f_3(v)}{\rho(0,v)f_1(v)}, \quad \forall v \in [0,1].$$

The geometrical meaning of these coefficients is given by:

$$\mathcal{R}_u(0,v) = a(v)L_u(1,v) + b(v)L_v(1,v), \forall v \in [0,1].$$

Theorem 2. *When the two conditions (1) and (2) are satisfied, the G^2 continuity between S_l and S_r along Γ is equivalent to the existence of four continuous real functions $\{h_1, h_2, h_3, h_4\}$, defined on $[0,1]$ such that*

$$h_1(v) \neq 0, \forall v \in [0,1]$$

and

$$h_1(v)\mathcal{A}(v) + h_2(v)\mathcal{L}_u(1,v) + h_3(v)\mathcal{L}_v(1,v) + h_4(v)\mathcal{L}(1,v) = 0 \tag{3}$$

where

$$\mathcal{A}(v) = \frac{\lambda(1,v)}{\rho(0,v)}\mathcal{R}_{uu}(0,v) - a^2(v)\mathcal{L}_{uu}(1,v) - 2a(v)b(v)\mathcal{L}_{uv}(1,v) - b^2(v)\mathcal{L}_{vv}(1,v).$$

The functions $\{f_1, f_2, f_3, f_4\}$ (resp. $\{h_1, h_2, h_3, h_4\}$) are called *coefficient-functions* of the G^1 continuity (resp. G^2 continuity) between the surfaces S_l and S_r. In case S_l and S_r are (SBR) surfaces, those functions are polynomial and some of their properties have already been studied in [1] and in [11] for rational Bézier patches. The main purpose of this paper is to illustrate Theorem 2 with two examples.

§3. First Example

A complete presentation of the first example can be found in [1].

Let $n = 2$, $p = 2$ and set $S_r = SBR[r_{ij}, 0 \leq i \leq 2, 0 \leq j \leq 2; [0,1]^2]$ with a massic net defined by $r_{ij} = \widehat{\Omega}^{-1}(\mathcal{R}_{ij})$, where the $(\mathcal{R}_{ij})$ are given in the following table:

j\i	0	1	2
2	(0,1,0,1)	$(-\frac{1}{2},1,0,1\ )$	(-1,1,1,1)
1	$(0,\frac{1}{2},-1,1)$	$(-\frac{1}{2},\frac{1}{2},-1,1)$	$(-1,\frac{1}{2},0,1)$
0	(0,0,-1,1)	$(-\frac{1}{2},0,-1,1)$	(-1,0,0,1)

For the second surface, let $n1 = 2$, $p1 = 2$, and set $S_l = SBR[l_{ij}, 0 \leq i \leq 2, 0 \leq j \leq 2; [0,1]^2]$ with a massic net defined by $l_{ij} = \widehat{\Omega}^{-1}(\mathcal{L}_{ij})$, where the $(\mathcal{L}_{ij})$ are given in the following table:

j\i	0	1	2
2	(6,3,-4,9)	(1,2,-2,3)	(0,1,0,1)
1	$(\frac{7}{2},1,-4,6)$	$(\frac{3}{4},\frac{3}{4}, -\frac{5}{2},\frac{5}{2})$	$(0,\frac{1}{2},-1,1)$
0	(2,0,-3,4)	$(\frac{1}{2},0,-2,2)$	(0,0,-1,1)

We observe that $L(1,v) = R(0,v)$, $\forall v \in [0,1]$, which means that we have G^0 continuity between S_l and S_r. The functions $\{f_1, f_2, f_3, f_4\}$ of equation (2) are obtained as solutions of a 4×4 linear system. After some calculations, we obtain

$$\begin{pmatrix} f_1 \\ f_2 \\ f_3 \\ f_4 \end{pmatrix} = \begin{pmatrix} -(1+v) \\ 1 \\ -v(1+v) \\ 2(1+v) \end{pmatrix} f_2.$$

We calculate $a(v) = -\frac{f_2}{f_1} = \frac{1}{1+v}$ and $b(v) = -\frac{f_3}{f_1} = -v$. Then, equation (3) leads to a 4×4 linear system, whose solution is

$$\begin{pmatrix} h_1 \\ h_2 \\ h_3 \\ h_4 \end{pmatrix} = \begin{pmatrix} (1+v)^2 \\ -(2+4v) \\ 2v(1+v)^2 \\ -2(1+v)^2 \end{pmatrix} \frac{h_1}{(1+v)^2}.$$

Thus, S_l and S_r are G^2 connected along $\Gamma = \{L(1,v), v \in [0,1]\}$.

G^2 continuity between S_l and S_r could be expected because S_r and S_l are two patches of the elliptic paraboloid $z = x^2 + y^2 - 1$ defined respectively on the domains $\{-1 \leq x \leq 0, 0 \leq y \leq 1\}$ and $\{0 \leq x, 0 \leq y, x+y \leq 1, 2x-y \leq 1\}$ and therefore are G^k connected along the parabola $\Gamma = \{z = y^2 - 1\}$ for all integer k (see [1,3]).

§4. A Second Example

Let $n1 = 1$, $p1 = 1$, and suppose $S_l = SBR[l_{ij}, 0 \leq i \leq 1, 0 \leq j \leq 1; [0,1]^2]$ with a massic net defined by $l_{ij} = \widehat{\Omega}^{-1}(\mathcal{L}_{ij})$, where the $(\mathcal{L}_{ij})$ are given by:

j\i	0	1
1	(-1,1,-1,1)	(0,1,0,1)
0	(-1,0,0,1)	(0,0,0,1)

Let $S_r = SBR[r_{ij}, 0 \leq i \leq 2, 0 \leq j \leq 2; [0,1]^2]$ with a massic net defined by $r_{ij} = \widehat{\Omega}^{-1}(\mathcal{R}_{ij})$, where the $(\mathcal{R}_{ij})$ are given by:

j\i	0	1	2
2	(0,1,0,1)	(1,2,1,3)	(6,3,2,9)
1	(0,1,0,2)	$(\frac{3}{2},\frac{3}{2},\frac{1}{2},5)$	(7,2,1,12)
0	(0,0,0,4)	(2,0,0,8)	(8,0,0,16)

Proposition 2. S_l and S_r are G^0 connected along the straight line $\Gamma = \{L(1,v), 0 \leq v \leq 1\}$.

Proof: $\mathcal{L}(1,v) = B_0^1(v)\mathcal{L}_{10} + B_1^1(v)\mathcal{L}_{11}$ i.e. $\mathcal{L}(1,v) = (0,v,0,1)$. Thus, $L(1,v) = (0,v,0)$. Moreover, $\mathcal{R}(0,v) = B_0^2(v)\mathcal{R}_{00} + B_1^2(v)\mathcal{R}_{01} + B_2^2(v)\mathcal{R}_{02}$ i.e. $\mathcal{R}(0,v) = (0, 2v - v^2, 0, (2-v)^2)$ and $R(0,v) = (0, \frac{v}{2-v}, 0)$.

Then $R(0,v) = L(1, \frac{v}{2-v})$, $\forall v \in [0,1]$, is a G^0 continuity relation between the surfaces S_l and S_r, more general than the usual one (1). (Liu [11] and Degen [6] have already made a remark about this.) ∎

Now we want to prove the G^2 continuity between S_l and S_r and, in order to apply the relation (2) we change the parametrization of S_l.

Let $w = \frac{v}{2-v}$. Then $B_0^1(w) = \frac{2B_0^1(v)}{2-v}$ and $B_1^1(w) = \frac{B_1^1(v)}{2-v}$.

Thus, $\mathcal{L}(u, w) = \sum_{i=0}^{1} \sum_{j=0}^{1} B_i^1(u) B_j^1(w) \mathcal{L}_{ij}$,

i.e. $\mathcal{L}(u, w) = \frac{1}{2-v} \sum_{i=0}^{1} \sum_{j=0}^{1} B_i^1(u) B_j^1(v) \mathcal{L}_{ij}^*$ with $\mathcal{L}_{ij}^* = 2^{1-j} \mathcal{L}_{ij}$.

We obtain a new surface in $\mathcal{F}$ whose S_l is the $\Pi\Omega$-projection, parametrized by $\mathcal{L}^*(u, v) = SBP[\mathcal{L}_{ij}^*, 0 \leq i \leq 1, 0 \leq j \leq 1; [0,1]^2](u,v)$, with $(\mathcal{L}_{ij}^*)$ given by the following table:

j\i	0	1
1	(-1,1,-1,1)	(0,1,0,1)
0	(-2,0,0,2)	(0,0,0,2)

We first verify the G^0 relation $L^*(1, v) = R(0, v)$, $\forall v \in [0, 1]$, and therefore we can apply Theorems 1 and 2. The fact that $L^*(1, v) = R(0, v)$, $\forall v \in [0, 1]$, means the equality of two (BR) curves although their lengths are not the same, parametrized by

$C_r(v) = BR[r_{0j}, 0 \leq j \leq 2; [0,1]](v)$ and $C_l(v) = BR[l_{1j}, 0 \leq j \leq 1; [0,1]](v)$. After a degree elevation we obtain $C_l(v) = BR[m_{1j}, 0 \leq j \leq 2; [0,1]](v)$ with $m_{1j} = \widehat{\Omega}^{-1}(\mathcal{M}_{1j})$ and $\mathcal{M}_{10} = \mathcal{L}_{10}^* = (0,0,0,2)$, $\mathcal{M}_{12} = \mathcal{L}_{11}^* = (0,1,0,1)$, $\mathcal{M}_{11} = 0.5(\mathcal{L}_{10}^* + \mathcal{L}_{11}^*) = (0, 0.5, 0, 1.5)$.

We observe that C_r and C_l define the same curve in $\mathcal{E}$ without having proportional massic polygons. An explanation is provided in [5] by the fact that $BP[\mathcal{R}_{0j}, 0 \leq j \leq 2; [0,1]](v) = (2 - v)BP[\mathcal{M}_{1j}, 0 \leq j \leq 2; [0,1]](v)$.

Proposition 3. *The (SBR) surfaces S_l and S_r are G^2 connected along the straight line $\Gamma = \{L(1, v), 0 \leq v \leq 1\}$.*

Proof: First, let us calculate four functions $\{f_1, f_2, f_3, f_4\}$ such that

$$f_1(v)\mathcal{R}_u(0, v) + f_2(v)\mathcal{L}_u^*(1, v) + f_3(v)\mathcal{L}_v^*(1, v) + f_4(v)\mathcal{L}^*(1, v) = 0.$$

Then the unknowns $\{f_1, f_2, f_3, f_4\}$ are solutions of the 4×4 linear system

$$\begin{bmatrix} 4 - 2v & 2 - v & 0 & 0 \\ 2v & 0 & 1 & v \\ 2v & v & 0 & 0 \\ 8 - 4v & 0 & -1 & 2 - v \end{bmatrix} \begin{bmatrix} f_1(v) \\ f_2(v) \\ f_3(v) \\ f_4(v) \end{bmatrix} = \begin{bmatrix} 0 \\ 0 \\ 0 \\ 0 \end{bmatrix}$$

and we obtain

$$\begin{bmatrix} f_1(v) \\ f_2(v) \\ f_3(v) \\ f_4(v) \end{bmatrix} = \begin{bmatrix} 1 \\ -2 \\ 2v - v^2 \\ v - 4 \end{bmatrix} f_1(v).$$

This leads to the following expressions of the functions a and b in (3):

$$-\frac{\lambda^*(1,v)f_2(v)}{\rho(0,v)f_1(v)} = -\frac{(2-v)f_2(v)}{(2-v)^2 f_1(v)}, \quad \text{i.e.} \quad a(v) = \frac{2}{(2-v)}.$$

$$-\frac{\lambda^*(1,v)f_3(v)}{\rho(0,v)f_1(v)} = -\frac{(2-v)f_3(v)}{(2-v)^2 f_1(v)}, \quad \text{i.e.} \quad b(v) = -v.$$

Now we are able to find four functions $\{h_1, h_2, h_3, h_4\}$, coefficients of the G^2 continuity relation (3) between the surfaces S_l et S_r. Let us first calculate

$$A(v) = (\frac{\lambda^*(1,v)}{\rho(0,v)}\mathcal{R}_{uu}(0,v) - a^2\mathcal{L}^*_{uu}(1,v) - 2ab\mathcal{L}^*_{uv}(1,v) - b^2\mathcal{L}^*_{vv}(1,v))$$

$$A(v) = \frac{2-v}{(2-v)^2}(8,0,0,8) + \frac{4v}{2-v}(-1,0,1,0) = (4,0,\frac{4v}{2-v},\frac{8}{2-v}).$$

The unknown functions $\{h_1, h_2, h_3, h_4\}$ are then solutions of the 4×4 linear system

$$\begin{bmatrix} 4 & 2-v & 0 & 0 \\ 0 & 0 & 1 & v \\ \frac{4v}{2-v} & v & 0 & 0 \\ \frac{8}{2-v} & 0 & -1 & 2-v \end{bmatrix} \begin{bmatrix} h_1(v) \\ h_2(v) \\ h_3(v) \\ h_4(v) \end{bmatrix} = \begin{bmatrix} 0 \\ 0 \\ 0 \\ 0 \end{bmatrix},$$

and we obtain :

$$\begin{bmatrix} h_1(v) \\ h_2(v) \\ h_3(v) \\ h_4(v) \end{bmatrix} = \begin{bmatrix} 2-v \\ -4 \\ 4v \\ -4 \end{bmatrix} \frac{h_1(v)}{2-v}.$$

The parametrizations $R(u,v) = \Pi\Omega(\mathcal{R}(u,v))$ and $L^*(u,v) = \Pi\Omega(\mathcal{L}^*(u,v))$ of the surfaces S_l and S_r verify (3). Thus, the two surfaces are G^2 connected along Γ. ∎

We can give a geometric interpretation of this G^2 example. The surface S_l is parametrized by $L : [0,1]^2 \longrightarrow \mathbb{R}^3$ with $L(u,v) = \Pi\Omega(\mathcal{L}(u,v))$. Let us calculate

$$\mathcal{L}(u,v) = \sum_{i=0}^{1}\sum_{j=0}^{1} B_i^1(u)B_j^1(v)\mathcal{L}_{ij}$$

$$= ((1-u)(1-v)\mathcal{L}_{00} + u(1-v)\mathcal{L}_{10} + (1-u)v\mathcal{L}_{01} + uv\mathcal{L}_{11}$$

Finally, $\Pi\Omega(\mathcal{L}(u,v)) = (-(1-u),v,-(1-u)v)$. Thus, S_l is the patch of the hyperbolic paraboloid $S(x=s, y=t, z=st)$ defined on the quadrangle $\mathcal{Q}_l = \{-1 \leq s \leq 0, 0 \leq t \leq 1\}$.

Proposition 4. *The surface S_r is the patch of the hyperbolic paraboloid S defined on the convex quadrangle $Q_r = \{0 \leq s, 0 \leq t, s + t \leq 1, 2s - t \leq 1\}$. Thus, S_l and S_r are G^k connected along $\Gamma = \{(0, v, 0), v \in [0, 1]\}$, for all integer k.*

Proof: It is shown in [8] that the projective transformation induced by the linear transformation of $\mathbb{R}^3$ whose matrix is:

$$M = \begin{bmatrix} -2 & 0 & 2 \\ 0 & 1 & 0 \\ -2 & -1 & 4 \end{bmatrix}$$

maps the square domain $[0, 1]^2$ into the quadrangle Q_r. The homogeneous coordinates of the point (s, t, st) of S are $(X = sw, Y = tw, Z = st, T = w^2)$. Replacing (s, t, w) by $(-2u_1 + 2w_1, v_1, -2u_1 - v_1 + 4w_1)$, and setting $u = 1 - \frac{u_1}{w_1}$, $v = \frac{v_1}{w_1}$ we find:

$$\begin{cases} (u, v) \in [0, 1]^2 \\ X = 4u^2 + 4u - 2uv \\ Y = 2uv - v^2 + 2v \\ Z = 2uv \\ T = 4u^2 + 8u - 4uv + v^2 - 4v + 4 \end{cases}$$

These coordinates in $\mathcal{F}$ define a (SBP) surface, whose S_r is the $\Pi\Omega$-projection; we can write them in the Bernstein basis and obtain that :
$$S_r = SBR[r_{ij}, 0 \leq i \leq 2, 0 \leq j \leq 2; [0, 1]^2],$$ where the massic net $\{r_{ij}, 0 \leq i \leq 2, 0 \leq j \leq 2\}$ is given above in this section. ∎

§ 5. Conclusion

The boundary curve Γ along which the surfaces S_l and S_r are G^2 connected is defined by two (BR) curves whose massic polygons are not proportional. Yet, in most CAD applications [7], the user wants to build a G^2 composite rational surface, i.e. from a given (SBR) surface S_l, he defines the massic net of another (SBR) surface S_r, G^2 connected with S_l. Usually, he begins identifying the first row of massic vectors of S_r with the last one of the massic net defining S_l and, then, using *convenient* G^1 functions coefficients $\{f_1, f_2, f_3, f_4\}$ (resp. G^2 functions coefficients) he obtains the second row (resp. the third row) of the massic net of S_r. This process is supported by the relations (1,2,3) used as *sufficient* conditions. The second example developed shows that those relations (1,2, 3) are more general. In fact, they are G^2 *necessary and sufficient* conditions of continuity between the (SBR) surfaces S_l and S_r. Similar conditions have been obtained by Canonne [4], for the C^k continuity between two (SBR) surfaces.

References

1. Bauchat, J. L., Etude du raccordement géométrique G^2 de surfaces rationnelles mises sous forme (SBR), Thèse Université des Sciences et Technologies de Lille, Laboratoire A.N.O., 1992.

2. Bauchat, J. L., J. C. Fiorot, P. Jeannin, G^1 and G^2 continuity of (SBR) surfaces in *Curves and Surfaces* , P.J. Laurent, A. Le Méhauté and L.L. Schumaker (eds.), Academic Press, New York, 1991, 33-36.

3. Bauchat, J. L., Aspects projectifs du raccordement G^2 entre deux surfaces, Journées "Courbes et Surfaces", C.I.R.M. Luminy, 25-29 March 1991.

4. Canonne, J. C., J. C. Fiorot, P. Jeannin, $\mathcal{C}^k$ continuity of (SBR) surfaces in *Curves and Surfaces* , P.J. Laurent, A. Le Méhauté and L.L. Schumaker (eds.), Academic Press, New York, 1991, 67-70.

5. Canonne, J. C., A Necessary and Sufficient Condition for the $\mathcal{C}^k$ Continuity of Triangular Rational Surfaces, in *Curves and Surfaces in Geometric Design*, P.-J. Laurent, A. Le Méhauté, and L. L. Schumaker (eds.), A K Peters, Wellesley, 1994, 75–82.

6. Degen, W. L. F., Explicit continuity conditions for adjacent Bézier surfaces patches, Computer Aided Geometric Design **7** (1990), 181-189.

7. Farin, G., A construction for visual C^1 continuity of polynomial surfaces patches, Computer Graphics and Image Processing, **20** (1982), 272-282 .

8. Fiorot, J. C., P. Jeannin, *Courbes et Surfaces Rationnelles. Applications à la CAO*, Coll R.M.A. 12, Masson, Paris, 1989. English version: *Rational Curves and Surfaces . Application to CAD*, J. Wiley and Sons, Chichester, 1992.

9. Fiorot, J. C., P. Jeannin, *Courbes Splines Rationnelles. Applications à la CAO*, Coll R.M.A.24, Masson, Paris, 1992.

10. Hahn, J., Geometric continuous patch complexes, Computer aided Geometric Design, **6** (1989), 55-67.

11. Liu, D., GC^1 continuity conditions between two adjacent rational Bézier surfaces patches, Computer Aided Geometric Design, **7** (1990), 151-163.

Jean-Luc Bauchat
Lab. de Conception et de réalisation mécaniques assistées par ordinateur.
ENSAM, 8 Bd Louis XIV, 59046 Lille
FRANCE
bauchat@ ensam.decnet.citilille.fr

A 3D Generalized Voronoi Diagram
for a Set of Polyhedra

E. Bertin and J.-M. Chassery

Abstract. This paper is a study of the 3D Generalized Voronoi Diagram for a set of polyhedra. Two main ideas have been developed. The first is a theoretical study of separator equations between 3D polyhedral objects. The second is a generalization of an algorithm to compute an approximate solution of the 3D Generalized Voronoi Diagram using the 3D Voronoi Diagram for points. A demonstration of the convergence of this approach is provided.

§1. Introduction

Computational geometry [10] is used in many domains. For example, image analysis and pattern recognition make use of 2D and 3D Voronoi Diagram for Points (VDP) [2,9]. This paper deals with the Generalized Voronoi Diagram (GVD). The generalized concept may be understood in various ways:

1) Metric: in image analysis, we can use discrete distance function d4, d8, chamfer distances, and Euclidean distance [4].

2) Dimension: 2D, 3D, and nD [2,5,10].

3) Elements also called *seeds*: points, segments, polygons [7,8], and polyhedra.

One application of the 3D GVD is to the computation of the skeleton, the exoskeleton and the SKIZ (SKeleton by Influence Zone) of a set of 3D objects [6,7]. These may be used in pattern recognition, image analysis and robotics. In this paper, the problem at hand is the element generalization and the dimension generalization (3D) of the Voronoi diagram. In the second section, we present some formal definitions. In the third section, a detailed study of the 3D GVD is realised. In the fourth section, an algorithm to compute an approximation of the 3D GVD is given using the 3D VDP. A proof is provided for the convergence of our algorithm. Finally we give a conclusion.

Curves and Surfaces in Geometric Design
P. J. Laurent, A. Le Méhauté, and L. L. Schumaker (eds.), pp. 43–50.
Copyright © 1994 by A K PETERS, Wellesley, MA
ISBN 1-56881-039-3.

§2. Definitions

Our goal is to find a decomposition of polyhedral objects into *simple elements* (such as points, open segments, and open polygons), and then to define a distance between those simple elements. The choice of the Euclidean distance allows bisectors, separators, and the 3D GVD to be defined.

Definition 1. *Points, open segments and open polygons are called simple elements.*

Definition 2. *Let p and q be two points of $\mathbb{R}^3$, and let $d(p,q)$ be the Euclidean distance between p and q. The distance between a point p and a nonempty set S is defined by $d(p,S) = inf_{q \in S} d(p,q)$.*

Definition 3. *An object Obj is defined as a connected set composed of simple elements.*

Definition 4. *The bisector $B(ei,ej)$ between two simple elements e_i and e_j is defined as the set of points equidistant from e_i and e_j:*

$$B(ei,ej) = \{q \in \mathbb{R}^3, d(q,e_i) = d(q,e_j)\}.$$

Definition 5. *The separator between two distinct objects Obj_1 and Obj_2, denoted by $Sep(Obj_1, Obj_2)$, is defined as the set of points equidistant from Obj_1 and Obj_2:*

$$Sep(Obj_1, Obj_2) = \{q \in \mathbb{R}^3, d(q, Obj_1) = d(q, Obj_2)\}.$$

Definition 6. *The region $H(e_i, e_j)$ associated to the element e_i with respect to the element e_j is the set of points nearer to element e_i than to element e_j:*

$$H(e_i, e_j) = \{q \in \mathbb{R}^3, d(q, e_i) \leq d(q, e_j)\}.$$

Definition 7. *Let S be a set of objects. The Voronoi region associated with an object Obj_1, denoted by $Cel(Obj_1)$, is the set of points nearer to object Obj_1 than to the other objects of S. For a simple element e_i, we have*

$$Cel(e_i) = \bigcap_{e_j \in S - e_i} H(e_i, e_j), \text{ and then, } Cel(Obj_1) = \bigcup_{e_i \in Obj_1} Cel(e_i).$$

Definition 8. *The Voronoi diagram, $Vor(S)$, of a set S of objects, is the union of all the Voronoi regions associated with each object:*

$$Vor(S) = \bigcup_{Obj_p \in S} Cel(Obj_p).$$

Remark. *If the set S is composed exclusively of points, we recognize the classical definition of the 3D VDP.*

§3. Equations of the Bisectors and the Separators

In this section, the equations of bisectors between simple elements are given in 3D. using the Euclidean distance (see Definition 2). Then a general theorem on the separator between polyhedral objects is presented. We consider several cases, but leave out the degenerate cases. Throughout this section, $P(x, y, z)$ denotes any point of $\mathbb{R}^3$.

Case 1: (2 points a and b.) $B(a, b)$ is the bisector plane between a and b. Then we have the following lemma.

Lemma 1. *The bisector between two points a and b is the bisector plane of points a and b.*

Case 2: (a point a and a segment $]b, c[$). The problem reduces to the 2D case considering the plane P defined by the three points a, b and c. In this plane, the bisector is divided into straight lines and a parabolic arc [7]. Consequently in 3D, the bisector becomes a parabolic surface and planes.

Lemma 2. *The bisector between a point a and a segment $]b, c[$ consists of planes and a parabolic surface.*

Case 3: (a polygon π and a point a). Firstly, consider the case of a plane Π and a point a. Let $z = 0$ be the plane Π and a the point of coordinates $(0, 0, 1)$. The orthogonal projection of P onto Π is $H(x, y, 0)$. Looking for points P situated at equal distance from Π and a, we obtain $x^2 + y^2 + 1 = 2.z$, which is the equation of a paraboloid of revolution (Fig. 1).

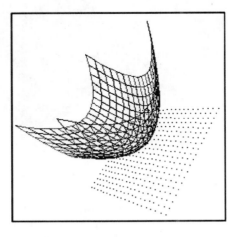

Fig. 1. Bisector between a point and a square.

If the equation of the plane Π is more complex, by reduction of the resulting quadric surface, we find the equation of a paraboloid of revolution. Let us examine the general case of a polygon π represented by vertices $(a_1, ..., a_n)$. We propose to decompose the problem into sub-problems. The points a, and a_i, $i \in [1...n]$, provide parts of plane (case 1). The point a and any segment $]a_i, a_{i+1}[$ generate parabolic surfaces (case 2). Finally the bisector between the point a and the polygon π is a paraboloid of revolution as was shown above.

Lemma 3. *The bisector between a point and a polygon is the union of parts of plane and parts of parabolic surfaces and a paraboloid of revolution.*

Case 4: (two segments $]a, b[$, and $]c, d[$). We suppose that the two segments are not coplanar. Firstly, before analysing the general case, let us consider two orthogonal and not coplanar straight lines D_1 and D_2 given by the following equations: $D_1 : \begin{cases} y = 0 \\ z = 1 \end{cases}$ and $D_2 : \begin{cases} x = 0 \\ y = -1 \end{cases}$. The orthogonal projection of P onto D_1 is $H_1(x, 0, 1)$, and the orthogonal projection of P onto D_2 is $H_2(0, y, -1)$. Looking for points P located at an equal distance from D_1 and D_2, we have $y^2 - x^2 = 4.z$, which is the equation of a hyperbolic paraboloid (Fig. 2).

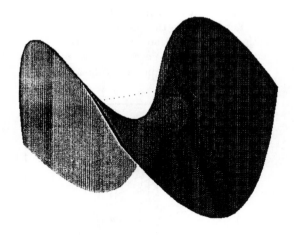

Fig. 2. Bisector between two segments.

The general case of noncoplanar straight lines D_1 and D_2 is obtained by reduction of quadratic form. Now, we are able to study the general case of two noncoplanar segments. The bisector includes 9 parts: one hyperbolic paraboloid generated by two segments $]a, b[$, and $]c, d[$, four parabolic surfaces generated by any extremal point a, b, c or d, and the complementary segment, and finally four parts of plane generated by any extremal bipoint (a, c), (a, d), (b, c) and (b, d).

Lemma 4. *The bisector between two segments is a union of half-planes, parts of a plane, parabolic surface, and an hyperbolic paraboloid.*

Case 5: (a segment $]a, b[$ and a polygon π). Firstly, consider a straight line D and a plane Π whose equations are $D : \begin{cases} x = 0 \\ y = 0 \end{cases}$ and $\Pi : \{ z = 0$. The orthogonal projection of P onto D is $H_1(0, 0, z)$, and the orthogonal projection of P onto Π is $H_2(x, y, 0)$. Points P equidistant from Π and D satisify $x^2 + y^2 - z^2 = 0$. We recognize here a cone of revolution around the z-axis (Fig. 3). The general case is obtained by reduction of a quadratic form. For a segment $]a, b[$ and a polygon π, by decomposition and with the previous lemmas, we have

Lemma 5. *The bisector between a segment and a polygon is a union of parts of a plane, parts of parabolic surfaces, parts of hyperbolic paraboloids, parts of paraboloids of revolution, and one cone.*

Case 6: (two polygons π_1 and π_2). It is well known that the bisector between two planes is a plane. Consequently, for two polygons, we have a part of plane generated by π_1 and π_2. The other bisectors result from the previous lemmas.

Lemma 6. *The bisector between two polygons is a union of pieces of a plane, parabolic surfaces, hyperbolic paraboloids, paraboloids of revolution and cones.*

Combining the previous cases leads to the following theorem.

Theorem 1. *Let S be a set of objects, for example polyhedra. Then the separators between objects of S consist of the following quadric surfaces: Paraboloid of revolution: $x^2 + y^2 + z = 1$, Hyperbolic paraboloid: $x^2 - y^2 = z$, Parabolic surface: $x^2 = z$, Cone: $x^2 + y^2 - z^2 = 0$, and Plane: $a.x + b.y + c.z + d = 0$.*

Proof: All the objects consist of unions of simple elements (see Definition 3). According to the above, the bisector equations between simple elements are exclusively planes, parabolic surfaces, hyperbolic paraboloids, paraboloids of revolution and revolution cones. According to Definitions 7 and 8, the resulting diagram is exclusively composed of those elementary quadrics, which completes the proof. ∎

§4. An Approximation of the 3D GVD by the 3D VDP

This section concerns the computation of the 3D GVD of a set O of polyhedra which is not an obvious problem of computational geometry. Since there are no methods to compute the 3D GVD directly, we propose to proceed by discretization of the polyhedra and use of the 3D VDP. This approach involves a definition of the dicretization.

Definition 9. *Let O be an object. A discretization of O is a finite set O_h of point seeds such as: for any point P inside O, there exists P_h inside O_h such that $d(P, P_h) < h$, where h is called the step of the dicretization.*

The algorithm has three main steps and may be described as follow:

1. Discretization of the objects of the set O (Such discretization provides a set O_h of points).
2. Computation of the 3D VDP based on the set O_h.
3. Deletion of the edges created by two points of the set O_h belonging to the same object (such edges are called *useless edges*).

The result obtained with this algorithm is an approximation of the 3D GVD of the set of objects (Fig. 4).

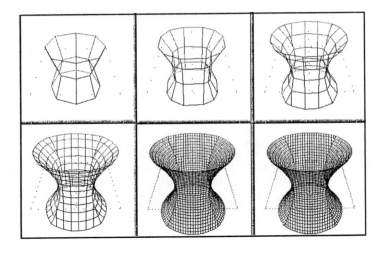

Fig. 3. Convergence of the algorithm.

Figure 5 shows the result obtained for the interior of a cube. This algorithm is a generalization of the 2D case [3,6,9]. The proof of this algorithm is given by Theorem 2 for the 3D case.

To avoid classical problems of computational geometry, let us assume that no five points are on the same sphere, and no four points are on the same circle.

Theorem 2. *Let S be a set of objects, and S_h a discretization of S. The 3D DVP of S_h converges toward the 3D GVD of S when the step of discretization h goes to zero.*

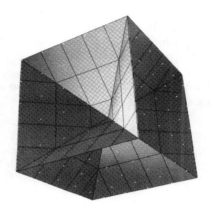

Fig. 4. Separator for the interior of a square.

Proof: Let O^1 (resp. O^2) be two objects, and O_h^1 (resp. O_h^2) their discretization. Let O_h be the union of O_h^1 and O_h^2. Let S_h be a Voronoi vertex of the 3D DVP based on O_h. S_h is created by four points P_1, P_2, P_3, and P_4 not belonging to the same object because of the linearity of O^1 and O^2. There are two cases. Case 1: P_1, P_2, P_3 are onto O^1 and P_4 onto O^2. (P_1, P_2, P_3, P_4) is a Delaunay tetrahedron, that is the ball B circumscribed by the four seeds P_1, P_2, P_3, and P_4 is empty. This is a classical property of the Delaunay tetrahedralization which is the dual partition of the Voronoi tesselation [10]. Furthermore, let H, (respec. H') be the projection onto O^1 (respec. O^2) of S_h. We have $d(H, P_i) < h$, for $i \in \{1, 2, 3\}$. Otherwise, if $d(H, P_i) \geq h$, then there exists P_n on O_h^1 such that $P_n \in B$ which gives a contradiction. For the same reason $d(H', P4) < h$. Then when h goes to zero, H, P_1, P_2, and P_3 are identical, H' and P_4 also. At the limit case, we have $d(H', S) = d(H, S)$, which completes the proof for this case. Case 2: P_1, P_2 are onto O^1 and P_3, P_4 are onto O^2. A similar reasoning completes the proof of this case. ■

This result is also established in [11].

§5. Conclusion

In this paper, we have given some general results about the 3D GVD. In particular, the 3D GVD of polyhedral objects turns out to be a union of parts of quadric surfaces: paraboloid of revolution, hyperbolic paraboloid, parabolic surface, cone and plane. Due to such properties, one would like to compute the exact 3D GVD with a specific approach like in 2D [7,8]. Unfortunately, many algorithmic difficulties appear in 3D: the computation of the intersection of two quadric surfaces, the construction of the convex hull in 3D, the problem of visibility of two objects, etc. Consequently, we have proposed an algorithm to compute an approximation of the GVD in 3D. It consists of a discretization of the objects (such discretization provides a set S of points), the computation of the 3D VDP based on the set S and the suppression of useless edges. This algorithm can be generalized in 2D for objects not necessary polygonal [1,6], and future research would be made in 3D for a set of objects not necessary polyhedral [1].

Acknowledgements. We would like to thank P. Moreau and D. Attali for enlightening discussions.

References

1. Attali D., E. Thiel, Du squelette discret ou continu. Colloque Géométrie Discrète en Imagerie-Fondement et applications. Strasbourg. (1993), 236-244.
2. Bertin E., F. Parazza, and J.-M. Chassery, Segmentation and measurement based on 3D Voronoi diagram: application to confocal microscopy, Computerized Medical Imaging and Graphics **17** (1993), 175-182.
3. Boissonnat J.-D. , O. D. Faugeras and E. Le Bras-Mehlman, Representing stereo data with the Delaunay triangulation. Rapport INRIA **788** (1988).
4. Chassery J.-M. and A. Montanvert, *Géométrie discrète en analyse d'images*. Ed. Hermes, 1991.
5. Bowyer A., Computing Dirichlet tesselation, The Computer Journal **24**, 2 (1981), 162-166.
6. Brandt J. W., Algazi V. R., Continuous Skeleton Computation by Voronoi Diagram, CVGIP, Image Understanding **55** (1992), 329-338.
7. Hu H. T., Diagramme de Voronoï Généralisé Pour un ensemble de Polygones. PhD. Thesis, University Joseph Fourier, Grenoble, France, 1991.
8. Lee D. T. and R. L. Drysdale, Generalization of Voronoi Diagrams in the plane, J. Comp. **10** (1981), 363–369.
9. Melkemi M., Approches géométriques par modèles de Voronoï en segmentation d'images. PhD. Thesis, University Joseph Fourier, Grenoble, France, 1992.
10. Preparata J. P. and M. I. S Shamos, *Computational Geometry, an Introduction*, Springer, New York, (1988).
11. Schmitt F., Some examples of algorithms analysis in computational geometry by means of mathematical morphology techniques. in *Lecture Note in Computer Science, Geometry and Robotics*, J.-D Boissonnat and J.-P. Laumond (eds.), Springer Verlag, Berlin, 1989, 225-246.

E. Bertin and J.-M. Chassery
TIMC-IMAG, Université Joseph Fourier, BP 53
38041 Grenoble cedex 9, FRANCE
Etienne.Bertin@imag.fr

Variational Design of
Rational Bézier Curves and Surfaces

G.-P. Bonneau and H. Hagen

Abstract. The design of curves and surfaces in CAD systems has many applications in the car, aeroplane and ship industries. Because they offer more flexibility, rational functions are often preferred to polynomial functions to model curves and surfaces. In this work, several methods to generate weights of rational Bézier curves and surfaces which minimize some functionals are proposed. The functionals measure a technical smoothness of the curves and surfaces, and are related to the energy of beams and plates in the sense of elasticity theory.

§1. Introduction

The purpose of Computer Aided Geometric Design (CAGD) is to define some mathematical modelling of free-form curves and surfaces, to study their properties, and to improve their quality. The Bézier and B-splines curves and surfaces were first based on polynomial parametric equations. The wish to give more freedom to the designer has led to the generalization of polynomial schemes to rational functions. The use of rational functions in CAGD began in the late 70's. Rational B-splines (also called NURBS, short for non-uniform rational B-splines) are becoming now a standard in CAD technology (see [1]). The flexibility of these curves and surfaces is achieved through the assignment of a scalar (called weight) to each control point. If a weight increases while the others remain constant, the curve or surface is pulled in the direction of the corresponding control point. The increase in flexibility (in particular the ability to exactly describe conics) is paid by an increased complexity of already known algorithms (e.g. the evaluation of derivatives), but has also brought new algorithms (for instance the reparameterization of rational curves , see [3]).

This paper deals with the rational counterpart of the Bézier schemes: the rational Bézier curves, the rational rectangular Bézier patches, and the rational triangular Bézier patches.

Curves and Surfaces in Geometric Design
P. J. Laurent, A. Le Méhauté, and L. L. Schumaker (eds.), pp. 51–58.
Copyright © 1994 by A K PETERS, Wellesley, MA
ISBN 1-56881-039-3.

In attempting to improve the quality of a mathematical modelling of curves and surfaces, algorithms based on the minimization of a functional are of great importance. The process of minimization is often constrained by interpolation conditions ([10, 12, 4, 13]). But a least square condition is sometimes more suitable than the interpolation condition. Hagen and Santarelli ([8]) use the minimization of the integral $\int \alpha \|X''(t)\|^2 + \beta \|X'''(t)\|^2 dt$, together with a least square constraint, to obtain Bézier and B-spline polynomial curves. They extend this result to the surface case .

In the present paper, new functionals are introduced, which can be used as minimization criteria to produce rational Bézier curves (§3), and rational tensor-product or triangular Bézier patches (§4).

We want to make maximal use of the additional parameters of rational schemes: the weights. Therefore we only allow the weights to vary during the process of minimization. Then the control points can be given directly by the user, and are not affected by the variational process.

To perform the minimization, we need to calculate the derivatives of the rational curves and surfaces as functions of the weights. An important idea here is to find an appropriate reparameterization for which the curve (or the surface) and all its derivatives become a polynomial function in the weights, at a particular parameter point. This result is presented in the case of rational Bézier curves in §2. It is quite independent of our other results, and could be used in other problems involving derivatives of rational curves and surfaces.

§2. Reparameterization of Rational Bézier Curves

We assume that the reader is familiar with the definition of rational Bézier curves (see [2, 11]). If the parameter value u of such a curve is changed by a properly chosen function of the type $\frac{au+b}{cu+d}$ (called here a rational linear function), then the control points are unchanged, and only the weights are changed (see [15, 3, 12]). For example, one can find a rational linear function such that the reparameterized curve has its first and last weights equal to one. The next theorem generalizes this result.

Theorem 1. *For any rational Bézier curve, and for any two parameter values a, b in its parameter interval* $[u_0, u_1]$, *there exists a rational linear reparameterization of the curve such that,*

  i) *the control points of the curve are unchanged*
  ii) *if* $\bar{\omega}_0, \cdots, \bar{\omega}_n$ *are the new weights, then*

$$\sum_{i=0}^{n} \bar{\omega}_i B_i^n \left(\frac{a - u_0}{u_1 - u_0} \right) = 1$$
$$\sum_{i=0}^{n} \bar{\omega}_i B_i^n \left(\frac{b - u_0}{u_1 - u_0} \right) = 1. \tag{1}$$

Proof: A rational linear reparameterization $\varphi(u)$ which preserve the parameter interval $[u_0, u_1]$ of the curve is of the form

$$\varphi(u) = \frac{\rho u_1 (u - u_0) + \hat{\rho} u_0 (u_1 - u)}{\rho (u - u_0) + \hat{\rho} (u_1 - u)},$$

where ρ and $\hat{\rho}$ are two non zero scalar values with the same sign, so that the denominator doesn't vanish. If $b_0, \cdots, b_n$ and $\omega_0, \cdots, \omega_n$ are respectively the control-points and control-weights of the curve X, then the reparametrized curve $\bar{X}$ has the following parametric equation :

$$\bar{X}(u) = X(\varphi(u)) = \frac{\sum_{i=0}^{n} \bar{\omega}_i b_i B_i^n \left(\frac{u - u_0}{u_1 - u_0} \right)}{\sum_{i=0}^{n} \bar{\omega}_i B_i^n \left(\frac{u - u_0}{u_1 - u_0} \right)},$$

with $\bar{\omega}_i = \rho^i \hat{\rho}^{n-i} \omega_i$, $i = 0, \ldots, n$. Dividing the two equations (1) by $\hat{\rho}^n$ and writing $\alpha = \rho / \hat{\rho}$, we find the following equivalent conditions to (1):

$$\exists\, \alpha > 0 /$$

$$\sum_{i=0}^{n} \alpha^i \omega_i \left[B_i^n \left(\frac{a - u_0}{u_1 - u_0} \right) - B_i^n \left(\frac{b - u_0}{u_1 - u_0} \right) \right] = 0 \qquad (2)$$

$$\frac{1}{\hat{\rho}^n} = \sum_{i=0}^{n} \alpha^i \omega_i B_i^n \left(\frac{a - u_0}{u_1 - u_0} \right)$$

$$\rho = \hat{\rho} \alpha.$$

If $a = b$ then $\rho = 0$ and $\hat{\rho} = \left(\omega_0 B_0^n (\frac{a - u_0}{u_1 - u_0}) \right)^{-\frac{1}{n}}$ is a solution.

We assume now $a \neq b$, and denote the left member of the equation (2) by f. Then $f(\alpha)$ is a polynomial of degree n, with first and last coefficients of opposite sign (because ω_0 and ω_n must have the same sign, B_0^n is a strictly decreasing function, B_n^n a strictly increasing function, and $a \neq b$). So f must have at least one positive root. ∎

Proposition 2. *If $a \neq b$, then for any two indices $i_0 \neq i_1$, the system of linear equations (1) has a unique solution in the weights $\bar{\omega}_{i_0}, \bar{\omega}_{i_1}$.*

The equations (1) assure that the denominator of the parametric equation of the curve "disappears" for the two parameter values a and b. Proposition 2 permits the solution of the system (1) for the weights $\bar{\omega}_{i_0}, \bar{\omega}_{i_1}$. At the parameter values a and b, the curve then becomes a polynomial function in the other weights. In other words, Theorem 1 states that, *for any two parameter values, there exists a reparameterization such that, in these two parameter values, the reparameterized curve and all its derivatives are polynomial functions of the weights.*

We illustrate Theorem 1 with a rational Bézier curve of degree 3, and control weights $(1, 1, 1, 20)$. We apply Theorem 1 with the parameter values $a = 0.8$ and $b = 1.0$. After the reparameterization, the new weights are $(16.20, 2.36, 0.34, 1.00)$. Fig. 1 shows the curve before and after the reparameterization. The points plotted on the curves are images of regularly-spaced parameter values in the parameter interval.

Fig. 1. Reparameterization of a rational Bézier curve.

§3. Weight Estimation of Rational Bézier Curves

Free-form objects are an essential part of powerful CAD systems. A major issue is the generation of smooth curves and surfaces which can be immediately supplied to the NC-process. The fundamental idea of our method is to minimize a certain functional that can be interpreted in the sense of physics/geometry.

In the case of curves, a thin elastic beam can serve as a model for a fair shape. Such a beam tends to take a position of least strain energy and the energy stored in the beam is proportional to the integral

$$\int \kappa^2(t)\|X'(t)\|dt. \tag{3}$$

As an approximation of the integral criterion (3), we can use

$$\int \|X''(t)\|^2 dt. \tag{4}$$

Following the physical analogy, we can add a jerk term to (4):

$$\int \alpha\|X''(t)\|^2 + \beta\|X'''(t)\|^2 dt. \tag{5}$$

Minimizing the integral (5) performs a blended optimization of energy and jerk. We adjust α and β in an interactive way. More information on the use of this integral for non-rational curves can be found in [9].

In the case of rational Bézier curves, the integral (5) is a transcendental function of the weights. Therefore we use in [5] a discretization of the integral (5) at the two endpoints of the curve:

$$\alpha(\|X''(u_0)\|^2 + \|X''(u_1)\|^2) + \beta(\|X'''(u_0)\|^2 + \|X'''(u_1)\|^2). \tag{6}$$

The results of §2 allow us to use a more general discretization: given any two parameter value a, b in the parameter interval $[u_0, u_1]$, we minimize the functional

$$\alpha(\|x''(a)\|^2 + \|x''(b)\|^2) + \beta(\|x'''(a)\|^2 + \|x'''(b)\|^2). \tag{7}$$

While the function (6) depends only on the first three and last three weights of the curve, function (7) involves all of the weights. Furthermore, a proper choice of the parameter values a and b in the function (7) can give better results.

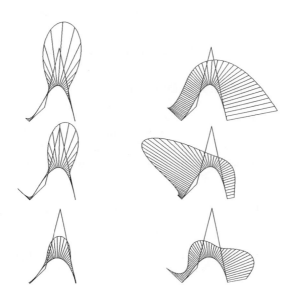

Fig. 2. Weight estimation of a rational cubic Bézier curve.

We illustrate the minimization of the functions (6) and (7) on a cubic Bézier curve parameterized on $[u_0 = 0, u_1 = 1]$, with $\alpha = 0.8$ and $\beta = 0.2$. For this example, we use $a = 0.2$ and $b = 0.64$ in the function (7) (0.64 is the parameter value in which the non-rational curve reaches its highest curvature value). Figure 2 shows the non-rational curve (top), the curve with the weights minimizing (6) (middle), and the curve with the weights minimizing (7) (bottom). Each curve $(u \rightarrow X(u))$ is represented twice, to the left together with the curve $(u \rightarrow X(u) + f(u)N(u))$ where $f(u) = \kappa^2\|X'(u)\|$, N is the normal vector, and to the right taking $f(u) = \alpha\|X''(u)\|^2 + \beta\|X'''(u)\|^2$. This gives a good idea of the minimization of both integrals (3) and (5).

§4. Weight Estimation of Rational Bézier Surfaces

In the case of surfaces, a thin elastic plate of small deflection can serve as a model for a fair shape. Such a plate tends to take a position of least strain energy of flexure and torsion. The energy stored in this plate is proportional to the integral

$$\int_S (\kappa_1^2 + \kappa_2^2) ds. \tag{8}$$

In [6], we use a quadrature of the integral (8) to find weights for rational tensor-product Bézier patches. In this paper we present another solution for both tensor-product and triangular Bézier patches.

In the case of *tensor-product Bézier patches*, following [14], we can use

$$\int \int \|X_{uu}\|^2 + 2\|X_{uv}\|^2 + \|X_{vv}\|^2 dudv \tag{9}$$

as an approximation of the integral criterion (8).

For general rational tensor-product Bézier patches, the integral (9) is a transcendental function of the weights. To get a polynomial function in the weights, we use a discretization of the integral (9) at the corners of the parameter domain $[u_0, u_1] \times [v_0, v_1]$. We assume the four corners weights to be equal to one, and using the other weights, we minimize the function

$$\sum_{i=0,1;j=0,1} \|X_{uu}(u_i, v_j)\|^2 + 2\|X_{uv}(u_i, v_j)\|^2 + \|X_{vv}(u_i, v_j)\|^2. \tag{10}$$

For a *triangular Bézier patch*, the parameter domain is the triangle of vertices $(u_{ijk})_{i+j+k=1;i,j,k>=0}$. In this case we prefer to use the following analog of the function (6):

$$\sum_{i+j+k=1,i,j,k>=0} \Big(\alpha\|X_{uu}(u_{ijk})\|^2 + \beta\|X_{uuu}(u_{ijk})\|^2 +$$
$$\alpha\|X_{vv}(u_{ijk})\|^2 + \beta\|X_{vvv}(u_{ijk})\|^2 + \tag{11}$$
$$\alpha\|X_{ww}(u_{ijk})\|^2 + \beta\|X_{www}(u_{ijk})\|^2 \Big).$$

To show the minimization of the criteria (10) and (11), we use the generalized focal surfaces tool (see [7]): *Each surface X in Figures 3, 4 and 5 is represented together with the surface $X + fN$, where f is an offset function, and N is the normal vector.*

We illustrate the minimization of the criterion (10) on a biquintic surface with four patches. The offset function for Figure 3 is $f(u, v) = \|X_{uu}\|^2 + 2\|X_{uv}\|^2 + \|X_{vv}\|^2$. This means that the distances between the two surfaces X and $X + fN$ along the normal vector is equal to the integrand of (9). Thus *Figure 3 shows both the minimization of the approximated energy integral (9), and of our criterion function (10).*

The surface with all weights equal to 1 is shown on the left of Figure 3, the surface with the new weights which minimize the function (10) on the right.

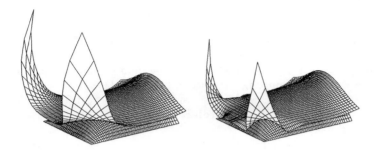

Fig. 3. Weight estimation for a biquintic rational Bézier surface.

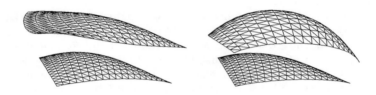

Fig. 4. Weight estimation for a cubic rational triangular Bézier patch.

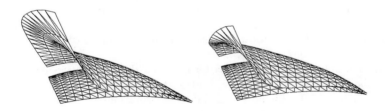

Fig. 5. Weight estimation for a cubic rational triangular Bézier patch.

We illustrate the minimization of the criterion (11) on a triangular cubic Bézier patch. For this example we choose $\alpha = 0.8$ and $\beta = 0.2$ in the function (11). The offset function for Figure 4 is

$$f = \alpha\|X_{uu}\|^2 + \beta\|X_{uuu}\|^2 + \alpha\|X_{vv}\|^2 + \beta\|X_{vvv}\|^2 + \alpha\|X_{ww}\|^2 + \beta\|X_{www}\|^2.$$

Therefore, *Figure 4 shows the minimization of our criterion function* (11). For

Figure 5, we choose

$$f = (\kappa_1^2 + \kappa_2^2)\sqrt{g},$$

where g is the determinant of the first fundamental form. Thus, *Figure 5 shows the minimization of the exact energy integral* (8). On the left of Figures 4 and 5 is the surface with all weights equal to one, and on the right is the surface with the weights minimizing our criteria function (11).

References

1. Farin, G., Algorithms for rational Bézier curves, Comput. Aided Design **15**, No. **2** (1983), 73–77.
2. Farin G., *Curves and Surfaces for Computer-Aided Geometric Design* 2nd. edition, Academic Press, New-York, 1988.
3. Farin, G. and A. Worsey, Reparameterization and degree elevation for rational Bézier curves, in *NURBS for Curve and Surface Design*, G. Farin (ed.), SIAM, 1991, 47–57.
4. Hagen, H., Geometric spline curves, Computer-Aided Geom. Design **2** (1985), 223–227.
5. Hagen, H. and G. P. Bonneau, Variational design of smooth rational Bézier curves, Computer-Aided Geom. Design **8** (1991) 393–399.
6. Hagen, H. and G. P. Bonneau, Variational design of smooth rational Bézier surfaces, Computing, Supplementum **8** (1993) 133–138.
7. Hagen, H. and S. Hahmann, Generalized focal surfaces: a new method for surface interrogation, in *Proceedings Visualization '92*, A.E. Kaufman and G.M. Nielson (eds.), IEEE, 1992, 70–75.
8. Hagen, H. and P. Santarelli, Variational design of smooth B-splines surfaces, in "Topics in Surface Modeling", H. Hagen ed., SIAM, 1992, 85–92.
9. Hagen, H. and G. Schulze, Variational principles in curve and surface design, in *Geometric Modeling - Methods and Applications*, Springer, Berlin, 1991, 161–184.
10. Holladay, J.C., Smoothest curve approximation, Math. Table Aids Comput., **11** (1957), 233–243.
11. Hoschek, J. and D. Lasser: *Grundlagen der Geometrischen Datenverarbeitung* 2nd. ed., Teubner, Stuttgart, 1992.
12. Lucian, M. L., Linear fractional transformations of rational Bézier curves, in *NURBS for Curve and Surface Design*, G. Farin ed., SIAM, 1991, 131–139.
13. Nielson, G., Some piecewise polynomial alternatives to splines under tension, in *Computer Aided Geometric Design*, R.E. Barnhill, W. Böhm (eds.), 1974, 209–235.
14. Nowacki, H., D. Reese, Design and fairing of ship surfaces, in *Surfaces in CAGD*, R.E. Barnhill, W. Böhm (eds.), 1983, 121–134.
15. Patterson, R. R., Projective transformations of the parameter of a Bernstein-Bézier curve, ACM Trans. Graphics **4**, No. **4**, (1985), 276–290.

An Introduction to Padé Approximations

C. Brezinski

Abstract. Padé approximants are rational approximations to functions defined as a formal power series expansion. Their power series expansion matches the series as far as possible. Padé approximants have many applications. This paper provides a short introduction to this subject.

The aim of this paper is to provide an introduction to Padé approximations. This is a wide subject connected to other important topics such as continued fractions and orthogonal polynomials and having many applications in numerical analysis and in applied mathematics. So we shall restrict our discussion here to the basic definitions and results, and refer the interested reader to [5] and the literature quoted here.

Let f be a formal power series

$$f(z) = c_0 + c_1 z + c_2 z^2 + \cdots$$

We are looking for a rational function

$$R(z) = \frac{a_0 + a_1 z + \cdots + a_p z^p}{b_0 + b_1 z + \cdots + b_q z^q}$$

such that its power series expansion in ascending powers of z (obtained by division) agrees with that of f *as far as possible*, that is

$$f(z) - R(z) = O(z^{p+q+1}). \tag{1}$$

Such a rational fraction is called a *Padé approximant* of f and it is denoted by

$$[p/q]_f(z).$$

Padé approximants are named after the French mathematician Henri Padé (1863–1953) who studied them in detail in his thesis in 1892 under the supervision of Charles Hermite (1821–1901). But in fact these approximants were already known. They were obtained in 1758 by Johann Heinrich

Curves and Surfaces in Geometric Design
P. J. Laurent, A. Le Méhauté, and L. L. Schumaker (eds.), pp. 59–65.
Copyright © 1994 by A K PETERS, Wellesley, MA
ISBN 1-56881-039-3.

Lambert (1728–1777) and also by Joseph Louis Lagrange (1736–1813) in 1776 via the continued fraction approach. An historical study can be found in [3]. See [11] for the biography of Padé.

Multiplying both sides of (1) by the denominator of $[p/q]$ and equating the coefficients of the identical powers of z, we obtain

$$a_0 = c_0 b_0$$
$$a_1 = c_1 b_0 + c_0 b_1$$
$$\vdots$$
$$a_p = c_p b_0 + c_{p-1} b_1 + \cdots + c_{p-q} b_q$$
$$0 = c_{p+1} b_0 + c_p b_1 + \cdots + c_{p-q+1} b_q$$
$$\vdots$$
$$0 = c_{p+q} b_0 + c_{p+q-1} b_1 + \cdots + c_p b_q$$

with the convention that $c_i = 0$ for $i < 0$.

Since a rational fraction is defined up to a multiplying factor, we can set $b_0 = 1$ and thus the last q equations give $b_1, \ldots, b_q$ and then $a_0, \ldots, a_p$ can be obtained directly from the $p + 1$ first equations.

Usually the approximants are displayed in a two dimensional array called the Padé table

$$
\begin{array}{cccc}
[0/0] & [0/1] & [0/2] & \cdots \\
[1/0] & [1/1] & [1/2] & \cdots \\
[2/0] & [2/1] & [2/2] & \cdots \\
\vdots & \vdots & \vdots & \ddots
\end{array}
$$

From the preceding system we see that $[p/q]$ exists and is unique if and only if the system giving the b_i's is nonsingular, that is, if and only if the Hankel determinant

$$
H_{q-1}^{(p-q+1)} =
\begin{vmatrix}
c_{p-q+1} & \cdots & c_p \\
\vdots & & \vdots \\
c_p & \cdots & c_{p+q-1}
\end{vmatrix}
$$

is different from zero. If this condition is satisfied for all p, q, the Padé table is said to be *normal*. It was proved by Padé in his thesis [11] that, in the non–normal case, adjacent identical approximants can only occur in square blocks in the Padé table.

Let us give a numerical example showing the effectiveness of Padé approximants. We consider the series

$$f(z) = \log(1 + z) = z - z^2/2 + z^3/3 + \cdots$$

Let us denote by $S_k(z)$ the partial sums of this series. For $z = 1$, we have $\log 2 = 0.6931471805599453 \ldots$ and we obtain

k	$S_{2k}(1)$	$[k/k]_f(1)$
1	0.830	0.7
2	0.783	0.6933
3	0.759	0.693152
4	0.745	0.69314733
5	0.736	0.6931471849
6	0.730	0.69314718068
7	0.725	0.693147180563
8	0.721	0.69314718056000
9	0.718	0.6931471805599485
10	0.716	0.6931471805599454

Exact digits are underlined.

We mention that to obtain an error of the order of 10^{-n} by computing the partial sums of the series, 10^n terms are needed. For $z = 2$, the series diverges and we have $\log 3 = 1.098612288668110\ldots$

k	$S_{2k}(2)$		$[k/k]_f(2)$
1	0.260	10^1	1.14
2	0.506	10^1	1.101
3	0.126	10^2	1.0988
4	0.375	10^2	1.098625
5	0.121	10^3	1.0986132
6	0.410	10^3	1.09861235
7	0.142	10^4	1.098612293
8	0.504	10^4	1.0986122890
9	0.181	10^5	1.098612288692
10	0.655	10^5	1.0986122886698

This example shows that Padé approximants can be used for analytic continuation of some series outside their disc of convergence.

We now discuss the algebraic theory of Padé approximants. For simplicity, let us consider only the case $p = k - 1$ and $q = k$, that is the approximants $[k - 1/k]$. The relations giving the coefficients b_i of the denominator are

$$c_0 b_k + \cdots + c_{k-1} b_1 + c_k = 0$$
$$\vdots$$
$$c_{k-1} b_k + \cdots + c_{2k-2} b_1 + c_{2k-1} = 0.$$

If we define the linear functional c on the space of complex polynomials by

$$c(x^i) = c_i, \qquad i = 0, 1, \ldots$$

and if we set

$$P_k(x) = b_k + b_{k-1} x + \cdots + b_1 x^{k-1} + x^k,$$

then the preceding relations can be written, for $i = 0, \ldots, k - 1$, as

$$b_k c(x^i) + \cdots + b_1 c(x^{i+k-1}) + c(x^{i+k}) = c(x^i P_k(x)) = 0.$$

Polynomials satisfying such conditions are called *formal orthogonal polynomials* with respect to the linear functional c. They enjoy all the properties of the usual orthogonal polynomials except some on their zeros. In particular, they satisfy a three–term recurrence relationship and the Christoffel–Darboux identity.

We have

$$f(z) = c\left(\frac{1}{1-xz}\right) = c(1 + xz + x^2z^2 + \cdots)$$

$$= c(1) + c(x)z + \cdots = c_0 + c_1z + c_2z^2 + \cdots$$

Let R_k be the Hermite interpolation polynomial of the function $x \to (1-xz)^{-1}$ at the zeros of P_k. Then, it can be proved that

$$c(R_k(x)) = [k-1/k]_f(z)$$

which shows that Padé approximants can be considered as formal Gaussian quadrature processes. We have

$$f(z) - [k-1/k]_f(z) = \frac{z^{2k}}{\widetilde{P}_k^2(z)} c\left(\frac{P_k^2(x)}{1-xz}\right)$$

with $\widetilde{P}_k(x) = x^k P_k(x^{-1}) = 1 + b_1 x + \cdots + b_k x^k$. Since Padé approximants are Gaussian quadratures, then the error $f(z) - [k-1/k]_f(z)$ can be estimated by Kronrod's procedure.

Using the theory of formal orthogonal polynomials, it is quite easy to see that the denominator $\widetilde{P}_k^{(n)}$ of the approximant $[n+k-1/k]$ is such that

$$c^{(n)}(x^i P_k^{(n)}(x)) = 0 \qquad \text{for} \qquad i = 0, \ldots, k-1,$$

where $P_k^{(n)}(x) = x^k \widetilde{P}_k^{(n)}(x)$, and where the linear functional $c^{(n)}$ is defined by

$$c^{(n)}(x^i) = c_{n+i}.$$

Thus the polynomials $P_k^{(n)}$ are orthogonal with respect to $c^{(n)}$. The families $\{P_k^{(n)}\}$ are called adjacent families of orthogonal polynomials. There exists many recurrence relationships between members of adjacent families of orthogonal polynomials. They provide relations for computing recursively any sequence of adjacent Padé approximants [2]. They have been extended to the non–normal case by Draux [7]. The corresponding FORTRAN subroutines can be found in [8].

Let us mention that Padé approximants are related to the ε–algorithm of Wynn [13] which is a procedure for implementing Shanks sequence transformation [12], a method for accelerating the convergence of some sequences by extrapolation [4].

We shall now discuss the convergence of Padé approximants that is the convergence of a sequence of Padé approximants. This is a quite difficult problem as illustrated by the following two examples.

Let $f(z) = (10 + z)/(1 - z^2)$. This series converges for $|z| < 1$. We have

$$[k/1]_f(z) = \sum_{i=0}^{k-1} c_i z^i + \frac{c_k z^k}{1 - \frac{c_{k+1}}{c_k} z}$$

with $c_{2i} = 10$ and $c_{2i+1} = 1$. Thus, when k is odd, $[k/1]$ has a pole at $z = 0.1$ and the sequence $([k/1]_f)$ cannot converge to f in $|z| < 1$ while the series does.

Thus poles can prevent convergence, not a surprising result. But zeros can also prevent it. As an illustration, let us consider the series $f(z) = (1 - z^2)/(10 + z)$ which is the reciprocal of the preceding one. It converges in $|z| < 10$. However

$$[1/2k + 1]_f(0.1) = 0 \qquad \text{while} \qquad f(0.1) \neq 0,$$

which shows that the sequence $([1/k]_f)$ cannot converge to f in $|z| < 10$ while the series does.

Because of the difficulty of the problem, there are not so many convergence results for Padé approximants. The first theorem concerns functions f analytic at the origin, meromorphic whith n poles $\alpha_1, \ldots, \alpha_n$ counted with their multiplicities in the open disc D_k of radius R. Then, as proved by R. Montessus de Ballore in 1902, the sequence $([k/n])$ converges to the function f uniformly on every compact subset of $D_k - \{\alpha_1, \ldots, \alpha_n\}$ when k goes to infinity and the poles of $[k/n]$ tend to those of f [10].

Quite complete convergence results are known for two classes of functions:

1) *Stieltjes series*, that is series such that

$$c_i = \int_0^\infty x^i d\alpha(x)$$

with α bounded and nondecreasing in $[0, \infty)$. For such series, uniform and geometric convergence occurs on every compact set of the complex plane cut along $[0, \infty)$.

2) *Pólya frequency series*, that is series of the form

$$f(z) = a_0 e^{\gamma z} \prod_{i \geq 0} (1 + \alpha_i z)/(1 - \beta_i z)$$

with $a_0 > 0$, $\gamma \geq 0$, $\alpha_i \geq 0$, $\beta_i \geq 0$ and $\sum_{i \geq 0}(\alpha_i + \beta_i) < +\infty$. Let (m_k) and (n_k) be two sequences of positive integers such that $\lim_{k \to \infty} m_k = \infty$ and $\lim_{k \to \infty} m_k/n_k = w \in [0, \infty)$. Then the sequence $([m_k/n_k]_f)$ converges to f uniformly on any compact subset of the complex plane. This result was obtained by Arms and Edrei [1].

There exists several generalizations of Padé approximants. The first one concerns the so–called Padé–type approximants. In Padé approximation, no free choice is left to the user. Thus it could be interesting to construct rational approximations in which the user can choose the denominator (that is the poles) and then the numerator is constructed in order to achieve the maximum order of approximation at the origin. Let v_k be an arbitrary polynomial of degree k and let w_k be defined by

$$w_k(z) = c \left(\frac{v_k(z) - v_k(x)}{z - x} \right),$$

where c acts on x. w_k is a polynomial of degree $k - 1$ in z and we set $\widetilde{v}_k(z) = z^k v_k(z^{-1})$ and $\widetilde{w}_k(z) = z^{k-1} w_k(z^{-1})$. Then it can be proved that

$$\widetilde{w}_k(z)/\widetilde{v}_k(z) - f(z) = O(z^k).$$

Such an approximant is called a Padé–type approximant and it is denoted by $(k - 1/k)_f(z)$ [2].

If some information on the poles of f are available, these approximants can give better results than the usual Padé approximants. If f is analytic in a domain D of the complex plane and if $\lim_{k \to \infty} v_k(x)/v_k(z^{-1}) = 0$ uniformly with respect to x and z on any compact subset of an open set A containing $\{(x, 0), x \in \mathbb{C}\}$ then, as proved by Eiermann [9], $\lim_{k \to \infty} (k - 1/k)_f(z) = f(z)$ uniformly on any compact subset of

$$\{z \in \mathbb{C}, \lim_{k \to \infty} v_k(t^{-1})/v_k(z^{-1}) = 0, \forall t \in \bar{\mathbb{C}} \backslash D\}.$$

Other generalizations concern Padé approximants for series of functions, for multiple series, for Laurent series, for vector and matrix series, etc. Moreover, Padé approximants approximating series in several points can be defined, and there are other extensions as well. All these generalizations can also be combined.

Of course, the main purpose of Padé approximants is to approximate functions given by a formal series expansion. In this respect they achieved great successes and they were much useful in solving problems in theoretical physics, fluid dynamics, numerical analysis, etc.

Padé approximants have also many other applications in the solution of various questions of numerical analysis not so closely related to power series approximation ; see [6], where such topics are developed. In particular, Padé approximants and their generalizations have applications in the construction of A-acceptable approximations to the exponential function, a question of primary importance in the numerical solution of differential equations. They have also been applied to the Laplace transform inversion and to the Borel transform. By their link with the ε–algorithm, as mentioned above, they have applications in the solution of systems of nonlinear equations where they provide a method with quadratic convergence (under certain assumptions) which does not require the computation of derivatives. In the case of a system of linear equations, they are related to the Lanczos method and to other projection methods. They also have applications in the computation of the eigenvalues of a matrix.

Padé approximation has many applications. It is of interest to pure mathematicians, numerical analysts, computer scientists, applied mathematicians, physicists, etc. It is the subject of a vast literature covering several centuries, and is still very much alive today.

References

1. Arms, R. J., and A. Edrei, The Padé tables and continued fractions generated by totally positive sequences, in *Mathematical Essays Dedicated to A. J. MacIntyre*, Ohio University Press, Athens, Ohio, 1970, 1–21.
2. Brezinski, C., *Padé–Type Approximation and General Orthogonal Polynomials*, Birkhäuser, Basel, 1980.
3. Brezinski, C., *History of Continued Fractions and Padé Approximants*, Springer Verlag, Berlin , 1991.
4. Brezinski, C. and M. Redivo Zaglia, *Extrapolation Methods. Theory and Practice*, North-Holland, Amsterdam, 1991.
5. Brezinski, C., and J. Van Iseghem, Padé approximations, in *Handbook of Numerical Analysis*, Vol. III, P. G. Ciarlet and J. L. Lions (eds.), North-Holland, Amsterdam, 1994.
6. Cuyt, A., and L. Wuytack, *Nonlinear Methods in Numerical Analysis*, North-Holland, Amsterdam, 1987.
7. Draux, A., *Polynômes Orthogonaux Formels. Applications*, LNM vol. 974, Springer Verlag, Berlin, 1983.
8. Draux, A., and P. Van Ingelandt, *Polynômes Orthogonaux et Approximants de Padé. Logiciels*, Technip, Paris, 1987.
9. Eiermann, M., On the convergence of Padé–type approximants to analytic functions, J. Comput. Appl. Math. **10** (1984), 219–227.
10. Montessus de Ballore, R., Sur les fractions continues algébriques, Bull. Soc. Math. France **30** (1902), 28–36.
11. Padé H., *Oeuvres*, C. Brezinski (ed.), Librairie Scientifique et Technique A. Blanchard, Paris, 1984.
12. Shanks, D., Non linear transformations of divergent and slowly convergent sequences, J. Math. phys. **34** (1955), 1–42.
13. Wynn, P., On a device for computing the $e_m(S_n)$ transformation, MTAC **10** (1956), 91–96.

Claude Brezinski
Laboratoire d'Analyse Numérique et d'Optimisation
U.F.R. I.E.E.A. - M3
Université des Sciences et Technologies de Lille
59655–Villeneuve d'Ascq Cedex, FRANCE

Discrete Curves and Curvature Constraints

Hermann G. Burchard

Abstract. In this paper, continuing work in [1], procedures are considered for computing *discrete plane curves* subject to constraints on the *numerical curvatures*, to be piecewise of one sign, monotone, and convex, in a strong sense. Both, linearized and nonlinear constraints for numerical curvatures are analyzed. Subject to such constraints, existence of interpolating G^1 and piecewise G^2–curves satisfying sufficiently strong curvature constraints is shown.

§1. Introduction

In this paper, continuing work in [1], procedures are considered for computing *discrete plane curves* subject to constraints on the *numerical curvatures*, to be piecewise of one sign, monotone, and convex, in a strong sense. Both, linearized and nonlinear constraints for numerical curvatures are analyzed. Subject to such constraints, existence of interpolating G^1 and piecewise G^2- curves satisfying sufficiently strong curvature constraints is shown. The discrete curves themselves are obtained by a nonlinear optimization algorithm which simultaneously enforces the constraints and ensures the curve deviates from data points by the least possible amount. Due to the constraints, a very small number of data points is required, so this procedure lends itself naturally to CAGD applications. Computing discrete curves (and surfaces) can be a convenient tool in CAGD for such operations as offsets, trimming, and quality control. If needed, many procedures are available for interpolating or approximating a discrete curve by a smooth parametric curve of smoothness class G^k for sufficiently large k [3]. As observed by several authors, constraints on the curvature are essential to guarantee quality of curves [1,3,5]. Given a discrete curve, existence of an interpolating or approximating G^1-curve satisfying sufficiently strong curvature constraints poses interesting and challenging mathematical problems, which seem to have been little explored. In the present report, the results of [1] are improved, strengthened, and generalized.

Curves and Surfaces in Geometric Design
P. J. Laurent, A. Le Méhauté, and L. L. Schumaker (eds.), pp. 67–74.
Copyright © 1994 by A K PETERS, Wellesley, MA
ISBN 1-56881-039-3.

67

It should be noted that there does not exist any straightforward relationship between constraints imposed on numerical curvatures of a discrete curve vs. the resulting properties which we are able to prove for the actual curvature of an interpolating G^1 piecewise G^2–curve.

Specifically, desirable properties of G^1 and piecewise G^2–curves are for the curvature $\kappa = \kappa(s)$ (s =arclength) to be piecewise of one sign, monotone, and convex (in a certain, strong sense). In [1] the significance of such constraints is established for the aesthetics of a plane curve, as relevant to designers, especially within the traditions of styling of exterior automobile surfaces. In that environment, there are known locations of inflection points (sign changes of $\kappa(s)$), as well as maxima and minima of $|\kappa(s)|$. In addition, the fairly strong condition of logarithmic convexity of $|\kappa(s)|$, *i.e.*,

$$\frac{d^2 \log |\kappa(s)|}{ds^2} \geq 0$$

was proposed in [1]. Note that at maxima of $|\kappa(s)|$, a discontinuity of $\frac{d|\kappa(s)|}{ds}$ is implied. The proposal is the basis of a code which has been in use at GMR for some time, and in turn is based in part on the empirical success of that code [1].

§2. Circular Splines

Our discussion concerns merely one of a family of possible models. Its basis is the existence of an interpolating *circular spline* possessing curvatures κ_i for the i-th circular arc, $i = 1, \ldots, n$, which on a typical segment (usually consisting of several circular arcs) satisfies curvature constraints as indicated above. See [4] for a recent account of circular splines. However, even for a circular spline, convexity of curvature with respect to a parameter t can be satisfied only in a discretized sense. Hence, to obtain satisfactory results, we must require the length of each circular arc to be sufficiently small. The jumps in curvature between the arcs are fundamentally undesirable, and can be accepted only because under conditions of small arclengths they would not be noticeable. A practical rule for the choice of arclength is to make the chordheight less than some tolerance ϵ. For small ϵ, arcs become visually indistinguishable from chords. An additional advantage of this restriction to small arclengths is that changes in the tangent angles from one arc to the next will likewise be small, hence one may linearize some of the relationships by approximating the cosines of these changes by unity.

A discrete plane curve is given by data points $(x_i, \; y_i)$, $i = 0, \ldots, n$, in a Cartesian (x, y)–coordinate system. We denote *chord vectors* $\mathbf{u}_i = (x_i - x_{i-1}, \; y_i - y_{i-1})$, and, using the Euclidean norm in $\mathbb{R}^2$, the chord lengths $L_i = \|\mathbf{u}_i\|$, $i = 1, \ldots, n$, double chord lengths $M_i = \|\mathbf{u}_i + \mathbf{u}_{i+1}\|$, and the angle between succesive chord vectors, $\alpha_i = \angle(\mathbf{u}_i, \mathbf{u}_{i+1})$, $i = 1, \ldots, n - 1$.

For any three successive points of the discrete curve we define the *numerical curvature*

$$\tilde{\kappa}_i = \frac{2}{M_i} \sin \alpha_i, \quad i = 1, \ldots, n - 1,$$

which is the usual curvature of the circle through three successive points (x_j, y_j), $j = i-1, i, i+1$.

We require circular arcs with curvatures κ_i, connecting any two successive points (x_j, y_j), $j = i-1, i$, $i = 1, \ldots, n$, of the discrete curve. The overall circular spline forms a G^1-curve, with the ith and $(i+1)$st circular arcs joining at the point (x_i, y_i) and forming a common tangent line including angles ϕ_i, ϕ_{i+1} respectively with the chord vectors $\mathbf{u}_i$, $\mathbf{u}_{i+1}$, $i = 1, \ldots, n-1$. Then $\kappa_i = \frac{2}{L_i} \sin \phi_i$, $\alpha_i = \phi_i + \phi_{i+1}$. Let us abbreviate

$$c_i = \cos \phi_i = \sqrt{1 - \frac{\kappa_i^2 L_i^2}{4}}, \quad i = 1, \ldots, n. \tag{1}$$

We will always work with the condition $0 < \phi_i < \frac{\pi}{2}$, and conditions usually exist that allow us to guarantee that $\phi_i << \frac{\pi}{2}$, so that $c_i \approx 1$. This is a consequence of using small arclengths. We now obtain as a G^1 condition for the circular spline

$$c_{i+1} L_i \kappa_i + c_i L_{i+1} \kappa_{i+1} = M_i \tilde{\kappa}_i, \quad i = 1, \ldots, n-1. \tag{2}$$

This is a system of nonlinear equations to be solved for the unknowns κ_i.

§3. Linearized Model

The most desirable procedure would be to specify constraints to be imposed on the curvatures κ_i. There is in principle no reason why this should not be done, except the number of variables to be carried in the resulting optimization problem can be reduced if any direct reference to the circular arcs and their curvatures κ_i can be avoided. The question arises of what properties of the numerical curvatures $\tilde{\kappa}_i$ are required to guarantee existence of a high–quality interpolating circular spline with spline curvatures κ_i satisfying constraints similar to or almost as strong to those imposed on the $\tilde{\kappa}_i$.

Our first result, Theorem 1 below, is proved in [1], and is the basis of a computer code that has been in use at GMR for some time. We impose constraints on the numerical curvatures $\tilde{\kappa}_i$ as follows. To simplify, we consider in the following a curve segment with positive, monotone curvatures, and impose an additional convexity constraint:

$$\begin{aligned} \tilde{\kappa}_i &> 0, \quad i = 1, \ldots, n-1; \\ \tilde{\kappa}_i &\geq \tilde{\kappa}_{i-1}, \quad i = 2, \ldots, n-1; \\ \frac{\tilde{\kappa}_{i+1} - \tilde{\kappa}_i}{t_{i+1} - t_i} &\geq \frac{\tilde{\kappa}_i - \tilde{\kappa}_{i-1}}{t_i - t_{i-1}}, \quad i = 2, \ldots, n-2. \end{aligned} \tag{3}$$

The parameter values t_i are yet to be determined. For reasons explained in [1], a good choice of parameters resulting in a strong convexity condition is

$$t_{i+1} - t_i = \frac{1}{L_{i+1}}, \quad i = 1, \ldots, n-1. \tag{4}$$

Under suitable conditions on the spacing of the points, this choice represents a discretized form of logarithmic convexity of the numerical curvatures with respect to arclength. The values in (4) result from spacing of points along the discrete curve subject to approximately constant chord height of the circular segments. This should ideally be done by an optimization procedure, but in practice it often suffices to estimate the chordheight from the data. In some cases we may require successive iterations.

While the first two inequalities (3) merely state the numerical curvatures should be positive and increasing, using (4), the third of the inequalities (3) may be written in the form

$$L_i \tilde{\kappa}_{i-1} + L_{i+1} \tilde{\kappa}_{i+1} \geq (L_i + L_{i+1}) \tilde{\kappa}_i, \quad i = 2, \ldots, n-2. \tag{5}$$

From this it is easy to see that it would be advantageous to make the approximation $c_i = 1$ in the G^1–condition (2) above. This procedure has worked well in practice, if the condition of small arclengths is satisfied. Theorem 1 is based on this approximation and the related linearization of (2),

$$L_i \kappa_i + L_{i+1} \kappa_{i+1} = (L_i + L_{i+1}) \tilde{\kappa}_i, \quad i = 2, \ldots, n-1. \tag{6}$$

Here we have linearized $M_i = L_i + L_{i+1}$. In this form it is apparent that (6) amounts to replacing the circles locally by parabolas.

Theorem 1. *Given points (x_i, y_i), $i = 0, 1, \ldots, n$, with positive, nondecreasing numerical curvatures $\tilde{\kappa}_i$, satisfying the discrete linearized convexity condition (5), we can solve the linearized system (6) interpolating data points so that circular spline curvatures κ_i satisfy*

$$0 < \kappa_1 \leq \tilde{\kappa}_1 \leq \kappa_2 \leq \tilde{\kappa}_2 \leq \cdots \leq \tilde{\kappa}_{n-2} \leq \kappa_{n-1} \leq \tilde{\kappa}_{n-1} \leq \kappa_n. \tag{7}$$

Proof: A special case must be made for $i = 1$, where we choose $\kappa_1 > 0$ subject to $\tilde{\kappa}_1 - \frac{L_2}{L_1}(\tilde{\kappa}_2 - \tilde{\kappa}_1) \leq \kappa_1 \leq \tilde{\kappa}_1$. The lower bound on κ_1 is consistent with the upper bound, and this choice of κ_1 implies $\tilde{\kappa}_1 \leq \kappa_2 \leq \tilde{\kappa}_2$. Having obtained $\tilde{\kappa}_{i-1} \leq \kappa_i \leq \tilde{\kappa}_i$ for $i \geq 2$ by induction, we solve (6) for κ_{i+1} with

$$\kappa_{i+1} = \tilde{\kappa}_i + \frac{L_i}{L_{i+1}}(\tilde{\kappa}_i - \kappa_i).$$

Hence $\tilde{\kappa}_i \leq \kappa_{i+1}$, advancing one half of the induction. On the other hand, subtracting (6) from (5) we obtain

$$L_i(\tilde{\kappa}_{i-1} - \kappa_i) + L_{i+1}(\tilde{\kappa}_{i+1} - \kappa_{i+1}) \geq 0.$$

But $\tilde{\kappa}_{i-1} \leq \kappa_i$ by induction hypothesis, therefore $\kappa_{i+1} \leq \tilde{\kappa}_{i+1}$. This advances the other half of the induction. ∎

Although the interlacing condition (7) does not imply convexity of the spline curvatures κ_i as in (5), the fact that the κ_i interlace the numerical curvatures $\tilde{\kappa}_i$, which do satisfy (5), implies that the resulting circular spline is a high-quality curvature constrained interpolant.

Computational methods for computing discrete constrained curves based on Theorem 1 are described in [1]. By its nature, computing a discrete curve to approximate data points subject to numerical curvature constraints is a non-linear optimization problem. One may conveniently use iterated linear programming, but today many good nonlinear optimization methods are known. To set up the optimization, in most cases data points are too far apart, hence additional points must be filled in to generate a useful discrete curve. Points may be filled reliably spaced along lines or circles. A practical scheme is to make the chord height of each circular segment less than a tolerance. Of course, filled points only serve to obtain adequate discretizations; it is not necessary that they improve the esthetic definition of the curve. This is taken care of by imposing constraints on the curvatures. Points are conveniently displaced along approximate normal vectors, the i-th point by an amount δ_i. The objective is to minimize the sum of absolute values of displacements of non-filled points, $\sum_{i \in J} |\delta_i|$, where J is the index set of those original points which are not filled. To be able to apply linear programming we must linearize the nonlinear dependence of $\tilde{\kappa}_i$ on δ_i. This amounts to a first degree Taylor formula:

$$\tilde{\kappa}_i^* \approx \tilde{\kappa}_i + T_{i1}\delta_{i-1} + T_{i2}\delta_i + T_{i3}\delta_{i+1}.$$

The superscript * refers to displaced points. The somewhat complicated formulas for the expansion coefficients T_{ij} are given in [1].

The convexity condition (5) of Theorem 1 is stronger than ordinary convexity if $L_{i+1} < L_i$. This would normally be the case for increasing curvatures because of the chord height condition. If the chordheight is made very nearly constant (and small), then it is shown in [1] that (5) becomes a discrete version of logarithmic convexity. The next result seeks to explore to what extent condition (5) can be weakened while still obtaining (7). We abbreviate

$$\eta_i = L_{i+1}(\tilde{\kappa}_{i+1} - \tilde{\kappa}_i) - L_i(\tilde{\kappa}_i - \tilde{\kappa}_{i-1}), \quad i = 2, \ldots, n-2.$$

Then the sufficient convexity condition of Theorem 1 amounts to $\eta_i \geq 0$. The next theorem gives a weaker condition which is actually necessary and sufficient.

Theorem 2. *Given points* (x_i, y_i), $i = 1, \ldots, n$, *with positive, nondecreasing numerical curvatures* $\tilde{\kappa}_i$, *then necessary and sufficient for the existence of a solution of the linearized system (6) by circular spline curvatures* κ_i *satisfying conditions (7) are the inequalities*

$$0 < L_1\tilde{\kappa}_1 + \eta_2 + \ldots + \eta_{2\nu},$$

and

$$0 \leq L_2(\tilde{\kappa}_2 - \tilde{\kappa}_1) + \eta_3 + \ldots + \eta_{2\nu-1}$$

for $2\nu \leq n-2$, *or* $2\nu - 1 \leq n - 2$ *respectively.*

We omit the proof which is a somewhat tedious mathematical induction.

§4. Nonlinear Models

We now drop all linearizations of the preceding section. If some $c_i << 1$, then the spline curvatures κ_i should satisfy the nonlinear equations (2) instead of the linearized condition (6). In this case the constraint (6) must be replaced by the stronger nonlinear convexity condition (10) of Theorem 3 below. A difficulty is that c_i is related to κ_i by (1). This is remedied in Theorem 4.

Lemma 1. *Given points* (x_i, y_i), $i = 0, 1, \ldots, n$, *and an interpolating circular spline with curvatures* κ_i *define cosines* c_i *by equation (1) above, then we obtain the inequality*

$$M_i \geq c_{i+1}L_i + c_iL_{i+1}. \tag{8}$$

Proof: This is easily proved as follows.

$$
\begin{aligned}
M_i^2 &= L_i^2 + L_{i+1}^2 + 2\cos\alpha_i L_i L_{i+1} \\
&= L_i^2 + L_{i+1}^2 + 2L_i L_{i+1}[c_i c_{i+1} - \sin\phi_i \sin\phi_{i+1}] \\
&= [c_{i+1}L_i + c_iL_{i+1}]^2 + [L_i\sin\phi_{i+1} - L_{i+1}\sin\phi_i]^2.
\end{aligned}
$$

From this we obtain (8). ∎

Remark. Inequality (8) becomes an equation in the special circumstance that $\kappa_i = \kappa_{i+1} = \tilde{\kappa}_i$ (this would be consistent with (2)).

Theorem 3. *Given a discrete curve* (x_i, y_i), $i = 0, \ldots, n$, *with positive and nondecreasing numerical curvatures* $\tilde{\kappa}_i$, *satisfying the discretized nonlinear convexity condition*

$$c_{i+1}L_i\tilde{\kappa}_{i-1} + c_iL_{i+1}\tilde{\kappa}_{i+1} \geq M_i\tilde{\kappa}_i, \quad i = 2, \ldots, n-2, \tag{10}$$

a solution of the nonlinear system (2) exists with circular spline curvatures κ_i *satisfying (7). The* c_i *are given in terms of* κ_i *by (1).*

Proof: The existence can be proved as in Theorem 4 below. For now we assume the c_i are given beforehand. The rest of the proof is similar that of Theorem 1. If by induction $\tilde{\kappa}_{i-1} \leq \kappa_i \leq \tilde{\kappa}_i$, then solve (2) for κ_{i+1}:

$$
\begin{aligned}
\kappa_{i+1} &= \frac{1}{c_iL_{i+1}}(M_i\tilde{\kappa}_i - c_{i+1}L_i\kappa_i) \\
&= \tilde{\kappa}_i + \frac{1}{c_iL_{i+1}}((M_i - c_iL_{i+1})\tilde{\kappa}_i - c_{i+1}L_i\kappa_i).
\end{aligned}
$$

Here, $M_i - c_iL_{i+1} \geq c_{i+1}L_i$ by Lemma 1, and $\kappa_i \leq \tilde{\kappa}_i$ by induction, hence $\tilde{\kappa}_i \leq \kappa_{i+1}$. For the other half of the induction, observe that subtraction of (2) from (10) results in

$$c_{i+1}L_i(\tilde{\kappa}_{i-1} - \kappa_i) + c_iL_{i+1}(\tilde{\kappa}_{i+1} - \kappa_{i+1}) \geq 0.$$

By induction $\tilde{\kappa}_{i-1} \leq \kappa_i$, hence we must have $\kappa_{i+1} \leq \tilde{\kappa}_{i+1}$. ∎

The c_i in the nonlinear constraint (10) depend on the κ_i; hence Theorem 3 may not be very practical. However, the constraints (10) would work well when we seek to optimize the circular spline simultaneously with the discrete curve.

We now state one possible fully discrete nonlinear version involving only the discrete curve itself. It is based on the relationship (1) for the c_i. Accordingly, in (11) we define *numerical cosines* $\tilde{c}_i$, which may be substituted in (10) to obtain a *stronger* constraint (12). The condition (13) of small arclengths also implies all cosines $c_i \approx 1$.

Theorem 4. *Given the discrete curve* (x_i, y_i), $i = 0, 1, \ldots, n$, *with nondecreasing numerical curvatures* $\tilde{\kappa}_i$, $i = 1, \ldots, n-1$, *define numerical cosines*

$$\tilde{c}_i = \sqrt{1 - \frac{\tilde{\kappa}_i^2 L_i^2}{4}}, \quad i = 1, \ldots, n-1. \tag{11}$$

On the numerical curvatures impose the nonlinear convexity constraint

$$\tilde{c}_{i+1} L_i \tilde{\kappa}_{i-1} + \tilde{c}_i L_{i+1} \tilde{\kappa}_{i+1} \geq M_i \tilde{\kappa}_i, \quad i = 2, \ldots, n-2. \tag{12}$$

Also require a small arclengths condition for some $\delta > 0$:

$$L_i \tilde{\kappa}_i \leq \delta \leq \frac{1}{8}, \quad i = 1, \ldots, n-1. \tag{13}$$

Subject to (12) and (13) a solution κ_i, $i = 1, \ldots, n$, *satisfying (7) of the nonlinear system (2) exists. In (2), the* c_i *are the exact cosines defined by the circular spline curvatures* κ_i, *i.e.,* $c_i = \cos\phi_i = \sqrt{1 - \frac{\kappa_i^2 L_i^2}{4}}$, $i = 1, \ldots, n$.

Proof: The proof proceeds again by induction. If κ_i has already been found and $\tilde{\kappa}_{i-1} \leq \kappa_i \leq \tilde{\kappa}_i$, then $L_i \kappa_i \leq \delta$ from (13). A univariate contraction argument is now applied to prove existence of κ_{i+1}. Abbreviate $u = L_{i+1}\kappa_{i+1}$ and write (2) in the form

$$u = g(u) := \frac{M_i \tilde{\kappa}_i}{c_i} - \frac{L_i \kappa_i}{c_i} \sqrt{1 - \frac{u^2}{4}} \geq 0. \tag{14}$$

On the interval $I = \{u : 0 \leq u \leq 0.5\}$, one proves the bound

$$|g'(u)| \leq \lambda = \frac{\delta}{4}.$$

Iterating (14) with $u_0 = 0$ and $u_1 = g(u_0)$ we estimate $r = \frac{\lambda}{1-\lambda}|u_1 - u_0| \leq \delta^2$. From this, the interval $I_r = \{u : 0 \leq u \leq u_1 + r\} \subset I$. Hence, the contraction mapping theorem applies to the interval I_r, and a solution of equation (14) exists and satisfies $u = L_{i+1}\kappa_{i+1} \in I_r$. Next, by induction (12) is first weakened to

$$\tilde{c}_{i+1} L_i \kappa_i + c_i L_{i+1} \tilde{\kappa}_{i+1} \geq M_i \tilde{\kappa}_i.$$

Subtracting (2) one gets $(\tilde{c}_{i+1} - c_{i+1})L_i\kappa_i + c_i L_{i+1}(\tilde{\kappa}_{i+1} - \kappa_{i+1}) \geq 0$. Substituting from (1) and (11), there follows after some manipulation

$$(\tilde{\kappa}_{i+1} - \kappa_{i+1})L_{i+1}\Big[c_i - \frac{1}{4}L_{i+1}\frac{\tilde{\kappa}_{i+1} + \kappa_{i+1}}{\tilde{c}_{i+1} + c_{i+1}}\Big] \geq 0.$$

This implies $\kappa_{i+1} \leq \tilde{\kappa}_{i+1}$, again using (13). Once this is shown, then (12) will imply (10) and $\tilde{\kappa}_i \leq \kappa_{i+1}$ follows as in the proof of Theorem 3. ∎

Curvature monotonicity and convexity are nonlinear forms of generalized convexity condition. This subject of approximation under higher order generalized convexity is treated in [2].

References

1. Burchard, H. G., Ayers, J. A., Frey, W. H., and Sapidis, N, Approximation with Aesthetic Constraints, GMR-7814, to appear in *Designing Fair Curves and Surfaces*, N. Sapidis (ed.), SIAM, Philadelphia, 1994.
2. Burchard, H. G. Extremal positive splines with applications to interpolation and approximation by generalized convex functions, Bull. Amer. Math. Soc. **79** (1973), 959–963.
3. Farin, G., *Curves and Surfaces for Computer Aided Geometric Design*, Academic Press, Boston, 1988.
4. Hoscheck, J., Circular Splines, Comput. Aided Design **94** (1992), 611–618.
5. Roulier, J., Rando, T. and Piper, B., Fairness and monotone curvature, in *Approximation Theory and Functional Analysis*, C. K. Chui (ed.), Academic Press, Boston, 1990.

Hermann G. Burchard
Dept. of Mathematics
Oklahoma State University
Stillwater, OK 74078-0613
burchard@ math.okstate.edu

A Necessary and Sufficient Condition
for the C^k Continuity
of Triangular Rational Surfaces

J. C. Canonne

Abstract. The rational surfaces in three dimensions can be defined as projections of polynomial surfaces in four dimensions. We can project discontinuous surfaces in four dimensions onto a smooth piecewise surface in three dimensions. Here we give a necessary and sufficient condition for C^k continuity between two rational surfaces in three dimensions, and explain why there are fewer interesting possibilities to construct a C^k piecewise surface from discontinuous surfaces in the triangular case than in the rectangular case.

§1. Introduction

Rational curves (resp. surfaces) in three dimensions can be defined as projections of curves (resp. surfaces) in four dimensions [10,11,12,13]. We can project discontinuous curves (resp. surfaces) in four dimensions onto a smooth piecewise curve (resp. surface) in three dimensions [10]. In [4,15] necessary and sufficient conditions were given for the C^k continuity of rational curves. Next, in [6], we gave a necessary and sufficient condition for the C^k continuity of the projections of two surfaces in four dimensions onto a space of three dimensions. This condition was afterwards written in terms of *massic vectors* for rectangular rational surfaces (SBR), as they have been defined by Fiorot and Jeannin [12,13,14]. As in the case of curves, we have many possibilities to construct a C^k piecewise surface, depending on free parameters, which we called ϕ_0, ϕ_1, ..., and ϕ_k. In [7], we applied our condition to rational triangular surfaces. We remarked on the difficulties originating from the elevation of degree used in our formulas.

Here we explain why there are fewer interesting possibilities than in the rectangular case and why a good choice of parameters in the triangular case is $\phi_1 = 0$, $\phi_2 = 0, \ldots, \phi_k = 0$.

Curves and Surfaces in Geometric Design
P. J. Laurent, A. Le Méhauté, and L. L. Schumaker (eds.), pp. 75–82.

§2. Notation

Let $\mathcal{E}$ (resp. $\mathcal{F}$) be a real affine space of dimension three (resp. four), and $\overrightarrow{\mathcal{E}}$ (resp. $\overrightarrow{\mathcal{F}}$) be its associated linear vector space so that $\mathcal{E}$ (resp. $\overrightarrow{\mathcal{E}}$) is a hyperplane of $\mathcal{F}$ (resp. $\overrightarrow{\mathcal{F}}$). We denote by $\tilde{\mathcal{E}}$ the real projective space associated to $\mathcal{E}$.

Let us consider a point $\Omega \in \mathcal{F} \setminus \mathcal{E}$. We work in $\mathcal{F}$ with cartesian coordinates relatively to $\mathcal{R} = (\Omega, \vec{e_1}, \vec{e_2}, \vec{e_3}, \vec{e_4})$, and in $\mathcal{E}$ with cartesian coordinates relatively to $\mathcal{R}_1 = (\Omega_1, \vec{e_1}, \vec{e_2}, \vec{e_3})$, where Ω_1 is defined by $\vec{e_4} = \overrightarrow{\Omega\Omega_1}$. We denote by $(\mathcal{A}_{(1)}, \mathcal{A}_{(2)}, \mathcal{A}_{(3)}, \alpha)$ the cartesian coordinates of a point $\mathcal{A}$ in $\mathcal{F}$ relatively to $\mathcal{R}$.

We denote by $\Pi\Omega : \mathcal{F} \setminus \{\Omega\} \longrightarrow \tilde{\mathcal{E}}$ the projection of apex Ω. If $\alpha \neq 0$, $\Pi\Omega(\mathcal{A})$ belongs to $\mathcal{E}$ and is denoted by A. Thus we get $\overrightarrow{\Omega A} = \alpha \overrightarrow{\Omega\mathcal{A}}$

Let $\mathcal{P}$ be a real affine space of dimension 2, and $\overrightarrow{\mathcal{P}}$ its linear vector space. Let σ be a triangle in $\mathcal{P}$ and τ be another triangle in $\mathcal{P}$ such that σ and τ have the common edge $W = [p_0, q_0]$.

We denote by $(\lambda_1, \lambda_2, \lambda_3)$ the barycentric coordinates of a point p in $\mathcal{P}$ relative to the nondegenerate triangle $\sigma = (s_1, s_2, s_3)$, where we choose $s_2 = p_0$, $s_3 = q_0$ (W is then described by the points $p(v)$ of coordinates $\lambda_1 = 0$,$\lambda_2 = 1 - v$, $\lambda_3 = v$ with v in $[0, 1]$), and by (μ_1, μ_2, μ_3) the barycentric coordinates in $\mathcal{P}$ relatively to the non degenerate triangle $\tau = (t_1, s_2, s_3)$. Let $t_1 = x_1 s_1 + x_2 s_2 + x_3 s_3$.

In the same way, we write $\vec{u} = u_1 s_1 + u_2 s_2 + u_3 s_3$ (resp. $\vec{u} = v_1 t_1 + v_2 s_2 + v_3 s_3$), with $u_1 + u_2 + u_3 = 0$ (resp. $v_1 + v_2 + v_3 = 0$) for $\vec{u}$ in $\overrightarrow{\mathcal{P}}$. We set $\vec{j} = \overrightarrow{s_2 s_3}$ and denote by $\vec{i}$ a vector such that $(\vec{i}, \vec{j})$ is a basis of $\overrightarrow{\mathcal{P}}$.

If g is a function defined at the point $p(v)$ of W, $v \in [0, 1]$, we set $g(v) = g(p(v))$. If f is a function defined on $\mathcal{P}$, we denote by $f^{(n)}$ the directional derivative of f of order n with respect to $\vec{i}$ when it exists.

§3. C^k Continuity for $\Pi\Omega$-projections of Surfaces

Let k be an integer. Let $\mathcal{B}$ (resp. $\mathcal{A}$) be a C^k function from an open set V_τ (resp. V_σ) containing τ (resp. σ) to $\mathcal{F}$. We set

$$F_a = \{v \in [0, 1] \ / \ \mathcal{A}(v) \neq \Omega \ and \ \Pi\Omega(\mathcal{A}(v)) \in \mathcal{E}\}$$
$$F_b = \{v \in [0, 1] \ / \ \mathcal{B}(v) \neq \Omega \ and \ \Pi\Omega(\mathcal{B}(v)) \in \mathcal{E}\}$$
$$F = F_a \cap F_b$$
$$\widetilde{W} = \{p(v) \in W \ / \ v \in F\}$$

Let $\mathcal{A}^{(o)}(v) = \overrightarrow{\Omega\mathcal{A}}(v)$ and $A^{(o)}(v) = \overrightarrow{\Omega A}(v)$.

Proposition 1. *There is C^k continuity of the $\Pi\Omega$-projections A and B along $\widetilde{W}$ if and only if there exist $k+1$ functions $\Phi_0, \Phi_1, \ldots, \Phi_k : [0,1] \longrightarrow \mathbb{R}$ such that*

$$\text{for} \quad v \in F, \quad \mathcal{B}^{(l)}(v) = \sum_{j=0}^{j=l} \binom{l}{j} \Phi_j(v) \mathcal{A}^{(l-j)}(v), \qquad (0 \le l \le k). \qquad (1.l)$$

Proof: We prove this proposition by induction on k. For $k = 0$, the proposition is obvious. Suppose the proposition is true for $k - 1$, i.e., there is C^{k-1} continuity between the $\Pi\Omega$-projections $A(v)$ and $B(v)$ along $\widetilde{W}$, and we have $(1.l)$ for $0 \le l \le k-1$.

At a point $A(p) \in \mathcal{E}$, we have $\overrightarrow{\Omega A}(p) = \alpha(p)\overrightarrow{\Omega A}(p)$. Using Leibniz's formula and multiplying by $\alpha(p)$, for $v \in F$ we get

$$\alpha^2 A^{(j)}(v) = \alpha \mathcal{A}^{(j)}(v) - \alpha^{(j)}\overrightarrow{\Omega A}(v) - \sum_{i=1}^{i=j-1} \alpha \binom{j}{i} \alpha^{(j-i)} A^{(i)}(v) \qquad (2.j)$$

Multiplying $(2.k)$ by $\Phi_0(v)$, we get

$$\Phi_0\alpha^2 A^{(k)}(v) = \Phi_0\alpha\mathcal{A}^{(k)}(v) - \Phi_0\alpha^{(k)}\overrightarrow{\Omega A}(v) - \Phi_0 \sum_{i=1}^{i=k-1} \alpha \binom{k}{i} \alpha^{(k-i)} A^{(i)}(v) \qquad (3)$$

In the same way, for $v \in F$ we have

$$\Phi_0\alpha^2 B^{(k)}(v) = -\beta^{(k)}\overrightarrow{\Omega A}(v) + \alpha B^{(k)}(v) - \sum_{i=1}^{i=k-1} \alpha \binom{k}{i} \beta^{(k-i)} A^{(i)}(v) \qquad (4)$$

According to (3) and (4), C^k continuity is then equivalent to

$$\alpha\big(\mathcal{B}^{(k)}(v) - \Phi_0\mathcal{A}^{(k)}(v)\big) - \big(\beta^{(k)} - \Phi_0\alpha^{(k)}\big)\mathcal{A}(v) = H_k(v), \qquad (5)$$

where

$$H_k(v) = \sum_{i=1}^{k-1} \Big(\alpha(v)\binom{k}{i}\big(\beta^{(k-i)}(v) - \Phi_0(v)\alpha^{(k-i)}(v)\big)A^{(i)}(v)\Big).$$

We now give a lemma which allows us to write $H_k(v)$ in another way:

Lemma 1. *For $1 \le \ell \le k$,*

$$H_k(v) = \sum_{i=1}^{k-\ell} \alpha(v)\binom{k}{i}\Big(\sum_{j=\ell}^{k-i} \binom{k-i}{j}\Phi_j(v)\alpha^{(k-i-j)}(v)\Big)A^{(i)}(v) \qquad (6.l)$$

$$+ \sum_{m=1}^{\ell-1} \binom{k}{k-m}\Phi_m(v)\Big(\alpha(v)\mathcal{A}^{(k-m)}(v) - \alpha^{(k-m)}(v)\mathcal{A}(v)\Big)$$

Proof: We give the proof of this lemma after the proof of Proposition 1. ∎

We now come back to the proof of Proposition 1. According to (6.k),

$$H_k(v) = \sum_{m=1}^{k-1} \binom{k}{k-m} \Phi_m(v) \Big(\alpha(v) \mathcal{A}^{(k-m)}(v) - \alpha^{(k-m)}(v) \mathcal{A}(v) \Big)$$

Thus, C^k continuity is equivalent to

$$\alpha(v) \Big(\mathcal{B}^{(k)} - \sum_{0}^{k-1} \binom{k}{k-m} \Phi_m(v) \mathcal{A}^{(k-m)}(v) \Big)$$

$$- \Big(\beta^{(k)} - \sum_{0}^{k-1} \binom{k}{k-m} \Phi_m(v) \alpha^{(k-m)}(v) \Big) \mathcal{A}(v) = 0$$

Therefore, C^k continuity holds if and only if there exists $\Phi_k : [0;1] \longrightarrow \mathbb{R}$ such that

$$\mathcal{B}^{(k)}(v) - \sum_{0}^{k-1} \binom{k}{k-m} \Phi_m(v) \mathcal{A}^{(k-m)}(v) = \Phi_k(v) \mathcal{A}(v)$$

which completes the proof for Proposition 1. ∎

Proof of Lemma 1. We proceed by induction on ℓ. The identity is true for $\ell = 1$. According to (1.k − i), for $1 \le i \le k - i$, we have

$$\mathcal{B}^{(k-i)}(v) - \Phi_0(v) \mathcal{A}^{(k-i)}(v) = \sum_{j=1}^{j=k-i} \binom{k-i}{j} \Phi_j(v) \mathcal{A}^{(k-i-j)}(v)$$

and

$$\beta^{(k-i)}(v) - \Phi_0(v) \alpha^{(k-i)}(v) = \sum_{j=1}^{j=k-i} \binom{k-i}{j} \Phi_j(v) \alpha^{(k-i-j)}(v),$$

which immediately gives (6.1).

Suppose now the identity is true for any ℓ with $1 \le \ell \le k$. Then taking the term corresponding to $i = k - \ell$ out of the first sum, we get

$$H_k(v) = \sum_{m=1}^{\ell-1} \binom{k}{k-m} \Phi_m(v) \Big(\alpha(v) \mathcal{A}^{(k-m)}(v) - \alpha^{(k-m)}(v) \mathcal{A}(v) \Big)$$

$$+ \sum_{i=1}^{k-\ell-1} \alpha(v) \binom{k}{i} \Big(\sum_{j=\ell}^{k-i} \binom{k-i}{j} \Phi_j(v) \alpha^{(k-i-j)}(v) \Big) \mathcal{A}^{(i)}(v)$$

$$+\binom{k}{k-\ell}\Phi_\ell(v)\alpha^2(v)A^{(k-\ell)}(v).$$

Now we use $(2.k-\ell)$ to get rid of $\alpha^2(v)A^{(k-\ell)}(v)$ in the last term, and remove the terms corresponding to $j=\ell$ in the second sum. We have

$$H_k(v) = \sum_{m=1}^{\ell-1}\binom{k}{k-m}\Phi_m(v)\Big(\alpha(v)\mathcal{A}^{(k-m)}(v)-\alpha^{(k-m)}(v)\mathcal{A}(v)\Big)$$

$$+\sum_{i=1}^{k-\ell-1}\alpha(v)\binom{k}{i}\Big(\sum_{j=\ell+1}^{k-i}\binom{k-i}{j}\Phi_j(v)\alpha^{(k-i-j)}(v)\Big)A^{(i)}(v)$$

$$+\sum_{i=1}^{k-\ell-1}\alpha(v)\binom{k}{i}\binom{k-i}{\ell}\Phi_l(v)\alpha^{(k-i-\ell)}(v)A^{(i)}(v)$$

$$+\binom{k}{k-\ell}\Phi_\ell(v)\big(\alpha(v)A^{(k-\ell)}(v)-\alpha^{(k-\ell)}(v)\overrightarrow{\Omega A}(v)\big)$$

$$-\sum_{i=1}^{k-\ell-1}\alpha(v)\binom{k-\ell}{i}\binom{k}{k-\ell}\Phi_\ell(v)\alpha^{(k-i-\ell)}(v)A^{(i)}(v),$$

which gives $(6.\ell+1)$, and completes the proof of Lemma 1. ∎

§4. Application to ΠΩ Projections of Polynomial Surfaces

Let $B_i^n(v) = \binom{n}{i}v^{n-i}(1-v)^i$ be the usual Bernstein basis polynomials of degree n. Let $\{Q_i, 0 \le i \le n\}$ be a polygon of points in a real affine space $\mathcal{Q}$. The curve described by the point $BP[Q_i,[0;1]](v) = \sum_{i=0}^n Q_i B_i^n(v)$ is called the *polynomial Bézier curve* and is denoted by $BP[Q_i,[0;1]]$. n is called the *length* of this curve.

Let $\{Q_\alpha, \alpha = (\alpha_1, \alpha_2, \alpha_3), |\alpha| = \alpha_1 + \alpha_2 + \alpha_3 = n\}$ be a triangular net of points in a real affine space $\mathcal{Q}$. We denote by $SBP[Q;\sigma]$ the triangular polynomial Bézier surface described by the point $Q(\lambda) = \sum_{|\alpha|=n} B_\alpha^n(\lambda)Q_\alpha$, for $\lambda_1 + \lambda_2 + \lambda_3 = 1, \lambda_i \ge 0$, where $B_\alpha^n(\lambda)$ are the usual Bernstein polynomials $B_\alpha^n(\lambda) = \frac{|\alpha|!}{\alpha_1!\alpha_2!\alpha_3!}\lambda^\alpha$.

The elements of $\hat{\mathcal{E}} = (\mathcal{E} \times \boldsymbol{R}^*) \cup \overrightarrow{\mathcal{E}}$ are called *massic vectors*. The isomorphism $\hat{\Omega}: (\hat{\mathcal{E}}, \oplus, *) \longrightarrow (\overrightarrow{\mathcal{F}}, +, .)$ is defined by $\hat{\Omega}(A,\alpha) = \alpha\overrightarrow{\Omega A}$, $\hat{\Omega}(\vec{u}) = \vec{u}$. We denote by Π the natural projection $(\Pi: \hat{\mathcal{E}} \setminus \{\overrightarrow{0}\} \longrightarrow \tilde{\mathcal{E}}: \Pi(A,\alpha) = A, \Pi(\vec{u}) = (\vec{u})_\infty)$.

If $a = \{a_i, 0 \le i \le n\}$ is a polygon of massic vectors, the curve defined by

$$BR[a;[0;1]](v) = \Pi\big(BP[a;[0;1]](v)\big)$$

or, equivalent,

$$BR[a;[0;1]](v) = \Pi\Omega\big(BP[A;[0;1]](v)\big),$$

where the polygon $\mathcal{A}$ is defined by $\hat{\Omega}(a_i) = \mathcal{A}_i, 0 \leq i \leq n$, is called the *(BR) curve*, and is denoted by $BR[a; [0; 1]]$.

In the same way, if $a = \{a_\epsilon, |\epsilon| = n\}$ is a triangular net of massic vectors, the surface defined by

$$SBR[a, \sigma](\lambda) = \Pi\big(SBP[a, \sigma](\lambda)\big)$$

or, equivalently,

$$SBR[a, \sigma](\lambda) = \Pi\Omega\big(SBP[\mathcal{A}, \sigma](\lambda)\big),$$

where the net $\mathcal{A}$ is defined by $\hat{\Omega}(a_\epsilon) = \mathcal{A}_\epsilon, |\epsilon| = n$, is called the *triangular (SBR) surface*, and is denoted by $SBR[a, \sigma]$. From Proposition 1 we have

Proposition 2. *Let $\mathcal{A}$ (resp. $\mathcal{B}$) be a polynomial surface in $\mathcal{F}$, and A (resp. B) be its $\Pi\Omega$ projection. There is C^k continuity of A and B along $\widetilde{W}$ if and only if there exist $k+1$ polynomials $P_0(v)$ (not always equal to 0), $P_1(v),...,P_k(v)$, so that*

$$\mathcal{P}_{\mathcal{A}}^{s+1}(v)\mathcal{B}^{(s)}(v) = \sum_{j=0}^{j=s} \binom{s}{j} P_j(v)\mathcal{P}_{\mathcal{A}}^{s-j}(v)\mathcal{A}^{(s-j)}(v), \qquad (0 \leq s \leq k), \qquad (7)$$

where $\mathcal{P}_{\mathcal{A}}(v)$ is a GCD of the four components $\mathcal{A}_{(1)}(v), \mathcal{A}_{(2)}(v), \mathcal{A}_{(3)}(v), \alpha(v)$ of $\mathcal{A}(v)$ in $\mathcal{F}$. Furthermore, $P_0(v)$ is then a GCD of the four components $\mathcal{B}_{(1)}(v), \mathcal{B}_{(2)}(v), \mathcal{B}_{(3)}(v), \beta(v)$ of $\mathcal{B}(v)$ in $\mathcal{F}$.

We now define the notion of a *well-written* (BR) curve. Let $A(v) = BR[\{a_i, 0 \leq i \leq n\}; [0; 1]](v)$ be a (BR) curve associated to the (BP) curve $\mathcal{A}(v) = BP[\{\mathcal{A}_i, 0 \leq i \leq n\}; [0; 1]](v)$.

Definition 2. *We define the degree of the (BR) curve A by*

$$deg\ A = \max\{deg\ \mathcal{A}_{(1)}(t), deg\ \mathcal{A}_{(2)}(t), deg\ \mathcal{A}_{(3)}(t), deg\ \alpha(v)\}$$

Definition 3. *We say that the (BR) curve A is a well-written (BR) curve if and only if*

$$deg\ \mathcal{P}_{\mathcal{A}} = 0$$
$$deg\ A = n.$$

Lemma 2. *Let $A(v) = BR[\{a_i, 0 \leq i \leq n\}; [0; 1]](v)$ be a well-written (BR) curve. Let $B(v) = BR[\{b_i, 0 \leq i \leq n\}; [0; 1]](v)$ be another (BR) curve of the same length n. If $A(v) = B(v)$ for all v in $[0; 1]$, then $deg\ B = n$, $B(v)$ is also a well-written (BR) curve, and moreover the massic polygons of $A(v)$ and $B(v)$ are proportional.*

These definitions allow us to give a practical description for C^k continuity of triangular (SBR) surfaces.

Let $A(\lambda) = SBR[\{a_\epsilon, |\epsilon| = n\}; \sigma](\lambda)$ be a (SBR) of heigth n, and let $\mathcal{A}(\lambda)$ be its associated (SBP). We denote by $A_F(v) = BR[\{a_{(0,n-j,j)}, 0 \leq j \leq n\}; [0; 1]](v)$ its border. Let $B(\mu) = SBR[\{b_\epsilon, |\epsilon| = n\}; \tau](\mu)$ be another triangular (SBR) of height n, and $\mathcal{B}(\mu)$ its associated (SBP). We denote by $B_F(v) = BR[\{b_{(0,n-j,j)}, 0 \leq j \leq n\}; [0; 1]](v)$ its border.

Proposition 3. *Suppose $A_F(v)$ is a well-written (BR) curve. Then there is C^0 continuity of $A(\lambda)$ and $B(\mu)$ along $\widetilde{W}$ if and only if the massic polygons of $A_F(v)$ and $B_F(v)$ are proportional. Moreover we can suppose these polygons are equal without losing generality. Now suppose the massic polygons of $A_F(v)$ and $B_F(v)$ are equal, so that there is C^0 continuity of $A(\lambda)$ and $B(\mu)$ along $\widetilde{W}$. Then we have C^k continuity ($k \geq 1$) of $A(\lambda)$ and $B(\mu)$ along $\widetilde{W}$ if and only if there is C^k continuity of $\mathcal{A}(\lambda)$ and $\mathcal{B}(\mu)$ along W.*

Comparing this result to those in [5,6], we see that there are in fact fewer interesting possibilities to construct a C^k piecewise surface in $\mathbb{R}^3$ from a C^j, $0 \leq j < k$, piecewise surface of $\mathbb{R}^4$ in the triangular case than in the rectangular case. Due to Proposition 3, we can obtain C^k continuity ($k \geq 1$) of $A(\lambda)$ and $B(\mu)$ along $\widetilde{W}$ by using the classical results on polynomial surfaces [3,8,9]. We do not study G^k continuity here; see [1,2].

Acknowledgements. I wish to thank Prof. J. C. Fiorot for inspiring discussions on some problems considered in this paper.

References

1. Bauchat, J. L., J. C. Fiorot and P. Jeannin, G^1 and G^2 continuity between (SBR) Surfaces, in *Curves and Surfaces*, P. J. Laurent, A. Le Méhauté, and L. L. Schumaker (eds.), Academic Press, Boston, 1991, 33 -36.
2. Bauchat, J. L., G^k continuity of (SBR) surfaces, to appear.
3. Boehm, M., G. Farin, and J. Kahman, A survey of curve and surface methods, Comp. Aided Geom. Design **1** (1984),1-60.
4. Canonne, J. C., J. C. Fiorot and P. Jeannin, Une condition nécessaire et suffisante de raccordement C^k de courbes rationnelles,C. R. Acad. Sci. Paris **312**,série I,(1991), 171-176.
5. Canonne, J. C., J. C. Fiorot, and P. Jeannin, C^k continuity of (SBR) Surfaces, in *Curves and Surfaces*, P. J. Laurent, A. Le Méhauté, and L. L. Schumaker (eds.), Academic Press, Boston, 1991, 67-70.
6. Canonne, J. C.,J. C. Fiorot,Une condition nécessaire et suffisante de raccordement C^k de surfaces rationnelles,C. R. Acad. Sci. Paris **312**,série I, (1991), 897-901.
7. Canonne, J. C., J. C. Fiorot, Des conditions suffisantes de raccordement C^k de surfaces rationnelles mises sous forme (SBR) triangulaire,C. R. Acad. Sci. Paris **313**,série I, (1991), 541-546.
8. de Boor, C., B-form basics, in *Geometric Modeling : Algorithms and New Trends*, G. Farin (Ed.),SIAM, Philadelphia, 1987, 131-148.

9. Farin, G., Triangular Bernstein-Bézier patches, Comp. Aided Geom. Design **3** (1986), 83-127.

10. Farin, G., From conics to nurbs: a tutorial and survey, IEEE Computer Graphics and Applications **12**, 5 (1992), 78-86.

11. Fiorot, J. C., and P. Jeannin, Courbes Bézier rationnelles, $XIX^{i\grave{e}me}$ Congrès National d'Analyse Numérique, Port-Barcarès,France, 26-30 Mai 1986.

12. Fiorot, J. C., and P. Jeannin, Nouvelle description et calcul des courbes rationnelles à l'aide de points et de vecteurs de contrôle,C. R. Acad. Sci.Paris **305**,série I,(1987), 435-440.

13. Fiorot, J. C., and P. Jeannin, *Courbes et Surfaces Rationnelles. Applications à la CAO*, Masson, Paris,1989.

14. Fiorot, J. C., and P. Jeannin, *Rational curves and surfaces. Application to CAO* ,J. Wiley and Sons, 1992.

15. Fiorot, J. C., and P. Jeannin, *Courbes Splines Rationnelles. Applications à la CAO*, Masson, Paris, 1992. English Version, to appear.

J. C. Canonne
LIMAV, I. U. T. G. E. I. I.
Univ. de Valenciennes
Le Mont Houy
59326 Valenciennes cedex FRANCE
canonne@univ-valenciennes.fr

Monotonicity Preserving Representations

J. M. Carnicer and J. M. Peña

Abstract. In this paper we deal with shape preserving properties related with monotonicity. A basis of blending functions is monotonicity preserving if the corresponding collocation matrices transform monotone increasing vectors into monotone increasing vectors. We characterize these matrices and other related classes of matrices in a simple way. We show that normalized totally positive bases provide monotonicity preserving representations of curves. Finally, we prove that monotonicity preserving representations are length diminishing.

§1. Introduction

In the past few years, there has been much work on shape preserving properties of curves obtained from control polygons (see for instance [3,2]). Given functions $u_0, \ldots, u_n$ defined on $[a, b]$ and $P_0, \ldots, P_n \in \mathbb{R}^k$, a curve $\gamma(t)$ may be defined by $\gamma(t) = \sum_{i=0}^n u_i(t) P_i$. The points $P_0, \ldots, P_n$ are usually called *control points*, because we expect to modify the shape of the curve by changing these points. The polygon with vertices $P_0, \ldots, P_n$ is called the *control polygon* of γ. In computer aided geometric design the functions $u_0, \ldots, u_n$ are usually nonnegative and $\sum_{i=0}^n u_i(t) = 1$, $\forall t \in [a, b]$, and in this case we call them *blending functions*.

One of the simplest shape properties is monotonicity. The sequence of control points $(P_0, \ldots, P_n)$ and the corresponding points of the curve $\gamma(t)$, $t \in [a, b]$, are projected on a line l. Let us denote by $\hat{P}_i$ the projection of P_i, $i = 0, \ldots, n$, and by $\hat{\gamma}(t)$ the projection of $\gamma(t)$ on l. Then we want the following property to be satisfied for any control polygon and any line l: if the projected points $\hat{P}_0, \ldots, \hat{P}_n$ are ordered, that is,

$$\hat{P}_i = \hat{P}_0 + \alpha_i v, \quad 0 \le \alpha_1 \le \cdots \le \alpha_n,$$

and v is a directional vector of the line l, then the projected points $\hat{\gamma}(t)$, $t \in [a, b]$, are also ordered, that is, $\hat{\gamma}(t) = \hat{P}_0 + \alpha(t)v$, where $\alpha(t)$ is an increasing function. This is equivalent to the following property:

$$v^T P_0 \le \cdots \le v^T P_n \Rightarrow v^T \gamma(t) \text{ is an increasing function} \tag{1}$$

Curves and Surfaces in Geometric Design
P. J. Laurent, A. Le Méhauté, and L. L. Schumaker (eds.), pp. 83–90.

for each $P_0, \ldots, P_n \in \mathbb{R}^k$, $v \in \mathbb{R}^k$.

Now, let $(u_0, \ldots, u_n)$ be a system of blending functions, and let γ be the curve defined by $\gamma(t) = \sum_{i=0}^{n} u_i(t) P_i$. Then we may write $v^T \gamma(t) = \sum_{i=0}^{n} (v^T P_i) u_i(t)$. Thus the property suggested by (1) motivates the following definition.

Definition 1.1. *A system of blending functions* $(u_0, \ldots, u_n)$ *is monotonicity preserving if for any increasing vector* $\lambda = (\lambda_0, \ldots, \lambda_n)^T$, $(\lambda_0 \leq \cdots \leq \lambda_n)$, *the function* $u(t) = \sum_{i=0}^{n} \lambda_i u_i(t)$ *is an increasing function.*

Now, let us describe a matrix interpretation of monotonicity preserving systems of blending functions. Let us recall that a vector $v = (v_0, \ldots, v_n)^T$ (resp. a matrix $A = (a_{ij})_{1 \leq i,j \leq n}$) is said to be nonnegative if $v_i \geq 0$, $i = 0, \ldots, n$ (resp. $a_{ij} \geq 0$, $i = 0, \ldots, n$, $j = 0, \ldots, n$). We shall denote by $v \geq 0$ a nonnegative vector (resp. $A \geq 0$ a nonnegative matrix).

Let us observe that a system of blending functions $(u_0, \ldots, u_n)$ defined on $[a, b]$ can be characterized in terms of its collocation matrices

$$M \begin{pmatrix} u_0, \ldots, u_n \\ t_0, \ldots, t_n \end{pmatrix} := (u_j(t_i))_{i,j=0,\ldots,n}, \quad a \leq t_0 < t_1 < \cdots < t_n \leq b,$$

which must be nonnegative and satisfy the property that the elements of each row add up to 1. That is, $(u_0, \ldots, u_n)$ is a system of blending functions if and only if all the collocation matrices are *stochastic*.

Definition 1.2. *A matrix* $A = (a_{ij})_{i,j=0,\ldots,n}$ *is said to be monotonicity preserving if for any increasing vector* $v = (v_0, \ldots, v_n)^T$, $v_0 \leq v_1 \leq \cdots \leq v_n$, *the vector* Av *is also increasing, that is,* $(Av)_0 \leq \cdots \leq (Av)_n$.

The next result shows that Definitions 1.1 and 1.2 are closely related.

Proposition 1.3. *A system of blending functions* $(u_0, \ldots, u_n)$ *is monotonicity preserving if and only if all the collocation matrices* $M \begin{pmatrix} u_0, \ldots, u_n \\ t_0, \ldots, t_n \end{pmatrix}$, $0 \leq t_0 < t_1 < \cdots < t_n \leq b$, *are monotonicity preserving.*

Proof: First, let us observe that $u(t)$ is an increasing function on $[a, b]$ if and only if for any $t_0 < \cdots < t_n$ in $[a, b]$, $u(t_0) \leq \cdots \leq u(t_n)$. Therefore $u_0(t), \ldots, u_n(t)$ is a monotonicity preserving system of blending functions if and only if for any increasing vector $\lambda = (\lambda_0, \ldots, \lambda_n)^T$, the vector $(u(t_0), \ldots, u(t_n))^T$ is also increasing for any $t_0 < \cdots < t_n$ in $[a, b]$, where $u(t) = \sum_{i=0}^{n} \lambda_i u_i(t)$. The result follows by taking into account the following relationship

$$(u(t_0), \ldots, u(t_n))^T = M \begin{pmatrix} u_0, \ldots, u_n \\ t_0, \ldots, t_n \end{pmatrix} (\lambda_0, \ldots, \lambda_n)^T. \quad \blacksquare$$

In the next section we shall characterize monotonicity preserving matrices. In [5, pp. 22-23] monotonicity preserving transformations were considered. This concept leads to a matrix interpretation. A monotonicity preserving matrix $A = (a_{i,j})_{i,j=0}^{n}$ in the sense of [5] is a matrix which transforms monotone (i.e. increasing or decreasing) vectors into monotone vectors. In fact, Proposition 3.1 of Chapter 1 of [5] states that if $A = (a_{i,j})_{i,j=0}^n$ is a matrix such that $\sum_{j=0}^n a_{ij} = 1$, $i = 0, \ldots, n$, and all the minors of order k of A have the same sign ε_k, $k = 0, 1$, then A transforms monotone vectors into monotone vectors. Our definition of monotonicity preserving matrices is more restrictive, but adequate for design purposes because it is convenient to preserve shape properties and even "orientation" properties.

In Section 3 we shall prove that totally positive transformations are monotonicity preserving. Some authors (cf. [3,4]) had already shown other shape properties which are preserved by totally positive transformations. It is well-known that the length of a curve generated by a totally positive system of blending functions is less than or equal to the length of the corresponding control polygon. In Section 4 we show that this property also holds for monotonicity preserving systems of blending functions.

§2. Characterizations of Monotonicity Preserving Matrices

In the previous section we have seen that monotonicity preserving matrices arise in the problem of describing monotonicity preserving representations. Here we are going to give a simple characterization of these kinds of matrices.

In the following, we shall use the notations: $e = (1, 1, \ldots, 1)^T$,

$$
E = \begin{pmatrix} 1 & 0 & \cdots & 0 \\ \vdots & \ddots & \ddots & \vdots \\ 1 & \cdots & 1 & 0 \\ 1 & \cdots & 1 & 1 \end{pmatrix}, \quad
E^{-1} = \begin{pmatrix} 1 & 0 & \cdots & \cdots & 0 \\ -1 & 1 & \ddots & & \vdots \\ 0 & -1 & 1 & \ddots & \vdots \\ \vdots & \ddots & \ddots & \ddots & 0 \\ 0 & \cdots & 0 & -1 & 1 \end{pmatrix}.
$$

Remark 2.1. *Taking into account that*

$$
E^{-1}(v_1, v_2, \ldots, v_n)^T = (v_1, v_2 - v_1, \ldots, v_n - v_{n-1})^T,
$$

we note that $v \in \mathbb{R}^n$ is an increasing vector if and only if the $n - 1$ last components of $E^{-1}v$ are nonnegative.

Remark 2.2. *Let us observe that if A is a monotonicity preserving matrix and v is a decreasing vector, then Av is a decreasing vector because $A(-v) = -Av$. This fact agrees with the name "monotonicity preserving" matrices in the sense that they transform increasing vectors into increasing vectors as well as they transform decreasing vectors into decreasing vectors. On the other hand, since e is an increasing and decreasing vector, Ae is necessarily a multiple of e.*

Proposition 2.3. *Let A be an $n \times n$ matrix. Then A is a monotonicity preserving matrix if and only if*

$$
E^{-1}AE = \begin{pmatrix} \beta & b^T \\ 0 & B \end{pmatrix}, \tag{2}
$$

where $B \geq 0$ is an $(n-1) \times (n-1)$ matrix, $\beta \in \mathbb{R}$ and b a vector with $n-1$ components.

Proof: Let us assume that A is a monotonicity preserving matrix. Taking into account that the vector columns of E are increasing, we deduce that AE is a matrix with increasing vector columns. By Remark 2.2, $Ae = \beta e$ and then the first column of AE is precisely βe. Clearly $E^{-1}AE$ is of the form (2) and, applying Remark 2.1 to the vector columns of AE, we have $B \geq 0$.

For the converse, let us assume that (2) holds, and let v be any increasing vector. Then by Remark 2.1, $E^{-1}v = \begin{pmatrix} w_1 \\ \tilde{w} \end{pmatrix}$, where $w_1 \in \mathbb{R}$, $\tilde{w} \in \mathbb{R}^{n-1}$ and $\tilde{w} \geq 0$. Therefore,

$$E^{-1}Av = E^{-1}AE(E^{-1}v) = \begin{pmatrix} \beta w_1 + b^T \tilde{w} \\ B\tilde{w} \end{pmatrix},$$

which implies that the $n-1$ last components of $E^{-1}(Av)$ are nonnegative and, again by Remark 2.1, Av is an increasing vector. ∎

In the following we shall often deal with matrices A such that $E^{-1}AE \geq 0$. This property can be interpreted as a "shape preserving" property.

Proposition 2.4. Let A be an $n \times n$ matrix. Then $E^{-1}AE \geq 0$ if and only if A transforms nonnegative and increasing vectors into nonnegative and increasing vectors.

Proof: $E^{-1}AE \geq 0$ if and only if it transforms nonnegative vectors into nonnegative vectors. The set of nonnegative vectors is given by the set of vectors $E^{-1}v$, where v is any nonnegative and increasing vector. Thus, $E^{-1}AE \geq 0$ if and only if $E^{-1}AE(E^{-1}v) \geq 0$ for any nonnegative and increasing vector v. Therefore $E^{-1}AE \geq 0$ if and only if $E^{-1}(Av) \geq 0$ for any nonnegative and increasing vector v, which is equivalent to saying that Av is nonnegative and increasing. ∎

The monotonicity preserving matrices which appeared in the previous section are always nonnegative. Then these matrices transform nonnegative vectors into nonnegative vectors and increasing vectors into increasing vectors. In particular, these matrices transform nonnegative and increasing vectors into vectors of the same kind. Thus, by Proposition 2.4, $E^{-1}AE \geq 0$ is always satisfied for the collocation matrices A of any monotonicity preserving system of blending functions.

Corollary 2.5. Let A be a nonnegative matrix. Then A is monotonicity preserving if and only if $E^{-1}AE \geq 0$ and $Ae = \beta e$ for some $\beta \geq 0$.

Proof: If $Ae = \beta e$, then $E^{-1}AE = \begin{pmatrix} \beta & b^T \\ 0 & B \end{pmatrix}$ and if $E^{-1}AE \geq 0$, then $B \geq 0$, which implies by Proposition 2.3 that A is monotonicity preserving. Conversely, if A is nonnegative and monotonicity preserving, A transforms nonnegative and increasing vectors into nonnegative and increasing vectors, and by Proposition 2.4 $E^{-1}AE \geq 0$. By Remark 2.2, Ae is necessarily a multiple of e. In fact $Ae = \beta e$, where β is the (1,1) entry of $E^{-1}AE(\geq 0)$, and so $\beta \geq 0$. ∎

Let us recall that a matrix T is said to be *stochastic* if $T \geq 0$ and $Te = e$.

Remark 2.6. *From the previous corollary we obtain that a nonnegative monotonicity preserving matrix A is a multiple of a stochastic matrix. In fact, $A = \beta T$, with T stochastic and monotonicity preserving and $\beta \geq 0$.*

In the previous section we have mentioned that the collocation matrices of any system of blending functions are always stochastic. The following obvious consequence of Corollary 2.5 allows us to characterize preservation of monotonicity in this case.

Corollary 2.7. *Let T be a stochastic matrix. Then T is monotonicity preserving if and only if $E^{-1}TE \geq 0$.*

§3. Total Positivity and Monotonicity Preserving Matrices

Let us recall that a *totally positive (TP) matrix* is a matrix such that all its minors are nonnegative. A system of blending functions $(u_0, \ldots, u_n)$ is totally positive if all its collocation matrices are totally positive. A totally positive system of blending functions, that is, $\sum_{i=0}^{n} u_i = 1$, is usually called a *normalized totally positive* (NTP) system. A system is NTP if all the collocation matrices are TP and stochastic.

Many common systems of blending functions are NTP. For instance, the Bernstein polynomials and the B-spline basis. Due to the variation diminishing properties of NTP systems, many shape properties of the control polygon are inherited by the corresponding curve (see for instance [3,2]). Now we are going to show that NTP systems are monotonicity preserving.

Proposition 3.1. *If A is a TP matrix such that $Ae = \beta e$, then $E^{-1}AE$ is TP.*

Proof: Clearly E is a TP matrix and then, by Theorem 3.1 of [1],

$$AE = \begin{pmatrix} \beta & \sum_{j=2}^{n} a_{1j} & \sum_{j=3}^{n} a_{1j} & \cdots & a_{1n} \\ \beta & \sum_{j=2}^{n} a_{2j} & \sum_{j=3}^{n} a_{2j} & \cdots & a_{2n} \\ \vdots & \vdots & \vdots & & \vdots \\ \beta & \sum_{j=2}^{n} a_{nj} & \sum_{j=3}^{n} a_{nj} & \cdots & a_{nn} \end{pmatrix}$$

is TP. Applying Proposition 3.2 of [2] to $(AE)^T$, we deduce that the matrix $B = (b_{ij})$ given by

$$b_{ij} := \begin{cases} \sum_{k=j}^{n} a_{1k}, & \text{if } i = 1, \\ \sum_{k=j}^{n} a_{ik} - \sum_{k=j}^{n} a_{i-1,k}, & \text{if } i > 1, \end{cases}$$

is TP. Finally let us observe that $B = E^{-1}AE$. ∎

Corollary 3.2. *Let A be a totally positive matrix. Then A is monotonicity preserving if and only if A is a multiple of a stochastic matrix.*

Proof: If A is totally positive and monotonicity preserving then, by Remark 2.6, A is a multiple of a stochastic matrix. Conversely, if A is totally positive and $Ae = \beta e$, we deduce from Proposition 3.1 that $E^{-1}AE$ is totally positive. In particular, $E^{-1}AE \geq 0$ and by Corollary 2.5 A is a monotonicity preserving matrix. ∎

Now we may derive the main result of this section, which is a straightforward consequence of Corollary 3.2.

Corollary 3.3. *A totally positive and stochastic matrix is monotonicity preserving.*

In fact, Corollary 3.3 provides the announced fact: each NTP system is monotonicity preserving.

Finally, let us state a result which allows us to characterize the totally positive matrices which transform nonnegative and increasing vectors into nonnegative and increasing vectors.

Proposition 3.4. *Let A be a totally positive matrix. Then $E^{-1}AE \geq 0$ if and only if Ae is an increasing vector.*

Proof: If $E^{-1}AE \geq 0$, then $E^{-1}AE(1, 0, \ldots, 0)^T = E^{-1}Ae$ is nonnegative, which implies by Remark 2.1 that Ae is an increasing vector.

Conversely, let us assume that $Ae = (d_1, \ldots, d_n)^T$ is increasing, that is

$$d_1 \leq \cdots \leq d_n, \tag{3}$$

and let $D = \text{diag}(d_1, \ldots, d_n)$. Then we may write $A = DT$, where T is a stochastic TP matrix. Since $E^{-1}AE = (E^{-1}DE)(E^{-1}TE)$ and by Proposition 3.1, $E^{-1}TE$ is TP, it is sufficient to see that

$$E^{-1}DE = \begin{pmatrix} d_1 & 0 & \cdots & 0 \\ d_2 - d_1 & d_2 & \ddots & \vdots \\ \vdots & \ddots & \ddots & 0 \\ d_n - d_{n-1} & \cdots & d_n - d_{n-1} & d_n \end{pmatrix}$$

is a nonnegative matrix, which follows directly from (3). ∎

§4. Monotonicity Preserving Representations and Length

In the previous section we have shown that normalized totally positive representations are monotonicity preserving. It is well-known (cf. [3]) that normalized totally positive representations generate curves whose length is less than or equal to the length of the corresponding control polygon. In this section we shall prove that this property holds for *any* monotonicity preserving representation.

The first result gives a characterization of monotonicity preserving systems of blending functions.

Proposition 4.1. *Let $(u_0, \ldots, u_n)$ be a system of blending functions. Then $(u_0, \ldots, u_n)$ is monotonicity preserving if and only if $v_1 := u_1 + \cdots + u_n, \ldots, v_n := u_n$ are increasing functions.*

Proof: Let us assume the $(u_0, \ldots, u_n)$ is monotonicity preserving. Since the vectors $(0, 1, \ldots, 1)^T$, $(0, 0, 1, \ldots, 1)^T$, $\ldots$, $(0, \ldots, 0, 1)^T \in \mathbb{R}^k$ are increasing, we derive from Definition 1.1, that $v_1, \ldots, v_n$ are increasing functions.

Conversely, let us assume that $v_1, \ldots, v_n$ are increasing functions and let $v_0 = u_0 + \cdots + u_n = 1$. If $(\lambda_0, \ldots, \lambda_n)^T$ is any increasing vector we obtain that

$$u(t) := \sum_{i=0}^{n} \lambda_i u_i(t) = \lambda_0 + \sum_{i=1}^{n} (\lambda_i - \lambda_{i-1}) v_i(t). \tag{4}$$

Since $\lambda_i - \lambda_{i-1} \geq 0$, $i = 1, \ldots, n$, and v_i, $i = 1, \ldots, n$, are increasing functions, we derive from (4) that u is an increasing function. $\blacksquare$

Let us observe that Corollary 2.7 can be seen as a discrete version of the previous Proposition.

Let us recall that if $\gamma : [a, b] \to \mathbb{R}^k$ is a vector valued function, the total variation of γ is defined by

$$TV(\gamma) := \sup_{\substack{t_0 < \cdots < t_n \in [a,b] \\ N \in \mathbb{N}}} \sum_{k=1}^{N} \|\gamma(t_k) - \gamma(t_{k-1})\|,$$

where $\| \cdot \|$ is any norm of $\mathbb{R}^k$. If $TV(\gamma) < +\infty$, then γ is said to be a vector valued function of bounded variation.

Proposition 4.2. *Let $(u_0, \ldots, u_n)$ be a monotonicity preserving system of blending functions defined on $[a, b]$ and $P_0, \ldots, P_n \in \mathbb{R}^k$. Then $\gamma(t) := \sum_{i=0}^{n} P_i u_i(t)$ satisfies*

$$TV(\gamma) \leq \sum_{i=1}^{n} \|P_i - P_{i-1}\|.$$

Proof: Let $v_i := \sum_{j=i}^{n} u_j$ for each $i = 1, \ldots, n$. By Proposition 4.1, $v_1, \ldots, v_n$ are increasing functions and clearly $v_0 = 1$. Then we may write

$$\gamma(t) = P_0 + \sum_{i=1}^{n} (P_i - P_{i-1}) v_i(t), \quad t \in [a, b]. \tag{5}$$

Since $v_1, \ldots, v_n$ are increasing, the total variation of these scalar functions is $TV(v_i) = v_i(b) - v_i(a)$, and taking into account that $0 \leq v_i(t) \leq v_0(t) = 1$ for all $t \in [a, b]$, we deduce that $TV(v_i) \leq 1$, $i = 0, 1, \ldots, n$. Now (5) allows us to compute $TV(\gamma)$:

$$TV(\gamma) = TV\left(P_0 + \sum_{i=1}^{n}(P_i - P_{i-1})v_i\right) = \sum_{i=1}^{n} \|P_i - P_{i-1}\| TV(v_i)$$

$$\leq \sum_{i=1}^{n} \|P_i - P_{i-1}\|. \quad \blacksquare$$

Now we may specialize Proposition 4.2 for the Euclidean norm in $\mathbb{R}^k$. In this case $TV(\gamma)$ is precisely the length of the curve γ and $\sum_{i=1}^{n} \|P_i - P_{i-1}\|$ is the length of the control polygon of γ with respect to $(u_0, \ldots, u_n)$.

Corollary 4.3. *For monotonicity preserving representations, the length of the curve is less than or equal to the length of its control polygon.*

Acknowledgements. Both authors were partially supported by DGICYT PS90-0121

References

1. Ando, T., Totally positive matrices, Linear Algebra Appl. **90** (1987), 165-219.
2. Carnicer, J. M. and Peña, J. M., Shape preserving representations and optimality of the Bernstein basis, Advances in Computational Mathematics **1** (1993), 173-196.
3. Goodman, T. N. T. , Shape preserving representations, in *Mathematical methods in Computer Aided Geometric Design*, T. Lyche and L.L. Schumaker (eds.), Academic Press, New York, 1989, 333-357.
4. Goodman, T. N. T., Inflections on curves in two and three dimensions, Computer-Aided Geom. Design **8** (1991), 37-50.
5. Karlin, S., *Total Positivity*, Stanford University Press, Stanford, 1968.

Departamento de Matemática Aplicada.
Universidad de Zaragoza.
Edificio de Matemáticas, Planta 1a.
50009 Zaragoza, SPAIN
carnicer@ cc.unizar.es

Splines Focales

Paul de Faget de Casteljau

Abstract. Rational physics is essentially of metric nature, and it would be utopian to associate an algebraic degree with any shape in Nature, as is easily seen from the example of a simple screw. The circular helix is not at all algebraic. The presentation here can be considered as a sample of a larger set of my ideas ranging from Descartes' ovals to quaternions, and on the consequences of a possible revival of Appolonius' viewpoint, using only the notions of distance and angle. The projective infinity and the imaginary numbers are examples of this aberration, which comes directly from the dogma attached to the notion of analytic degree. In order not to neglect the existence of a differential aspect, we conclude the paper with the interesting problem of conjugate aplanetic mirrors. The highly Hamiltonian vision of the association between point and tangent, where each appear to be the derivative of the other, is also presented. Moreover, the solution of this problem depends only on a unique curve, the intermediate caustic, i.e, the envelope of the mean radius. Without taking account of the duality between the orthogonal circle and two tangents to the same circle, we start by adjoining a variable third tangent. Step by step, we obtain the figure of two pairs of tangents intersected by a fifth one. We also present a possible generalization. The presentation here is too short to develop all the ideas evoked here. The author hopes to write a new book on this subject in the near future.

Résumé. La Physique rationnelle est résolument métrique, et il serait utopique d'imposer un degré algébrique aux formes de la Nature ainsi que le prouve le simple système vis-écrou ; l'hélice circulaire n'est pas algébrique. Cette étude n'est qu'un échantillonnage des réflexions, s'étendant aux Ovales de Descartes ou aux Quaternions, que mène l'auteur sur les conséquences d'un retour au point de vue d'Appolonius, en évacuant tout ce qui est contradictoire avec les notions de distance ou d'angle. L'infini projectif, ou les imaginaires sont des exemples de ces aberrations qui proviennent des dogmes rattachés à la notion de degré analytique. Pour ne pas oublier l'existence d'un aspect différentiel, cet article semble conclure en amorçant le problème très passionnant des miroirs aplanétiques conjugués. La vision très Hamiltonnienne de l'association point-tangente, où chacun semble apparaître comme la dérivée de l'autre, y est présente ; de plus, malgré les apparences, la solution ne dépend que d'une courbe unique, la caustique intermédiaire (enveloppe du rayon

Curves and Surfaces in Geometric Design
P. J. Laurent, A. Le Méhauté, and L. L. Schumaker (eds.), pp. 91–103.
Copyright © 1994 by A K PETERS, Wellesley, MA
ISBN 1-56881-039-3.

moyen). Délaissant la dualité avec le cercle orthogonal, par échange avec les rayons, des deux tangentes à un cercle, cet exposé commence avec l'adjonction d'une troisième, variable : progressivement, on aboutit à la figure de deux paires coupées par une cinquième ; les généralisations ne sont pas oubliées. Le cadre trop restreint de cet article ne permet pas de développer convenablement toutes les idées évoquées ici, dont la synthèse pourrait faire l'objet d'un livre plus complet, que l'auteur ambitionne toujours de rédiger un jour.

§1. Géométrie métrique

Les formes rencontrées en Mécanique comme en Physique font intervenir les distances mutuelles entre points, ou les angles et par suite appartiennent à la Géométrie Métrique. Ce n'est ni le cas des formes à Pôles, ou B-Splines qui sont affines, ni a fortiori des NURBS, projectives. On peut se demander s'il existe des algorithmes simples, rapides et efficaces, compatibles avec les exigences de la Géométrie Métrique. Rien ne vaut la règle et le compas, instruments d'usage courant, pour aborder l'étude de la Géométrie Métrique.

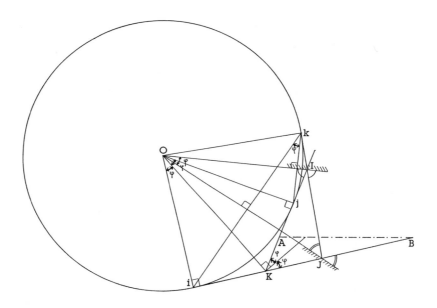

Fig. 1. Une tangente variable à un cercle recoupe deux tangentes fixes.

§2. Trois Tangentes au Cercle

Une des propriétés caractéristiques du Cercle, est de présenter une symétrie de figure, par rapport à tout diamètre.

La longueur des deux tangentes issues d'un point, ainsi que les deux angles qu'elles forment avec le diamètre passant par le point, sont égaux.

Ainsi si une tangente variable kIJ, recoupe en I et J deux tangentes fixes Aj et Bi, qui se coupent entre elles en K, on a :

$$Ij = Ik, \quad Jk = Ji$$

et par suite la longueur de la ligne brisée $BJIj$, formée de segments de tangentes (en nombre quelconque) est identique à la longueur de la tangente Bi, *quantité essentiellement positive*, encore appelée distance du point B au cercle.

La longueur du trajet $AIJB$ est donc constante, quel que soit la position de la tangente en k, et quel que soit la position des points A et B sur les tangentes fixes. L'égalité des angles montre qu'il s'agit d'un rayon, qui se réfléchirait en I et J sur des miroirs orientés suivant OI et OJ, et comme on le voit par l'étude des triangles KIJ et OIJ l'angle $\widehat{IOJ}$ est moitié de l'angle $\widehat{IKJ}$.

Il est clair que suivant les cas de figure, la somme peut être remplacée par une différence, mais surtout si un rayon devient virtuel, il y a aussitôt inversion des rôles entre la tangente et la normale, qui sont les bissectrices intérieures ou extérieures de l'angle formé par les rayons.

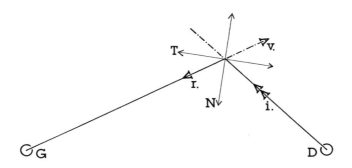

Fig. 2. Où tangente et normale échangent leur rôle.

Il en résulte que le cercle O, ex-inscrit au triangle IJK peut alors être remplacé par un des trois autres cercles inscrits, et on retrouverait ainsi de nombreuses propriétés de Géométrie élémentaire, cercle d'Euler Feuerbach et autres. Notons aussi que la considération des cercles centrés en I, J ou K et de rayon égal à leur distance au cercle (cercles orthogonaux) conduit à de multiples propriétés.

La tangente variable qui découpe sur les deux tangentes fixes les points I et J, vus sous un angle constant, découpe sur les deux droites des divisions homographiques, c'est à dire que les abscisses des points I ou J ne figurent qu'au premier degré dans l'équation qui les relie ($mxy + nx + py + q = 0$).

§3. Sections Coniques

Suivant que les deux cercles sont du même coté, ou de part et d'autre, d'une tangente commune on obtient deux valeurs possibles $2a$ ou $2c$ de leur "distance".

Si on fait tourner la figure autour d'un axe O_1O_2 leur ligne des centres on obtient deux sphères, et pour la première tangente un tronc de cône : on peut encore définir ce cône comme lieu des points dont la somme des distances aux deux sphères, est égale à $2a$. Il s'agit encore de la longueur de n'importe quelle ligne brisée formée de tangentes. Pour une longueur comprise entre $2c$ et $2a$, on aurait obtenu un Hyperboloïde à une nappe, et pour une longueur supérieure à $2a$ un ellipsoïde. Pour n'importe quelle section plane de la figure, l'intersection sera le lieu des points dont la somme des distances aux deux cercles, supposés réels, intersection avec les deux sphères, sera constante. Si le plan de section est le plan Π qui se projette suivant l'autre tangente commune, on obtient la définition focale d'une ellipse (Théorème de Dandelin), sinon on arrive à la définition par cercles focaux.

Notons que pour la même raison, les autres quadriques tangentes à ces deux sphères sont coupées suivant un système de coniques homofocales. Par différence, on aurait obtenu une section Hyperbole ; la Parabole s'obtient par un passage à la limite (foyer à l'infini).

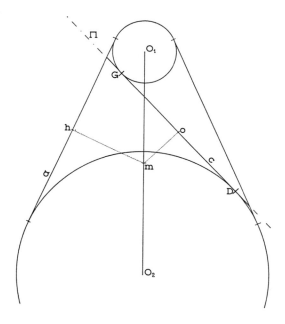

Fig. 3. Section conique d'un cône de révolution.

§4. Quadrilatère Circonscrit à un Cercle

Il est dès lors facile de définir un arc d'ellipse (ou d'hyperbole, par différence)
en utilisant la somme des distances de deux points à un cercle.

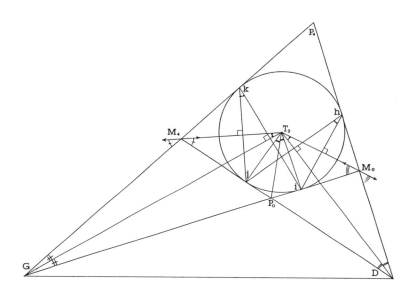

Fig. 4. Définition métrique d'un arc de conique.

En additionnant les distances d'un point G, à gauche $Gi = Gk = GM_4 +$
M_4j et d'un point D, à droite $DM_0 + M_0i = Dh = Dj$ à un même cercle
centré en T_2, on obtient

$$GM_0 + DM_0 = GM_4 + DM_4.$$

Les points M_0, M_4 définissent un arc d'ellipse (P_0, P_4 un arc d'hyperbole) de
foyers G et D.

Les tangentes M_0T_2 et M_4T_2 à l'ellipse, sont des diamètres du cercle, ce
qui implique l'égalité des angles des tangentes avec les deux rayons focaux.
Les diamètres DT_2 et GT_2 sont aussi les bissectrices des rayons focaux, issus
des foyers. On retrouve les angles des diamètres DT_2 avec M_0T_2 ou GT_2
avec M_4T_2 égaux à ceux des cordes rectangulaires ihj, ou ikj, angles inscrits
dans l'arc ij donc tous égaux! Il faudrait y ajouter l'angle sous lequel on
verrait un segment découpé par un tangente variable, sur les tangentes fixes
Gi, Dj, comme on l'a vu précédemment. On a ainsi démontré les théorèmes
de Poncelet, et même davantage par la seule utilisation de la symétrie du
cercle par rapport à un diamètre!

Cette figure est très riche : Ainsi comme tout quadrilatère circonscrit à une conique possède le même triangle conjugué que le quadrilatère inscrit, le diamètre qui relie un des sommets de ce triangle au centre, est perpendiculaire, dans le cas du cercle, au coté opposé, qui en est la polaire. La considération des cercles orthogonaux centrés en G, D, M_0, M_4 permet de trouver encore de nombreuses propriétés intéressantes.

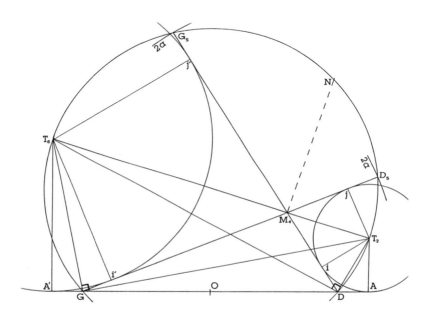

Fig. 5. Cas particulier de la tangente à l'extrémité de l'axe focal.

Le cas particulier où une des extrémités de l'arc est l'extrémité A du grand axe mérite d'être souligné. Le cercle tangent en A à l'axe focal est ex-inscrit au triangle GM_4D. Comme O est le centre de l'ellipse

$$2a = 2OA = GA + DA = Gj + Di = GM_4 + DM_4.$$

Par suite l'arc complémentaire M_4A' est défini par l'autre cercle ex-inscrit GT_2 est perpendiculaire à GT_6, comme le sont DT_2 et DT_6. De nombreuses propriétés liées à la symétrie diamétrale se trouveraient sur la figure du cercle de diamètre T_2T_6 appartenant au faisceau de cercles passant par les points G et D. Outre les points évidents d'intersection avec les cercles directeurs, citons les points d'intersection avec la Normale, situés à des distances $a+b$ et $a-b$ du centre O $(b^2 = a^2 - c^2)$.

§5. Fractionnement de l'Arc

Soit l'arc $M_0 T_2 M_4$ à fractionner, DT_2 étant la bissectrice de l'angle focal DM_0, DM_4.

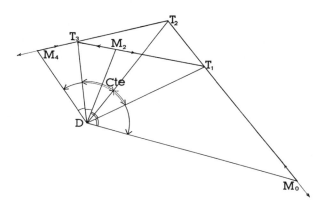

Fig. 6. Fractionnement de l'arc.

Les bissectrices d'un angle DM_2, droite quelconque, avec DM_0 ou DM_4 donneront les points T_1 et T_3 d'intersection avec les tangentes $M_0 T_2$ et $M_4 T_2$. $T_1 T_3$ est la tangente en M_2, ainsi déterminé, et les arcs $M_0 T_1 M_2$ et $M_2 T_3 M_4$. Cette construction s'applique aussi à l'arc complémentaire de $M_0 M_4$, sans être à proprement parler une extrapolation. L'angle DT_1, DT_3 est évidemment moitié de l'angle $M_0 D M_4$ et par suite égal aux angles DM_0, DT_2 ou DT_2, DM_4 et donc par suite constant.

Les cercles tangents aux rayons focaux extrêmes DM_0, DM_2 centré en T_1 et DM_2, DM_4 centré en T_3 achèvent la construction de l'arc fractionné en deux, chacun étant défini par deux points et leurs tangentes associés au cercle centré en leur intersection.

§6. Généralisation aux degrés supérieurs

Si on prend les précautions essentielles, imposées par l'identité des droites définies à π près, et en particulier en limitant les amplitudes des arcs à π, il est possible de répéter cette construction, sur plusieurs étages. Un point, origine polaire, étant donné : on considère n points pilotes $P_0, P_1, P_2, \ldots, P_n$ dont les angles polaires sont en progression arithmétique $\omega, \omega + \theta, \cdots \omega + n\theta$. Il en sera encore de même des points $Q_0, \ldots, Q_{n-1}$, situés dans les directions $\omega + \phi, \omega + \phi + \theta, \cdots, \omega + \phi + (n-1)\theta$. Opération réitérée, jusqu'à l'obtention d'un point unique à l'angle $\omega + n\phi$.

C'est l'algorithme des courbes à pôles simples, appliqué à l'angle polaire. Notons que si le centre polaire s'éloigne vers l'infini, on retrouve, tout au moins au degré 2, l'algorithme qui permet de construire l'arc de parabole, l'axe focal étant dans la direction de la médiane du triangle. La condition d'avoir une direction fixe est restrictive, ce qui inciterait à chercher des courbes focales, où ce foyer ne serait plus fixe.

§7. Pôles Généralisés

Dès lors, il est élémentaire, d'écrire, en coordonnées polaires, l'alignement de trois points, en écrivant que la surface du triangle P_iFP_j est égal à la somme des triangles P_iFP_θ et $P_\theta FP_j$. Cette surface obtenue par le produit vectoriel des rayons vecteurs, conduit à la relation

$$\frac{\vec{U}_i \wedge \vec{U}_j}{\rho} = \frac{\vec{U}_i \wedge \vec{U}}{\rho_i} + \frac{\vec{U} \wedge \vec{U}_j}{\rho_j},$$

$\vec{U}$ désignant le vecteur unitaire correspondant, en ayant divisé par le produit $\rho\rho_i\rho_j$ l'identité. On en déduit un équivalent en coordonnées polaires de l'algorithme de Cox-de Boor

$$\frac{\sin(\theta_j - \theta_i)}{\rho} = \frac{\sin(\theta_j - \theta)}{\rho_i} + \frac{\sin(\theta - \theta_i)}{\rho_j}.$$

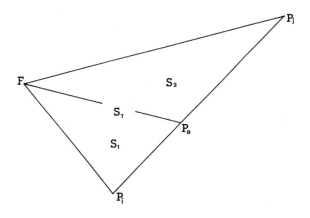

Fig. 7. Algorithme d'insertion du paramètre angulaire.

§8. Insertion d'indices

Avec une logique en inverse du rayon polaire, dont il ne faut pas s'étonner, quand on se souvient des formules de l'optique $\frac{1}{p'} + \frac{1}{p''} = \frac{1}{f}$, cette insertion d'indice, permet de construire une courbe à pôles généralisés, en coordonnés polaires, encore appelée "Blossoming" aux Etats-Unis.

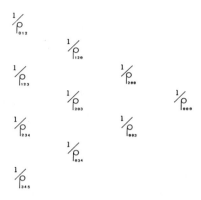

Fig. 8. Insertion d'indice : Points de construction.

§9. Addition Vectorielle

Elle reste toujours compatible ; en particulier si on prend plusieurs progressions angulaires ω, ϕ, θ proportionnelles à des quantités m, n, p la position du vecteur $\overrightarrow{ST}$ pourra être connue pour l'angle indicé $\omega_i + \omega_j + \omega_k(\overrightarrow{ST}_{ijk})$ de même $\overrightarrow{TL}_{ijk}$ avec $\phi_i + \phi_j + \phi_k$ et $\overrightarrow{LM}_{ijk}$ avec $\theta_i + \theta_j + \theta_k$ et ainsi on aura une excellente interpolation, (puisqu'au second degré on peut décrire une orbite képlerienne, tout à fait rigoureuse) d'une trajectoire d'un mobile M gravitant autour de la lune L, qui tourne autour de la terre T, et cette dernière autour du soleil S.

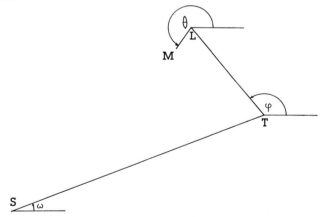

Fig. 9. Epicycles elliptiques.

Il est bon de noter seulement qu'il convient d'effectuer un changement de variable. Ici on décrit une orbite képlerienne à angles constants, ce qui est une excellente définition en projection sur la voûte céleste, alors que le paramètre habituel reste le temps.

Cette extension des épicycles, à des coniques, avec une préférence pour l'ellipse, permet la génération des rosaces les plus complexes. Il est bon de repenser la mécanique, en différences finies, par exemple en remarquant que $\frac{1}{2}(\vec{V}_2^2 - \vec{V}_1^2) = \frac{1}{2}(\vec{V}_2 + \vec{V}_1) \cdot (\vec{V}_2 - \vec{V}_1)$ produit scalaire de la vitesse moyenne par l'accélération, évidemment dirigée dans la direction de la moyenne des forces.

Afin de montrer que les calculs de Géométrie Métrique, n'ont rien à voir avec ceux de la Géométrie Analytique, paralysée par la notion de degré, c'est à dire de Groupes projectifs, nous allons aborder l'étude de miroirs conjugués. Comme pour les orbites képleriennes, les coniques sont à la base de l'étude de l'Optique Géométrique, sous leur définition focale. L'énumération des cas de figure, y compris les cas particuliers serait passionnante et je m'excuse auprès du lecteur de me borner à l'étude, utilisable en pratique d'une combinaison Grégory ; elle génère des formes plus régulières que le système Cassegrain. Et on se limitera à des rayons réels.

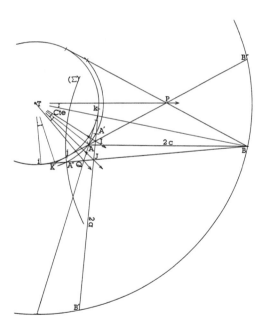

Fig. 10. Miroirs aplanétiques I: Principe de Fermat.

On veut faire converger au foyer A, suivant la direction $B''A$, un rayon BA'' issu de l'objet B, hors de portée, identifié comme une normale à la surface d'onde Σ, cercle directeur relatif au foyer B, dont on ne connait que la portion accessible. Notons qu'on obtient la continuité métrique à l'infini suivant le sens de la concavité de cette surface d'onde, en passant progressivement d'une ellipse ($AB = 2c, AB'' = A''B = 2a = A'B'$) à la parabole, puis à l'hyperbole. Au début de cet exposé, on a appris à définir l'arc PQ d'ellipse, par les points extrêmes, et un cercle centré à l'intersection des tangentes PT et QT. La longueur $2a$, est la longueur constante des tangentes comprises entre les deux cercles centrés en T et passant par A ou B, que l'on peut retrouver sous forme de ligne brisée $AIJB$, IJ étant une tangente variable au cercle initial. Si 2ϕ est l'angle des rayons AB'' et $A''B$, qui se coupent en K, on voit IJ, PB, AQ ainsi que Ki ou Kj sous l'angle ϕ.

En fait le point T n'est pas connu : c'est le point de la médiatrice de AA'', d'où l'on voit ce segment sous l'angle 2ϕ.

Pour déterminer la tangente intermédiaire $A'B'$, symétrique par rapport à TI de AB'', et par rapport à TJ de $A''B$, on utilisera l'angle 2γ qu'elle fait avec l'axe. On a ainsi a résoudre un problème différentiel, calculant la valeur de γ en fonction de la position du point K, quand celui-ci décrit une courbe.

Pour une combinaison aplanétique, respectant donc la convergence, pour des points se projettant en A ou B sur l'axe, il faut respecter la condition des sinus, donnée par $ABBE$:

$$\frac{\sin 2\beta}{\sin 2\alpha} = \frac{KA}{KB} = \frac{RA}{RB} = \frac{AE}{BF} = \frac{AS}{SB}$$

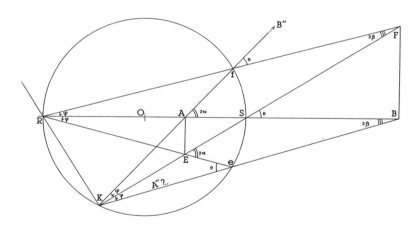

Fig. 11. Miroirs aplanétiques II: Relation d'ABBE.

Les points R et S, extrémités d'un diamètre d'un cercle centré en O sont les intersections avec l'axe des bissectrices des rayons AB'' et $A''B$. La division

AB, RS est harmonique. On voit encore que si E et F sont les intersections des rayon inclinés de l'angle ϕ, en plus ou en moins, issus de R avec la droite SK, ils se projettent en A et B, et on a encore $2\alpha = \theta + \phi$, $2\beta = \theta - \phi$, ($\theta =$ angle BSF) et $\frac{\tan\theta}{\tan\phi} = \frac{RA}{AS} = \frac{RB}{SB}$ (segments AE et BF vus de R ou S).

Il est dès lors facile de construire la figure de la zone accessible (intérieur du cercle), à partir du point E, par exemple ; en déplaçant le point A, on obtient la continuité à l'infini, en O parabole, au delà hyperbole.

Si $AI = u$, $IJ = w$, $JB = v$, $AB = 2c$, $u + v + w = 2a$, $u_0 + v_0 = a + c$, $w_0 = a - c$, $e = c/a$ et $\tan\phi = q\tan\theta$, la variable qui permet l'intégration étant $t = \frac{KA}{RA} = \frac{\cos\theta}{\cos\phi}$, $\frac{t}{q} = \frac{KA}{AS} = \frac{\sin\theta}{\sin\phi}$, le calcul algébrique conduit à calculer le rapport BB'/AA' et on trouve, *dans le cas de la figure*

$$\frac{BB'}{AA'} = \frac{v_0 \left(\frac{\cos(\beta)}{\cos(\alpha)}\right)^{\frac{e(1-q^2)}{1-e^2q^2}}}{u_0 \left(\frac{\frac{a}{\cos(\theta)} - \frac{c}{\cos(\varphi)}}{a-c}\right)^{\frac{q(1-e^2)}{1-e^2q^2}}}$$

$$\frac{(a+c)\cos(\gamma)}{AA'\cos(\alpha) + BB'\cos(\beta)} = \frac{(a-c)\sin(\gamma)}{AA'\sin(\alpha) + BB'\sin(\beta)} = \frac{1}{2}$$

les angles d'incidences en I ou J étant $\gamma - \alpha$ ou $\gamma - \beta$, les normales en I et J font les angles $\gamma + \alpha$ et $\gamma + \beta$, il est dès lors possible de calculer la figure : si le sommet de l'ellipse est en R, $e = q$ entraine quelques simplifications. Si $A = A'$ le miroir I se réduit en un point, comme le miroir J, si $B = B'$.

En s'écartant de ces cas extrêmes, on obtient les formes les plus pures, avec une obstruction centrale, voisine de $1/3$, valeur que les astronomes jugeront excessive. Les zones physiquement significatives restent voisines des coniques; au delà, le miroir I est piriforme (forme d'une poire), et le miroir J connait deux rebroussements, et ressemble à un croissant pour rejoindre à la fin la queue de la poire.

On peut aussi définir ces formes par une double récurrence, avec une précision stupéfiante. Connaissant le trajet $B'I_{2n-1}J_{2n}A'$ reliant deux points voisins de A et B, et sa longueur, il est élémentaire de réfléchir en J_{2n} le rayon issu de B'', symétrique, pour aller en b'' : la médiatrice de $B''b''$ coupera ce rayon réfléchi en I_{2n+1}, et donnera la tangente. Il est facile de recommencer depuis B' et ainsi de suite. Plus difficile est l'initialisation du problème. On peut obtenir les points image et objet respectivement par réflexion de rayons initiaux $J_{-3}I_{-1}$ et $J_{-1}I_1$ pour le premier et I_1J_{-1} et I_3J_1 pour le second (ou plutôt pour des indices divisés par 2).

Si on fait tourner la figure autour de l'axe, on peut même obtenir une interpolation continue de l'arc $I_{2n-1}I_{2n+1}$; de même J. En écrivant que les normales rencontrent l'axe, on obtient une généralisation en éléments finis de la condition d'$ABBE$ (respect de la focale sagittale) qui conditionne d'ailleurs la stabilité de la récurrence.

Ce n'est qu'au voisinage de la queue de la poire que la récurrence peut s'affoler, très rapidement.

Ces formes dépendent de quatre paramètres, de grandeur $2c$, e, q et la constante d'intégration $\frac{v_0}{u_0}$ contre deux $2c$, e aux coniques. On peut aussi les étendre au cas de la réfraction avec, en sus, l'indice comme paramètre.

Les esprits curieux pourront encore étudier le problème pour des couples de points $A'B'$ et $A''B''$, qui ne seraient plus symétriques par rapport à un axe (problème plan).

References

1. de Casteljau P., *Formes à Pôles*, Hermès, Paris, 1985.
2. de Casteljau P., *Les quaternions*, Hermès, Paris, 1987.
3. de Casteljau P., *Le lissage*, Hermès, Paris, 1990.
4. Deltheil R., et D. Caire, *Géométrie*, 4e édition, J. Gabay, Paris, 1950.
5. Deltheil R., et D. Caire, *Compléments*, J. Gabay, Paris, 1951.
6. Farin G., *Curves and Surfaces for Computer Aided Geometric Design, a Practical Guide*, Academic Press, New York, 1988.

Paul de Faget de Casteljau
4 Avenue du Commerce
78000 Versailles, FRANCE

A Surface-Surface Intersection Algorithm with a Fast Clipping Technique

M. Daniel and A. Nicolas

Abstract. We present a surface-surface intersection algorithm suitable for B-splines and NURBS surfaces. Computation of singular cases (tangency, partially identical surfaces) is provided. The local modeling allows a clipping which defines, in the two parametric planes, corresponding free-form regions where there are potential intersections. After this clipping, the regions are recursively split, without subdivision if possible, until no more closed curve exists. Areas with singular cases are simultaneously detected and separately processed. Open curves are computed with a marching technique. All continuous curve segments are finally connected.

§1. Introduction

Intersection problems are fundamental in CAGD, and continue to be an important topic of research especially for surfaces. At least two approaches are possible. The first one is to consider general parametric surfaces, see [1,6]. Only a surface evaluator and a derivative evaluator have to be provided. The second one is to consider specific surface modeling and take advantage of geometric properties. This principally concerns Bézier or B-splines (rational or not) and their well-known convex hull property. In our method, we consider B-splines or NURBS because it corresponds to a local scheme modeling. Notice that numerous CAD systems are based on these modelings.

Many different methods have been developed. Subdivision methods (for example [6]) are reliable, but the numerous surfaces subdivisions entail long computations and proliferation of data. On the other hand, the main and real problem of marching methods is the determination of a good number of starting points for marching only once along all the intersection curves. These methods are however faster than the previous ones. So recent trends are now to develop reliable techniques for finding starting points. The use of vector fields is studied in [2,7]. An hybrid method combining subdivisions and marching is proposed in [1]. Another approach can be found in [9].

Curves and Surfaces in Geometric Design
P. J. Laurent, A. Le Méhauté, and L. L. Schumaker (eds.), pp. 105–112.

The result of a surface-surface intersection is usually a set of curves, closed or not. But singular cases can occur: isolated points or pieces of surfaces. In the last case, the two surfaces are partially identical. The limit case is the tangency one. Approaches for tangency processing exist, see [1,7,10]. These methods find a curve segment. In fact, the two surfaces are numerically identical around the theoretical intersection curve in a narrow region which evidently depends on the tolerance used. It is as if the curve segment has a thickness depending on the tolerance. This explains why we have a continuous variation from the general case to the limit one.

Our method is also an hybrid method which computes the usual set of curves and the singular cases. The local scheme modeling allows a clipping technique. It precisely locates the potential intersection regions which are free-form regions. A marching process is only achieved for open curves. When closed curves exist, they are transformed into open curve segments by splitting, or if necessary subdividing, the corresponding regions. Starting points have then to be found on region boundary curves. Connecting the continuous segments produces the final result.

After a short review on NURBS in the following section, in Section 3 we describe the clipping process and in Section 4 the suppression of closed curves and the detection of singular cases. How we compute the intersections is explained in Section 5, and we end with examples. Additional information is available in [11].

§2. B-splines and NURBS Surfaces

NURBS studies can be found in [5,12]. We only propose a short review of the background necessary for our method.

Notation 1. *Let*

1) $(u_i)_{i=0,\ldots,nu+ku}$, $(v_j)_{j=0,\ldots,nv+kv}$ *be two nondecreasing knot sequences, ku and kv being the surface orders in the two parametric directions (order=degree+1)*
2) $P_{i,j}$ *the control points* $(0 \leq i \leq nu; \quad 0 \leq j \leq nv)$
3) ω_{ij} *the associated positive weights.*

Due to the local modeling, on each elementary parametric rectangle $[u_i, u_{i+1}] \times [v_j, v_{j+1}]$, the NURBS surface S, associated with these data is defined by

$$S(u,v) = \frac{\displaystyle\sum_{m=i-ku+1}^{i} \sum_{n=j-kv+1}^{j} \omega_{m,n}\, P_{m,n}\, N_{m,ku}(u)\, N_{n,kv}(v)}{\displaystyle\sum_{m=i-ku+1}^{i} \sum_{n=j-kv+1}^{j} \omega_{m,n}\, N_{m,ku}(u)\, N_{n,kv}(v)},$$

where the functions $N_{m,ku}(u)$ and $N_{n,kv}(v)$ are the classical B-spline functions defined over the two knot vectors.

Usually, the extreme knots are repeated with a multiplicity equal to the order $(u_0 = \cdots = u_{ku-1}, u_{nu+1} = \cdots = u_{nu+ku})$ and $(v_0 = \cdots = v_{kv-1}, v_{nv+1} = \cdots = v_{nv+kv})$ which generalizes the Bézier case. To simplify our implementation, we assume there are no other multiple knots, but this is not a constraint on the method. If all the weights are equal, the NURBS surface is a B-spline surface.

Our clipping technique is based on the well-known convex hull property: the surface patch defined over an elementary parametric rectangle is located in the convex hull of $ku * kv$ neighbouring control points.

Normal vectors are computed for the detection of closed curves. They are defined by the cross product of the two first derivatives. In fact, only the direction of each vector is necessary. The direction of each derivative is included in the cone defined by the vectors of the hodograph control points or, for NURBS, the hodograph of the surface obtained by setting all the weights to one [13]. The local definition is evidently always true.

§3. Location of Intersections and Clipping

Our idea is to quickly localize the regions on both surfaces where there are potential intersections. For this, we developed an iterative clipping process. We first consider the control points of surface one and the minmax box of surface 2. The respective position of these control points and the three pairs of parallel slabs of the bounding box indicates, according to property 1, which elementary patches defined over elementary parametric rectangles potentially intersect the box ([3,11]). All these critical rectangles of surface 1 are combined into free-form parametric regions as described in the following. We now consider these regions and their minmax boxes (computed from the associated control points). We study the potential intersection between each box and surface 2, as described above for surface 1. We obtain on both surfaces a set of corresponding regions where there are potential intersections. These regions can be reduced by iterating this clipping process. The convergence is quickly obtained (3 or 4 iterations in practice).

At each step, all the neighbouring critical rectangles are linked together to define free-form regions with potential intersections. Each region is disjoint from all others. The considered relationship is a 4-connex one. It provides smaller regions than an 8-connex one. There is a difficulty if two regions have precisely one common corner as in Figure 1. In that case, we check if this corner is an intersection point. If it is not, the two regions are completly independent. If it is an intersection point, we still independently process the two regions but we store the information that the two curves found on the two regions have to be connected in that point. All the informations upon the regions are stored in a tree.

Notice that this localization step can be made very fast through a careful study [11]. The regions are only potential intersection regions. No subdivision has been computed. The next step is to analyse the type of intersection on each region.

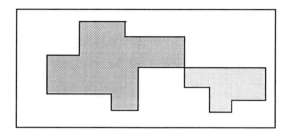

Fig. 1. Two parametric free-form regions with a common corner.

§4. Closed Curve Suppression and Singular Case Detection

There are four basic cases of intersection on each region: no intersection, open curve, closed curve, singular case. The last three cases can appear simultaneously, and with more than one element of each type. The open curves are obtained with a marching method. Starting points are found on the boundary curves as explained in Section 5. There is a problem to determine first points to follow closed curves and singular cases. Thus, we have to suppress closed curves by splitting them into several open curve segments and to detect singular cases. We use the test proposed in [15] to detect potential closed curve: if there is a closed intersection curve, there are two colinear normal vectors on the two surfaces. Notice that this property can be extended if there is a singular case since the normal vectors are then colinear for every intersection points. Two normal bounding cones are constructed for both surfaces (a bounding pyramid is proposed in [8]). The control points taken into account are only those associated with the region. If the cones do not overlap, then no closed curve and no singular case can exist. On the contrary, the surfaces have to be split and the technique is recursively processed, only if potential intersections remain. Other tests have been proposed [8,14]. Before splitting the surfaces, we check if the "middle point" of one region belongs to a singular case: the regions are then studied as explained in the following section. If a singular case exist without including this point, it will be found after some splittings [11].

The technique for splitting depends on the number of elementary parametric rectangles defining the region. If this number is greater that one, an immediate decomposition in up to four new regions is possible by separating the rectangles (and the corresponding control points) from the "middle" of the parametric domain. Otherwise, the region is reduced to a Bézier patch and a subdivision is necessary to produce four new regions. The clipping process is applied again to limit the studies.

§5. Computation of Intersections

We have to compute two kinds of intersections: open curves and singular cases. We use a marching method to compute the open curves. Our interest in this paper concerns finding start points. An open curve reaches the boundaries of at least one region. So the first points must only be searched on the boundary curves of both regions. If no first point is found, no intersection exists for these regions. Notice that the boundary curves of the regions cannot be (partially) included in the other surface. A study of respective position of the control points of each boundary curve and the min-max box of the other region allows us to eliminate the curve (or pieces of it) when there is no intersection [3]. We need to study the problem of curve-surface intersection under favorable conditions due to the initial clipping and the splitting for closed curve suppression, which increase the flatness of the regions, and also to the clipping applied to each boundary curve. In practice, generally at most one point has to be found which evidently simplifies the problem. But reliability has constrained us to consider more general approaches. Notice that the "degree of flatness" of each surface can be estimated from the normal bounding volumes previously computed. This provides information concerning the suitable approach. We have developed the following method based on a bisection strategy. We compute the oriented distance of a point on the curve to the surface. If the curve segment intersects the surface, the sign of this oriented distance changes when it crosses the surface, as illustrated in Figure 2. This is a *local* necessary condition, not a sufficient one, but it allows a bisection approach. Finally, we consider n points on each curve segment. This implicitly defines $n - 1$ intervals where we study, with the oriented distance, the respective position of the extreme points and the surface. With this technique, we can find up to $n - 1$ first points for each curve segment. The value n is a parameter which can also be controlled by the flatness of the regions.

After all the open curves have been obtained, some curve segments have to be connected to define the closed curves, if any. Branching is not detected with a marching technique. It is necessary to postprocess as in [1], or to consider another approach as in [9].

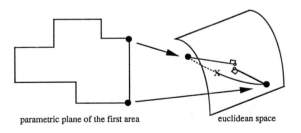

parametric plane of the first area   euclidean space

Fig. 2. Starting point detection.

To compute the singular cases, we consider grids covering both surface regions, constructed with a set of isoparametric curves in one direction. The accuracy of the solution depends on the distance between lines. We apply a

3D curve-curve algorithm [3] to each pair of curves. This algorithm has been developed to compute the singular cases (a curve segment can be found). We finally obtain for each isoparametric curve, when they exist, two opposite points on the boundary of the singular case as illustrated in Figure 3. The boundary is defined by joining these successive points.

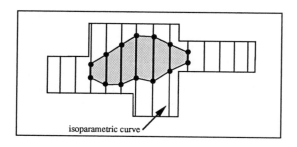

Fig. 3. Boundary of a singular case in one parametric domain.

§6. Examples

The following examples concern B-splines surfaces. The two surfaces are plotted. Bold lines indicate the intersection curves, or singular cases, found. The result in the two parametric planes is also given in the right part of each figure. The intersection of two approximate cylinders is show in Figure 4. Figure 5 illustrates the case of two nonsymmetric surfaces which are partially identical. Figure 6 deals with the limit case of a tangency. Notice that the curves in the parametric planes have a thickness as explained in Section 1, which depends on the number of isoparametric curves computed (see Figure 2).

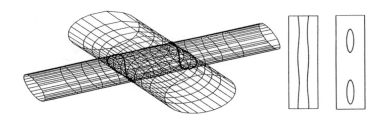

Fig. 4. Intersection of two cylinders.

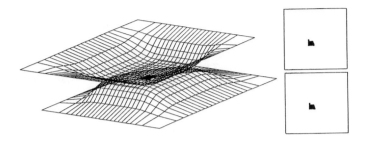

Fig. 5. Singular case: partially identical surfaces.

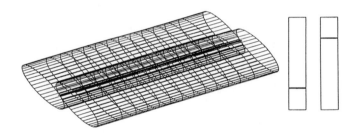

Fig. 6. Singular case: a tangency.

§7. Conclusions

The hybrid method described in this paper computes the intersection between two B-splines or NURBS surfaces. Our goal was to provide a reliable method which finds all the intersection curves and the singular cases (tangencies, partially identical surfaces) as illustrated with the previous examples. A clipping technique quickly locates the regions with potential intersections. The method can be slowed down for singular cases processing or when many closed curves have to be found because numerous splittings are necessary. We also developed a surface-plane intersection algorithm with a similar method [4]. In that particular case, the clipping step is not iterative, and is very accurate.

A specific method for branches or cusp detection must be included during the marching step. A parallel implementation of this method is studied.

References

1. Barnhill, R. E. and S. N. Kersey, A marching method for parametric surface/surface intersection, Computer-Aided Geom. Design **7** (1990), 257–280.
2. Cheng, K. P., Using plane vector fields to obtain all the intersection curves of two general surfaces, in *Theory and practice of geometric Modeling*, W. Strasser and H.P. Seidel (eds.), Springer Verlag, 1989, 187–280.
3. Daniel, M., A curve intersection algorithm with processing of singular cases: Introduction of a clipping technique, in *Mathematical Methods in Computer Aided Geometric Design II*, T. Lyche and L. L. Schumaker (eds.), Academic Press, 1992, 161–170.
4. Daniel, M. and N. Nicolas, An hybrid suface-plane algorithm using a clipping technique, Graphicon'93, Third Conference on computer graphics and visualization in Russia, September 93, St Petersburg.
5. Farin, G., *Curves and Surfaces for Computer Aided Geometric Design, a Practical Guide*, Academic Press, Boston, 1988.
6. Houghton, E. G., R. F. Hemnett, J.D. Factor and C.L. Sabharwal, Implementation of a divide and conquer method for intersection of parametric surfaces, Computer-Aided Geom. Design **2** (1985), 173–183.
7. Kriezis, G. A., N. M. Patrikalakis and F. E. Wolter, Topological and differential equation methods for surface intersections, Comput. Aided Design **24** (1992), 41–55.
8. Kriezis, G. A. and N. M. Patrikalakis, Rational polynomial surface intersections, 17th ASME Design Automation Conference, september 1991, Miami, FL.
9. Manosha, D. and J. F. Canny, A new approach for surface intersection, Int. J. Comp. Geom. Applic. **1**, 4 1991, 491–516.
10. Markot, R. P. and R. L. Magedson, Solutions of tangential surface and curve intersection, Comput. Aided Design **21**, 7 (1989), 421–429.
11. Nicolas, A., Localisation et décomposition en éléments simples de l'intersection de deux surfaces, Research Report IRIN-24, february 1993.
12. Piegl, L., On NURBS: a Survey, IEEE Computer Graphics and Applications **11**, 1 (1991), 55–71.
13. Sederberg, T. W. and X. Wang, Rational hodographs, Computer-Aided Geom. Design **4** (1987), 333-335.
14. Sederberg, T. W., H. N. Christiansen, and S. Katz, Improved test for closed loops in surface intersections, Comput. Aided Design **21**, 8 (1989), 505–508.
15. Sinha, P., E. Klassen and K.K. Wang, Exploiting topological and geometric properties for selective subdivision, ACM Symposium on Computational Geometry (1985), 39–45.

M. Daniel, A. Nicolas, Ecole Centrale de Nantes, I.R.I.N.
1 rue de la Noë, 44072 Nantes cedex 03 FRANCE
(Marc.Daniel Alain.Nicolas) @ ec-nantes.fr

Zonoidal Surfaces

O. Daoudi, B. Lacolle, N. Szafran, and P. Valentin

Abstract. Zonoids are particular convex sets in $\mathbb{R}^n$ studied by many authors. In this paper we study some local regularity properties of the boundary of some particular zonoids related to an actual application in Chemical Engineering.

§1. Introduction

We address here the problem of the structure of a special class of zonoids in the affine space $\mathbb{R}^3$. Zonoids have been introduced as useful tools in various fields (see the reviews by Schneider & Weil [3]). In Chemical Engineering, zonoids represent the separation state associated with a time or space distribution of n chemical species (or any conservative quantities such as energy or impulsion) [5,6,7]. Such a distribution can arise through a separation process, e.g. batch distillation, chromatography, whose output consists of the flowrates of different species in the course of time, that is, by a vector function $F(t)$ which generates a (convex) set of mixtures Z.

A typical practical problem consists in finding the maximal blend of a given composition x which can be made from such an output. This means geometrically that we look for the intersection of Z, which is a zonoid, with a given straight line of direction x:

$$\max\{\phi: \ \phi \in \mathbb{R}^+, M = \phi x \in Z\}.$$

Since the point M is on ∂Z, the sensitivity of the solution to x is obviously linked with the structure and regularity of ∂Z

A *zonotope* is a special kind of polytope, a Minkowski sum of segments in an affine space of finite dimension. A *zonoid* is a limit, relative to Hausdorff metric, of a sequence of zonotopes.

In his paper [1], Bolker gave several equivalent definitions. Let us quote two of them:

Curves and Surfaces in Geometric Design
P. J. Laurent, A. Le Méhauté, and L. L. Schumaker (eds.), pp. 113–120.

Definition 1. *A zonoid is the convex hull of the range of a finite vector measure.*

Definition 2. *A zonoid is a convex set K with symmetry center x_o if a positive measure μ exists on the unit sphere S such that the supporting function of the translated $\tilde{K} = K - x_0$ is*

$$H_{\tilde{K}}(\xi) = \frac{1}{2} \int_S |\langle x, \xi \rangle| d\mu(x).$$

Some particular zonoids are associated to parametric curves in $\mathbb{R}^n$ or referring to Def. 2, to positive measures μ whose supports are curves.

Although zonoids, as convex sets, have been rather thoroughly studied [1,3], their boundary structure, as well as their differentiability properties remain rather obscure. Furthermore, discretization by itself does not give clear answers to these questions.

In this paper, we define a *zonoidal surface* to be any surface which happens to be the boundary of some zonoid associated to a parametric curve, and we study the relationship between the regularity of this curve and the zonoidal surface.

This paper is divided into three parts: Section 1 gives general definitions, Section 2, some more precise definitions and tools, Section 3, a local regularity theorem of zonoidal surfaces, and Section 4, an example.

§2. Zonoids

2.1. The zonoid associated with a regular parametric curve

Let F be a mapping defined on a real compact interval $I = [a, b]$ with values in $E = \mathbb{R}^n$, satisfying the following *conditions* $\mathcal{H}_1$:

1) F is one-to-one,

2) F is C^p, $p \geq 1$ on I,

3) $F'(t) \neq 0$, $\forall t \in I$.

F is a regular parametric representation of the C^p curve $F(I) = \Gamma$.

Let Λ be the subset of the functions of $L_\infty(I)$ with values in $[0, 1]$.

Definition 3. *The set defined by*

$$Z(F) = \{M \in E : \ M = \int_I \lambda(t) F(t) dt, \ \lambda \in \Lambda\},$$

is called the zonoid associated with Γ and may be denoted alternately by $Z(\Gamma)$ or $Z(F)$. Clearly, $Z(\Gamma)$ does not depend on the regular parametric representation of Γ.

Refering to Bolker [1], we have

Property 1. *$Z(F)$ is a centrally symmetric convex compact set.*

In the sequel, we take $E = \mathbb{R}^3$, the three dimensional Euclidean space.

2.2. The first projective diagram of $Z(F)$

This is a natural generalization of the first projective diagram of a zonohedron introduced by Coxeter [2]. We further suppose that there exists a vector μ such that $\langle F(t), \mu \rangle > 0$. The first projective diagram is the conic projection of Γ onto the plane $P_\mu = \{ x : \langle x, \mu \rangle = 1 \}$, and so it is defined by

$$F_p(t) = \frac{F(t)}{\langle F(t), \mu \rangle}, \qquad t \in I,$$

which is not necessarily a regular parametric curve in P_μ.

In the sequel, we denote the homogeneous plane with oriented normal vector η by Π_η, and the line $\Pi_\eta \cap P_\mu$ by Δ_η.

§3. Zonoidal Surfaces

Definition 4. *Using the definitions in Section 1, we define the zonoidal surface $S(F)$ as the boundary of $Z(F)$.*

Hypothesis $\mathcal{H}_2$.
1) the number of intersection points of $\Gamma = F(I)$ with any plane is less than a fixed integer m.
2) there exists $\mu \in E$ such that: $\forall t \in I, \langle F(t), \mu \rangle > 0$.

Without loss of generality we may take $\mu = (0, 0, 1)$. In the following, we assume the hypothesis $\mathcal{H}_1$ and $\mathcal{H}_2$ hold true.

Proposition 2. *There exists a supporting plane to $Z(F)$ through any point of $S(F)$.*

Proof: This holds true since $Z(F)$ is a closed convex set. ∎

3.1. Supporting planes to $Z(F)$

Proposition 3. *The unique intersection point between the supporting plane with outward normal u and the zonoid $Z(F)$ is given by*

$$\int_{\omega_u} F(t)dt, \tag{4}$$

where $\omega_u = \{ t \in I : \langle F(t), u \rangle \geq 0 \}$.

Proof: Let u be a non zero vector of $\mathbb{R}^3$, and $H_{Z(F)}$ the supporting function. We have

$$H_{Z(F)}(u) = \max\{\langle x, u \rangle : x \in Z(F)\}$$

$$= \max\{\int_I \lambda(t)\langle F(t), u \rangle dt : \lambda \in \Lambda\}$$

$$= \int_I \langle F(t), u \rangle_+ dt,$$

$$= \int_{\omega_u} \langle F(t), u \rangle dt.$$

By hypothesis $\mathcal{H}_2$, we have a unique intersection point given by (4). ∎

3.2. Parametrization

From now on, we use the usual spherical coordinates (θ, φ), although any plane projective coordinates could be used. Let $\eta(\theta, \varphi)$ be the unit vector

$$\eta(\theta, \varphi) = (cos\theta \, cos\varphi, sin\theta \, cos\varphi, sin\varphi).$$

Let M be the mapping defined by

$$M : \mathbb{R}/_{2\pi\mathbb{Z}} \times [-\frac{\pi}{2}, \frac{\pi}{2}] \longrightarrow S(F),$$

$$(\theta, \varphi) \longrightarrow \int_{\omega_{\eta(\theta,\varphi)}} F(t)dt.$$

The above properties lead directly to the following two propositions:

Proposition 4. M is a surjective mapping from $\mathbb{R}/_{2\pi\mathbb{Z}} \times [-\frac{\pi}{2}, \frac{\pi}{2}]$ onto $S(F)$.

Proposition 5. If the homogeneous plane of outward normal $\eta(\theta, \varphi)$ has q intersections with the curve $\Gamma = F([a, b])$, whose parameters are denoted by $a \leq t_1 < t_2 < \cdots < t_q \leq b$, with $\langle F'(t_i), \eta(\theta, \phi) \rangle \neq 0$, $i = 1, \ldots, q$, then

$$M(\theta, \varphi) = \epsilon_0 \int_a^{t_1} F(t)dt + \sum_{i=1}^{q-1} \epsilon_i \int_{t_i}^{t_{i+1}} F(t)dt + \epsilon_q \int_{t_q}^b F(t)dt, \qquad (5)$$

where

$$\epsilon_i = \begin{cases} 1 & \text{if } \langle F'(t_i), \eta(\theta, \varphi) \rangle > 0 \\ 0 & \text{elsewhere,} \end{cases} \qquad i = 1, 2, .., q,$$

$$\epsilon_0 = \begin{cases} 1 & \text{if } \langle F'(t_1), \eta(\theta, \varphi) \rangle < 0 \\ 0 & \text{elsewhere.} \end{cases}$$

Remarks.
1) The ϵ_i above alternatively take the values 0 and 1.
2) The intersection point between the supporting plane of normal μ
   (resp. $-\mu$) and $Z(F)$ is $\int_I F(t)dt$ (resp. 0). These points are called poles.

3.3. Regularity of the parametrization

We recall that F is a regular parametric representation of the curve $\Gamma = F(I)$, $I = [a, b]$, such that hypotheses $\mathcal{H}_1$ and $\mathcal{H}_2$ hold true.

Theorem. *Consider* $(\theta_0, \varphi_0) \in \Omega = \mathbb{R}/_{2\pi\mathbb{Z}} \times (-\frac{\pi}{2}, \frac{\pi}{2})$, $t \in \overset{\circ}{I} = (a, b)$. *Let*
$G : \Omega \times \overset{\circ}{I} \longrightarrow \mathbb{R}$ *be defined by* $G(\theta, \varphi, t) = \langle \eta(\theta, \varphi), F(t) \rangle$, *and let* $t_1 < t_2 < \cdots < t_q$ *be the* q *roots of the equation* $G(\theta_0, \varphi_0, t) = 0$, $t \in I$. *Suppose*

1) $\frac{\partial G}{\partial t}(\theta_0, \varphi_0, t_i) \neq 0$, *for* $i = 1, 2, \ldots, q$,

2) *there exists* $V_{(\theta_0, \varphi_0)}$, *an open neighborhood of* (θ_0, φ_0) *such that for every* $(\theta, \varphi) \in V_{(\theta_0, \varphi_0)}$, *the homogeneous plane with oriented normal* $\eta(\theta, \varphi)$ *and the curve* Γ *have exactly* q *intersections.*

Then

i) *if* F *is of class* C^p *on* I, *then* M *is of class* C^p *in an open neighborhood of* (θ_0, φ_0).

ii) $\frac{\partial M}{\partial \theta}(\theta, \varphi)$ *et* $\frac{\partial M}{\partial \varphi}(\theta, \varphi)$ *are linearly independent if and only if all the intersection points between the curve and the homogeneous plane with normal* $\eta(\theta, \varphi)$, *do not lie on a same line through* 0.

Remark. *In the first projective diagram, condition (ii) of the theorem can be expressed by the following equivalent property:*

ii') $\frac{\partial M}{\partial \theta}(\theta, \varphi)$ *and* $\frac{\partial M}{\partial \varphi}(\theta, \varphi)$ *are linearly independent if and only if the number of intersection points between the conic projection curve of* Γ *and the line* $\Delta_{\eta(\theta, \varphi)}$ *is greater or equal than* 2.

To prove the theorem, we need the following lemma:

Lemma. *Let* G *be the function* $G : \Omega \times \overset{\circ}{I} \longrightarrow \mathbb{R}$,

$$G(\theta, \varphi, t) = \langle \eta(\theta, \varphi), F(t) \rangle, \tag{6}$$

and assume that $(\theta_0, \varphi_0, t_0)$ *is such that*

$$G((\theta_0, \varphi_0, t_0) = 0 \quad \text{and} \quad \frac{\partial G}{\partial t}(\theta_0, \varphi_0, t_0) \neq 0. \tag{7}$$

If F *is a mapping of class* C^p *in* I, *then there exists an open neighborhood* U_0 *of* (θ_0, φ_0), *an open neighborhood* T_0 *of* t_0 *and a* C^p *function* $g : U_0 \longrightarrow T_0$ *so that*

$$t_0 = g(\theta_0, \varphi_0) \quad \text{and} \quad G(\theta, \varphi, g(\theta, \varphi)) = 0, \quad ((\theta, \varphi) \in U_0).$$

Moreover, $g(\theta, \varphi)$ *is the unique solution of* $G(\theta, \varphi, t) = 0$ *for* $t \in T_0$, *and partial derivatives with respect to* θ *(or* φ*), exist for* $(\theta, \varphi) \in U_0$:

$$\frac{\partial g}{\partial \theta}(\theta, \varphi) = -\frac{\langle \frac{\partial \eta}{\partial \theta}(\theta, \varphi), F(g(\theta, \varphi)) \rangle}{\langle \eta(\theta, \varphi), F'(g(\theta, \varphi)) \rangle}. \tag{8}$$

Proof: This is a straightforward application of the implicit function theorem. ∎

Proof of the theorem: Let $a < t_1 < t_2 < \cdots < t_q < b$ be the parameters of the q intersections.

From the above lemma, for each t_i, $i = 1, \ldots, q$,
- there exists U_i, an open neighborhood of (θ_0, φ_0),
- there exists T_i, an open neighborhood of t_i,
- there exists a C^p function $g_i : U_i \longrightarrow T_i$ such that

$$t_i = g_i(\theta_0, \varphi_0) \text{ and } G(\theta, \varphi, g_i(\theta, \varphi)) = 0, \quad ((\theta, \varphi) \in U_i).$$

Let $U = \bigcap_{i=1}^{q} U_i$ be an open neighborhood of (θ_0, φ_0). Since the t_i's are distinct, we can further restrict U so that the $g_i(U)$ are disjoint sets.

A straightforward change of variables $t_i = g_i(\theta, \varphi)$ in the bounds of integrals in Eqn (5) gives the expression of $M(\theta, \varphi)$. It follows that M is a C^p mapping of $U \cap V_{(\theta_0, \varphi_0)}$ into $S(F)$.

In order to prove (ii), we evaluate the partial derivatives of M, then apply the second part of the lemma to get

$$\frac{\partial M}{\partial \theta}(\theta, \varphi) = \sum_{i=1}^{q} F(g_i(\theta, \varphi)) \frac{\langle \frac{\partial \eta}{\partial \theta}(\theta, \varphi), F(g_i(\theta, \varphi)) \rangle}{|\langle \eta(\theta, \varphi), F'(g_i(\theta, \varphi)) \rangle|}, \tag{9}$$

$$\frac{\partial M}{\partial \varphi}(\theta, \varphi) = \sum_{i=1}^{q} F(g_i(\theta, \varphi)) \frac{\langle \frac{\partial \eta}{\partial \varphi}(\theta, \varphi), F(g_i(\theta, \varphi)) \rangle}{|\langle \eta(\theta, \varphi), F'(g_i(\theta, \varphi)) \rangle|}. \tag{10}$$

Suppose that the vectors $F(g_i(\theta, \varphi))$, $i = 1, \ldots, q$ are all collinear. Then, for all i with $i = 1, \ldots, q$ there exist α_i so that $F(g_i(\theta, \varphi)) = \alpha_i F(g_1(\theta, \varphi))$. Taking into account these expressions, we deduce from equations (9) and (10) that $\frac{\partial M}{\partial \theta}(\theta, \varphi)$ and $\frac{\partial M}{\partial \varphi}(\theta, \varphi)$ are linearly dependent.

Conversely, assume that for a value (θ, φ), $\frac{\partial M}{\partial \theta}(\theta, \varphi)$ and $\frac{\partial M}{\partial \varphi}(\theta, \varphi)$ are linearly dependent. Then there exist $(\alpha, \beta) \neq (0, 0)$ so that

$$0 = \alpha \frac{\partial M}{\partial \theta}(\theta, \varphi) + \beta \frac{\partial M}{\partial \varphi}(\theta, \varphi) = \sum_{i=1}^{q} \langle v(\theta, \varphi), F(g_i(\theta, \varphi)) \rangle c_i F(g_i(\theta, \varphi)),$$

where

$$c_i = \frac{1}{|\langle \eta(\theta, \varphi), F'(g_i(\theta, \varphi)) \rangle|},$$

and

$$v(\theta, \varphi) = \alpha \frac{\partial \eta}{\partial \theta}(\theta, \varphi) + \beta \frac{\partial \eta}{\partial \varphi}(\theta, \varphi).$$

A simple verification shows that $v(\theta, \varphi) \neq 0$ since $\varphi \in (-\frac{\pi}{2}, \frac{\pi}{2})$. Then

$$0 = \sum_{i=1}^{q} c_i \langle v(\theta, \varphi), F(g_i(\theta, \varphi)) \rangle F(g_i(\theta, \varphi)),$$

and

$$0 = \langle 0, v(\theta, \varphi) \rangle = \sum_{i=1}^{q} c_i \langle v(\theta, \varphi), F(g_i(\theta, \varphi)) \rangle^2.$$

All the values c_i are positive, and therefore $v(\theta, \varphi)$ is orthogonal to $F(g_i(\theta, \varphi))$, $i = 1, \ldots, q$. Furthermore, the vectors $F(g_i(\theta, \varphi))$ and v are in the same plane $\Pi_{\eta(\theta, \varphi)}$. This proves that for $i = 1, \ldots, q$, $F(g_i(\theta, \varphi))$ are collinear. ∎

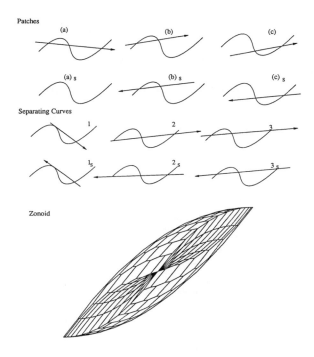

Fig. 1. First projective diagram and boundary structure of a zonoid.

§4. Example

For the sake of simplicity, let Γ be a curve lying in the plane P_μ with $\mu = (0, 0, 1)$. Then Γ and its conic projection onto P_μ are one and the same curve and any intersection between Γ and a plane Π_η can be seen as an intersection between Γ and the corresponding oriented line Δ_η. Consider

$$F(t) = (t, t^3, 1), \ t \in [-1, 1].$$

From the main theorem, the local regularity of our parametrization of $S(F)$ can be obtained in any of the cases "Patches" shown on Figure 1. In fact, a direct work on the algebraic equations shows that the local regularity can be extended in the large to the set of all the values of the parameters corresponding to the same type of intersection in the previous cases. This leads to a decomposition of $S(F)$ into patches, separated by some smooth curves called "separating curves" in Figure 1. The topological arrangement of "patches" and "separating curves" on $Z(F)$ are drawn at the bottom of Figure 1.

§5. Conclusions

It has been shown that zonoidal surfaces inherit almost everywhere the regularity of the generating curve.

It is believed that the present results can be extended to a more complete description of the topological structure and regularity properties of zonoidal surfaces. To this end, we will need a generalization of the Second Projective Diagram, introduced by Coxeter [2] for zonotopes.

Practical applications of the present results stem from the fact that zonoids are generated by models of processes, mainly ODE systems or PDE systems of conservation laws and such a classification of separation processes would be extremely useful.

Another way of getting zonoids is by using interpolating formulas wich generate a parametric curve from a set of control points, e.g. Bézier polynomials. Therefore it may be quite useful to establish a direct link between control points, zonoid and zonoidal surfaces.

Acknowledgements. This work was supported by Université Joseph Fourier de Grenoble (France), and Société Nationale ELF-FRANCE.The authors gratefully acknowledge Société Nationale ELF-FRANCE for kind permission to publish it.

References

1. E.D. Bolker, A class of convex bodies, Trans. Amer. Math. Soc. **145** (1969).
2. H.S. Coxeter, The classification of zonohedra by means of projective diagrams, Journ. de Math, tome XLI., Fasc. 2, 1962.
3. R. Schneider, W. Weil, Zonoids and related topics, in *Convexity and its Application*, Birkhäuser Verlag, Basel, 1983.
4. N. Szafran, Zonoèdres : de la géométrie algorithmique à la théorie de la séparation, Thèse de l'Univ. Joseph Fourier, Grenoble, Oct., 1991.
5. P. Valentin, Is chromatography a separation process ? The zonoid answer, J. of Chromatog. **556** (1991), 25–80.
6. P. Valentin, Theory of Zonoids : I, A Mathematical Summary, NATO ASI, Advances in Theory of Chromatography, Ferrare, Aug. 1991, F. Dondi, G. Guiochon Ed., Kluwer, 1992.
7. P. Valentin, Theory of Zonoids : II, Application to Chromatography, NATO ASI, Advances in Theory of Chromatography, Ferrare, Aug. 1991, F. Dondi, G. Guiochon eds., Kluwer, 1992.

O. Daoudi, B. Lacolle, N. Szafran
Laboratoire LMC-IMAG, BP 53
F38041 Grenoble cedex 9, FRANCE

P. Valentin
ELF-ANTAR-FRANCE, Centre de Recherche ELF-Solaize, BP 22
F69360 Saint Symphorien d'Ozon, FRANCE

Spline Conversion: Existing Solutions and Open Problems

T. Dokken and T. Lyche

Abstract. In this paper we address some of the mathematical problems resulting from practical conversion issues. We separate the conversion process into two distinct problems, exact conversion, and approximate conversion. For approximate conversion we recommend a method based on knot removal. This method is tested on three different surfaces.

§1. Introduction

The increased use of splines in industry, and the multitude of representation formats (for example the power form, the Bernstein/Bézier form, and the B-spline form), have resulted in a need for efficient and accurate methods to convert from one surface representation to another.

It is convenient to consider two aspects of the conversion process: exact conversion and approximate conversion. *Exact conversion* simply means to change representation, for example, increasing the degree, using more polynomial patches, or just use a different representation. In *approximate conversion* we would like to compute a piecewise polynomial approximation to a given piecewise polynomial surface. The approximation is normally required to have a certain degree (which could be either higher, equal, or lower), and then the number and location of the patches for the approximation should be adjusted so that the error is within a given tolerance. In degree reduction we would normally have to increase the number of patches, while in degree raising the number of patches can sometimes be reduced.

To structure the concept of conversion in §2 we will introduce different surface representation forms and the appropriate spline spaces. Existing surface transfer formats are discussed in §3. In §4 we discuss exact conversion problems while in §5 we will address briefly approximative conversion problems. The results from using these conversion concepts on three test surfaces are given in §6. We conclude the paper in §7 with some general remarks and list some open problems.

For additional references related to the conversion problem see [8,14].

Curves and Surfaces in Geometric Design
P. J. Laurent, A. Le Méhauté, and L. L. Schumaker (eds.), pp. 121–130.
Copyright ⊕ 1994 by A K PETERS, Wellesley, MA
ISBN 1-56881-039-3.

§2. Surface Forms

To fix notation we review the surface representations we are interested in and define the mathematical approximation problem underlying the conversion problem. We restrict attention to parametric surfaces on rectangular grids. For surfaces on triangular grids see [9,25].

We denote B-splines of degree d on a nondecreasing knot vector $t = (t_i)_{i=1}^{n+d+1}$, by $B_{i,d,t}$ or $B(x|t_i, \cdots, t_{i+d+1})$. Normally we assume that $t_{i+d+1} > t_i$ for all i and for a suitable interval $[a, b]$ we have $a = t_1 = \cdots = t_{d+1}$ and $b = t_{n+1} = \cdots = t_{n+d+1}$. We call such t a d-extended partition of $[a, b]$. The B-splines form a basis for a space $\mathcal{S}_{d,t}$ of univariate piecewise polynomials on $[a, b]$.

We consider parametric B-spline surfaces $f : \Omega \to \mathbb{R}^3$ where $\Omega = [a^1, b^1] \times [a^2, b^2]$ is a planar rectangle. We say that $t = (t^1, t^2)$ is a $d = (d^1, d^2)$ extended partition of Ω if t^i is a d^i extended partition of $[a^i, b^i]$ for $i = 1, 2$. We define the polynomial spline space

$$\mathcal{S}_{d,t} = \left\{ \sum_{i=1}^{n_1} \sum_{j=1}^{n_2} c_{i,j} \phi_i(u) \psi_j(v) : \quad c_{i,j} \in \mathbb{R}^s \right\}. \tag{2.1}$$

Here $\phi_i(u) = B_{i,d^1,t^1}(u)$ and $\psi_j(v) = B_{j,d^2,t^2}(v)$ are B-splines of degree d^1 and d^2 on t^1 and t^2. We are mostly interested in the case $s = 3$ of parametric surfaces. We let $S_d(\Omega)$ be the space of all functions $f : \Omega \to \mathbb{R}^s$ such that $f \in S_{d,t}$ for some d-extended partition t of Ω.

A spline will simply mean a function which for a suitable parametrization can be written in the form (2.1).

A spline can be represented in a number of ways. In addition to (2.1) two other surface representations will be considered. To define them let $\Omega_\mu = [t_{\mu_1}^1, t_{\mu_1+1}^1) \times [t_{\mu_2}^2, t_{\mu_2+1}^2)$, where $\mu = (\mu_1, \mu_2)$, denote a nonempty subrectangle of Ω. The function f_μ representing a spline f on Ω_μ can be written

$$f_\mu(u, v) = f(t_{\mu_1}^1 + u(t_{\mu_1+1}^1 - t_{\mu_1}^1), t_{\mu_2}^2 + v(t_{\mu_2+1}^2 - t_{\mu_2}^2)), \quad 0 \le u, v \le 1. \tag{2.2}$$

We then have three ways of representing f_μ

(i) The *power form* (P).

$$f_\mu(u, v) = \sum_{i=0}^{d^1} \sum_{j=0}^{d^2} a_{i,j} \frac{u^i v^j}{i! \, j!}, \quad a_{i,j} = a_{i,j}^\mu \in \mathbb{R}^s. \tag{2.3}$$

(ii) The *Bernstein/Bézier form* (B). With $B_i^d(t) = \binom{d}{i} t^i (1 - t)^{d-i}$

$$f_\mu(u, v) = \sum_{i=0}^{d^1} \sum_{j=0}^{d^2} b_{i,j} B_i^{d^1}(u) B_j^{d^2}(v), \quad b_{i,j} = b_{i,j}^\mu \in \mathbb{R}^s. \tag{2.4}$$

(iii) The *spline form* (S) given by (2.1).

The knot vector $t = (t^1, t^2)$ where each t_j^i is repeated $d^i + 1$ times, for $i = 1, 2$ is called the *Bernstein knot vector*. The S-form reduces to the B-form on such a t.

With $\mathcal{S}$ denoting a spline spaces $\mathcal{S}_{d,t}$ and F one of the surface forms S, P or B we let

$$C(f, \mathcal{S}, F)$$

denote the coefficients of a spline $f \in \mathcal{S}$ written in the form F. We can now define the following conversion problem.

Conversion Problem. *Suppose we are given two spline spaces $\mathcal{S}_1$ and $\mathcal{S}_2$ with $\mathcal{S}_1 \subset \mathcal{S}_2$ defined over the same rectangle Ω, two surface forms F_1 and F_2, and the coefficients $C(f, \mathcal{S}_1, F_1)$ of some $f \in \mathcal{S}_1$. Find $C(f, \mathcal{S}_2, F_2)$.*

Since $\mathcal{S}_1 \subset \mathcal{S}_2$ this problem involves exact conversion. We simply want to rewrite f using another representation. With $\mathcal{S}_1 = \mathcal{S}_{d,\tau}$ and $\mathcal{S}_2 = \mathcal{S}_{e,t}$ the problem is called *knot insertion* if $e = d$ and $\tau \subset t$, and more generally *degree raising* if $e > d$. See [16].

In applications there is also a need for approximate conversion. Thus the input to the conversion problem will often be the solution of the following approximation problem:

Approximation Problem. *Suppose we are given degrees d, e, a spline $f \in \mathcal{S}_d(\Omega)$, a tolerance $\epsilon > 0$, and a norm $||\cdot||$ on $C(\Omega)$. Find a space $\mathcal{S}_{e,\tau} \subset \mathcal{S}_e(\Omega)$ of smallest possible dimension, an element $\tilde{f} \in \mathcal{S}_{e,\tau}$, and a reparametrization $p : \Omega \to \Omega$ such that $||f - \tilde{f} \circ p|| \le \epsilon$.*

For our purpose a reparametrization of Ω is any C^0 function $p : \Omega \to \Omega$ which is one-to-one and onto, and increasing in the x and y direction. With $d > e$ the problem is called *degree reduction*, while if $d < e$ we refer to the problem as *degree raising*.

In certain cases the surface f is C^0 and G^1, to support this case we allow for reparametrizations with only C^0 continuity.

§3. Some Geometry Transfer Formats and their Surface Forms

Different parts of the industry use different transfer formats for surfaces. VDA-FS originated in the German automotive industry and is a transfer format for piecewise nonrational curves and surfaces in power form. Unless some continuity information can be decided the underlying spline space must be defined using the Bernstein knots. The coefficients $C(f, \mathcal{S}, P)$ in this format are given in the power form (2.3).

IGES and the similar French standard SET are transfer formats that cover a much broader range than geometry and include NURBS. Thus the spline space used is $\mathcal{S}_{d,t}$, with $d = (d^1, d^2) \ge 0$, and t being nondecreasing knot sequences. The coefficients $C(f, \mathcal{S}_{d,t}, S)$ are given by (2.1). Because of the added complexity the success rate for transfer with IGES has been reported to be as low as 50%. Since IGES and SET contains B-splines and NURBS, it is a much better transfer format for geometry than VDA-FS. There are two reasons for this, one is that NURBS can represent a much broader range of geometry than the power basis. The other is that Bernstein and B-spline representation utilize the representation possibilities of float numbers on digital computers better than the power basis [10]. Trimmed surface can be transferred by IGES.

ISO 10303 (STEP) is a standard for communicating between technical data bases in CAD/CAE. The standard is currently under development. The STEP standard contains NURBS and is well suited for geometry transfer. STEP is covering the functionality of IGES, SET and VDA-FS, and is intended to replace these transfer formats.

§4. Conversion between Different Surface Forms

The conversion problem for tensor product surfaces can be stated in terms of univariate conversion. Using lower case c for the coefficients of a univariate spline f in some univariate spline spaces $\mathcal{S}_1 \subset \mathcal{S}_2$, we can write

$$c(f, \mathcal{S}_2, F_2) = Ec(f, \mathcal{S}_1, F_1),$$

where E is a matrix transforming one vector of coefficients into another. (If $c = (c^1, \cdots, c^s)$ has s components then we define the product Ec as $(Ec^1, \cdots, Ec^s)$.) For tensor product surfaces we now have

$$C_2 = E_u^T C_1 E_v, \qquad (2.5)$$

where $C_i = C(f, S_i, F_i)$ for $i = 1, 2$, and E_u, E_v are univariate conversion matrices in the u and v direction, respectively, see [17]. We discuss next some special instances of (2.5).

4.1. Conversion from P-form to B-form

There is a simple explicit formula for this conversion. Let $A = (a_{i,j})$ and $B = (b_{i,j})$ be the matrices with $d^1 + 1$ rows and $d^2 + 1$ columns containing the coefficients of the P and B-form on one subrectangle. Then $B = E_{d^1}^T A E_{d^2}$ where E_d transforms the Bernstein basis to the power basis $(1, t, \cdots, t^d/d!)^T = E_d(B_0^d(t), \cdots, B_d^d(t))^T$. The matrix E_d is nonnegative and upper triangular with elements $e_{i+1,j+1} = \binom{j}{i}/(i!\binom{d}{i})$ for $0 \leq i \leq j \leq d$, see [2,10]. These entries can be generated either as exact rational numbers or recursively from $e_{i+1,j+1} = e_{i+1,j} + e_{i,j}/(d-i+1)$, $1 \leq i \leq j \leq d$, starting with a first row of ones.

4.2. Conversion from B-form to P-form

Using the notation of §4.1 we have $A = F_{d^1}^T B F_{d^2}$ where $F_d = (f_{i,j})_{i,j=1}^{d+1}$ is an upper triangular matrix (the inverse of E_d) with $f_{i+1,j+1} = (-1)^{j-i}\binom{d}{j}\binom{j}{i}j!$, for $j \geq i \geq 0$, see [10]. A recursive formula is $f_{i+1,j+1} = (d-j+1)(f_{i,j}-f_{i+1,j})$. As d increases these elements grow quickly in size and extended precision might be used for large values of d.

4.3. Conversion from S-form to P-form

Algorithms for this conversion problem are given already in the books [3,24]. Since this is a generalization of the conversion in §4.2, numerical problems can be encountered for high degrees.

4.4. Conversion from S-form to B-form

Using knot insertion to ensure that the elements of the knot vector t^1 and t^2 are repeated a maximal number of times at each location reduces the S-form to the B-form. With $C = (c_{i,j})_{i=\mu_1-d^1, j=\mu_2-d^2}^{\mu_1, \mu_2}$ the S-form coefficients on the subrectangle Ω_μ, and B the B-form coefficients on this rectangle the conversion can be written $B^\mu = G_u^T C G_v$ with G_u, G_v the 1D transformation matrices inserting knots at the boundary of Ω_μ. Various knot insertion algorithms have recently been tested and compared for timing and accuracy, see [11].

4.5. Conversion from B-form to S-form

One approach is to first represent the spline in S-form on the Bernstein knots and then using the knot removal technique in [18] with a small tolerance. The remaining knots determines automatically the continuity of f. This method does not work well for parametric surfaces that are C^0, and G^1, but not C^1. For such surfaces we can obtain C^1 continuity by a reparametrization. The procedure is quite straightforward. Indeed, suppose two parts $f_-(u, v)$ and $f_+(u, v)$ of a tensor product surface meet at a knot line $z = t_{\mu_1}^1$ say. Having C^0 and G^1 continuity mean that $f_-(z, v) = f_+(z, v)$ and $\partial f_-(z, v)/\partial u = \beta \partial f_+(z, v)/\partial u$ for all $v \in [c, d]$ and some $\beta \in \mathbb{R}$. If $\beta = 1$ then we already have C^1 continuity. Suppose $\beta \neq 0, 1$. Define $\tilde{t}^1$ by $\tilde{t}_i^1 = t_i^1$, $\tilde{t}_{i+1}^1 - \tilde{t}_i^1 = t_{i+1}^1 - t_i^1$ for $i \neq \mu_1 - 1$, and $\tilde{t}_{i+1}^1 - \tilde{t}_i^1 = (t_{i+1}^1 - t_i^1)/\beta$ for $i = \mu_1 - 1$. On the parametrization defined by the knots $(\tilde{t}^1, t^2)$ the two surfaces f_- and f_+ will now meet with continuity C^1 at $z = t_{\mu_1}^1$. This procedure can be repeated for all interior knot lines.

If the continuity of the surface is known from other considerations, we can use the generalized Oslo algorithm to convert from the B-form to the S-form, see [19].

4.6. Conversion from P-form to S-form

One method is to first convert to B-form and then use one of the methods in §4.5. Alternatively, suppose we have a suitable knot vector for the S-form. A simple conversion method is to determine the unknown coefficients by solving an interpolation problem on a rectangular grid. The inversion problem gets ill conditioned as the polynomial degree grows. To improve the situation we have used local spline projectors that can be constructed to reproduce a required polynomial behavior [20].

4.7. Degree raising

Suppose now that the degree of S_2 is larger than the degree of S_1. Such a conversion can be carried out in two steps. First we raise the degree without changing the form, and then we change the form if necessary. The second step has already been discussed, so consider the first step. Degree raising is trivial for the power form. We simply include the correct number of zero coefficients. For the B-form we can repeatedly increase the degree by one. The formula for increasing the degree from d to $d+1$ is ([13])

$$B_i^d = \frac{d+1-i}{d+1} B_i^{d+1} + \frac{i+1}{d+1} B_{i+1}^{d+1}.$$

For the S-form we can either convert to B-form and raise the degree of the B-form, or use one of the methods described in [4,21,22]. For raising the degree from d to e we need to add $e-d$ extra knots at each distinct knot location. A numerically stable recurrence relation for directly finding the transformation matrix E in this case, is given in [21].

§5. Approximate Conversion

We consider now the approximation problem defined in §2.

5.1. The nonparametric case

Typically some kind of least squares method is used both to find the space and the approximation. To find the space there are two main strategies. Either starting with few knots, and inserting knots where the error is large [6] or starting with some simple approximation using lots of knots and then removing the knots which are determined not to be important for the approximation [18]. It is also possible to include shape constraints in the approximation [1,6].

5.2. The parametric case

The traditional way of approximating parametric spline represented geometries is to approximate the x-, y- and z-coordinates independently, thus keeping the parametrization of the original geometry. Thus most existing methods use a nonparametric method for each space component [6,18]. A few methods work directly with the parametric case [6,13].

During the last years research has turned in the direction of reparametrization of the geometry to get more freedom in the parametrization. In [5], Degen introduces the concept of normal distance between an original curve and an approximating curve. For certain classes of Bézier curves he shows that the error of the best approximant has the equioscillating property as in the case of Chebyshev approximation. The clue of this approximation method is to allow the approximating curve to have a different parametrization than the curve to be approximated. However, there is a one to one relation between points on the two curve segments. For some explicit approximation schemes see [7,12,15,23].

5.3. Degree Reduction and Degree Raising

For degree reduction we can use the knot removal technique of [18]. We construct a bilinear interpolant to the surface by sampling at a sufficinetly dense set of points, raise the degree, and then remove knots. Other initial approximations can also be used. (See the SURFB example in §6 where bicubic Hermite interpolation is used to construct the initial approximation.) Other approaches can be found in [8].

For degree raising the sampling technique combined with knot removal as described for degree reduction can be used.

5.4. Approximating a rational surface

Suppose

$$f = \frac{\sum_{i=1,j=1}^{n_1,n_2} w_{i,j} c_{i,j} \phi_i(u) \psi_j(v)}{\sum_{i=1,j=1}^{n_1,n_2} w_{i,j} \phi_i(u) \psi_j(v)} = \frac{g(u,v)}{w(u,v)}, \quad c_{i,j} \in \mathbb{R}^s$$

is a surface written in NURBS form. Here ϕ_i and ψ_j are B-splines as in (2.1), say of degrees $e = (e_1, e_2)$, and the $w_{i,j}$'s are all positive scalars. Conversion methods for this problem can be found in [14]. We can also approximate f by a polynomial spline in some spline space $S_d(\Omega)$ to within a tolerance ϵ. We can formulate this problem as a slight generalization of the approximation problem in §2. After multiplying by w we simply replace $\|f - \tilde{f} \circ p\|$ by $\|g - w\tilde{f} \circ p\|$. This problem could be solved by least squares introducing suitable discrete coefficient norms as in [18].

§6. Test Results and Methods Used

The test surfaces BRODE, SEITE 1 and SURFB were all originally represented in P-form and converted to B-form. For the surfaces SEITE 1 and SURFB a conversion to S-form combined with a reparametrization as described in §4.5 was first applied. This gave surfaces with C^1 continuity (see [14] for plots). For the results of the tests see the tables. Column two in the tables gives the number of polynomial segments in the approximation. These numbers are determined automatically by the algorithm. The compression factor is calculated as the ratio between the storage space needed for the original B-form representation of a surface (3 x number of vertices) and the storage needed for the storage space of the B-spline approximation to the surface (3 x number of vertices + length of knot vectors).

BRODE consists of one polynomial patch of degree 9×9. To degree reduce this surface, we first made a bilinear approximation represented in the S-form based on sampling the surface at 50×200 points. Then the polynomial degrees were raised to respectively 3×3 and 5×5. As a final step knot line removal as described in [18] was applied. SEITE 1 was given as 9×7 bicubic segments. We applied the knot line removal on the C^1 S-form version. For degrees 5×5 we first had to apply a degree raising transformation to the

surface: BRODE ; degree: 9 × 9 ; segments: 1 × 1						
degree	segm.	minimal contin.	compr. factor	error toler.	meas. error	CPU HP 9000/735
3×3	6×6	C^2, C^2	1.1	0.01	0.009	3.6 sec
	3×3	C^2, C^2	2.3	0.1	0.048	1.5 sec
5×5	5×3	C^4, C^4	1.1	0.01	0.007	5.7 sec
	1×1	C^5, C^5	2.3	0.1	0.057	3.7 sec

surface: SEITE 1 ; degree: 3 × 3 ; segments: 9 × 7						
degree	segm.	minimal contin.	compr. factor	error toler.	meas. error	CPU HP 9000/735
3×3	8×7	C^2, C^2	8.4	0.01	0.005	0.03 sec
	4×7	C^2, C^2	12.9	0.1	0.033	0.04 sec
5×5	5×7	C^2, C^2	3.3	0.01	0.009	0.16 sec
	3×7	C^4, C^2	5.1	0.1	0.086	0.14 sec

surface: SURF B ; degree: 5 × 5 ; segments: 17 × 63						
degree	segm.	minimal contin.	compr. factor	error toler.	meas. error	CPU HP 9000/735
3×3	17×27	C^2, C^2	62.3	0.01	0.004	29.0 sec
	17×18	C^2, C^2	88.4	0.1	0.033	22.5 sec
5×5	17×37	C^2, C^3	20.5	0.01	0.005	11.9 sec
	15×19	C^2, C^4	48.8	0.1	0.052	9.04 sec

S-form, and then use knot line removal. The surface called SURFB consisted of 17 × 63 biquintic patches and should be converted to a bicubic surface and a biquintic surface with fewer patches. For degrees 5 × 5 knot line removal could be applied directly to the C^1 S-from version. For degrees 3 × 3 we first made an approximation to within half the tolerance using piecewise bicubic Hermite interpolation. We then removed knots from this approximation.

The software used is contained in SISL - SI Spline Library written in C. An earlier version of the data reduction software written in FORTRAN is available from NETLIB.

§7. Remarks and Open Problems

1. The new methods so far have addressed approximation of planar parametric curves and constructing approximations using reparametrization. The problem gets more complex when 3D parametric surfaces are to be addressed. Much research still remains here.

2. Another approach is approximation of parametric curves and surfaces by algebraic curves and surfaces. Existing methods find the exact algebraic representation and are based on symbolic computation and infinite curves. In CAGD there is a need for numerical methods that can find an algebraic approximation to a parametric curve segment or a parametric surface patch.

Acknowledgements. This work was support by the Research Council of Norway through program STP. 28402.

References

1. Arge, E., M. Dæhlen, T. Lyche, and K. Mørken, Constrained spline approximation of functions and data based on constrained knot removal, in *Algorithms for Approximation II*, J. C. Mason and M. G. Cox (eds.), Chapman and Hall, London, 1990, 4–20.
2. Barry, P. J., and R. N. Goldman, Algorithms for progressive curves: Extending B-spline and blossoming techniques to the monomial, power, and Newton dual bases, in *Knot Insertion and Deletion Algorithms for B-spline Curves and Surfaces*, R. N. Goldman and T. Lyche (eds.), SIAM, Phil., 1993, 11–63.
3. de Boor, C., *A Practical Guide to Splines*, Springer Verlag, New York, 1978.
4. Cohen, E., T. Lyche, and L. L. Schumaker, Algorithms for degree raising of splines, ACM Trans. Graph. **4** (1985), 171–181.
5. Degen, W. L. F., Best Approximation of parametric curves by splines, in *Mathematical Methods in Computer Aided Geometric Design II*, T. Lyche and L. L. Schumaker(eds.) Academic Press, Boston, 1992, 171–184.
6. Dierckx, P., *Curve and Surface Fitting with Splines*, Clarendon Press, Oxford, 1993.
7. Dokken, T., M. Dæhlen, T. Lyche, and K. Mørken, Good Approximation of circles by curvature-continuous Bézier curves, Computer-Aided Geom. Design **7** (1990), 33–41.
8. Eck, M., and J. Hadenfeld, A stepwise algorithm for converting B-Splines, in *Curves and Surfaces in Geometric Design*, P.-J. Laurent, A. Le Méhauté, and L. L. Schumaker (eds.), A K Peters, Wellesley, 1994, 131–138.
9. Farin, G. *Curves and Surfaces for Computer Aided Geometric Design, A practical Guide*, Academic Press, San Diego 1990.
10. Farouki, R. T., and V. T. Rajan, On the numerical condition of polynomials in Bernstein form, Computer-Aided Geom. Design **4** (1987), 191–216.
11. Fugelli, P., Knot insertion for B-splines — evaluation of new and established algorithms, Master Thesis (in Norwegian) , University of Oslo, Norway, 1993.
12. Goult, R. J., Parametric curve and surface approximation, in *The Mathematics of Surfaces III*, D. C. Handscomb (ed), Clardon Press, Oxford, 1989, 331–346.

13. Hoschek, J., and D. Lasser, *Fundamentals of Computer Aided Geometric Design*, AKPeters, Boston, 1993.
14. Hoschek, Josef, and Franz-Josef Schneider Approximate conversion and data compression of integral and rational B-spline surfaces, in *Curves and Surfaces in Geometric Design*, P.-J. Laurent, A. Le Méhauté, and L. L. Schumaker (eds.), A K Peters, Wellesley, 1994, 241–250.
15. Lachance, M. A., Chebyshev economization for parametric surfaces, Computer Aided Geometric Design **3** (1988) 195–208..
16. Lyche, T., Discrete B-splines and conversion problems, in *Computations of Curves and Surfaces*, M. Gasca (ed.), Kluwer Academic Publishers, Dordrecht, 1990, 117–134.
17. Lyche, T, E. Cohen, and K. Mørken, Knot line refinement algorithms for tensor product B-spline surfaces, Computer-Aided Geom. Design **2** (1985), 133–139.
18. Lyche, T. and K. Mørken, Knot removal for parametric B-spline curves and surfaces, Computer-Aided Geom. Design **4** (1987), 217–230.
19. Lyche, T, K. Mørken, and K. Strøm, Conversion between B-spline bases using the generalized Oslo algorithm, in *Knot Insertion and Deletion Algorithms for B-spline Curves and Surfaces*, R. N. Goldman and T. Lyche (eds.), SIAM, Phil., 1993, 135–153.
20. Lyche, T. and L. L. Schumaker, Local spline approximation methods, J. Approx. Th. , **15** (1975), 294–325.
21. Mørken, K. Some identities for products and degree raising of splines, Constr. Approx. **7** (1991), 195–208.
22. Prautzsch, H., Degree elevation of B-spline curves, Computer-Aieded Geometric Design **1** (1984), 193–198.
23. Schaback, R., Rational geometric curve interpolation, in *Mathematical Methods in Computer Aided Geometric Design II*, T. Lyche and L. L. Schumaker, (eds.), Academic Press, Boston, 1992, 517–535.
24. Schumaker, L. L., *Spline functions: Basic Theory*, Wiley, New York, 1981.
25. Schumaker, L. L., Applications of multivariate splines, preprint.

Tor Dokken
SINTEF SI
P.O. Box 124, Blindern
0314 Oslo, NORWAY

Tom Lyche
Institutt for Informatikk
P.O. Box 1080, Blindern
0316 Oslo, NORWAY
tom@ifi.uio.no

A Stepwise Algorithm
for Converting B-Splines

M. Eck and J. Hadenfeld

Abstract. We present a new solution to convert polynomial B-splines. This stepwise operating algorithm can be understood as an appropriate combination of a degree reduction method for Bézier curves (resp. surfaces) and a knot removal strategy for B-splines. These two basic methods were developed separately in recent papers of the authors [3,4,5].

§1. Problem Statement

It is a frequent problem in Computer Aided Design systems (see [8,11] for several examples) to represent a given B-spline curve $\mathbf{x}(t)$ of arbitrary order k as a B-spline curve $\bar{\mathbf{x}}(t)$ of any prescribed order $\bar{k}$ by applying suitable methods for basis conversion. The same task naturally appears for tensor product B-spline surfaces if the polynomial order has to be changed from (k_1, k_2) to $(\bar{k}_1, \bar{k}_2)$.

However, it might be an additional wish (or even constraint) of the user in this particular situation that the resulting new B-spline curves or surfaces are based on knot vectors which consist of relative small numbers of knots. Here, the two basic ideas are that the results should be smooth as well as consisting of only few spline segments.

Both requirements can obviously be fulfilled if the deviation between the input and output spline is not of interest. Nevertheless, in most applications this deviation also has to be limited by a prescribed tolerance δ. Here, often the maximal absolute deviation d according to the parametrization is an appropriate measure for this deviation since it is very easy to bound by the control points.

Therefore, we always have to take account of the following constraint during the conversion algorithm for B-spline curves (and similarly for surfaces):

$$d(\mathbf{x}, \bar{\mathbf{x}}) := \max\{|\mathbf{x}(t) - \bar{\mathbf{x}}(t)|\} \leq \delta. \tag{1}$$

Curves and Surfaces in Geometric Design
P. J. Laurent, A. Le Méhauté, and L. L. Schumaker (eds.), pp. 131–138.
Copyright © 1994 by A K PETERS, Wellesley, MA
ISBN 1-56881-039-3.

However, we note that the following, more special requirements can also be solved with help of the restriction (1) if the error tolerance δ is set close to the accuracy of the computing machine:

- Remove all knots which have no influence on the shape of the B-spline.
- Reduce the polynomial degree of the B-spline as much as possible.

In the literature a lot of schemes can be found which deal with the above problems [2,7,8,10–12]. Some of them are also part of the comparison of different conversion methods in the current proceedings.

In the present paper we will introduce a stepwise approach for carrying out the basis conversion for polynomials (not for rationals) which is mainly based on recent results of the authors on the topics of degree reduction of Bézier curves (resp. surfaces) [3,4] and knot removal for B-splines [5].

The advantages of these basic methods are, first, that the new control points are obtained by fast and explicitly known constructions only from the given control points and, secondly, that an upper bound of the maximal deviation is known a priori in every step. Unfortunately, the methods cannot be described here in detail because of space limitations.

§2. Stepwise Strategy

In the following scheme the stepwise structure of the conversion algorithm is obvious since four (partly) known operations are gone through sequentially:

input B-spline curve (resp. surface) $\mathbf{x}$ of order k (resp. (k_1, k_2))

1. knot insertion (exactly)

Bézier spline of degree $k - 1$ (resp. $(k_1 - 1, k_2 - 1)$)

2. degree elevation (exactly) or degree reduction (approximately)

Bézier spline of degree $\bar{k} - 1$ (resp. $(\bar{k}_1 - 1, \bar{k}_2 - 1)$)

3. creating knot vectors with multiple knots

B-spline curve (resp. surface) of order $\bar{k}$ (resp. $(\bar{k}_1, \bar{k}_2)$)

4. knot removal (approximately)

output B-spline curve (resp. surface) $\bar{\mathbf{x}}$ of order $\bar{k}$ (resp. $(\bar{k}_1, \bar{k}_2)$)

The first step is the (exact) conversion of the input B-spline curve or surface into a Bézier spline of the same polynomial degree. This is done with the help of the knot insertion formula of Boehm [1] whereby the control points of the Bézier segments are achieved by raising the multiplicity of all knots to its maximum.

The second step is the adjustment of the polynomial degree of the input Bézier spline to the target degree $\bar{k} - 1$ (resp. $(\bar{k}_1 - 1, \bar{k}_2 - 1)$). Here, the cases $k \leq \bar{k}$ (resp. $k_1 \leq \bar{k}_1$ and $k_2 \leq \bar{k}_2$) can be carried out exactly by applying the well-known degree elevation formula for Bernstein polynomials to every Bézier spline segment. On the other hand, the remaining cases need a first approximate treatment. As mentioned above this is done with the help of methods derived in [3,4].

In the third step, we transform the actual spline back into the B-spline basis. Doing so, the freedom is left over to choose new knot vectors which only have to contain knots with multiplicity of the respective target order of the spline. Here, we always use the proposal described in [6] whereby the relations of neighbouring knot spacings and the relation of neighbouring control point differences at the respective segments boundaries are (nearly) equal. This produces a C^1 parametrization of the spline if it is possible at all.

The final step is to reduce the (artificial) large number of (multiple) knots from the knot sequences just created. Again this knot removal process cannot be executed exactly, since in general, the shape of the spline changes if reversal knot insertion formulas are applied; see [11,13]. Thus, we will use the approximate methods for knot removal developed in [5].

Furthermore, we have to be very careful within the two approximate steps of degree reduction and knot removal in order to ensure the error is less than the tolerance δ according to restriction (1). Thus these two main parts must be combined appropriately as explained in the following two sections.

§3. Degree Reduction

The degree reduction method which we use in our algorithm reduces the polynomial degree of every Bézier spline segment separately, and so allows a local sequential treatment. Moreover, for each spline segment the polynomial degree is again lowered iteratively. Thus, in every iteration step the degree is reduced by one (in each parameter direction).

The method to perform this single polynomial degree reduction for Bézier curves is based on *constrained* Chebyshev polynomials which are also used in [10]. The main advantages of these polynomials are that in every iteration step the uniform error norm is minimized, the actual approximation error is known explicitly, and preselected continuity constraints at the boundaries are satisfied.

A geometric construction is given in [3] which allows a very stable and fast computation of the new control points of the Bézier curve with degree reduced by one. This fully determined, unique construction can be understood as a particular combination of two different solutions obtained from inverting the degree elevation formulas for Bézier curves.

The case of tensor product Bézier surfaces, moreover, is handled straight-forwardly by applying the above curve construction to every row and every column of the control net [4]. So we do not minimize the uniform error norm, but make it only sufficiently small. This fact represents a gradual difference to the case of curves.

Now, coming back to the entire degree reduction, first in the curve case, we have to carry out $k - \bar{k}$ steps if the polynomial degree should be reduced from $k - 1$ to $\bar{k} - 1$:

$$\underbrace{k - 1}_{\mathbf{x}^*} \xrightarrow{\bar{\varepsilon}_1} k - 2 \xrightarrow{\bar{\varepsilon}_2} k - 3 \xrightarrow{\bar{\varepsilon}_3} \ldots \xrightarrow{\bar{\varepsilon}_{k-\bar{k}}} \underbrace{\bar{k} - 1}_{\bar{\mathbf{x}}^*} \ . \tag{2}$$

Here, the known approximation errors $\bar{\varepsilon}_i$ in every step depend only on the Bézier points which have been created in the previous degree reduction step. Moreover, the total error of the entire stepwise process (2) has the following simple bound ε^*:

$$d(\mathbf{x}^*, \bar{\mathbf{x}}^*) \leq \bar{\varepsilon}_1 + \bar{\varepsilon}_2 + \ldots + \bar{\varepsilon}_{k-\bar{k}} := \varepsilon^* \tag{3}$$

where $\mathbf{x}^*$ (resp. $\bar{\mathbf{x}}^*$) denotes the Bézier curve of degree $k - 1$ (resp. $\bar{k} - 1$).

In the case of degree reduction of Bézier spline surfaces we perform a similar stepwise strategy, namely

$$(k_1 - 1, k_2 - 1) \xrightarrow{\bar{\varepsilon}_1} (k_1 - 2, k_2 - 2) \xrightarrow{\bar{\varepsilon}_2} \ldots \longrightarrow (\bar{k}_1 - 1, \bar{k}_2 - 1) \tag{4}$$

whereby again the sum $\bar{\varepsilon}^*$ of all known partial errors $\bar{\varepsilon}_i$ can be used as an appropriate bound for the maximal deviation d.

Now with respect to the total conversion process, the sum ε^* has to be smaller than a certain part of the global error tolerance δ. Here, we observed in several tests that the requirement $\varepsilon^* \leq \frac{1}{5}\delta$ leads to good results.

Thus in order to satisfy this restriction, it might be necessary to carry out a segmentation of the input Bézier spline with the help of the de Casteljau algorithm. Doing so, a very careful treatment of this subdivision procedure is required, since otherwise the number of spline segments increases very rapidly.

Here again the curve and the surface cases are different in a special way. For curves, the number of segmentations can be determined a priori by a simple inversion of the stepwise error formula. The similar formula for surfaces is more complicated so that the number of segmentation has to be found by a suitable searching strategy. For large sets of adjacent Bézier surfaces, this part of the algorithm is very time-consuming as we will see in Section 5.

§4. Knot Removal

Similar to the above degree reduction process, the knot removal procedure described now is also of stepwise nature. In every iteration step we remove only one knot from the knot vectors of the polynomial B-spline curves or surfaces.

In the case of univariate B-splines, we apply a method recently presented in [5] which carries out this single knot removal by minimizing the uniform norm. In particular, it is shown that the new control points of the curve with one knot removed can be computed by a local geometric construction if a simple one-dimensional Remes-typed algorithm is first applied.

However, the advantages of this single knot deleting algorithm can be summarized as follows: the local computation of the control points is fast and stable, the approximation error known a priori is non-zero only in a local interval with the removed knot in it, and prescribed continuity constraints at the two end points can be fulfilled.

Further, the extension of this univariate solution to tensor product B-spline surfaces is again simply done by applying the curve construction to all rows (resp. columns) of the control net if a single knot of the u-direction (resp. v-direction) has to be removed. And although this method does not generally minimize the uniform norm, the results are convenient for our purposes since at least a simple upper bound of the uniform error is directly obtained. This bound then depends only on the control points, too.

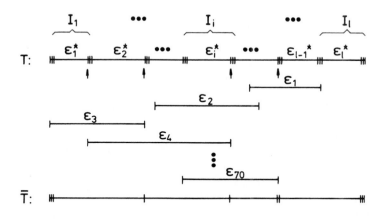

Fig. 1. An example of stepwise knot removal for curves.

We now discuss the entire (stepwise) knot removal process for B-spline curves. The method works analogously for surfaces. In the example shown in Figure 1, the input knot vector T consists of l non-vanishing knot intervals $\{I_j\}_{j=1}^{l}$ with $\bar{k}$-fold knots in between. Here, on each of these intervals I_j, a degree reduction error $\varepsilon_j^* \leq \frac{1}{5}\delta$ is possible from the previous stage of degree adapting. Then the knot removal process is carried out stepwise. Thus, 70 knots could be removed in the example shown in Figure 1, where the local errors $\{\epsilon_r\}_{r=1}^{70}$ are also illustrated.

During this algorithm the total conversion error on each interval $\{I_j\}_{j=1}^{l}$ can be controlled directly by summing up all local errors ε_r influencing I_j. In Figure 1 we get on the typical interval I_i the following bound of d:

$$d(\mathbf{x}, \bar{\mathbf{x}}) \leq \varepsilon_i^* + \varepsilon_2 + \varepsilon_4 + \cdots + \varepsilon_{70}. \tag{5}$$

Moreover, this bound is used in every step to decide whether a selected knot can be removed or not. Here, we simply ensure that the bound is smaller than the prescribed tolerance δ on every interval $\{I_j\}_{j=1}^{l}$. The algorithm then terminates when no further knots can be removed at all.

Obviously, another important aspect of the algorithm is the order in which the knots are removed, since different orders yield different approximating curves or surfaces. Here, we use the partial error ε_r as the ranking number of the considered knot. Then we order the knots with respect to their ranking number, whereby multiple knots are taken into consideration only once.

Then the knots are removed according to this ordered list. When the list is worked through, a new list is created based on the new knot vectors. This is repeated until only simple knots are left. Afterwards, the ranking list is updated after every removed knot, and this guarantees a convenient distribution of the knots in the final knot vectors in most cases.

§5. Test Examples

For comparison of our conversion methods with others, we use the three benchmark surfaces in the paper of Hoschek & Schneider [7] in this proceedings. Each of these surfaces is already represented as a Bézier spline surface since the VDAFS file format is used for storage. Therefore the first stage of our strategy described in Section 2 is left out.

The results of our computations done on a HP 9000/735 machine can be found in Tables 1–3 below. The degree and the segment numbers of the output B-spline surfaces are listed in the first two columns.

In the third column the minimal inner continuities in both parameter directions are entered. These continuity orders cannot be prescribed in our method but result from the knot removal stage. Nevertheless, in all cases at least inner C^2 continuity is achieved. In the four corners of the parameter domain we always have enforced only C^0 continuity. Here, also higher orders would be possible.

The fourth column contains the data compression factor meaning the quotient of the number of all input Bézier points over the number of all B-spline control points plus knot vectors of the output spline. However, the compression effect is more impressive in the second and third example.

In the fifth and sixth column the approximation error is considered. Here, the first one represents the prescribed tolerance δ whereas the second one gives the *measured* error $\max_i\{|\mathbf{x}(u_i, v_i) - \bar{\mathbf{x}}(u_i, v_i)|\}$ obtained by comparing the two surfaces at sampled parameter locations (u_i, v_i). Comparing these two entries we notice that our added bounds are in most cases two or three times too large.

In the final column the needed computing time is listed whereby IO operations are not included. Moreover, we obtain that the degree reduction needs two times more CPU time than knot removal because of the mentioned search of possible segmentations.

We believe that the tables speak for themselves, and so we do not comment on them further. Two pecularities which cannot be found in the tables should be mentioned:

- The second surface SEITE 1 already has high continuity at the inner boundaries, and thus a lot of knots can be removed without errors.
- The third surface SURF B has a *true* degree $(3, 3)$, and thus the degree can be reduced without errors.

degree	segm.	minimal contin.	compr. factor	error toler.	meas. error	CPU
(3,3)	8 × 6	C^2, C^2	0.9	0.01	0.003	1.0 sec
	4 × 3	C^2, C^2	2.0	0.1	0.036	0.3 sec
(5,5)	3 × 3	C^3, C^4	1.2	0.01	0.004	0.3 sec
	2 × 1	C^4, C^5	2.0	0.1	0.036	0.2 sec

Table 1. Input surface: BRODE ; degree: (9,9) ; segments: 1 × 1

degree	segm.	minimal contin.	compr. factor	error toler.	meas. error	CPU
(3,3)	8 × 7	C^2, C^2	8.4	0.01	0.004	0.3 sec
	5 × 7	C^2, C^2	11.4	0.1	0.068	0.1 sec
(5,5)	4 × 7	C^2, C^2	3.1	0.01	0.003	1.5 sec
	2 × 7	C^3, C^2	5.3	0.1	0.041	1.6 sec

Table 2. Input surface: SEITE 1 ; degree: (3,3) ; segments: 9 × 7

degree	segm.	minimal contin.	compr. factor	error toler.	meas. error	CPU
(3,3)	17 × 24	C^2, C^2	69.1	0.01	0.005	17.8 sec
	17 × 19	C^2, C^2	84.4	0.1	0.074	17.6 sec
(5,5)	17 × 45	C^2, C^3	18.5	0.01	0.007	21.0 sec
	17 × 31	C^3, C^4	43.4	0.1	0.045	20.2 sec

Table 3. Input surface: SURF B ; degree: (5,5) ; segments: 17 × 63

References

1. Boehm, W., Inserting new knots into B-spline curves, Comp. Aided Design **12** (1980), 199–201.
2. Dannenberg, L. and H. Nowacki, Approximate conversion of surface representations with polynomial bases, Comp. Aided Geom. Design **2** (1985), 123–132.
3. Eck, M., Degree reduction of Bézier curves, Comp. Aided Geom. Design **10** (1993), 237–251.
4. Eck, M., Degree reduction of Bézier surfaces, to appear in *The Mathematics of Surfaces V*, R. Fisher (ed.), Oxford University Press, 1994.
5. Eck, M., and J. Hadenfeld, Knot removal for B-spline curves, submitted to Comp. Aided Geom. Design.
6. Farin, G., *Curves and Surfaces for Computer Aided Geometric Design. A Practical Guide. Third Edition*, Academic Press, New York, 1993.
7. Hoschek, J., and F.-J. Schneider, Approximate Conversion and Data Compression of Integral and Rational B-spline Surfaces, in *Curves and Surfaces in Geometric Design*, P.-J. Laurent, A. Le Méhauté, and L. L. Schumaker (eds.), A K Peters, Wellesley, 1994, 241–250.
8. Hoschek, J., and F. J. Schneider, Spline conversion for trimmed rational Bézier- and B-spline surfaces, Comp. Aided Design **22** (1990), 580–590.
9. Hoschek, J., and D. Lasser, *Fundamentals of Computer Aided Geometric Design*, AK Peters, Wellesley, 1993.
10. Lachance, M. A., Chebyshev economization for parametric surfaces, Comp. Aided Geom. Design **5** (1988), 195–208.
11. Lyche, T., and K. Mørken, Knot removal for parametric B-spline curves and surfaces, Comp. Aided Geom. Design **4** (1987), 217–230.
12. Patrikalakis, N. M., Approximate conversion of rational splines, Comp. Aided Geom. Design **6** (1989), 155–166.
13. Tiller, W., Knot-removal algorithms for NURBS curves and surfaces, Comp. Aided Design **24** (1992), 445–453.

Matthias Eck, Jan Hadenfeld
Department of Mathematics
University of Technology and Science
Schlossgartenstrasse 7
D-64289 Darmstadt, GERMANY
eck, hadenfeld @ mathematik.th-darmstadt.de

Best Constrained Approximations of Planar Curves by Bézier Curves

Eberhard F. Eisele

Abstract. Given a parametric planar curve **p** and any Bézier curve **x** of degree n such that **p** and **x** have contact of order k at the common end points, the normal vector field of **p** is used to measure the distance of corresponding points of **p** and **x**. Using alternation theorems of uniform approximation theory, we characterize the Bézier curve **x** which (locally) minimizes the uniform norm of this distance function $\rho_{\mathbf{x}}$ under certain constraints. We consider the cases that each Bézier curve interpolates **p** at certain prescribed points, or that each function $\rho_{\mathbf{x}}$ has non-negative range.

§1. Introduction

In CAD systems methods for good approximation of a given curve by a certain class of spline curves are frequently needed. For instance, a method for approximate conversion of spline representations is required for exchanging data between different CAD systems. Also, the problem arises to approximate offset curves by spline curves of a certain kind.

Given a parametric planar curve **p** and any Bézier curve **x** of degree n such that **p** and **x** have contact of fixed order k at the common end points, Degen [3] has used the normal vector field of **p** to measure the distance of corresponding points of **p** and **x**. Then the Bézier curve **x** is called a best approximation to **p** if the maximum norm $\|\rho_{\mathbf{x}}\|_{\infty}$ of this distance (or deviation) function $\rho_{\mathbf{x}}$ is minimal for **x**. The (locally) best approximation **x** has been characterized by an alternation property of $\rho_{\mathbf{x}}$ in [3] and [6].

This paper addresses the problem of characterizing a (locally) best approximation **x** *satisfying certain constraints*. We consider the cases that either each **x** interpolates **p** at certain prescribed points (see §3), or that each deviation function $\rho_{\mathbf{x}}$ has non-negative range (see §4). In both cases we characterize a locally best constrained approximation **x** by an alternation property of $\rho_{\mathbf{x}}$, applying results of Chebyshev approximation with side conditions. Finally, some best approximations to circle segments are calculated.

Curves and Surfaces in Geometric Design
P. J. Laurent, A. Le Méhauté, and L. L. Schumaker (eds.), pp. 139–146.

§2. The Normal Distance of Planar Curves

We only consider planar curves defined on the unit interval $I := [0, 1]$. Given $k \in \mathbb{N}^+$, any k-times continuously differentiable map $\mathbf{x} : I = [0, 1] \to \mathbb{R}^2$ is called a (*parametric*) C^k *curve*, and any k-times continuously differentiable bijective function $\alpha : I \to I$ such that $\alpha'(t) > 0$ for every $t \in I$ is called a C^k *reparametrization*. Any C^1 curve $\mathbf{x}$ satisfying $\dot{\mathbf{x}}(s) := \frac{d\mathbf{x}}{dt}(s) \neq (0, 0)$ for every $s \in I$ is called a *regular curve*.

The following parametrization–independent concept of contact of two curves at a common end point, also known as geometric continuity, is crucial in CAGD when fitting together several curve segments. It will be used in Sections 3 and 4.

Definition 1. *Let $s \in \{0, 1\}$ and $k \in \mathbb{N}_0$. Two C^k curves $\mathbf{x}$ and $\mathbf{y}$ have contact of order k at s if and only if $\mathbf{x}(s) = \mathbf{y}(s)$, and in case $k > 0$, there are C^k reparametrizations α and β such that $\frac{d^j(\mathbf{x} \circ \alpha)}{dt^j}(s) = \frac{d^j(\mathbf{y} \circ \beta)}{dt^j}(s)$, $j = 1, \ldots, k$.*

For the remainder of this paper, let $\mathbf{p} = (p_1, p_2) : I \to \mathbb{R}^2$ be a fixed regular curve (injectivity is not supposed), to be approximated. Using the normal vector field of $\mathbf{p}$, we now introduce the class of admissible curves with respect to $\mathbf{p}$ for which the normal distance to $\mathbf{p}$ can be defined (see [3,6]):

Definition 2. *The regular curve $\mathbf{x} : I \to \mathbb{R}^2$ is said to be admissible with respect to $\mathbf{p}$ if and only if:*

1) $\mathbf{p}(s) = \mathbf{x}(s)$, $s = 0, 1$.

2) *There exists a unique strictly increasing bijective map $\varphi_{\mathbf{x}} : [0, 1] \to [0, 1]$ such that for each $s \in [0, 1]$, the point $\mathbf{x}(\varphi_{\mathbf{x}}(s))$ lies on the normal $N(s) := \{\mathbf{p}(s) + t \cdot \mathbf{n_p}(s) : t \in \mathbb{R}\}$ of $\mathbf{p}$ at s, where $\mathbf{n_p}(s)$ denotes the unit normal vector $(-\dot{p}_2/\|\dot{\mathbf{p}}\|_2, \dot{p}_1/\|\dot{\mathbf{p}}\|_2)$ of $\mathbf{p}$ at*

3) *The tangent vector $\dot{\mathbf{x}}(\varphi_{\mathbf{x}}(s))$ of $\mathbf{x}$ is not parallel to $\mathbf{n_p}(s)$ for every $s \in [0, 1]$.*

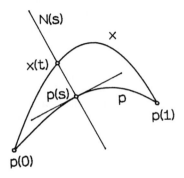

Fig. 1. Admissible parabola x.

Given any admissible curve $\mathbf{x} : I \to \mathbb{R}^2$, there exists a unique continuous map $\rho_{\mathbf{x}} : [0, 1] \to \mathbb{R}$ satisfying:

$$\mathbf{x}(\varphi_{\mathbf{x}}(s)) = \mathbf{p}(s) + \rho_{\mathbf{x}}(s) \cdot \mathbf{n_p}(s), \qquad s \in [0, 1]. \tag{1}$$

Therefore $\mathbf{x} \circ \varphi_{\mathbf{x}}$ is a *variable–distance* offset curve. From $\varphi_{\mathbf{x}}(0) = 0$ and $\varphi_{\mathbf{x}}(1) = 1$, it follows $\rho_{\mathbf{x}}(0) = \rho_{\mathbf{x}}(1) = 0$, and (1) implies that for $s \in [0, 1]$, $\| \mathbf{x}(\varphi_{\mathbf{x}}(s)) - \mathbf{p}(s) \|_2 = |\rho_{\mathbf{x}}(s)|$, where $\| \cdot \|_2$ denotes the Euclidean norm. Hence we define (cf. [3,6]):

Definition 3. *For each admissible curve* $\mathbf{x}$, *the map* $\rho_{\mathbf{x}}$ *is called the deviation function of* $\mathbf{x}$, *and* $\mathrm{d_N}(\mathbf{x}) := \| \rho_{\mathbf{x}} \|_\infty := \max_{s \in I} |\rho_{\mathbf{x}}(s)|$ *the normal distance from* $\mathbf{x}$ *to* $\mathbf{p}$.

Given any set $\mathcal{A}$ of admissible parametric curves, for instance a certain class of Bézier curves, a curve $\mathbf{x}_0 \in \mathcal{A}$ satisfying

$$\mathrm{d_N}(\mathbf{x}_0) \leq \mathrm{d_N}(\mathbf{x}) \text{ for all } \mathbf{x} \in \mathcal{A} \tag{2}$$

is called a *best approximation to* $\mathbf{p}$ *from* $\mathcal{A}$. Because $\mathrm{d_N}(\mathbf{x}) = \| \rho_{\mathbf{x}} - 0 \|_\infty$, the curve $\mathbf{x}_0$ is a best approximation to $\mathbf{p}$ from $\mathcal{A}$ if and only if $\rho_{\mathbf{x}_0}$ is a "best uniform approximation" to $f := 0$ from $\{ \rho_{\mathbf{x}} : \mathbf{x} \in \mathcal{A} \}$ in the sense of approximation theory.

Remarks 1a. The normal distance is parametrization–independent, because for any reparametrization α, $\rho_{\mathbf{x} \circ \alpha} = \rho_{\mathbf{x}}$, and replacing $\mathbf{p}$ by $\mathbf{p} \circ \alpha$, $\rho_{\mathbf{x}}$ has to replaced by $\rho_{\mathbf{x}} \circ \alpha$, where $\| \rho_{\mathbf{x}} \|_\infty = \| \rho_{\mathbf{x}} \circ \alpha \|_\infty$.

b. Applying the implicit function theorem to (1), it follows that $\varphi_{\mathbf{x}}$ and $\rho_{\mathbf{x}}$ are at least of class C^{k-1} ($k \geq 2$) if $\mathbf{p}$ is a C^k curve and $\mathbf{x}$ a C^{k-1} curve. Moreover, $\varphi'_{\mathbf{x}}(s) > 0$ for some $s \in I$ if $k \geq 2$ and $\rho_{\mathbf{x}}(s) \cdot \kappa(s) < 1$, where $\kappa(s)$ denotes the curvature of $\mathbf{p}$ at s (see [3] Proposition 1).

c. The normal distance $\mathrm{d_N}(\mathbf{x})$ is equal to the Hausdorff distance $\mathrm{d_H}(\mathbf{p}, \mathbf{x})$ of the point sets of $\mathbf{p}$ and $\mathbf{x}$ if the line segments $\overline{\mathbf{p}(s), \mathbf{x}(\varphi_{\mathbf{x}}(s))}$, $s \in [0, 1]$, are pairwise disjoint, and if $\rho_{\mathbf{x}} \cdot \kappa < 1$ (see [3] Theorem 1). Moreover, we always have $\mathrm{d_H}(\mathbf{p}, \mathbf{x}) \leq \mathrm{d_N}(\mathbf{x})$.

§3. Best Approximations Satisfying Interpolatory Constraints

First we introduce some classes of admissible Bézier curves. In the sequel let $n \geq 2$ be a fixed natural number and $k \in \{0, \ldots, n-2\}$. Furthermore, let $\mathcal{B}$ be the set of all admissible Bézier curves $\mathbf{x} : I \to \mathbb{R}^2$ of minimal (algebraic) degree n for which the resultant (see [11] p.24) of the two polynomial coordinate functions of the derivative $\dot{\mathbf{x}}$ of $\mathbf{x}$ does not vanish (cf. [6] (1.1)). This condition implies the regularity of each $\mathbf{x} \in \mathcal{B}$ and is needed for the proofs of our theorems. For each $\mathbf{x} \in \mathcal{B}$ we have a unique Bézier representation with control points $\mathbf{b}_0, \ldots, \mathbf{b}_n \in \mathbb{R}^2$:

$$\mathbf{x}(t) = \sum_{i=0}^{n} B_i^n(t) \cdot \mathbf{b}_i, \ t \in [0, 1], \tag{3}$$

where $B_i^n(t)$ denotes the i-th Bernstein polynomial of degree n, and by Definition 2(1), $\mathbf{b}_0 = \mathbf{p}(0)$ and $\mathbf{b}_n = \mathbf{p}(1)$. Using Definition 1 we put:

$$\mathcal{B}_k := \{\mathbf{x} \in \mathcal{B} : \mathbf{x} \text{ and } \mathbf{p} \text{ have contact of order } k \text{ at } 0 \text{ and } 1\}. \tag{4}$$

Remark 2. $\mathcal{B}_0 = \mathcal{B}$ is a $(2n - 2)$-dimensional manifold if each $\mathbf{x} \in \mathcal{B}$ is identified with the sequence of its variable Bézier points $\mathbf{b}_1, \ldots, \mathbf{b}_{n-1}$. Furthermore, it follows from [6] Corollary 2.8 and [10] Theorem 11–1 that $\mathcal{B}_k$ is a manifold of dimension:

$$\alpha := 2 \cdot (n - k - 1). \tag{5}$$

Now we prescribe r interpolation points in $(0, 1)$. Suppose we are given $r \in \{0, \ldots, \alpha - 1\}$ and r real values $s_i \in (0, 1)$ satisfying

$$0 < s_1 < \ldots < s_r < 1. \tag{6}$$

Then we define

$$\mathcal{B}_k^r := \{\mathbf{x} \in \mathcal{B}_k : \mathbf{x}(\varphi_{\mathbf{x}}(s_i)) = \mathbf{p}(s_i), \ i = 1, \ldots, r\}. \tag{7}$$

Note that $\mathbf{x}(\varphi_{\mathbf{x}}(s_i)) = \mathbf{p}(s_i)$ is equivalent to $\rho_{\mathbf{x}}(s_i) = 0$, and that $\mathcal{B}_k^0 = \mathcal{B}_k$. In the following theorem we call $\mathbf{x} \in \mathcal{B}_k^r$ a *locally best approximation* to $\mathbf{p}$ from $\mathcal{B}_k^r$ if there exists a neighborhood U of $\mathbf{x}$ in the manifold $\mathcal{B}_0$ (cf. Remark 2) such that $d_N(\mathbf{x}) \leq d_N(\mathbf{y})$ for every $\mathbf{y} \in \mathcal{B}_k^r \cap U$. The proof of the following characterization theorem is based on the local results in [6], and it is unknown whether there actually can be several locally best approximations from $\mathcal{B}_k^r$. At least for parabolas and cubics with $k = 1$ the latter is not the case [3].

Theorem 1. *The Bézier curve* $\mathbf{x}$ *is a locally best approximation to* $\mathbf{p}$ *from* $\mathcal{B}_k^r$ *if and only if there are (at least)* $\alpha - r + 1$ *points* $0 < t_1 < \ldots < t_{\alpha-r+1} < 1$ *such that*

$$|\rho_{\mathbf{x}}(t_i)| = \|\rho_{\mathbf{x}}\|_\infty = d_N(\mathbf{x}), \tag{8}$$

and if γ_i *denotes the number of points* $s_j \in (t_i, t_{i+1})$,

$$\rho_{\mathbf{x}}(t_{i+1}) = (-1)^{\gamma_i+1} \cdot \rho_{\mathbf{x}}(t_i), \ i = 1, \ldots, \alpha - r. \tag{9}$$

Moreover, any such locally best approximation is the only one in some neighborhood of $\mathbf{x}$.

Note that for $r = 0$ we obtain the usual alternation property: $-\rho_{\mathbf{x}}(t_i) = \rho_{\mathbf{x}}(t_{i+1}) = \pm\|\rho_{\mathbf{x}}\|_\infty$.

Proof: In case $r = 0$ this theorem has been proved in Eisele [6] Theorem 2.9. Moreover, we have shown in [6] Corollary 2.8 that the family $\{\rho_{\mathbf{x}} : \mathbf{x} \in \mathcal{B}_k\}$ of deviation functions is varisolvent (= unisolvent in [1] Definition 3) of fixed degree α in some neighborhood $U_{\mathbf{x}}$ of each $\mathbf{x} \in \mathcal{B}_k$, where the fixed endpoints 0 and 1 must be excluded in the definition of solvency (see [6] Section 2).

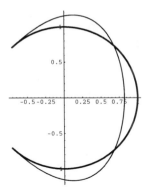

Fig. 2. Best tangent-continuous cubic to p from $\mathcal{B}_1$.

Therefore we may apply (the adapted) Theorem 3 in [1] to the family $\{\rho_\mathbf{x} : \mathbf{x} \in \mathcal{B}_k^r \cap U_\mathbf{x}\}$ to obtain the desired result. ∎

Remark 3. It is possible to show that the above theorem remains valid if coinciding interpolating points are allowed in the following sense: if $s_{i-1} < s_i = s_{i+1} = \ldots = s_{i+m} < s_{i+m+1}$, then assume instead of $\mathbf{x}(\varphi_\mathbf{x}(s_i)) = \mathbf{p}(s_i)$ that $\mathbf{x}$ and $\mathbf{p}$ have contact of order m at $\mathbf{p}(s_i)$.

To illustrate the alternation property of best approximations, we have approximated large segments of the unit circle (cf. [4] and [9]). In the following examples the best approximation could be calculated explicitly, but for higher degrees n it is necessary to use a nonlinear Remes type algorithm (cf. [6] 2.11 and [2]).

Examples 1a. Figure 2 shows the best approximation $\mathbf{x}$ to a segment $\mathbf{p}$ of the unit circle of length $\frac{3\pi}{2}$ from $\mathcal{B}_1^0 = \mathcal{B}_1$ with $n := 3$. The Bézier curve $\mathbf{x}$ is symmetric with respect to the x-axis, and its deviation function $\rho_\mathbf{x}$ has $3 = \alpha - 0 + 1$ extreme points with alternating sign.

b. Figure 3 shows the best (symmetric) approximation $\mathbf{x}$ to $\mathbf{p}$ by tangent-continuous Bézier cubics having one interpolating point • in $(0, \frac{1}{2})$, i.e., $n = 3$ and $k = r = 1$. The deviation function $\rho_\mathbf{x}$ has $2 = (\alpha - 1) + 1$ extreme points with the same sign. If the interpolating point is the midpoint of $\mathbf{p}$, the best approximation to $\mathbf{p}$ is tangent to $\mathbf{p}$ at the midpoint (see Figure 4).

§4. Best Non-negative Approximations

We define

$$\mathcal{B}_k^+ := \{\mathbf{x} \in \mathcal{B}_k : \rho_\mathbf{x} \geq 0\}, \tag{10}$$

and call a (locally) best approximation to $\mathbf{p}$ from $\mathcal{B}_k^+$ a *(locally) best non-negative approximation* to $\mathbf{p}$.

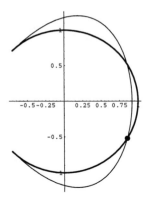

Fig. 3. Best cubic to p from $\mathcal{B}_1^1$, one interpolating point in $(0, \frac{1}{2})$.

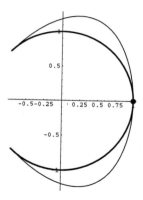

Fig. 4. Best cubic to p from $\mathcal{B}_1^1$, interpolating point at $s_1 = \frac{1}{2}$.

Definition 4. Let $\rho_{\mathbf{x}} \in \mathcal{B}_k^+$ and $t \in [0,1]$. Then t is a *plus point* of $\rho_{\mathbf{x}}$ if $\rho_{\mathbf{x}}(t) = +\|\rho_{\mathbf{x}}\|_\infty$; furthermore, t is a *minus point* of $\rho_{\mathbf{x}}$ if $\rho_{\mathbf{x}}(t) = 0$ and in case $t \in \{0,1\}$, in addition, $\mathbf{p}$ and $\mathbf{x}$ have contact of order $k+1$ at t.

Theorem 2. *The Bézier curve* $\mathbf{x} \in \mathcal{B}_k^+$ *is a locally best approximation to* $\mathbf{p}$ *from* $\mathcal{B}_k^+$ *if and only if there exist* $\alpha + 1 = 2(n - k - 1) + 1$ *points* t_i *with* $0 \le t_1 < \ldots < t_{\alpha+1} \le 1$ *which are alternately plus or minus points. Moreover, any such locally best non-negative approximation is the only one in some neighborhood of* $\mathbf{x}$.

Proof: Using the unisolvency of the family $\{\rho_{\mathbf{x}} : \mathbf{x} \in \mathcal{B}_k\}$ in some neighborhood of each $\mathbf{x} \in \mathcal{B}_k$, the statement can be derived from the characterization theorem in [5], but several modifications are necessary at the endpoints. In [7] the reader can find a detailed proof for Bézier curves having an arbi-

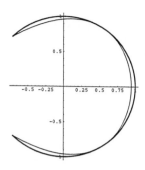

Fig. 5. Best non-negative cubic to p from $\mathcal{B}_0^+$.

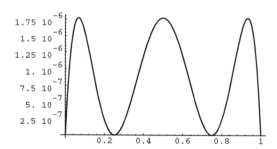

Fig. 6. Deviation function of the best cubic to a quarter circle from $\mathcal{B}_0^+$.

trary restricted range, i.e., given continuous functions u and v from $[0,1]$ into $\mathbb{R} \cup \{\pm\infty\}$ with $u < v$ and $u \leq 0 \leq v$, a locally best approximation from $\mathcal{B}_k^{u,v} := \{\mathbf{x} \in \mathcal{B}_k : u \leq \rho_{\mathbf{x}} \leq v\}$ is characterized (here: $u := 0$ and $v := +\infty$). ∎

Clearly, an analogous characterization theorem is valid for best non-positive approximations from $\mathcal{B}_k^- := \{\mathbf{x} \in \mathcal{B}_k : \rho_{\mathbf{x}} \leq 0\}$. Every non-negative or non-positive approximation lies on one side of $\mathbf{p}$, which depends on the orientation of $\mathbf{p}$.

Examples 2a. Figure 5 shows the best non-negative cubic $\mathbf{x}$ to the circle segment $\mathbf{p}(s) := (\cos(\frac{3\pi}{2}(s-\frac{1}{2})), \sin(\frac{3\pi}{2}(s-\frac{1}{2})))$, $s \in [0,1]$, from $\mathcal{B}_0^+$. The Bézier curve $\mathbf{x}$ is symmetric with respect to the x-axis and its deviation function $\rho_{\mathbf{x}}$ has alternately 3 plus points and 2 minus points in $(0,1)$, where $\alpha = 2 \cdot (3 - 0 - 1) = 4$. For the calculation of $\mathbf{x}$ we have used a non-linear Remes type algorithm (cf. [11] and [6] 2.11). Figure 6 shows the deviation function of the best non-negative cubic to a quarter circle from $\mathcal{B}_0^+$.

b. Figure 2 shows the best non-negative cubic to $\bar{\mathbf{p}}(s) := \mathbf{p}(1 - s)$, $s \in [0, 1]$, from $\mathcal{B}_0^+$ and $\mathcal{B}_1^+$. Since $\mathbf{x}$ and $\bar{\mathbf{p}}$ have contact of order 1 at 0 and 1, 0 and 1 are minus points of $\rho_{\mathbf{x}}$ in case $k = 0$, and $\rho_{\mathbf{x}}$ has two plus points in $(0, 1)$ and the minus point $\frac{1}{2}$.

References

1. Barrar, R. and H. L. Loeb, Best non-linear approximation with interpolation, Arch. Rational Mech. Anal. **33** (1969), 231–237.
2. Chalmers, B. L. and G. D. Taylor, Uniform approximation with constraints, Jber. d. Dt. Math.-Verein. **81** (1979), 49–86.
3. Degen, W. L. F., Best approximations of parametric curves by splines, in *Mathematical Methods in CAGD II*, T. Lyche and L. L. Schumaker (eds.), Academic Press, Boston, 1992, 171–184.
4. Dokken, T., M. Dæhlen, T. Lyche, and K. Mørken, Good approximation of circles by curvature–continuous Bézier surfaces, Comp. Aided Geom. Design **7** (1990), 33–41.
5. Dunham, C. B., Alternating minimax approximation with unequal restraints, Journal of Approx. Theory **10** (1974), 199–205.
6. Eisele, E. F., Chebyshev approximation of plane curves by splines, J. Approx. Theory **76** (1994), to appear.
7. Eisele, E. F., Best approximations of planar curves by Bézier curves having restricted ranges, preprint, 1993.
8. Hoschek, J., F.–J. Schneider, and P. Wassum, Optimal approximate conversion of spline surfaces, Computer Aided Geometric Design **6** (1989), 293–306.
9. Mørken, K., Best approximation of circle segments by quadratic Bézier curves, in *Curves and Surfaces*, Laurent, P. J., A. Le Méhaute, and L. L. Schumaker (eds.), Academic Press, Boston, 1990, 331–336.
10. Rice, J. K., *The Approximation of Functions, Vol. II – Advanced Topics*, Addison Wesley, London, 1969.
11. Taylor, G. D. and M. J. Winter, Calculation of best restricted approximations. Siam J. Numer. Anal. **7**, No. 2 (1970), 248–255.
12. Walker, R. J., *Algebraic Curves*, Springer, New York, 1950.

Eberhard F. Eisele
Mathematisches Institut B
Universität Stuttgart
70550 Stuttgart, GERMANY
e.eisele@ rus.uni-stuttgart.de

Projective Blossoms and Derivatives

G. Farin

Abstract. In projective space, derivatives of polynomial curves are related to blossom values at infinity. For polynomial surfaces, we employ the concept of a fundamental line and then apply the same principle.

§1. Projective Basics

In this paper, we are using projective two- or three-space with points $\mathbf{a} = [x, y, z]^T$ or $\mathbf{a} = [x, y, z, w]^T$. Points whose coordinates differ by a constant factor denote the same location.

Lines (in two-space) are defined as $\mathbf{L} = [l, m, n]$. Two points $\mathbf{a}, \mathbf{b}$ define a line $\mathbf{L}$; its coordinates are given by $\mathbf{L} = \mathbf{a} \wedge \mathbf{b}$, where $\wedge$ denotes the familiar vector product. Similarly, two lines $\mathbf{L}, \mathbf{M}$ define a point of intersection: $\mathbf{a} = \mathbf{L} \wedge \mathbf{M}$.

Once four points, no three of them collinear, have been determined, and coordinates have been assigned to them, the coordinates of every point in the projective plane are determined: we say that the four points constitute a *projective reference frame* for the projective plane. Without loss of generality, we may assign coordinates $\mathbf{e1} = [1, 0, 0]^T, \mathbf{e2} = [0, 1, 0]^T, \mathbf{e3} = [0, 0, 1]^T$, and $[1, 1, 1]^T$ to these four points. Fig. 1 shows how they generate a coordinate system: the lines of constant parameter values are shown. Notice that these lines form three pencils, each pencil containing one edge of the triangle. (A pencil is a set of straight lines through one common point, the center of the pencil.) The centers of these pencils are collinear, corresponding to the coordinates $[-\infty, \infty, 0]^T, [-\infty, 0, \infty]^T$, and $[0, -\infty, \infty]^T$. We shall call the line that contains them the *fundamental line* $\mathbf{F}$ of our coordinate system. For more details, see [2,3,6].

Curves and Surfaces in Geometric Design
P. J. Laurent, A. Le Méhauté, and L. L. Schumaker (eds.), pp. 147–152.

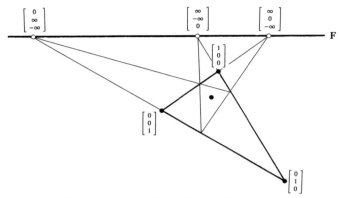

Fig. 1. A projective reference frame.

§2. Conic Blossoms

In projective two-space, a conic is represented by a quadratic parametric curve $\mathbf{b}(t)$. It may be written in Bernstein form and may be evaluated using the de Casteljau algorithm.

Let $\mathbf{b}_0$ and $\mathbf{b}_2$ be two distinct points on a conic Γ. Let the corresponding tangents $\mathbf{B}_0$ and $\mathbf{B}_2$ intersect in the point $\mathbf{b}_1$. Let

$$q_0 = \frac{1}{2}\mathbf{b}_0 + \frac{1}{2}\mathbf{b}_1 \quad \text{and} \quad q_1 = \frac{1}{2}\mathbf{b}_1 + \frac{1}{2}\mathbf{b}_2$$

be two points on these tangents, such that their connection is another tangent to the conic; it is called the *shoulder tangent*.

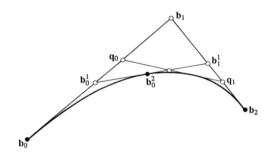

Fig. 2. The Bézier form of conics.

Using quadratic Bernstein polynomials B_i^2, we can write the conic as a *projective quadratic Bézier curve*:

$$\mathbf{b}_0^2(t) = \sum_{i=0}^{2} \mathbf{b}_i B_i^2(t).$$

The de Casteljau algorithm for conics reads

$$\mathbf{b}_i^r(t) = (1-t)\mathbf{b}_i^{r-1}(t) + t\mathbf{b}_{i+1}^{r-1}(t); \quad r = 1, 2; i = 0, ..., 2 - r.$$

This is a two-level procedure: at the first leve ($r = 1$), we interpolate (twice) with respect to t, and at the second level ($r = 2$), we interpolate with respect to t again. If we allow two *different* parameters s, t at the two levels, we obtain a point that depends on two variables and denote it by $\mathbf{b}[s, t]$. Of course, if $s = t$, we have $\mathbf{b}[t, t] = \mathbf{b}_0^2(t)$. We call $\mathbf{b}[s, t]$ the *blossom* of the conic; this terminology was introduced by L. Ramshaw in 1987, see [4] or [5].

The blossom $\mathbf{b}[s, t]$ is a *symmetric function*: this means that $\mathbf{b}[s, t] = \mathbf{b}[t, s]$ and is verified by straightforward algebra. In particular, $\mathbf{b}[r, t]$ is the intersection of the tangents at $\mathbf{b}[r, r]$ and $\mathbf{b}[t, t]$:

$$\mathbf{b}[r, t] = \mathbf{T}(r) \wedge \mathbf{T}(t),$$

which gives a geometric argument for the symmetry of the blossom function.

In classical geometry, one also sees the definition that the point $\mathbf{b}[r, t]$ is the *pole* to the *polar*

$$\mathbf{B}[r, t] = \mathbf{b}[r, r] \wedge \mathbf{b}[t, t].$$

§3. Derivatives

In projective geometry, there are no vectors, and hence the derivative $\dot{\mathbf{x}}(t)$ of a parametric curve $\mathbf{x}(t)$ is itself a point. It is located on the tangent $\mathbf{T}(t)$, which is thus given by

$$\mathbf{T} = \mathbf{x} \wedge \dot{\mathbf{x}}.$$

For conics, we have that

$$\dot{\mathbf{b}}(t) = 2(1-t)\Delta\mathbf{b}_0 + 2t\Delta\mathbf{b}_1,$$

or, equivalently,

$$\dot{\mathbf{b}}(t) = (1-t)\Delta\mathbf{b}_0 + t\Delta\mathbf{b}_1,$$

where Δ is the forward difference operator. Since both $\Delta\mathbf{b}_0$ and $\mathbf{b}_1$ are points, it follows that $\dot{\mathbf{b}}(t)$ describes a line, and we have that *all derivatives of a conic are collinear.*

If we rewrite the conic as

$$\mathbf{b}(t) = \mathbf{b}_0 + t\Delta\mathbf{b}_0 + \frac{1}{2}t^2\Delta^2\mathbf{b}_0,$$

we see that

$$\mathbf{b}(\infty) := \lim_{t \to \infty} \mathbf{b}(t) = \Delta^2\mathbf{b}_0.$$

Using the same limit argument again, we obtain

$$\mathbf{b}(\infty) = \dot{\mathbf{b}}(\infty) = \ddot{\mathbf{b}}(\infty),$$

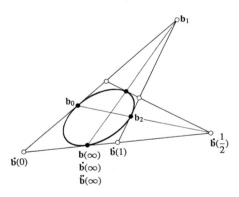

Fig. 3. Derivatives of a conic.

where $\ddot{\mathbf{b}}$ stands for the second derivative.

Since $\Delta \mathbf{b}_0, \Delta^2 \mathbf{b}_0$, and $\Delta \mathbf{b}_1$ are collinear points, it follows that all first derivatives to $\mathbf{b}$ lie on the tangent corresponding to $t = \infty$:

$$\dot{\mathbf{b}}(t) = \mathbf{T}(\infty) \wedge \mathbf{T}(t).$$

For an illustration, see Fig. 3.

We can write derivatives of a conic in blossom form as

$$\frac{\mathrm{d}^r}{\mathrm{d}t^r}\mathbf{b}^2(t) = \mathbf{b}[\infty^{<r>}, t^{<2-r>}]; \quad r \in \{0, 1, 2\},$$

where the superscript $< x >$ denotes x-fold repetition of the corresponding argument. Thus taking derivatives amounts to evaluating at $t = \infty$, and all derivatives to a conic are on the line $\mathbf{B}[\infty, t]$. Fig. 4 shows several blossom values related to derivatives. Except for the labeling, it is identical to Fig. 3.

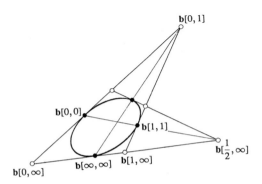

Fig. 4. Tangents and derivatives in blossom form.

§4. Projective Bézier Triangles

Consider an arbitary (nondegenerate) triangle in the projective plane, which, together with a fourth point, defines a projective reference frame and thus projective coordinates $\mathbf{u} = [u, v, w]^T$. We give the triangle vertices the coordinates $\mathbf{e1} = [1, 0, 0]^T, \mathbf{e2} = [0, 1, 0]^T, \mathbf{e3} = [0, 0, 1]^T$, and the fourth point the coordinates $[1/3, 1/3, 1/3]^T$. Then a point $\mathbf{x}$ on an triangular patch $\mathbf{b}(\mathbf{u})$ is expressed as

$$\mathbf{b}^n(\mathbf{u}) = \mathbf{b}_0^n(\mathbf{u}) = \sum_{|\mathbf{j}|=n} \mathbf{b_j} B_{\mathbf{j}}^n(\mathbf{u}),$$

where the $B_{\mathbf{i}}^n(\mathbf{u})$ are trivariate Bernstein polynomials:

$$B_{\mathbf{i}}^n(\mathbf{u}) = B_{i,j,k}^n(u, v, w) = \frac{n!}{i!j!k!} u^i v^j w^k.$$

The *blossom* of a triangular patch is obtained by feeding n possibly different arguments into the de Casteljau algorithm; it is denoted by $\mathbf{b}[\mathbf{u}_1, \ldots, \mathbf{u}_n]$. Note that each argument consists of three components, denoting a point in the projective plane. If all n arguments agree, we obtain the point on the surface corresponding to that argument.

§5. Derivatives

A projective reference frame implies the existence of a *fundamental line* $\mathbf{F}$, see Section 1. Let $\mathbf{f}$ be a point on $\mathbf{F}$, whereas the vertices of the domain triangle are not on it. In complete analogy to our development of the curve case, we may use $\mathbf{f}$ to compute *derivatives* of $\mathbf{b}(\mathbf{u})$ with respect to $\mathbf{f}$:

$$D_{\mathbf{f}}^r \mathbf{b}(\mathbf{u}) = \mathbf{b}[\mathbf{f}^{<r>}, \mathbf{u}^{<n-r>}].$$

If we set $r = n$, then we see that the nth derivatives of $\mathbf{b}$ with respect to $\mathbf{f}$ are on the image of $\mathbf{F}$, which is a curve on the surface.

We shall consider the case $r = 1$ in more detail. Setting $\mathbf{f} = [f, g, h]^T$, we have

$$D_{\mathbf{f}} \mathbf{b}(\mathbf{u}) = \mathbf{b}[\mathbf{f}, \mathbf{u}^{<n-1>}] = \sum_{|\mathbf{i}|=n-1} [f\mathbf{b_{i+e1}} + g\mathbf{b_{i+e2}} + h\mathbf{b_{i+e3}}] B_{\mathbf{i}}^{n-1}(\mathbf{u}).$$

Thus the coefficients of the derivative surface (of degree $n - 1$) are formed by the terms $f\mathbf{b_{i+e1}} + g\mathbf{b_{i+e2}} + h\mathbf{b_{i+e3}}$, obtained by application of one level of the de Casteljau algorithm to the original control points. Figure 5 illustrates the case $n = 2$.

If $\mathbf{f}$ and $\mathbf{g}$ are two points on $\mathbf{F}$, we may define *mixed derivatives* in the same way:

$$D_{\mathbf{f},\mathbf{g}}^{r,s} \mathbf{b}(\mathbf{u}) = \mathbf{b}[\mathbf{f}^{<r>}, \mathbf{g}^{<s>}, \mathbf{u}^{<n-r-s>}].$$

Let us investigate the relationship between blossoms and derivatives in more detail for the quadratic case $n = 2$. The fundamental line $\mathbf{F}$ is mapped

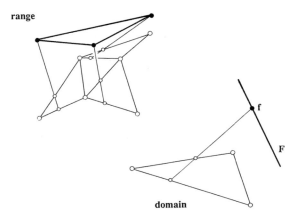

Fig. 5. Derivatives of a triangular patch.

to a conic c_f on the surface. Let $\mathbf{u}$ be a point in the domain, and $\mathbf{b}(\mathbf{u})$ its corresponding surface point. With $\mathbf{f}$ a point on $\mathbf{F}$, the line $\mathbf{u} \wedge \mathbf{f}$ is mapped to a conic $\mathbf{c}$ on the surface. The derivative of this conic at $\mathbf{u}$ is the intersection of the conic's tangents at $\mathbf{b}(\mathbf{u})$ and at $\mathbf{b}(\mathbf{f})$. Thus all derivatives to $\mathbf{b}$ lie on $\mathbf{b}$'s tangent planes at c_f.

In the degenerate case where c_f collapses to a single point $\mathbf{p}$, all tangent planes at $\mathbf{p}$ agree, and all first derivatives of $\mathbf{b}$ lie in this plane. Quadratic surfaces for which $\mathbf{F}$ is mapped to a single point $\mathbf{p}$ are obtainable by a *stereographic projection* through $\mathbf{p}$. In this special case, the quadratic surface is actually part of a *quadric*, see [1].

Acknowledgements. I wish to thank D. Hansford for proofreading. This research was partially supported by NSF grant DDM 9123527 to Arizona State University.

References

1. W. Boehm and D. Hansford, Bézier patches on quadrics, in: *NURBS for Curve and Surface Design*, G. Farin (ed.), SIAM, 1991, 1-14.
2. W. Boehm and H. Prautzsch, *Geometric Foundations of Geometric Design*, AK Peters, Boston, 1994.
3. G. Farin, *NURBS for rational Curve and Surface Design*, AK Peters, Boston, 1994.
4. L. Ramshaw, Blossoming: a connect-the-dots approach to splines, Technical Report, Digital Systems Research Center, Palo Alto, Ca, 1987.
5. L. Ramshaw, Blossoms are polar forms, Computer Aided Geometric Design 6(**4**), 323–359, 1989.
6. D. Struik, *Analytic and Projective Geometry*, Addison-Wesley, Cambridge, MA, 1953.

Characterizations of the Set of Rational Parametric Curves with Rational Offsets

J. C. Fiorot and Th. Gensane

Abstract. Rational curves do not generally have rational offsets. We give an analytic characterization of rational curves with rational offsets. We also propose a geometric characterization of these curves after showing that the set of rational curves with rational or piecewise rational arc length is equal to the set of caustics of the rational curves.

§1. Introduction

Let $r(t)$ be a parametric curve. The offsets to $r(t)$ are defined as the curves $r_d(t) = r(t) + d\, n(t)$, where $d \in \mathbb{R}$, and $n(t)$ is the unit normal vector to $r(t)$.

Offsets of rational curves are not necessarily rational themselves. A lot of papers in the eigthies deal with approximating offsets by piecewise polynomial or piecewise rational curves [4,11,12,14,16,20]. We will denote by $I_{\mathcal{R}}$ the *set of rational parametric curves with rational or piecewise rational offsets*. The aim of this paper is to characterize $I_{\mathcal{R}}$. Recently Farouki and Sakkalis [6,7] have been interested in polynomial curves with rational offsets. They approximate a curve by a special polynomial curve with rational offsets.

More generally, this problem was studied more than a century ago. Humbert was interested by the algebraic curves with an algebraic arc length [13]. In the complex field, Raffy characterized the curves of $I_{\mathcal{R}}$ by considering them as envelope of their tangent lines [19], but the difficulty is to pull out the real parametric curves. Whereas it is easy to characterize polynomial curves with rational offsets (Kubota [15]), this is no longer the case if one considers rational curves. In this paper we give two answers to this problem, on the one hand giving their explicit parametric form as well as the one of their offsets (Section 2), on the other hand characterizing them as involute of the caustics of the rational parametric curves (Section 3). In the same section, we show that the set of the rational parametric curves with rational arc length is identical to the set of the caustics of the rational parametric

Curves and Surfaces in Geometric Design
P. J. Laurent, A. Le Méhauté, and L. L. Schumaker (eds.), pp. 153–160.

curves. These geometrical characterizations were already presented at the French 24th Congress of Numerical Analysis in Vittel [10].

This study has been motivated by the use of offsets in manufacturing areas for tool-paths, CAGD problems, and by the fact that all rational curves are easily described as (BR) curves by means of a finite set of massic vectors as defined in Fiorot and Jeannin [9]. For CAGD and CAM purposes we can use this possibility to represent the curves of $I_\mathcal{R}$ and their rational offsets. The (BR) curves include the polynomial [1,3] and the rational [5,8] Bézier curves. Throughout this paper we use the following notation:

$\overline{\mathbb{R}} = \mathbb{R} \cup \{-\infty, \infty\}$,

$\mathbb{R}[X] = $ the ring of polynomials with real coefficients,

$\mathbb{R}(X) = $ the field of the rational functions,

$\mathcal{D} = $ the set of rational parametric curves with support on a line,

$\mathcal{C} = $ the set of rational parametric curves with support on a circle.

§2. An Analytic Characterization of $I_\mathcal{R}$

Let $r(t) = (x(t), y(t))$ be a parametric curve on $[a, b]$, with $a, b \in \overline{\mathbb{R}}$. The curves $r_d(t) = r(t) + dn(t)$ with $d \in \mathbb{R}$ and $n(t) = \left(-y'(t)/\sqrt{x'^2(t) + y'^2(t)}\right.$, $\left. x'(t)/\sqrt{x'^2(t) + y'^2(t)}\right)$ the unit normal vector to $r(t)$, are called *offsets* to $r(t)$. Offsets of rational parametric curves are rational or piecewise rational iff the Gauss map $t \longrightarrow n(t)$ is rational, i.e.,

$$\text{iff the map } t \longrightarrow |r'(t)| \text{ is rational,} \quad \text{or} \tag{1}$$

$$\text{iff } \exists C \in \mathbb{R}(X), \ x'^2(t) + y'^2(t) = C^2(t). \tag{2}$$

To give a characterization of $I_\mathcal{R}$, we use the following theorem by Kubota [15]

Theorem 2.1. *Let $a, b, c \in \mathbb{R}[X]$. Then $a^2 + b^2 = c^2$ if and only if there exist $K, M, N \in \mathbb{R}[X]$ such that*

$$a = K(M^2 - N^2)$$
$$b = K(2MN).$$

Consequently, we obtain

Corollary 2.1. *The set $I_\mathcal{R}$ is determined by*

$$I_\mathcal{R} = \{r(t) = (x(t), y(t))^T : \ x(t) = \int K(t)\frac{M^2(t) - N^2(t)}{R^2(t)}\, dt,$$

$$\text{and } y(t) = \int K(t)\frac{2M(t)\,N(t)}{R^2(t)}\, dt \ \text{ are both rational}\}.$$

Example 2.1. The parameterization of the unit circle without the point $(-1, 0)$ given by $r(t) = ((1 - t^2)/(1 + t^2), (2t/(1 + t^2)))$ can be obtained from

Corollary 2.1 with $K(t) = -1$, $M(t) = 1 + t$, $N(t) = 1 - t$, $R(t) = 1 + t^2$. The parameterization of the astroid $r(t) = ((\frac{1-t^2}{1+t^2})^3, (\frac{2t}{1+t^2})^3)$ is given by $K(t) = -12t(1 - t^2)$, $M(t) = 1$, $N(t) = t$, $R(t) = (1 + t^2)$.

The characterization given by Corollary 2.1 is not an explicit one: If we consider $K, M, N, R \in \mathbb{R}[X]$, we need a costly check to verify if the two primitives are rational. We will avoid this problem by considering the curves of $I_\mathcal{R}$ as envelopes of their tangent lines. The general form of the tangent lines of the curves of $I_\mathcal{R}$ being

$$D_t \equiv 2M(t)N(t)\,x - (M^2(t) - N^2(t))\,y = f(t) \qquad (3)$$

with $f(t)$ a rational function, we find after calculation the following (see [17,18], for a similar result and the use of the curves of $I_\mathcal{R}$ in the dual form).

Theorem 2.2. *A curve $r(t) = (x(t), y(t))$ belongs to $I_\mathcal{R}$ iff*

$$\begin{cases} x(t) &= \frac{f'(M^2-N^2)-f(M^2-N^2)'}{2(MN'-M'N)(M^2+N^2)}(t) \\[2mm] y(t) &= \frac{f'(2MN)-f(2MN)'}{2(MN'-M'N)(M^2+N^2)}(t), \end{cases} \qquad (4)$$

where $f \in \mathbb{R}(X)$ and $M, N \in \mathbb{R}[X]$.

Example 2.2. Setting $f(t) = \lambda(M^2(t) + N^2(t))$ in (4) we get the explicit parameterized form of the curves of $\mathcal{C}$:

$$x(t) = \lambda \frac{-2M(t)N(t)}{M^2(t) + N^2(t)} \text{ and } y(t) = \lambda \frac{M^2(t) - N^2(t)}{M^2(t) + N^2(t)},$$

which can also be obtained directly from Theorem 2.1.

Another example is the following

Proposition 2.1. *The polynomial cubics of $I_\mathcal{R}$ which do not belong to $\mathcal{D}$ can be obtained by (4), setting $f(t) = \lambda(M^2(t) + N^2(t))^2$ with $M(t) = at + b$, $N(t) = ct + d$, $(a, c) \neq (0, 0)$ and $ad - bc \neq 0$.*

We now give the parameterizations of the offsets to a curve $r(t)$ of $I_\mathcal{R}$.

Corollary 2.2. *Let $r(t)$ be a curve defined by (4) with $f \in \mathbb{R}(X)$ and $M, N \in \mathbb{R}[X]$. The offsets $r_d(t)$ to $r(t)$ are defined by*

$$r_d(t) = r(t) + d\,\varepsilon(t)(\frac{-2M(t)N(t)}{M^2(t) + N^2(t)}, \frac{M^2(t) - N^2(t)}{M^2(t) + N^2(t)}),$$

where $\varepsilon(t)$ is the sign of the determinant

$$\begin{vmatrix} f(t) & 2M(t)N(t) & M^2(t) - N^2(t) \\ f'(t) & (2M(t)N(t))' & (M^2(t) - N^2(t))' \\ f''(t) & (2M(t)N(t))'' & (M^2(t) - N^2(t))'' \end{vmatrix}. \qquad (5)$$

Sketch of the proof: One can show that a curve $r(t)$ defined by (4) can be put in the form

$$r(t) = \left(\int g(t) \frac{M^2(t) - N^2(t)}{R^2(t)} \, dt \,, \int g(t) \frac{2M(t)N(t)}{R^2(t)} \, dt \right) \qquad (6)$$

with $R(t) = 2(M(t)N'(t) - M'(t)N(t))(M^2(t) + N^2(t))$ and $g(t)$ the determinant (5). The result follows. ∎

Remark 2.1. The curves of $I_{\mathcal{R}}$ are given by (4) and their offsets by Corollary 2.2. For use in CAGD or CAM, we can put them into the (BR) form given by [9]. Hence they are stored by a finite number of massic vectors.

Remark 2.2. It is obvious that the arc length of a rational curve can be put into the form of an hyperelliptic integral $S(t) = \int \sqrt{P(t)}/Q(t) \, dt$, where $P, Q \in \mathbb{R}[X]^2$. In the case of a curve in $I_{\mathcal{R}}$, we get $S(t) = \int |K(t)|(M^2(t) + N^2(t))/R^2(t) \, dt$. Generally, this integral is not (piecewise) rational, and in the following section, we characterize the subset of $I_{\mathcal{R}}$ consisting of the rational parametric curves with rational or piecewise rational arc length.

§3. A Geometric Characterization of $I_{\mathcal{R}}$ by the Caustics

A *caustic* is the solution of an old problem in optics: it is the envelope of light rays reflected by a mirror where the source of light is supposed to be at infinity (see Brieskorn and Knoerer [2]). More precisely, let $r(t) = (x(t), y(t))$ be a parametric curve considered as a mirror. Given a source of light at infinity, we suppose the light rays are all parallel to the line $y = 0$, coming from one side of the plane or the other depending on the convexity of the curve.

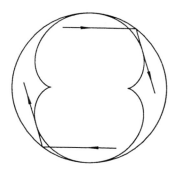

Fig. 3.1. The caustic of the circle is the nephroid.

We are now going to show that offsets of caustics of a rational parametric curve belong to $I_{\mathcal{R}}$.

Lemma 3.1. *The equation of the caustic of a parametric curve $r(t) = (x(t), y(t))$ is*

$$\begin{pmatrix} X(t) \\ Y(t) \end{pmatrix} = \begin{pmatrix} x(t) \\ y(t) \end{pmatrix} - \frac{y'(t)}{2[x''(t)\,y'(t) - x'(t)\,y''(t)]} \begin{pmatrix} x'^2(t) - y'^2(t) \\ 2x'(t)y'(t) \end{pmatrix}$$

Moreover,

$$\begin{pmatrix} X'(t) \\ Y'(t) \end{pmatrix} = k(t) \begin{pmatrix} x'^2(t) - y'^2(t) \\ 2x'(t)y'(t) \end{pmatrix},$$

where $k : \mathbb{R} \longrightarrow \mathbb{R}$. If in addition the curve $r(t)$ is rational, the caustic is also rational.

Proof: The caustic of $r(t)$ is the envelope of the family L_t of reflected lines on $r(t)$. The equation of the straight line L_t is

$$(y - y(t))(y'^2(t) - x'^2(t)) + (x - x(t))(2x'(t)y'(t)) = 0.$$

We find the results by a classical calculation on the envelopes of curves. ∎

Proposition 3.1. *Let $C_\mathcal{R}$ be the set of caustics of rational parametric curves. We have*

$$C_\mathcal{R} \subset I_\mathcal{R}.$$

Proof: Let $R(t) = (X(t), Y(t))$ be the caustic of the rational curve $r(t) = (x(t), y(t))$. By Lemma 3.1 we have $X'^2(t) + Y'^2(t) = k^2(t)(x'^2(t) + y'^2(t))^2$, where $k \in \mathbb{R}(X)$ because $R'(t)$ is rational. Hence by (1), $R(t)$ belongs to $I_\mathcal{R}$. ∎

One might think that each curve of $I_\mathcal{R}$ is the caustic of a rational parametric curve. Proposition 3.2 illustrates that it is not the case.

Theorem 3.1. *Let $A_\mathcal{R}$ be the set of rational parametric curves for which the arc length is a rational or piecewise rational map. Then*

$$C_\mathcal{R} = A_\mathcal{R}.$$

H. Pottmann [17] gives the explicit parametric form of the curve in $A_\mathcal{R}$. To prove Theorem 3.1, we need the following

Lemma 3.2. *Let $R(t)$ be a curve of $I_\mathcal{R}$. $R(t)$ is the caustic of a rational parametric curve if and only if its arc length is a rational or piecewise rational map.*

Proof: Let $R(t) = (X(t), Y(t)) \in I_\mathcal{R}$; we denote it by $Z(t) = X(t) + iY(t)$ in the complex plane. By Corollary 2.1, the derivative of $Z(t)$ can be put into the form

$$Z'(t) = K(t) \left(\frac{M(t) + i\,N(t)}{R(t)} \right)^2. \tag{7}$$

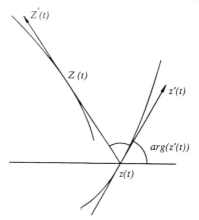

Fig. 3.2. Is there some rational mirror $z(t)$?

We are looking for $z(t) = x(t) + iy(t)$ which has $Z(t)$ as caustic. Since the point $z(t)$ is on the tangent line of $Z(t)$, $2 \arg z'(t) \equiv \arg Z'(t)(\pi)$,

$$\begin{cases} z(t) & = & Z(t) + g(t)\,Z'(t) \qquad \text{(a)} \\ z'^2(t) & = & f(t)\,Z'(t), \qquad\qquad \text{(b)} \end{cases} \qquad (8)$$

where $(f, g) \in \mathbb{R}(X)^2$. Squaring the derivative of (8-a) and substituting the expression (7) of $Z'(t)$ in the result, we obtain $g(t)$ as solution of a differential equation:

$$g(t) = \left[|K(t)| \frac{L^2(t)}{R^2(t)} \right]^{-1} \left[\lambda - \int |K(t)| \frac{L^2(t)}{R^2(t)}\, dt \right],\ \lambda \in \mathbb{R}, \qquad (9)$$

where $L = M$ or N. Hence, we find two classes of mirrors

$$z(t) = Z(t) + g(t)\,Z'(t) \qquad (10)$$

having $Z(t)$ as caustic. Let us investigate under which condition they are rational. Suppose first $R(t)$ is regular, i.e., $K(t) \neq 0$ for each t. We suppose in addition $K \geq 0$. Then, from (9) and using the rational expression $X(t) = \int K(t)(M^2(t) - N^2(t))/R^2(t)dt$ we find that the mirrors $z(t)$ are rational iff $\int K(t)(M^2(t) + N^2(t))/R^2(t)dt$, the arc length of $R(t)$, is rational. Let now $R(t)$ be singular. In contrast to the first case, where the constant λ was arbitrary, we need to take $\lambda = 0$ in (9). Then

$$g(t) = - \left[K(t) \frac{L^2(t)}{R^2(t)} \right]^{-1} \left[\int K(t) \frac{L^2(t)}{R^2(t)}\, dt \right],$$

and we are back in the regular case. ∎

Remark 3.1. Notice that a nonsingular curve of $A_\mathcal{R}$ has two classes of infinity many rational mirrors which have this curve as caustic. If the curve is singular (i.e. has a piecewise rational arc length), there are two rational mirrors, the others being piecewise rational.

Proof of Theorem 3.1: $M(t) \in C_\mathcal{R}$ implies by Proposition 3.1 that $M(t) \in I_\mathcal{R}$. We get $M(t) \in A_\mathcal{R}$ by Lemma 3.2, hence $C_\mathcal{R} \subset A_\mathcal{R}$. Conversely, $M(t) \in A_\mathcal{R}$ implies by (1) that $M(t) \in I_\mathcal{R}$, we get $M(t) \in C_\mathcal{R}$ by Lemma 3.2 and finally $A_\mathcal{R} \subset C_\mathcal{R}$. ∎

We now give an example of a curve which has rational offsets but not a rational arc length.

Proposition 3.2. *Let $r(t) \in A_\mathcal{R}$. An offset $r_d(t)$ of $r(t)$ with $d \neq 0$ belongs to $I_\mathcal{R} \backslash A_\mathcal{R}$.*

Before giving the geometrical characterization of $I_\mathcal{R}$, let us recall the definitions of evolute and involute. For a parametric curve $r(t) = (x(t), y(t))$, we define the *evolute* of $r(t)$ to be the curve $ev(t) = r(t) - \rho(t)n(t)$, where $\rho(t)$ is the radius of curvature. The evolute is the locus of the centers of curvature of $r(t)$. The curve $inv(t) = R(t) + S_{t_0}(t) \frac{R'(t)}{|R'(t)|}, t_0 \in \mathbb{R}$, is called the *involute* of the curve $R(t)$, where as usual S_{t_0} denotes the arc length of $R(t)$. A curve of $A_\mathcal{R}$ has rational involutes which are parallel to each other; they belong to $I_\mathcal{R}$. Conversely, consider a curve $r(t) \in I_\mathcal{R} \backslash \mathcal{D}$. The evolute of $r(t)$ is rational and has a rational arc length (considering the length of an arc as difference of the radius of curvature), then the evolute of $r(t)$ belongs to $A_\mathcal{R}$. So, the curve $r(t)$ is the involute of a curve of $A_\mathcal{R}$. The curves of $\mathcal{C}$ belong to $I_\mathcal{R}$, but their evolutes are reduced to a point (which is the caustic of a parabola). By Theorem 3.1, we get

Proposition 3.3. *The set $I_\mathcal{R}$ is equal to the union of the involutes of the caustics of rational parametric curves and of $\mathcal{D} \cup \mathcal{C}$.*

References

1. Bézier, P., *Procédé de définition numérique des courbes et surfaces non mathématiques*, Système Unisurf, Automatisme **13** (1968), 189–196.
2. Brieskorn, E., H. Knorrer, *Plane Algebraic Curves*, Birkhauser Verlag, 1986.
3. de Casteljau, P., *Outillage, Méthode de Calcul*, André Citrœn Automobile S.A., Paris 1959.
4. Farin, G., Curvature continuity and offsets for piecewise conics, ACM Trans. on Graphics, **8** (1989), 88–99.
5. Farin, G., Algorithms for rational Bézier curves, Comput. Aided Design **15** (1983), 73–77.
6. Farouki, R. T., T. Sakkalis, Pythagorean hodographs, IBM, Journal of Research and Development, **34**, 5 (1990), 736–752.

7. Farouki, R. T., Pythagorean hodograph curves in practical use, in *Geometry Processing for Design and Manufacturing*, R.E.Barnhill (ed.), SIAM Publications, Philadelphia, 1992, 3–33.
8. Faux , I. E., M. J. Pratt, *Computational Geometry for Design and Manufacture*, Ellis Horwood, Chistester, 1979.
9. Fiorot, J. C., P. Jeannin, *Courbes et Surfaces Rationnelles, Applications à la CAO*, Masson, RMA **12**, Paris, 1989. English version: *Rational Curves and Surfaces, Applications to CAD*, J.Wiley and Sons, Chichester, 1992.
10. Gensane, T., Etude des courbes parallèles rationnelles, 24^{ieme} Congrès National d'Analyse Numérique, 25-28 mai 1992, Vittel, France.
11. Hoschek, J., N. Wissel, Optimal approximate conversion of spline curves and spline approximation of offset curves, Comput. Aided Design **20** (1988), 475–483.
12. Hoschek, J., Offset curves in the plane, Comput. Aided Design **17** (1985), 33–40.
13. Humbert, G., Sur les courbes algébriques rectifiables, C.R.A.S. de Paris, **104** (1887), 1051–1052.
14. Klass, R., An offset spline approximation for plane cubic splines, Comput. Aided Design **15** (1983), 297–299.
15. Kubota, K. K., Pythagorean triples in unique factorization domains, Amer. Math. Monthly, **79** (1972), 503–505.
16. Pham, B., Offset approximation of uniform B-splines, Comput. Aided Design **20** (1988), 471-474.
17. Pottmann, H., Rational curves and surfaces with rational offsets, Computer Aided Geom. Design, to appear.
18. Pottmann, H., Applications of the dual Bézier representation of rational curves and surfaces, in *Curves and Surfaces in Geometric Design*, P.-J. Laurent, A. Le Méhauté, and L. L. Schumaker (eds.), A K Peters, Wellesley, 1994, 377–384.
19. Raffy, L., Sur la rectification des courbes planes unicursales, C.R.A.S. de Paris, **104** (1887), 892–893.
20. Tiller, W., E. G. Hanson, Offset of two dimensitional profiles, IEEE Computer Graphics Appl. **4** (1984), 36–46.

J. C. Fiorot
Université de Valenciennes et du Hainaut-Cambraisis
ENSIMEV, Laboratoire IMAV B.P. 311
F-59304 Valenciennes - cedex, FRANCE

Th. Gensane
Université des Sciences et Technologies de Lille
Laboratoire d'Analyse Numérique et d'Optimisation
UFR IEEA - M3
59655 Villeneuve d'Ascq - cedex, FRANCE

A Necessary and Sufficient Condition
for Joining B-Rational Curves
with Geometric Continuity G³

J. C. Fiorot and P. Jeannin

Abstract. We give a necessary and sufficient condition for joining rational curves in the (BR) form with geometric continuity G^3. It reduces to considering two polynomial Bézier curves of length three whose control polygons are determined in terms of the massic polygons of the (BR) curves.

§1. Introduction

Any polynomial curve of an affine space $\mathcal{E}$ defined on a finite interval can be written in the Bézier form. This curve is controlled by a polygon of points of $\mathcal{E}$. In a similar way, any rational curve of $\mathcal{E}$ can be written in (BR) form. So this rational curve is controlled by a polygon of massic vectors [5,6,7,8]. A massic vector is either a weighted point of $\mathcal{E}$, or a vector of $\vec{\mathcal{E}}$ where $\vec{\mathcal{E}}$ is the associated vector space to $\mathcal{E}$.

To link simple curves in order to obtain splines with more or less regularity is an outstanding problem. It is a common concern for piecewise polynomial curves. The segments can be pieced together using different concepts of continuity : C^k-continuity, geometric continuity, continuity of the Frenet frame, and higher order curvatures GF^k, TC^r-continuity. For a complete study, bibliography and historical details, see [1,4,12, 13,14,15,17,19].

These different kinds of continuity have been studied in the case of rational Bézier curves [2,10,11,15,16,18] or in a more theoretical way in [14]. More recently [3,9], we have examined the C^k-continuity of rational curves defined by (BR) curves. In terms of massic vectors, we gave a necessary and sufficient condition for two (BR) curves to be C^k-continuous at a junction point.

In this paper we deal with the geometric continuity of two (BR) curves. For the sake of simplicity, we limit ourselves to geometric continuity G^3. Our

Curves and Surfaces in Geometric Design
P. J. Laurent, A. Le Méhauté, and L. L. Schumaker (eds.), pp. 161–168.
Copyright © 1994 by A K PETERS, Wellesley, MA
ISBN 1-56881-039-3.

approach consists in approximating the (BR) curves by Bézier curves of length three with C^3-continuity. Exploiting the result of [3,9], the control polygon of the Bézier curves is obtained as a function of four massic vectors of the (BR) curves. Finally thanks to a simple lemma we are brought back to joining two Bézier curves with geometric continuity G^3. We discuss two problems to illustrate the results.

§2. Geometric Continuity, Definition and Notation

Definition 2.1. *Let $\mathcal{E}$ be a real affine space, $\mathbb{R}^2, \mathbb{R}^3$, for instance. Let $f : (t_0 - h, t_0] \to \mathcal{E}$ and $g : [t_0, t_0 + h)$ be two parametric curves of class C^3 supposed to be regular at $t = t_0 : f'(t_0) \neq 0, g'(t_0) \neq 0$. f and g are joined with geometric continuity G^3 at t_0 and we denote by $f \cong g$ $(G^3; t_0)$ when there exists $\varphi : (t_0 - h, t_0] \to \mathbb{R}$ of class C^3 such that*

(i) $\varphi(t_0) = t_0$

(ii) $\varphi'(t_0) > 0$

(iii) *$f \circ \varphi$ and g are C^3-continuous at t_0, we denote by $f \circ \varphi \cong g$ $(C^3; t_0)$.*

Proposition 2.1. *$f \cong g$ $(G^3; t_0) \Leftrightarrow$ there exist real constants $\beta_1 > 0, \beta_2, \beta_3$ such that*

$$\begin{pmatrix} 1 & 0 & 0 & 0 \\ 0 & \beta_1 & 0 & 0 \\ 0 & \beta_2 & \beta_1^2 & 0 \\ 0 & \beta_3 & 3\beta_1\beta_2 & \beta_1^3 \end{pmatrix} \begin{pmatrix} f(t_0) \\ f'(t_0) \\ f''(t_0) \\ f'''(t_0) \end{pmatrix} = \begin{pmatrix} g(t_0) \\ g'(t_0) \\ g''(t_0) \\ g'''(t_0) \end{pmatrix} \tag{1}$$

Proof: For instance, see Proposition 2.2 in [12]. ∎

Notation 2.1. *The matrix in (1) with $\beta_1 > 0$ is usually called a G^3 connection matrix ; it is denoted by $K(\beta_1, \beta_2, \beta_3)$ or briefly by K. We write $(D_3 f)(t_0) = (f(t_0), f'(t_0), f''(t_0), f'''(t_0))^t$.*

Now equation (1) can be written as $K(D_3 f)(t_0) = (D_3 g)(t_0)$.

§3. Some Lemmas on Continuity

Lemma 3.1. *Let $f, f_1 : (t_0 - h, t_0] \to \mathcal{E}$ and $g, g_1 : [t_0, t_0 + h] \to \mathcal{E}$ be four parametric curves of class C^3 satisfying the conditions : $f \cong f_1$ $(C^3; t_0)$ and $g \cong g_1$ $(C^3; t_0)$. Then if f and g are regular at t_0 and join with geometric continuity G^3 at t_0, the same applies to f_1 and g_1, and conversely.*

Proof: By hypothesis, $(D_3 f)(t_0) = (D_3 f_1)(t_0)$ and $(D_3 g)(t_0) = (D_3 g_1)(t_0)$. The result is a straightforward consequence of Proposition 2.1. ∎

Notation 3.1. *We denote by $BP[P_0, \ldots, P_n; [a, b]](t) = \sum_{i=0}^n B_i^n[a, b](t)P_i$ the Bézier curve related to the interval $[a, b]$ whose control polygon is $P = (P_0, \ldots, P_n)$. n is called the length. Let*

$$B_i^n[a, b](t) = \binom{n}{i} \left(\frac{b-t}{b-a}\right)^{n-i} \left(\frac{t-a}{b-a}\right)^i .$$

More briefly, we write $BP[P; [a, b]](t)$.

Lemma 3.2. *Let $BP[P_0, P_1, \ldots, P_n; [a, b]]$ and $BP[Q_0, Q_1, \ldots, Q_p; [b, c]]$ be two Bézier curves with $P_{n-1} \neq P_n$ and $Q_0 \neq Q_1$ (regularity at $t = b$). They are joined with geometric continuity G^3 at b if and only if there exist a G^3 connection matrix K satisfying the relation :*

$$K \begin{pmatrix} P_n \\ n\Delta P_{n-1} \\ n(n-1)\Delta^2 P_{n-2} \\ n(n-1)(n-2)\Delta^3 P_{n-3} \end{pmatrix} = \begin{pmatrix} Q_0 \\ p\Delta Q_0 \\ p(p-1)\Delta^2 Q_0 \\ p(p-1)(p-2)\Delta^3 Q_0 \end{pmatrix}$$

Proof: We use Proposition 2.1 and the expression of the derivatives of a Bézier curve. ■

Remark 3.1. The previous conditions are independent from a, b, c and d. It was what we expect since the relation $BP[P; [a, b]](t) = BP[P; [0, 1]] \left(\dfrac{t - a}{b - a} \right)$ implies that the supports of $BP[P; [a, b]]$ and $BP[P; [a', b']]$ are the same.

Remark 3.2. When $\beta_1 = \dfrac{c - b}{b - a}, \beta_2 = 0, \beta_3 = 0$, the corresponding conditions are those of C^3-continuity.

§4. C^k-Continuity of (BR) Curves

Any polynomial curve of $\mathcal{E}$ defined on a finite interval can be written in the Bézier form ((BP) form) : $BP[P_0, P_1, \ldots, P_n; [a, b]]$ with $P_i \in \mathcal{E}$ (see Notation 3.1). In a similar way any rational curve of $\mathcal{E}$ can be written in (BR) form : $BR[\theta_0, \theta_1, \ldots, \theta_n; [a, b]]$ with $\theta_i \in \hat{\mathcal{E}}$. $\hat{\mathcal{E}} = (\mathcal{E} \times \mathbb{R}^*) \cup \vec{\mathcal{E}}$ being the linear space of massic vectors. Then a massic vector θ_i is either a weighted point : $\theta_i = (P_i; \beta_i) \in \mathcal{E} \times \mathbb{R}^*$, $i \in I$ or a vector $\theta_i = \vec{V}_i \in \vec{\mathcal{E}}$, $i \in \overline{I}$ with $I \cup \overline{I} = \{0, 1, \ldots, n\}, I \cap \overline{I} = \emptyset$. Let $\theta = (\theta_0, \theta_1, \ldots, \theta_n)$. For a point at finite distance,

$$BR[\theta; [a, b]](t) = \frac{\sum_{i \in I} \beta_i B_i^n[a, b](t) P_i + \sum_{i \in \overline{I}} B_i^n[a, b](t) \vec{U}_i}{\sum_{i \in I} \beta_i B_i^n[a, b](t)}$$

The (BR) curves were introduced and studied in [5,6,7,8].

We give a preliminary result concerning with the C^k-continuity of (BR) curves. This result is a consequence of Theorem 7.2.1 of [9].

Proposition 4.1. *Given two (BR) curves : $BR[\omega_0, \omega_1, \ldots, w_q; [a - h_2, a]]$ and $BR[\theta_0, \theta_1, \ldots, \theta_p; [a - h_1, a]]$, $h_1, h_2 > 0$ with $\omega_q = \theta_p$, ω_q and $\theta_p \neq 0$, an integer $k \geq 1$. A necessary and sufficient condition that*

$$BR[\omega_0, \omega_1, \ldots, w_q; [a - h_2, a]] \cong BR[\theta_0, \theta_1, \ldots, \theta_p; [a - h_1, a]] \ (C^k; a)$$

is that there exist $Min(k, p+q)$ reals $\mu_1, \mu_2, \ldots, \mu_{Min(k,p+q)}$ such that

$$\binom{q}{\ell} \Delta^\ell \omega_{q-\ell} - \left(\frac{h_2}{h_1}\right)^\ell \binom{p}{\ell} \Delta^\ell \theta_{p-\ell} = \sum_{i=1}^{\ell} \mu_i \binom{q}{\ell - i} \Delta^{\ell-i} \omega_{q-\ell+i}$$

for $\ell = 1, 2, \ldots, Min(k, p+q)$. The real numbers $\mu_1, \mu_2, \ldots, \mu_{Min(k,p+q)}$ are unique. Interchanging the role of θ and ω, this condition takes the form: there exist $Min(k, p+q)$ reals $\mu_1', \mu_2', \ldots, \mu_{Min(k,p+q)}'$ such that

$$\binom{p}{\ell} \Delta^\ell \theta_{p-\ell} - \left(\frac{h_1}{h_2}\right)^\ell \binom{q}{\ell} \Delta^\ell \omega_{q-\ell} = \sum_{i=1}^{\ell} \mu_i' \binom{p}{\ell - i} \Delta^{\ell-i} \theta_{p-\ell+i}$$

for $\ell = 1, 2, \ldots, Min(k, p+q)$.

Proof: This follows from Theorem 7.2.1 of [9] via the Proposition 1.5.4 (γ) of [9]. ∎

§5. The Main Result

The following theorem gives a necessary and sufficient condition for two (BR) curves to join with geometric continuity G^3. To that end, to each (BR) curve we determine a Bézier curve of length 3, C^3-continuous with it. Using Lemma 3.1, it suffices to join the two Bézier curves with geometric continuity G^3.

Notation 5.1. The linear form $\chi : \hat{\mathcal{E}} \to R$ defined by $\chi(P; \alpha) = \alpha$, $(P; \alpha) \in \mathcal{E} \times \mathbb{R}^*$ and $\chi(\vec{U}) = 0$, $\vec{U} \in \vec{\mathcal{E}}$ is called the mass (Proposition 1.2.2.8 of [7,8]).

With $\tilde{\mathcal{E}}$ the projective completion of $\mathcal{E}$, we denote by $\Pi : \hat{\mathcal{E}} - \{\vec{0}\} \to \tilde{\mathcal{E}}$, the natural projection :

$$\Pi(P; \alpha) = P \quad \Pi(\vec{U}) = (\vec{U})_\infty,$$

where $(\vec{U})_\infty$ designates the point at infinity of $\mathcal{E}$ in the direction of $\vec{U}$ (Notation, Proposition 1.2.2.3 of [7,8]).

Theorem 5.1. A necessary and sufficient condition that

$$BR[\omega_0, \omega_1, \ldots, \omega_q; [a, b]] \cong BR[\theta_0, \theta_1, \ldots, \theta_p; [b, c]] \quad (G^3; b),$$

where $\omega_q = (R_q; \alpha_q), \alpha_q \neq 0$, $\Pi(\omega_{q-1}) \neq \Pi(\omega_q)$, $\theta_0 = (S_0; \gamma_0)$, $\gamma_0 \neq 0$, $\Pi(\theta_0) \neq \Pi(\theta_1)$), is that

$$BP[P_0, P_1, P_2, P_3; [a, b]] \cong BP[Q_0, Q_1, Q_2, Q_3; [b, c]] \quad (G^3; b),$$

where the control polygons $P = (P_0, P_1, P_2, P_3)$ and $Q = (Q_0, Q_1, Q_2, Q_3)$ are respectively determined by the equalities (2) and (3) :

$$\alpha_q \begin{pmatrix} (P_3; 1) \\ 3\Delta P_2 \\ 3\Delta^2 P_1 \\ \Delta^3 P_0 \end{pmatrix} = \begin{pmatrix} 1 & 0 & 0 & 0 \\ -\mu_1 & q & 0 & 0 \\ -\mu_2 & -q\mu_1 & \binom{q}{2} & 0 \\ -\mu_3 & -q\mu_2 & -\binom{q}{2}\mu_1 & \binom{q}{3} \end{pmatrix} \begin{pmatrix} \omega_q \\ \Delta\omega_{q-1} \\ \Delta^2\omega_{q-2} \\ \Delta^3\omega_{q-3} \end{pmatrix} \quad (2)$$

$$\gamma_0 \begin{pmatrix} (Q_0; 1) \\ 3\Delta Q_0 \\ 3\Delta^2 Q_0 \\ \Delta^3 Q_0 \end{pmatrix} = \begin{pmatrix} 1 & 0 & 0 & 0 \\ -\lambda_1 & p & 0 & 0 \\ -\lambda_2 & -p\lambda_1 & \binom{p}{2} & 0 \\ -\lambda_3 & -p\lambda_2 & -\binom{p}{2}\lambda_1 & \binom{p}{3} \end{pmatrix} \begin{pmatrix} \theta_0 \\ \Delta\theta_0 \\ \Delta^2\theta_0 \\ \Delta^3\theta_0 \end{pmatrix}. \tag{3}$$

The constants μ_1, μ_2, μ_3 (resp. $\lambda_1, \lambda_2, \lambda_3$) depend on the masses $\chi(\omega_{q-i}) = \alpha_{q-i}$, $i = 0,1,2,3$ (resp. $\chi(\theta_i) = \gamma_i$, $i = 0,1,2,3$). They are given by

$$\mu_1 = \frac{q\Delta\alpha_{q-1}}{\alpha_q} \qquad \mu_2 = -\left(\frac{q\Delta\alpha_{q-1}}{\alpha_q}\right)^2 + \binom{q}{2}\frac{\Delta^2\alpha_{q-2}}{\alpha_q}$$

$$\mu_3 = \left(\frac{q\Delta\alpha_{q-1}}{\alpha_q}\right)^3 - 2q\binom{q}{2}\frac{\Delta\alpha_{q-1}\Delta^2\alpha_{q-2}}{\alpha_q^2} + \binom{q}{3}\frac{\Delta^3\alpha_{q-3}}{\alpha_q} \tag{4}$$

$$\lambda_1 = \frac{p\Delta\gamma_0}{\gamma_0} \qquad \lambda_2 = -\left(\frac{p\Delta\gamma_0}{\gamma_0}\right)^2 + \binom{p}{2}\frac{\Delta^2\gamma_0}{\gamma_0}$$

$$\lambda_3 = \left(\frac{p\Delta\gamma_0}{\gamma_0}\right)^3 - 2p\binom{p}{2}\frac{\Delta\gamma_0\Delta^2\gamma_0}{\gamma_0^2} + \binom{p}{3}\frac{\Delta^3\gamma_0}{\gamma_0} \tag{5}$$

Proof: We apply Lemma 3.1., i.e. we determine $BP[P_0, P_1, P_2, P_3; [a,b]]$ and $BP[Q_0, Q_1, Q_2, Q_3; [b,c]]$ such that

$$BR[\omega_0, \omega_1, \ldots, \omega_q; [a,b]] \cong BP[P_0, P_1, P_2, P_3; [a,b]] \quad (C^3; b)$$
$$BR[\theta_0, \theta_1, \ldots, \theta_p; [b,c]] \cong BP[Q_0, Q_1, Q_2, Q_3; [b,c]]] \quad (C^3; b).$$

We now discuss the determination of P_0, P_1, P_2, P_3. Define $\alpha_i = \chi(\omega_i)$, for $i = 0, 1, \ldots, q$, the mass of ω_i and $P_3 = R_q$. We can write

$$BP[P_0, P_1, P_2, P_3; [a,b]](t) = BR[\theta_0', \theta_1', \theta_2', \theta_3'; [a,b]](t)$$

with $\theta_i' = (P_i; \alpha_q)$. By applying Proposition 4.1 (first group of relations) with $p = 3$, $h_1 = h_2 = b - a$, θ_i' in place of θ_i, we obtain with $\theta_3' = \omega_q$:

$$\ell = 1 \quad \binom{3}{1}\Delta\theta_2' = \binom{q}{1}\Delta\omega_{q-1} - \mu_1\omega_q$$

$$\ell = 2 \quad \binom{3}{2}\Delta^2\theta_1' = \binom{q}{2}\Delta^2\omega_{q-2} - \mu_1\binom{q}{1}\Delta\omega_{q-1} - \mu_2\omega_q$$

$$\ell = 3 \quad \binom{3}{3}\Delta^3\theta_0' = \binom{q}{3}\Delta^3\omega_{q-3} - \mu_1\binom{q}{2}\Delta^2\omega_{q-2} - \mu_2\binom{q}{1}\Delta\omega_{q-1} - \mu_3\omega_q$$

We have $\Delta\theta_2' = \theta_3' - \theta_2'$ i.e. $\Delta\theta_2' = (P_3; \alpha_q) - (P_2; \alpha_q)$ then $\Delta\theta_2' = \alpha_q\Delta P_2$. And similarly $\Delta^2\theta_1' = \alpha_q\Delta^2 P_1$, $\Delta^3\theta_0' = \alpha_q\Delta^3 P_0$. The previous relations $\ell = 1, 2, 3$ with $\theta_3' = \omega_q$ give (2).

By noticing that $\chi(\Delta^i P_{3-i}) = 0, i = 1, 2, 3$ and $\chi(\Delta^i \omega_{q-i}) = \Delta^i \alpha_{q-i}, i = 1, 2, 3$ and by passing to the masses in the above equalities $\ell = 1, 2, 3$, we obtain a particular triangular system giving μ_1, μ_2, μ_3 (4). The determination of Q_0, Q_1, Q_2, Q_3 is similar to that of P_0, P_1, P_2, P_3 by applying now the first group of relations of Theorem 7.2.1 of [9]. .

From (3) and (5) we can verify that $\Pi(\theta_0) \neq \Pi(\theta_1)$ implies $Q_0 \neq Q_1$ and similarly $P_2 \neq P_3$. ∎

§6. Two Applications

Application 6.1. *Let* $BR[\omega_0, \omega_1, \ldots, \omega_q; [a, b]]$ *and* $BR[\theta_0, \theta_1, \ldots, \theta_p; [b, c]]$ *be given. Do these two (BR) curves join with geometric continuity* G^3 *at* $t = b$?

The solution is given by applying the following steps:

1) From $\omega_q, \omega_{q-1}, \omega_{q-2}, \omega_{q-3} (\Pi(\omega_q) \neq \Pi(\omega_{q-1}))$ we determine the constants μ_1, μ_2, μ_3 following (4). Recall that $\chi(\omega_i) = \alpha_i$.

2) We define $P_3, \Delta P_2, \Delta^2 P_1, \Delta^3 P_0$ by equalities (2).

3) From $\theta_0, \theta_1, \theta_2, \theta_3$ $(\Pi(\theta_0) \neq \Pi(\theta_1))$ we determine the constants $\lambda_1, \lambda_2, \lambda_3$ following (5). Recall that $\chi(\theta_i) = \gamma_i$.

4) We define $Q_0, \Delta Q_0, \Delta^2 Q_0, \Delta^3 Q_0$ by equalities (3).

5) If $BP[P_0, P_1, P_2, P_3; [a, b]]$ and $BP[Q_0, Q_1, Q_2, Q_3; [b, c]]$ join with geometric continuity G^3 the answer is "yes". To this end, using Lemma 3.2 with $n = p = 3$, we check if there exist constants $\beta_1 > 0, \beta_2, \beta_3$ satisfying the relations

$$\begin{aligned}
P_3 &= Q_0 \quad (= \Pi(\omega_q) = \Pi(\theta_0)) \\
\beta_1 \Delta P_2 &= \Delta Q_0 \\
\beta_2 \Delta P_2 &= 2\Delta^2 Q_0 - 2\beta_1^2 \Delta^2 P_1 \\
\beta_3 \Delta P_2 &= 2\Delta^3 Q_0 - 6\beta_1 \beta_2 \Delta^2 P_1 - 2\beta_1^3 \Delta^3 P_0.
\end{aligned}$$

Corollary 6.2. *Let* Q_0, Q_1, Q_2, Q_3 *be given points, a necessary and sufficient condition that*

$$BR[\theta_0, \theta_1, \ldots, \theta_p; [b, c]] \cong BP[Q_0, Q_1, Q_2, Q_3; [b, c]] \quad (C^3; b)$$

is that $\theta_0, \theta_1, \theta_2, \theta_3$ *are determined by* $\theta_0 = (Q_0; \gamma_0)$, γ_0 *an arbitrary non zero constant,*

$$p\Delta\theta_0 = 3\gamma_0 \Delta Q_0 - \lambda_1'(Q_0; \gamma_0)$$

$$\binom{p}{2}\Delta^2\theta_0 = 3\gamma_0 \Delta^2 Q_0 - 3\lambda_1'\gamma_0 \Delta Q_0 - \lambda_2'(Q_0; \gamma_0) \tag{6}$$

$$\binom{p}{3}\Delta^3\theta_0 = \gamma_0 \Delta^3 Q_0 - 3\lambda_1'\gamma_0 \Delta^2 Q_0 - 3\lambda_2'\gamma_0 \Delta Q_0 - \lambda_3'(Q_0; \gamma_0)$$

with $\lambda_1', \lambda_2', \lambda_3'$ *arbitrary constants. The massic vectors* θ_i, $i \geq 4$ *are arbitrary.*

Proof: We write $BP[Q_0, Q_1, Q_2, Q_3; [b, c]] = BP[(Q_0; \gamma_0), (Q_1; \gamma_0), (Q_2; \gamma_0),$ $(Q_3; \gamma_0); [b, c]]$ with $\gamma_0 \neq 0$ an arbitrary constant. By applying Theorem 7.2.1 of [9] (second group of relations for $\ell = 1, 2, 3$) with $q = 3$, $h_1 = h_2 = c - b$ and $\omega_i' = (Q_i; \gamma_0)$, $i = 0, 1, 2, 3$, in place of ω_i, and remarking that $\Delta^i \omega_0' = \gamma_0 \Delta^i Q_0$, $i = 1, 2, 3$, we obtain the result. ∎

Application 6.2. *Given $BR[\omega_0, \omega_1, \ldots, \omega_q; [a, b]]$ with $\Pi(\omega_q) \neq \Pi(\omega_{q-1})$, determine $BR[\theta_0, \theta_1, \ldots, \theta_p; [b, c]]$ such that*

$$BR[\omega; [a, b]] \cong BR[\theta; [b, c]] \quad (G^3; b).$$

The massic polygon θ is obtained by the following steps:

1) Carry out steps 1 and 2 as for Application 6.1 to get $P_3, \Delta P_2, \Delta^2 P_1, \Delta^3 P_0$.

2) Let constants $\beta_1 > 0, \beta_2, \beta_3$ be given arbitrarily. Use Lemma 3.2 with $n = p = 3$ to determine $Q_0, \Delta Q_0, \Delta^2 Q_0, \Delta^3 Q_0 : Q_0 = P_3$, $\Delta Q_0 = \beta_1 \Delta P_2$, $2\Delta^2 Q_0 = \beta_2 \Delta P_2 + 2\beta_1^2 \Delta^2 P_1$, $2\Delta^3 Q_0 = \beta_3 \Delta P_2 + 6\beta_1 \beta_2 \Delta^2 P_1 + 2\beta_1^3 \Delta^3 P_0$.

3) $\theta_0 = (Q_0; \gamma_0)$, γ_0 an arbitrary constant, determine $\theta_1, \theta_2, \theta_3$ following (6) with $\lambda_1', \lambda_2', \lambda_3'$ arbitrary, θ_i, $i \geq 4$ are arbitrary too.

In step 3, a particular solution is obtained with $\theta_0 = (Q_0; \gamma_0)$, γ_0 arbitrary, $\theta_i = (S_i; \gamma_0)$, $i = 1, 2, 3$, S_1, S_2, S_3 being determined by $p\Delta S_0 = 3\Delta Q_0$, $\binom{p}{2}\Delta^2 S_0 = 3\Delta^2 Q_0$, $\binom{p}{3}\Delta^3 S_0 = \Delta^3 Q_0$. It suffices to take $\lambda_1' = \lambda_2' = \lambda_3' = 0$ in (6) which gives $\theta_0 = (Q_0; \gamma_0), p\Delta\theta_0 = 3\gamma_0\Delta Q_0$, $\binom{p}{2}\Delta^2\theta_0 = 3\gamma_0\Delta^2 Q_0$, $\binom{p}{3}\Delta^3\theta_0 = \gamma_0\Delta^3 Q_0$. By passing to the masses in each previous equality we deduce $\chi(\theta_i) = \gamma_0$, $i = 1, 2, 3$. Then $\theta_i = (S_i; \gamma_0)$, $i = 1, 2, 3$ and consequently by $\Delta^i\theta_0 = \gamma_0\Delta^i S_0$, $i = 1, 2, 3$, yields the result.

Remark 6.1. The use of (6) in step 3 above avoids solving the triangular linear system (3).

References

1. Barsky B. A., The beta-spline : a local representation based on shape parameters and fundamental geometric measures, dissertation, Utah University, 1981.

2. Boehm W., Rational geometric splines, Comp. Aided Geom. Design **4** (1987), 67–77.

3. Canonne J. C., J. C. Fiorot and P. Jeannin, Une condition nécessaire et suffisante de raccordement C^k de courbes rationnelles, C.R. Acad. Sci. Paris **312**, Série I (1991), 171–176.

4. Dyn N. and C. A. Micchelli, Piecewise polynomial spaces and geometric continuity of curves, Numerische Mathematik **54** (1988), 319–337.

5. Fiorot J. C. and P. Jeannin, Courbes Bézier rationnelles, XIXème Congrès National d'Analyse Numérique, Port-Barcarès, France, 26-30 Mai 1986.

6. Fiorot J. C. and P. Jeannin, Nouvelle description et calcul des courbes rationnelles à l'aide de points et vecteurs de contrôle, C.R. Acad. Sci. Paris **305**, Série I (1987), 435–440.

7. Fiorot J. C. and P. Jeannin, *Courbes et Surfaces Rationnelles, Applications à la CAO* , RMA 12, Masson, Paris, 1989.
8. Fiorot J. C. and P. Jeannin, *Rational Curves and Surfaces, Applications to CAD*, J. Wiley and sons, Chichester, 1992.
9. Fiorot J. C. and P. Jeannin, *Courbes Splines Rationnelles, Applications à la CAO*, RMA 24, Masson, Paris, 1992.
10. Goodman T. N. T., Constructing piecewise rational curves with Frenet frame continuity, Comp. Aided Geom. Design **7** (1990), 15–31.
11. Goodman T. N. T., Joining rational curves smoothly, Comp. Aided Geom. Design **8** (1991), 443–464.
12. Gregory J. A., Geometric continuity, in *Mathematical Methods in Computer Aided Geometric Design*, T. Lyche and L. L. Schumaker (eds.) Academic Press, Boston, 1989, 353–371.
13. Hagen H., Bézier curves with curvature and torsion continuity, Rocky Mountain J. Math. **16** (1986), 629–638.
14. Hohmeyer M. E. and B. A. Barsky , Rational continuity : parametric, geometric, and Frenet frame continuity of rational curves, ACM Trans. Graph. **8** (1989), 335–359.
15. Joe B, Multiple-knot and rational cubic beta-splines, ACM Trans. Graph. **8** (1989), 100–120.
16. Lasser D. and A. Purucker, B-spline Bézier representation of rational geometric spline curves, quartics and quintics, in *Nurbs for Curve and Surface Design*, G. Farin (ed.), SIAM publications, Philadelphia, 1991, 115–130.
17. Pottmann H., Projectively invariant classes of geometric continuity for CAGD, Comp. Aided Geom., Design **6** (1989), 307–321.
18. Pottmann H., A projective algorithm for curvature continuous rational splines in *Nurbs for Curve and Surface Design*, G. Farin (ed.), SIAM publications, Philadephia, 1991, 141–148.
19. Seroussi G. and B. A. Barsky, An explicit derivation of discretely shaped beta-spline basis functions of arbitrary order, in *Mathematical Methods in Computer Aided Geometric Design II*, T. Lyche and L. L. Schumaker (eds.), Academic Press, Boston, 1992, 567–584.

J. C. Fiorot
Université de Valenciennes et du Hainaut-Cambraisis
ENSIMEV, Laboratoire IMAV B.P. 311
F-59304 Valenciennes - cedex, FRANCE

P. Jeannin
Département de Mathématiques Appliquées
Université du Littoral
Bâtiment Henri Poincaré, B.P 699
62228 Calais cedex, FRANCE

Generalizations of Bézier Curves and Surfaces

I. Gânscă, Gh. Coman, and L. Ţâmbulea

Abstract. A Bézier curve (or surface) is defined by using the Bernstein basis and a set of points, so its shape depends only on the point positions. In this paper we generalize the Bernstein basis in order to obtain more flexible curves and surfaces corresponding to a set of fixed points.

§1. Introduction

A Bézier curve associated to $n+1$ given points $b_k \in \mathbb{R}^3$, $k = \overline{0,n}$, is represented by the following vectorial equation in terms of the Bernstein basis:

$$B(t) = \sum_{k=0}^{n} b_{n,k}(t) b_k, \qquad (1.1)$$

$$b_{n,k}(t) = \binom{n}{k}(1-t)^{n-k} t^k, \quad k = \overline{0,n}, \quad t \in [0,1]. \qquad (1.2)$$

A Bézier surface has the equation

$$B(u,v) = \sum_{i=0}^{m} \sum_{j=0}^{n} b_{m,i}(u) b_{n,j}(v) b_{ij}, \quad (u,v) \in [0,1] \times [0,1], \qquad (1.3)$$

and $b_{ij} \in \mathbb{R}^3$; $i = \overline{0,m}$; $j = \overline{0,n}$ are $(m+1)(n+1)$ given points.

In this paper we investigate in more detail the generalizations of Bézier curves and surfaces which were introduced in the papers [3,4,5] in order to create flexible curves and surfaces without changing control points.

§2. First Generalization of Bézier Curves

In [3] Bézier curves were generalized by replacing the Bernstein basis by the following new basis, created by D. D. Stancu in [7] p.213,

$$w_{n,k}(t;r) = \begin{cases} \binom{n-r}{k} t^k (1-t)^{n-r-k+1}, & 0 \le k < r \\ \binom{n-r}{k} t^k (1-t)^{n-r-k+1} + \\ \quad + \binom{n-r}{k-r} t^{k-r+1}(1-t)^{n-k}, & r \le k \le n-r \\ \binom{n-r}{k-r} t^{k-r+1}(1-t)^{n-k}, & n-r < k \le n \end{cases} \qquad (2.1)$$

Curves and Surfaces in Geometric Design
P. J. Laurent, A. Le Méhauté, and L. L. Schumaker (eds.), pp. 169–176.
Copyright © 1994 by A K PETERS, Wellesley, MA
ISBN 1-56881-039-3.

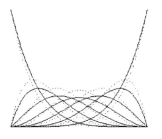
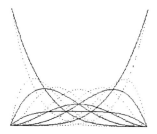

Figure 1.   Figure 2.

where r is a nonnegative integer which satisfies condition $2r \leq n + 1$. One easily checks that

$$w_{n,k}(t; 0) = w_{n,k}(t; 1) = b_{n,k}(t), \quad k = \overline{0, n}, \quad t \in [0, 1].$$

Now, we remark that the functions $w_{n,k}$ can be written in the form

$$w_{n,k}(t; r) = \begin{cases} (1 - t)b_{n-r,k}(t), & 0 \leq k < r \\ (1 - t)b_{n-r,k}(t) + t\, b_{n-r,k-r}(t), & r \leq k \leq n - r \\ t\, b_{n-r,k-r}(t), & n - r < k \leq n. \end{cases} \tag{2.2}$$

From (2.2) one deduces that $w_{n,k}$ have the same properties as the Bernstein functions $b_{n,k}$ have, that is

$$a) \quad w_{n,k}(t; r) \geq 0, \quad t \in [0, 1]; \quad k = \overline{0, n}$$

$$b) \quad \sum_{k=0}^{n} w_{n,k}(t; r) = 1. \tag{2.3}$$

The first property is evident if we use (2.2) and the fact that $b_{n,k} \geq 0$. For the proof of the second property, taking account that $\sum_{j=0}^{n-r} b_{n-r,j}(t) = 1$, we have successively

$$\sum_{k=0}^{n} w_{n,k}(t; r) = (1 - t) \sum_{k=0}^{n-r} b_{n-r,k}(t) + t \sum_{k=r}^{n} b_{n-r,k-r}(t) =$$

$$= (1 - t) \sum_{k=0}^{n-r} b_{n-r,k}(t) + t \sum_{i=0}^{n-r} b_{n-r,i}(t) = 1.$$

These properties ensure that the corresponding generalized Bézier curve lies within the convex hull of points $b_i, i = \overline{0, n}$ (property a) and is invariant under any affine transformation (property b).

Figure 1 shows the graphs of the polynomials $w_{6,k}(t; 2)$ and $b_{6,k}(t)$ (dotted curves), $k = \overline{0, 6}$, $t \in [0, 1]$, and Figure 2 presents the polynomials $w_{6,k}(t; 3)$ and $b_{6,k}(t)$.

By (2.3), the generalized Bézier curve corresponding to points b_k, $k = \overline{0, n}$ and basis $w_{n,k}$, $k = \overline{0, n}$, can be represented in the following form

$$S(t; r) = \sum_{k=0}^{n} w_{n,k}(t; r) b_k = \sum_{k=0}^{n-r} b_{n-r,k}(t)[(1 - t)b_k + t\, b_{r+k}], \quad t \in [0, 1]. \quad (2.4)$$

One observes that the degree of $S(t; r)$ is $n - r + 1$ and $S(t; 0) = S(t; 1) = B(t)$.

From (2.4) we deduce that the de Casteljau algorithm is valid, but for starting points $\overline{b_k}(t) = (1 - t)b_k + t\, b_{r+k}$, $k = \overline{0, n - r}$, $t \in [0, 1]$. For a given value of $t \in [0, 1]$, we obtain an intermediate de Casteljau point by using formula

$$b_j^{[p]} = \sum_{k=0}^{p} \overline{b}_{j+k}(t) b_{p,k} = \sum_{k=0}^{p} [(1 - t)b_{j+k} + t\, b_{r+j+k}] b_{p,k}(t),$$
$$p = \overline{0, n - r}, \quad j = \overline{0, n - r - p}. \quad (2.5)$$

In the special case, $r = 1$, the points $\overline{b}_j$, $j = \overline{0, n - 1}$, coincide with the first intermediate de Casteljau points $b_j^{(1)}(t)$, $j = \overline{0, n - 1}$, see [2], p.8.

If $p = n - r$ then the corresponding de Casteljau point is

$$b_0^{(n-r)}(t) = \sum_{k=0}^{n-r} [(1 - t)b_k + t\, b_{r+k}] b_{n-r,k}(t),$$

that is $b_0^{(n-r)}(t) = S(t; r)$, $t \in [0, 1]$.

In order to obtain the s–th derivative of $S(t; r)$ one uses the Leibniz's formula and formula for the s–th derivative of Bézier curve (see [2], p.9)

$$B^{(s)}(t) = \frac{n!}{(n - s)!} \sum_{k=0}^{n-s} b_{n-s,k}(t) \Delta^s b_k, \quad (2.6)$$

where

$$\Delta^s b_k = \sum_{j=0}^{s} (-1)^{s-j} \binom{s}{j} b_{k+j}. \quad (2.7)$$

Finally we get

$$S^{(s)}(t; r) = \frac{(n - r)!}{(n - r - s)!} \left\{ \sum_{j=0}^{n-r-s} b_{n-r-s,j}(t)[(1 - t)\Delta^s b_j + t\, \Delta^s b_{r+j}] + \right.$$
$$\left. + \frac{s}{n - r - s + 1} \sum_{j=0}^{n-r-s+1} b_{n-r-s+1,j}(t)(\Delta^{s-1} b_{r+j} - \Delta^{s-1} b_j) \right\}$$
$$(2.8)$$

From here we deduce the particular cases

$$S^{(s)}(0;r) = \frac{(n-r)!}{(n-r-s)!}\left[\Delta^s b_0 + \frac{s}{n-r-s+1}\Delta^{s-1}(b_r - b_0)\right] \qquad (2.9)$$

and

$$S^{(s)}(1;r) =$$
$$\frac{(n-r)!}{(n-r-s)!}\left[\Delta^s b_{n-s} + \frac{s}{n-r-s+1}\Delta^{s-1}(b_{n-s-1} - b_{n-r-s+1})\right].$$
$$(2.10)$$

Taking account of (2.7), these formulas become

$$S^{(s)}(0;r) =$$
$$\frac{(n-r)!}{(n-r-s)!}\left[b_s + s\sum_{k=0}^{s-1}(-1)^{s-k}\binom{s-1}{k}\left(\frac{b_k}{s-k} + \frac{b_k - b_{r+k}}{n-r-s+1}\right)\right]$$
$$(2.11)$$

and

$$S^{(s)}(1;r) =$$
$$\frac{(n-r)!}{(n-r-s)!}\left[(-1)^s b_{n-s} + \sum_{k=0}^{s-1}\binom{s-1}{k}\left(\frac{b_{n-k}}{s-k} + \frac{b_{n-k} - b_{n-r-k}}{n-r-s+1}\right)\right]$$
$$(2.12)$$

For $s = 1$, from (2.11) and (2.12), one deduces

$$S'(0;r) = (n-r)(b_1 - b_0) + b_r - b_0 \qquad (2.13)$$

and

$$S'(1;r) = (n-r)(b_n - b_{n-1}) + b_n - b_{n-r}. \qquad (2.14)$$

A consequence of formula (2.13) is that if the points $b_r, b_1, b_0, b_0 \neq b_1$ are collinear, then the seqment $b_0 b_1$ is tangent to the curve $S(t;r)$ in b_0. Analogously, if the points $b_{n-r}, b_{n-1}, b_n, b_{n-1} \neq b_n$ are collinear, then the segment $b_{n-1} b_n$ is tangent to the curve $S(t;r)$ in b_n (in view of formula (2.14)).

Figure 3 shows the graphs of Bézier $B(t)$ (dotted curve) and generalized Bézier curves $S(t;2)$ corresponding to the points

$$b_0(-1,3); \ b_1(-1,6); \ b_2(4,6); \ b_3(4,3); \ b_4(0,2); \ b_5(0,-1). \qquad (2.15)$$

In Figure 4 we present graphs of the curves $B(t)$ (dotted curve), $S(t;2)$ and $S(t;3)$ associated with the set of points

$$b_0(-1,3); \ b_1(-1,6); \ b_2(4,6); \ b_3(4,3); \ b_4(0,2) = b_5; \ b_6(0,-1). \qquad (2.16)$$

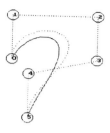

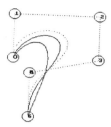

Figure 3.   Figure 4

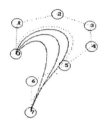

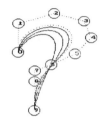

Figure 5.   Figure 6

From the above figures one observes that the generalized Bézier curves $S(t;r)$ follow the control polygon less closely than the Bézier curve does, so we will make some successive degree elevations, which affect only the curve $S(t;r)$; the Bézier curve remains unchanged (see [4], p.9). In this case, the equation of $S(t;r)$ is

$$S(t;r) = \sum_{k=0}^{n+p} w_{n+p,k}(t;r)b_k^{(p)}, \qquad (2.17)$$

where

$$b_k^{(p)} = \frac{k}{n+p}b_{k-1}^{(p-1)} + \left(1 - \frac{k}{n+p}\right)b_k^{(p-1)}, \quad p = 1,2,\ldots; \quad k = \overline{0,n+p}$$
$$(2.18)$$

and $b_k^0 = b_k$. Figure 5 shows the curves $S(t;r), r = 2,3,4, \ p = 2$ with starting points (2.15). Figure 6 presents $S(t;r), r = 2,3,4, p = 3$ with the starting points (2.16) (The dotted curves are Bézier curves).

§3. Second Generalization of Bézier Curves

As another generalization of Bézier curves, we introduce a real and positive parameter α in the basis functions (2.2) as follows:

$$w_{n,k}(t;r,\alpha) = \begin{cases} (1 - t^\alpha)b_{n-r,k}(t), & 0 \le k < r \\ (1 - t^\alpha)b_{n-r,k}(t) + t^\alpha b_{n-r,k-r}(t), & r \le k \le n-r \\ t^\alpha b_{n-r,k-r}(t), & n-r < k \le n. \end{cases} \qquad (3.1)$$

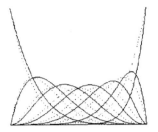   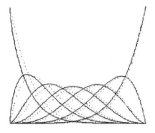

Figure 7.   Figure 8

Remarks

1. The integer parameter r, in this case, takes values in the interval $[1, \frac{n+1}{2}]$.

2. One easily checks that $w_{n,k}(\cdot; r, \alpha)$, $k = \overline{0, n}$, have the properties (2.3).

Figures 7 and 8 show the graphs of the functions $w_{n,k}(t; r, \alpha)$, $t \in [0, 1]$, $k = \overline{0, 6}$ for particular values of the parameters r and α. The new generalized Bézier curve has the equation

$$
\begin{aligned}
G(t; r, \alpha) &= \sum_{k=0}^{n} w_{n,k}(t; r, \alpha) b_k = \\
&= \sum_{k=0}^{n-r} b_{n-r,k}(t) \left[(1 - t^\alpha) b_k + t^\alpha b_{r+k} \right], \quad t \in [0, 1]
\end{aligned}
\tag{3.2}
$$

and the integer $r \in [1, \frac{n+1}{2}]$.

For the derivatives we write (3.2) in the form

$$
G(t; r, \alpha) = \sum_{k=0}^{n-r} b_{n-r,k}(t) b_k + t^\alpha \sum_{k=0}^{n-r} b_{n-r,k}(t) \left(b_{r+k} - b_k \right)
$$

and using (2.6) and the Leibniz's formula one obtains

$$
\begin{aligned}
G^{(p)}(t; r, \alpha) &= \frac{(n-r)!}{(n-r-p)!} \sum_{k=0}^{n-r-p} b_{n-r-p,k}(t) \Delta^p b_k + \\
&+ \sum_{i=0}^{p} \binom{p}{i} (t^\alpha)^{(i)} \frac{(n-r)!}{(n-r-p+i)!} \sum_{k=0}^{n-r-p+i} b_{n-r-p+i,k}(t) \Delta^{p-i} \left(b_{r+k} - b_k \right).
\end{aligned}
\tag{3.3}
$$

In the special cases $t = 0$ and $t = 1$ we have

$$
G^{(p)}(0; r, \alpha) =
$$

$$
= \begin{cases}
\frac{(n-r)!}{(n-r-p)!} \Delta^p b_0, & \alpha > p \geq 0 \\
\frac{(n-r)!}{(n-r-p)!} \Delta^p b_0 + \frac{p!}{(p-\alpha)!} \frac{(n-r)!}{(n-r-p+\alpha)!} \Delta^{p-\alpha} (b_r - b_0), & \alpha \in \mathbb{N}, 1 < \alpha \leq p \\
\text{is collinear with } b_r - b_0, & 0 < \alpha < 1
\end{cases}
\tag{3.4}
$$

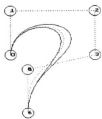

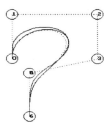

Figure 9.   Figure 10.

and

$$G^{(p)}(1; r, \alpha) = \frac{(n-r)!}{(n-r-p)!} \Delta^p b_{n-r-p} +$$

$$+ \sum_{i=0}^{p} \alpha(\alpha-1)\dots(\alpha-i+1)\frac{(n-r)!}{(n-r-p+i)!} \Delta^{p-i} \left(b_{n-p+i} - b_{n-r-p+i}\right),.$$

for all $\alpha > 0$.

Figures 9 and 10 present the curves $G(t; 2, \alpha)$, ($\alpha = 2$, $\alpha = 8$) and $G(t; 1, \alpha)$, ($\alpha = 0.25$, $\alpha = 9$) respectively, corresponding to control points (2.16); the Bézier curves are dotted.

§4. Generalizations of Bézier Surfaces

Extending the ideas from Bézier curves to Bézier surface (1.3) we are able to control the shape of a surface corresponding to the fixed points $b_{ij} \in \mathbb{R}^3$, $i = \overline{0, m}$, $j = \overline{0, n}$.

The first generalization of a Bézier surface has been made in [4] by replacing the Bernstein functions $b_{p,k}(\cdot)$ with the functions $w_{p,k}(\cdot, q)$; $p = m$, $q = r$ and $p = n$, $q = s$ respectively, so the equation (1.3) becomes

$$S(u, v; r, s) = \sum_{i=0}^{m-r} \sum_{j=0}^{n-s} b_{m-r,i}(u) b_{n-s,j}(v) \left[b_{ij} + u \left(b_{i+r,j} - b_{ij} \right) + \right.$$

$$\left. + v \left(b_{i,j+s} - b_{ij} \right) + uv \left(b_{i+r,j+s} - b_{i,j+s} - b_{i+r,j} + b_{ij} \right) \right], \tag{4.1}$$

$(u, v) \in [0, 1] \times [0, 1]$ and non-negative integers r, s satisfying the conditions $2r \leq m + 1$ and $2s \leq n + 1$. For $(r, s) \in \{0, 1\} \times \{0, 1\}$ we have $S(u, v; r, s) = B(u, b)$, that is (4.1) generalizes (1.3).

Another generalization of the Bézier surface was made in [5], using the basis of the form (3.1). The vectorial equation in this case is

$$G(u, v; r, s; \alpha, \beta) = \sum_{i=0}^{m-r} \sum_{j=0}^{n-s} b_{m-r,i}(u) b_{n-r,j}(v) [b_{ij} + +u^\alpha (b_{i+r,j} - b_{ij}) +$$

$$+ v^\beta (b_{i,j+s} - b_{ij}) + u^\alpha v^\beta ((b_{i+r,j+s} - b_{i,j+s} - b_{i+r,j} + b_{ij})], \tag{4.2}$$

$(u, v) \in [0,1] \times [0,1]$ and $\alpha,\ \beta \in (0, \infty)$.

One observes that $G(\cdot, \cdot; r, s; 1, 1) = S(\cdot, \cdot; r, s)$. The dependence of the shape of the generalized Bézier surfaces (4.1) and (4.2) by the parameters r, s and α, β respectively is illustrated in figures 4.1-4.4, corresponding to the following set of points $b_{i,j}$; $i = \overline{0,5}$, $j = \overline{0,6}$.

$$
\begin{array}{ccccccc}
(0,0,2) & (0,2,1) & (0,3,3) & (0,5,4) & (0,9,3) & (0,9,3) & (0,6,2) \\
(2,0,2) & (3,2,2) & (2,3,3) & (1,4,4) & (1,5,3) & (1,5,3) & (2,5,3) \\
(3,0,2) & (4,2,0) & (4,3,0) & (4,5,2) & (3,6,2) & (3,6,2) & (3,7,1) \\
(5,0,2) & (5,2,3) & (6,4,1) & (6,6,1) & (6,8,5) & (6,8,5) & (4,7,3) \\
(6,0,2) & (6,1,5) & (6,3,0) & (7,5,0) & (5,4,3) & (5,4,3) & (6,6,3) \\
(8,0,2) & (6,2,2) & (6,4,0) & (9,4,2) & (9,5,4) & (9,5,4) & (7,7,4)
\end{array}
$$

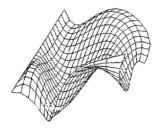 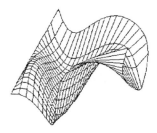

Figure 11. G(u,v;1,1;1,1).   Figure 12. G(u,v;2,1;0.25,8)

References

1. Barnhill R. R., Representation and approximation of surfaces, in *Math. Software III*, Academic Press, New York, 1977.
2. Boehm W., G. Farin, J. Kahmann, A survey of curves and surfaces methods in CAGD, Computer Aided Geometric Design **1** (1984), 1–60.
3. Gânscă I., A generalization of Bézier curves, in Babes-Bolyai University, Faculty of Mathematics, Research Seminaries, Seminar on Numerical and Statistical Calculus, Preprint no.9, 1987, 69–72.
4. Gânscă I., A generalization of Bézier surfaces, Studia Universitatis Babeş-Bolyai, Mathematica **34**, 4 (1989), 40–44.
5. Gânscă I., Gh. Coman, L. Ţâmbulea, On the shape of Bézier surfaces, Studia Universitatis Babeş-Bolyai, Mathematica **35**, 3 (1990), 37–42.
6. Riesenfeld R. F., Applications of B-spline Approximation to Geometric Problems of Computer-Aided Design, Computer Science, University of Utah, 1973.
7. Stancu D. D., Approximation of functions by means of a new generalized Bernstein operator, Calcolo **XX**, II (1983), 211–229.

Corner Cutting Algorithms and
Totally Positive Matrices

M. Gasca and J. M. Peña

Abstract. A totally positive, nonsingular, stochastic matrix of order $n + 1$ can be factored as the product of n lower triangular and n upper triangular matrices, all them bidiagonal, totally positive and stochastic. The geometric interpretation of this factorization is a corner cutting algorithm which transforms a control polygon of a curve into another control polygon of the same curve related to a different basis. Here we prove that there exist several different factorizations of this type and that some conditions on the ordering and form of the factors should be added in order to get uniqueness.

§1. Introduction

A matrix is totally positive (TP) if all its minors are nonnegative. If they are strictly positive the matrix is called strictly totally positive (STP). Properties of these matrices can be found, for example, in [10,1] and some recent papers by the authors ([3–6] among others).

We say that a sequence of functions $(u_0, u_1, \ldots, u_n)$ is *totally positive* on an interval I of $\mathbb{R}$ if for any $t_0 < t_1 < \cdots < t_m$ in I the collocation matrix $(u_j(t_i))_{0 \le i \le m, 0 \le j \le n}$ is TP. If in addition one has

$$\sum_{i=0}^{n} u_i(t) = 1, \qquad \forall t \in I,$$

the sequence is said to be *normalised totally positive* (NTP). For an NTP sequence of linearly independent functions $(u_0, u_1, \ldots, u_n)$ on I and a sequence of points $(P_0, P_1, \ldots, P_n)$ in $\mathbb{R}^k$, we may define a curve

$$\gamma(t) = \sum_{i=0}^{n} P_i u_i(t), \qquad t \in I. \tag{1}$$

Curves and Surfaces in Geometric Design   177
P. J. Laurent, A. Le Méhauté, and L. L. Schumaker (eds.), pp. 177–184.
Copyright ⓒ 1994 by A K PETERS, Wellesley, MA
ISBN 1-56881-039-3.

The functions u_i are usually called *blending functions* and the points P_i *control points*. The polygonal arc $P_0 P_1 \cdots P_n$ will be referred to as the *control polygon* of the curve. This construction is relevant in CAGD because the curve γ preserves many shape properties of the control polygon. See for example [7,9,2]. The most well-known example of NTP sequences is the Bernstein basis of polynomials of degree not greater than n on the interval $[0,1]$

$$u_i(t) = \binom{n}{i} t^i (1-t)^{n-i}, \qquad 0 \le i \le n.$$

The curve γ is said to be, in this case, a *Bézier curve* and the control polygon the *Bézier polygon*.

A matrix with nonnegative entries is called *stochastic* if the row sums are equal to one. Matrices which are nonsingular, stochastic and totally positive are of particular interest for us. Observe that a nonsingular, totally positive matrix $A = (a_{ij})_{0 \le i,j \le n}$ can be factored in the form

$$A = DB \qquad (2)$$

with $D = (d_{ij})_{0 \le i,j \le n}$ a diagonal matrix with positive diagonal entries $d_{ii} = 1/(\sum_{i=0}^n a_{ij})$ and B a nonsingular, totally positive, stochastic matrix. The factorization is obviously unique. By using (2) for each factor, a product of nonsingular, totally positive matrices $A_1, A_2, \ldots, A_n$ can be written

$$A_1 A_2 \cdots A_n = D B_1 B_2 \cdots B_n \qquad (3)$$

with $B_1, B_2, \ldots, B_n$ totally positive and stochastic and D a diagonal matrix with positive diagonal.

An *elementary corner cutting* is a transformation which maps any polygon $P_0 P_1 \cdots P_n$ into another one $Q_0 Q_1 \cdots Q_n$ defined by

$$\begin{aligned} Q_j &= P_j \quad j \neq i \\ Q_i &= \lambda P_i + (1-\lambda) P_{i+1} \end{aligned} \qquad (4)$$

for some $0 \le i \le n-1$ or

$$\begin{aligned} Q_j &= P_j \quad j \neq i \\ Q_i &= \lambda P_i + (1-\lambda) P_{i-1} \end{aligned} \qquad (5)$$

for some $1 \le i \le n$.

A *corner cutting algorithm* is any composition of elementary corner cutting tranformations. An elementary corner cutting is defined by a bidiagonal, nonsingular, totally positive and stochastic matrix, which is upper (respectively lower) triangular in the case (4) (resp. (5)). A corner cutting algorithm is described by a matrix which is nonsingular, totally positive and stochastic, as a product of the previous bidiagonal matrices.

The following proposition is straightforward taking into account that a nonnegative matrix A is stochastic if and only if its product by $(1, 1, \ldots, 1)^T$ is again $(1, 1, \ldots, 1)^T$.

Proposition 1.1. Let $(u_0, u_1, \ldots, u_n)$ be a totally positive basis of a space U and $(v_0, v_1, \ldots, v_n)$ another basis of U. Let $M = (m_{ij})_{0 \leq i,j \leq n}$ be the nonsingular matrix such that

$$(v_0, v_1, \ldots, v_n) = (u_0, u_1, \ldots, u_n)M. \tag{6}$$

If M is totally positive, then $(v_0, v_1, \ldots, v_n)$ is a TP basis of U. If the basis $(u_0, u_1, \ldots, u_n)$ is NTP and M is totally positive and stochastic, then $(v_0, v_1, \ldots, v_n)$ is NTP.

Let $\gamma(t)$ be a curve defined by (1). If $(v_0, v_1, \ldots, v_n)$ is a basis of U related to $(u_0, u_1, \ldots, u_n)$ by (5), we have

$$\gamma(t) = \sum_{i=0}^{n} P_i u_i(t) = \sum_{i=0}^{n} Q_i v_i(t) \tag{7}$$

with

$$\begin{pmatrix} P_0 \\ P_1 \\ \vdots \\ P_n \end{pmatrix} = M \begin{pmatrix} Q_0 \\ Q_1 \\ \vdots \\ Q_n \end{pmatrix}. \tag{8}$$

By Proposition 1.1, if M is totally positive and stochastic and $(u_0, u_1, \ldots, u_n)$ is NTP then $(v_0, v_1, \ldots, v_n)$ is NTP. In other words, if the control polygon $P_0 P_1 \cdots P_n$ corresponding to an NTP basis can be obtained by a corner cutting algorithm from another polygon $Q_0 Q_1 \cdots Q_n$, as in (8), then the basis $(v_0, v_1, \ldots, v_n)$ of U defined by (6) is NTP. Some years ago, Goodman and Micchelli ([8], Theorem 1) proved that any nonsingular, totally positive and stochastic matrix can be factored in such a way that it describes a corner cutting algorithm. The relevance of corner cutting algorithms with respect to the geometrical optimality of NTP bases can be seen, for example, in [2,9]. Our aim is to prove that there exist several similar factorizations and that some condition on the ordering and form of the factors should be added in order to get uniqueness.

§2. Decomposition of a Totally Positive Matrix

The following theorem is a reformulation of some results of [6].

Theorem 2.1. *A nonsingular matrix M of order $n + 1$ is totally positive if and only if it can be factored in the form*

$$M = D L_{n-1} L_{n-2} \cdots L_0 K_0 K_1 \cdots K_{n-1}, \tag{9}$$

where D is a diagonal matrix with positive diagonal entries, and where for

$l = 0, 1, \ldots, n - 1$, L_l, K_l are the bidiagonal matrices

$$
L_l = \begin{pmatrix}
1 & & & & & & \\
0 & 1 & & & & & \\
& & \ddots & \ddots & & & \\
& & & 0 & 1 & & \\
& & & & \lambda_{l+1}^{(l)} & 1 - \lambda_{l+1}^{(l)} & \\
& & & & & \ddots & \ddots \\
& & & & & & \lambda_n^{(l)} & 1 - \lambda_n^{(l)}
\end{pmatrix}
\tag{10}
$$

$$
K_l = \begin{pmatrix}
1 & 0 & & & & & \\
& \ddots & \ddots & & & & \\
& & 1 & 0 & & & \\
& & & 1 - \mu_l^{(l)} & \mu_l^{(l)} & & \\
& & & & \ddots & \ddots & \\
& & & & & 1 - \mu_{n-1}^{(l)} & \mu_{n-1}^{(l)} \\
& & & & & & 1
\end{pmatrix}
\tag{11}
$$

with $0 \leq \lambda_j^{(l)}, \mu_j^{(l)} \leq 1$ and

$$
\begin{aligned}
\lambda_r^{(l)} = 0 \quad (r > l) &\Longrightarrow \lambda_t^{(l)} = 0 \quad \forall t > r, \\
\mu_s^{(l)} = 0 \quad (s > l) &\Longrightarrow \mu_t^{(i)} = 0 \quad \forall t > s.
\end{aligned}
\tag{12}
$$

Moreover this factorization is unique.

Proof: Let us compare this decomposition with that of Theorem 4.1' of [6]. Here the matrices L_l, K_l are stochastic, while in [6] they were unit diagonal. Moreover the diagonal matrix D is the first factor of the decomposition (9), while in [6] it was placed between L_0 and K_0. Nevertheless these differences are not essential, and the proof of Theorem 4.1' can be easily adapted to the present conditions taking into account (3) to transform one of the decompositions into the other one.

The uniqueness of the decomposition (9) under the conditions of the theorem follows from the uniqueness of the right-hand side of (3) for prescribed matrices $A_1 A_2 \cdots A_n$ and from the uniqueness of the decomposition of Theorem 4.1' of [6] (see Remark 4.6 of that paper). ∎

Remark 2.2. The matrix M of Theorem 2.1 will be stochastic if and only if the matrix D of (9) is the identity matrix.

If M is a nonsingular, stochastic, totally positive matrix, the factorization (9) describes a corner cutting algorithm consisting of $2n$ steps, respectively associated to the matrices $K_{n-1}, K_{n-2}, \ldots, K_0, L_0, L_1, \ldots, L_{n-1}$. The matrix $K_l(n - 1 \geq l \geq 0)$, when applied to a control polygon with $n + 1$ vertices leaves invariant at least the vertices labelled with $0, 1, \ldots, l - 1$ and n. The matrix $L_l (0 \le l \le n-1)$ leaves invariant at least the vertices labelled with $0, 1, \ldots l$. Condition (12) means that if the matrix K_l (respectively L_l) is not the identity matrix, it cuts some *consecutive* vertices, starting with the one labelled with l (resp. $l+1$).

In summary, this corner cutting algorithm cuts at most i vertices in the ith step for $1 \le i \le n$ and at most $2n + 1 - i$ vertices in the ith step for $n + 1 \le i \le 2n$. This is not the usual form of corner cutting algorithms, because for example in the case of an upper triangular matrix M the number of changing vertices increases at each step. In the next theorem we prove in a different way the characterization of nonsingular totally positive matrices obtained in [8] in terms of factorizations similar to (9), and the uniqueness of such factorizations under conditions analogous to (12).

Theorem 2.3. *A nonsingular matrix M of order $n + 1$ is totally positive if and only if it can be factored in the form*

$$M = D L_{n-1} L_{n-2} \cdots L_0 U_{n-1} U_{n-2} \cdots U_0, \tag{13}$$

where D is a diagonal matrix with positive diagonal and, for $l = 0, 1, \ldots, n-1$,

$$L_l = \begin{pmatrix} 1 & & & & & & \\ 0 & 1 & & & & & \\ & & \ddots & \ddots & & & \\ & & 0 & 1 & & & \\ & & & \lambda_{l+1}^{(l)} & 1 - \lambda_{l+1}^{(l)} & & \\ & & & & \ddots & \ddots & \\ & & & & & \lambda_n^{(l)} & 1 - \lambda_n^{(l)} \end{pmatrix} \tag{14}$$

$$U_l = \begin{pmatrix} 1 - \beta_0^{(l)} & \beta_0^{(l)} & & & & & \\ & \ddots & \ddots & & & & \\ & & 1 - \beta_{n-l-1}^{(l)} & \beta_{n-l-1}^{(l)} & & & \\ & & & 1 & 0 & & \\ & & & & \ddots & \ddots & \\ & & & & & & 1 \end{pmatrix} \tag{15}$$

with $0 \le \lambda_j^{(l)}, \beta_j^{(l)} \le 1$ and

$$\begin{aligned} \lambda_r^{(l)} = 0 \quad (r > l) &\Longrightarrow \lambda_t^{(l)} = 0 \quad \forall t > r, \\ \beta_s^{(l)} = 0 \quad (s \le l) &\Longrightarrow \beta_t^{(l)} = 0 \quad \forall t < s. \end{aligned} \tag{16}$$

Moreover this factorization is unique.

Proof: By Theorem 2.1 the nonsingular matrix M is TP if and only if it can be factored in the form

$$M = D L_{n-1} L_{n-2} \cdots L_0 K_0 K_1 \cdots K_{n-1} \tag{17}$$

with (12). Let us define $U = K_0 K_1 \cdots K_{n-1}$. The *converse* $A^\#$ of a matrix $A = (a_{ij})_{0 \le i,j \le n}$ is the matrix whose (i,j) entry $(0 \le i, j \le n)$ is $a_{n-i,n-j}$ (see [6],[1]). It is easy to prove that $(AB)^\# = A^\# B^\#$. The matrix $U^\# = K_0^\# K_1^\# \cdots K_{n-1}^\#$ is lower triangular, nonsingular, TP and stochastic. As in Theorem 2.1 it can be written in the form $U^\# = \tilde{L}_{n-1} \tilde{L}_{n-2} \cdots \tilde{L}_0$, where

$$
\tilde{L}_l = \begin{pmatrix}
1 & & & & & & \\
0 & 1 & & & & & \\
& \ddots & & \ddots & & & \\
& & 0 & & 1 & & \\
& & & & \gamma_{l+1}^{(l)} & 1 - \gamma_{l+1}^{(l)} & \\
& & & & & \ddots & \ddots \\
& & & & & & \gamma_n^{(l)} \quad 1 - \gamma_n^{(l)}
\end{pmatrix}
$$

with $0 \le \gamma_j^{(l)} \le 1$ and

$$
\gamma_s^{(l)} = 0 \quad (s > l) \quad \Longrightarrow \gamma_t^{(l)} = 0 \quad \forall t > s.
$$

If we denote $\tilde{L}_l^\# = U_l$ and $\gamma_j^{(l)} = \beta_{n-j}^{(l)}$ we get

$$
M = D L_{n-1} L_{n-2} \cdots L_0 U_{n-1} U_{n-2} \cdots U_0
$$

with U_l as in (15). Uniqueness follows from Theorem 2.1. ∎

Remark 2.4. *As in Remark 2.2, M is stochastic if and only if $D = I$.*

By using converses we can state other theorems similar to the previous ones providing new factorizations. First we can proceed with the product $L_{n-1} L_{n-2} \cdots L_0$ in Theorems 2.1 and 2.3 as we have done with the product $K_0 K_1 \cdots K_{n-1}$ of Theorem 2.1 to get $U_{n-1} U_{n-2} \cdots U_0$ of Theorem 2.3. The corresponding factorizations are

$$
M = D H_0 H_1 \cdots H_{n-1} K_0 K_1 \cdots K_{n-1} \tag{18}
$$

$$
M = D H_0 H_1 \cdots H_{n-1} U_{n-1} U_{n-2} \cdots U_0, \tag{19}
$$

with

$$
H_l = \begin{pmatrix}
1 & & & & & & \\
\nu_1^{(l)} & 1 - \nu_1^{(l)} & & & & & \\
& \ddots & \ddots & & & & \\
& & \nu_{n-l}^{(l)} & 1 - \nu_{n-l}^{(l)} & & & \\
& & & 0 & 1 & & \\
& & & & & \ddots & \ddots \\
& & & & & & 0 \quad 1
\end{pmatrix}.
$$

On the other hand, we can consider $M^{\#}$ instead of M, apply any of the four preceding factorizations (9),(13),(18) and (19), and take converses again. Then we get four similar factorizations with the roles of the lower and upper triangular matrices interchanged:

$$M = \hat{D}\hat{U}_{n-1}\hat{U}_{n-2}\cdots\hat{U}_0\hat{H}_0\hat{H}_1\cdots\hat{H}_{n-1},$$

$$M = \hat{D}\hat{U}_{n-1}\hat{U}_{n-2}\cdots\hat{U}_0\hat{L}_{n-1}\cdots\hat{L}_1\hat{L}_0,$$

$$M = \hat{D}\hat{K}_0\hat{K}_1\cdots\hat{K}_{n-1}\hat{H}_0\hat{H}_1\cdots\hat{H}_{n-1},$$

$$M = \hat{D}\hat{K}_0\hat{K}_1\cdots\hat{K}_{n-1}\hat{L}_{n-1}\cdots\hat{L}_1\hat{L}_0.$$

§3. Example

As an example, let us interprete de Casteljau's algorithm as a corner cutting algorithm transforming the control polygon $P_0 P_1 \cdots P_n$ of a curve for the Bézier basis on $[0,1]$ into the control polygon $Q_0 Q_1 \cdots Q_n$ of the curve for the Bézier basis on $[t_0, 1]$. For $n = 3$ we have

$$\begin{pmatrix} Q_0 \\ Q_1 \\ Q_2 \\ Q_3 \end{pmatrix} = \begin{pmatrix} 1-t_0 & t_0 & 0 & 0 \\ 0 & 1 & 0 & 0 \\ 0 & 0 & 1 & 0 \\ 0 & 0 & 0 & 1 \end{pmatrix} \begin{pmatrix} 1-t_0 & t_0 & 0 & 0 \\ 0 & 1-t_0 & t_0 & 0 \\ 0 & 0 & 1 & 0 \\ 0 & 0 & 0 & 1 \end{pmatrix}$$

$$\times \begin{pmatrix} 1-t_0 & t_0 & 0 & 0 \\ 0 & 1-t_0 & t_0 & 0 \\ 0 & 0 & 1-t_0 & t_0 \\ 0 & 0 & 0 & 1 \end{pmatrix} \begin{pmatrix} P_0 \\ P_1 \\ P_2 \\ P_3 \end{pmatrix},$$

that is

$$\begin{pmatrix} Q_0 \\ Q_1 \\ Q_2 \\ Q_3 \end{pmatrix} = \begin{pmatrix} (1-t_0)^3 & 3t_0(1-t_0)^2 & 3t_0^2(1-t_0) & t_0^3 \\ 0 & (1-t_0)^2 & 2t_0(1-t_0) & t_0^2 \\ 0 & 0 & 1-t_0 & t_0 \\ 0 & 0 & 0 & 1 \end{pmatrix} \begin{pmatrix} P_0 \\ P_1 \\ P_2 \\ P_3 \end{pmatrix}.$$

According to Theorem 2.1 the same result is obtained with the factorization

$$\begin{pmatrix} Q_0 \\ Q_1 \\ Q_2 \\ Q_3 \end{pmatrix} = \begin{pmatrix} 1-d & d & 0 & 0 \\ 0 & 1-e & e & 0 \\ 0 & 0 & 1-f & f \\ 0 & 0 & 0 & 1 \end{pmatrix} \begin{pmatrix} 1 & 0 & 0 & 0 \\ 0 & 1-b & b & 0 \\ 0 & 0 & 1-c & c \\ 0 & 0 & 0 & 1 \end{pmatrix}$$

$$\times \begin{pmatrix} 1 & 0 & 0 & 0 \\ 0 & 1 & 0 & 0 \\ 0 & 0 & 1-a & a \\ 0 & 0 & 0 & 1 \end{pmatrix} \begin{pmatrix} P_0 \\ P_1 \\ P_2 \\ P_3 \end{pmatrix}$$

with $a = (t_0/3)/(1 - (2t_0/3))$, $b = (t_0 - (2t_0^2/3))/(1 - t_0 + (t_0^2/3))$, $c = (t_0/3)/(1-(t_0/3))$, $d = 1 - (1-t_0)^3$, $e = t_0 - (t_0^3/3)$ and $f = t_0/3$.

Acknowledgements. Both authors were partially supported by DGICYT Research Grant PS90-0121

References

1. Ando, T., Totally positive matrices, Linear Algebra Appl. **90** (1987), 165-219.
2. Carnicer, J.M., and J.M.Peña, Shape preserving representations and optimality of the Bernstein basis, Advances in Computational Mathematics **1** (1993), 173-196.
3. Gasca, M., and J.M.Peña, Total positivity and Neville elimination, Linear Algebra Appl. **165** (1992), 25-44.
4. Gasca, M., and J.M.Peña, On the characterization of totally positive matrices, in *Approximation Theory, spline functions and applications*, S.P.Singh (ed.), Kluwer Pub., 1992, 357-364.
5. Gasca, M., and J.M.Peña, Total positivity, QR factorization and Neville elimination, SIAM J. Matrix Anal. Appl. **14** (1993), 1132–1140.
6. Gasca, M., and J.M.Peña, A matricial description of Neville elimination, with applications to total positivity. To appear in Linear Algebra Appl.(1993).
7. Goodman, T.N.T., Shape preserving representations, in *Mathematical methods in CAGD*, T.Lyche and L.L. Schumaker (eds.), Academic Press, New York, 1989, 333-357.
8. Goodman, T.N.T., and C.A.Micchelli, Corner cutting algorithms for the Bézier representation of free form curves, Linear Alg. Appl. **99** (1988), 225-252.
9. Goodman, T.N.T., and H.B.Said, Shape preserving properties of the generalized Ball basis, Computer Aided Geometric Design **8** (1991), 115-121.
10. Karlin, S., *Total Positivity*, Stanford U.P., Stanford, 1968.

M. Gasca and J. M. Peña
Departamento de Matemática Aplicada,
Universidad de Zaragoza,
50009 Zaragoza, SPAIN
gasca@cc.unizar.es

Piecewise Polynomial Approximation of Spheres

S. Glærum

Abstract. We present a polynomial scheme for the approximation of spheres. The scheme is a simple extension of circle approximation and gives good results with regard to error bounds and continuity properties.

§1. Introduction

In this paper we will look into the problem of approximating spherical surfaces. The approximation of circles and spheres has received some attention in the past, see for instance [3,4,8,10]. Most of the schemes presented to date use some form of rationals. What we will present below is a piecewise polynomial scheme based on circle approximations. The reason for using polynomials is that there are many existing applications, e.g. CAD systems, that require that all surfaces should be represented as parametric polynomials.

There have lately been produced a number of papers on polynomial Hermite approximation of circle segments, for instance [2,5,9], all of which form a suitable basis for the approximation scheme presented below.

§2. The Approximation

We want to find an approximation to the spherical surface segment given by $\mathbf{R}_{\alpha,\beta}(\gamma, \delta) = (\cos\gamma\cos\delta, \sin\gamma\cos\delta, \sin\delta)$, where $(\gamma, \delta) \in [-\alpha, \alpha] \times [-\beta, \beta]$. With $\mathbf{r}_\alpha(\gamma) = (\cos\gamma, \sin\gamma, 1)^T$ and $\mathbf{s}_\beta(\delta) = (\cos\delta, \cos\delta, \sin\delta)^T$ we have

$$\mathbf{R}_{\alpha,\beta}(\gamma, \delta) = \text{diag}\left[\mathbf{s}_\beta(\delta)\right] \cdot \mathbf{r}_\alpha(\gamma) \tag{1}$$

where $\text{diag}[\mathbf{x}]$ is the diagonal matrix with diagonal $\mathbf{x}$.

Now, this is in fact an expression involving two circle segments. The vector $\mathbf{r}_\alpha$ is a circle segment of arc length 2α in the plane $z = 1$. Furthermore, we have that while the $\mathbf{s}_\beta$ is not a circle segment in itself, it is the *projection* of one, namely a circle segment of arc length 2β in the plane $y = 0$ (or $x = 0$) projected along the y-axis (x-axis) onto the plane $x = y$.

Curves and Surfaces in Geometric Design
P. J. Laurent, A. Le Méhauté, and L. L. Schumaker (eds.), pp. 185–192.
Copyright © 1994 by A K PETERS, Wellesley, MA
ISBN 1-56881-039-3.

These observations are the basis for our approximation of spherical surfaces. Since the two circular arcs are independently parameterized, by using some 2D scheme or other, we may approximate each of the arcs separately. We let

$$\bar{\boldsymbol{p}}_\alpha(u) = \left(p_{1\alpha}(u), p_{2\alpha}(u)\right)^T, \qquad u \in [-1, 1];$$
$$\bar{\boldsymbol{q}}_\beta(v) = \left(q_{1\beta}(v), q_{2\beta}(v)\right)^T, \qquad v \in [-1, 1], \tag{2}$$

be regular parametric curves in $\mathbb{R}^2$, where $\bar{\boldsymbol{p}}_\alpha(u)$ is some approximation to the circular arc $\{(\cos\gamma, \sin\gamma), \gamma \in [-\alpha, \alpha]\}$, and where $\bar{\boldsymbol{q}}_\beta(v)$ is some approximation to the circular arc $\{(\cos\delta, \sin\delta), \delta \in [-\beta, \beta]\}$.

Based on the components of these planar approximations, we define

$$\mathbf{p}_\alpha(u) = \left(p_{1\alpha}(u), p_{2\alpha}(u), 1\right)^T, \qquad u \in [-1, 1];$$
$$\mathbf{q}_\beta(v) = \left(q_{1\beta}(v), q_{1\beta}(v), q_{2\beta}(v)\right)^T, \qquad v \in [-1, 1]. \tag{3}$$

These curves are approximations to the curves swept by the two vectors in (1), and we can now construct an approximation $\mathbf{P}_{\alpha,\beta}$ to the spherical surface defined as

$$\mathbf{P}_{\alpha,\beta}(u, v) = \mathrm{diag}\left[\mathbf{q}_\beta(v)\right] \cdot \mathbf{p}_\alpha(u), \qquad (u, v) \in [-1, 1] \times [-1, 1]. \tag{4}$$

§3. The Approximation Error

The Euclidian distance between a point $\{x, y, z\}$ and the unit sphere is given by the absolute value of the function $\epsilon(\{x, y, z\}) = \sqrt{x^2 + y^2 + z^2} - 1$. Instead of this, we will use the function $e(\{x, y, z\}) = x^2 + y^2 + z^2 - 1$ as a signed error. The justification is that e is easier to handle and that ϵ and e are proportional for small errors. Similarly, we will use $\bar{e}(\{x, y\}) = x^2 + y^2 - 1$ as a signed error in the planar case of approximating the unit circle. In both 2D and 3D we will use the max-norm for measuring the approximation error. That is, the errors $E_\alpha = E(\mathbf{p}_\alpha) = \sup_u |\bar{e}(\mathbf{p}_\alpha(u))|$, $E_\beta = E(\mathbf{q}_\beta) = \sup_v |\bar{e}(\mathbf{q}_\beta(v))|$ and $E_{\alpha,\beta} = E(\mathbf{P}_{\alpha,\beta}) = \sup_{(u,v)} |e(\mathbf{P}_{\alpha,\beta}(u, v))|$.

Proposition 1. *If the errors for the planar circle approximations $\mathbf{p}_\alpha$ and $\mathbf{q}_\beta$ are given as E_α and E_β, then the error $E_{\alpha,\beta}$ for the spherical approximation $\mathbf{P}_{\alpha,\beta}$ defined in (4) satisfies*

$$E_{\alpha,\beta} \leq E_\alpha + E_\beta + E_\alpha E_\beta.$$

Proof: We have that $E_\alpha = \sup_u |p_{1\alpha}(u)^2 + p_{2\alpha}(u)^2 - 1|$ and that $E_\beta =$

$\sup_v |q_{1\beta}(v)^2 + q_{2\beta}(v)^2 - 1|$. Using this and (4) we get that

$$
\begin{aligned}
E_{\alpha,\beta} &= \sup_{(u,v)} |\mathbf{P}_{\alpha,\beta}(u,v)^T \mathbf{P}_{\alpha,\beta}(u,v) - 1| \\
&= \sup_{(u,v)} |q_{1\beta}(v)^2 p_{1\alpha}(u)^2 + q_{1\beta}(v)^2 p_{2\alpha}(u)^2 + q_{2\beta}(v)^2 - 1| \\
&= \sup_{(u,v)} |q_{1\beta}(v)^2 (p_{1\alpha}(u)^2 + p_{2\alpha}(u)^2 - 1) + q_{1\beta}(v)^2 + q_{2\beta}(v)^2 - 1| \\
&\leq \sup_v q_{1\beta}(v)^2 \sup_u |p_{1\alpha}(u)^2 + p_{2\alpha}(u)^2 - 1| + \sup_v |q_{1\beta}(v)^2 + q_{2\beta}(v)^2 - 1| \\
&= \sup_v q_{1\beta}(v)^2 E_\alpha + E_\beta
\end{aligned}
$$

For all v, we have $q_{1\beta}(v)^2 + q_{2\beta}(v)^2 - 1 \leq E_\beta$ so that $q_{1\beta}(v)^2 \leq E_\beta - q_{2\beta}(v)^2 + 1 \leq E_\beta + 1$. Inserting this into the above we get

$$
E_{\alpha,\beta} \leq (E_\beta + 1)E_\alpha + E_\beta = E_\alpha + E_\beta + E_\alpha E_\beta,
$$

which concludes the proof. ∎

§4. An Arbitrary Spherical Patch

In order to avoid degeneracy, we place some constraints on the end points of the circle approximations. We will require that the end points should lie on the edge of the sector swept by the circle segment, e.g. $\overline{\mathbf{p}}_\alpha(1) = r(\cos\alpha, \sin\alpha)$ for some $r \in \mathbb{R}^+$. If these requirements are satisfied we have that $\mathbf{P}_{\alpha,\beta}$ converges in Hausdorff distance to the spherical segment spanning longitude $-\alpha$ to α and latitude $-\beta$ to β whenever E_α and E_β both go to zero.

Our next problem is to approximate *any* spherical surface restricted by longitude and latitude. We will do this by transforming the original patch $\mathbf{P}_{\alpha,\beta}$ as depicted in Figure 1 below.

The way we obtain this is by applying 'rotation' matrices to $\mathbf{P}_{\alpha,\beta}$. These matrices are defined by

$$
A_\psi = \begin{pmatrix} \cos\psi & -\sin\psi & 0 \\ \sin\psi & \cos\psi & 0 \\ 0 & 0 & 1 \end{pmatrix}, \qquad B_\mu = \begin{pmatrix} \cos\mu & 0 & -\sin\mu \\ 0 & \cos\mu & -\sin\mu \\ \sin\mu & 0 & \cos\mu \end{pmatrix}. \tag{5}
$$

With the help of these matrices we define another, more general, approximation $\mathbf{Q}_{\alpha,\beta}^{\psi,\mu}$ which is, loosely stated (see below), a rotation of our original approximation $\mathbf{P}_{\alpha,\beta}$,

$$
\mathbf{Q}_{\alpha,\beta}^{\psi,\mu}(u,v) = \mathrm{diag}[B_\mu \mathbf{q}_\beta(v)] A_\psi \mathbf{p}_\alpha(u). \tag{6}
$$

In order to understand what these matrices are doing, it is important to note that a point on $\mathbf{Q}_{\alpha,\beta}^{\psi,\mu}$ is *not* obtained by a simple rotation of a point on

$\mathbf{P}_{\alpha,\beta}$. Since we are applying A_ψ and B_μ on each vector separately and then multiplying the result, the effect is somewhat different.

The effect of applying the unitary rotation matrix A_ψ is straightforward; it is the rotation of the curve $\mathbf{p}_\alpha$ through an angle ψ about the z-axis. The same thing happens to the patch as a whole, a unitary rotation about the z-axis.

Applying B_μ is, however, not so straightforward. First of all, B_μ is not unitary. The best way to perceive what is happening is to look at the projection of $\mathbf{q}_\beta$ in the xz-plane. Remember that $\mathbf{q}_\beta$ is the projection of $(q_{1\beta}, 0, q_{2\beta})^T$ along the y-axis onto the plane $x = y$. Applying B_μ to $\mathbf{q}_\beta$ is equivalent to performing the *unitary* rotation of $(q_{1\beta}, 0, q_{2\beta})^T$ of angle μ about the y-axis and then project the result along the y-axis onto the plane $x = y$.

In Figure 1 below, we give an example of such rotations, using a bi-quadratic approximation $\mathbf{P}_{\frac{\pi}{8},\frac{\pi}{8}}$.

Figure 1: The patches $\mathbf{P}_{\pi/8,\pi/8}$ and $\mathbf{Q}_{\pi/8,\pi/8}^{3\pi/16,5\pi/16}$.

If $\mathbf{P}_{\alpha,\beta}$ is a perfect spherical segment spanning $[-\alpha, \alpha]$ in longitude and $[-\beta, \beta]$ in latitude, applying A_ψ and B_μ yields a perfect spherical segment $\mathbf{Q}_{\alpha,\beta}^{\psi,\mu}$ that spans $[-\alpha+\psi, \alpha+\psi]$ in longitude and $[-\beta+\mu, \beta+\mu]$ in latitude, as expected.

The error estimate given in Proposition 1 is valid also for a rotated patch. The proof can be found in [5].

Proposition 2. *The approximation error* $E_{\alpha,\beta}^{\psi,\mu} = \sup_{(u,v)} |(\mathbf{Q}_{\alpha,\beta}^{\psi,\mu})^T \mathbf{Q}_{\alpha,\beta}^{\psi,\mu} - 1|$ *satisfies*

$$E_{\alpha,\beta}^{\psi,\mu} \le E_\alpha + E_\beta + E_\alpha E_\beta.$$

§5. Geometric Continuity

In this section we will examine what type of conditions we must place on the planar circle approximations in order to ensure geometric continuity of a given order between two or more spherical approximations.

The surface is formed by rotations of the original approximation, so that $\mathbf{Q}_{\alpha,\beta}^{\psi,\mu}(u, v)$ and $\mathbf{Q}_{\alpha,\beta}^{\psi+2\alpha,\mu}(u, v)$ will form a common border along $\mathbf{Q}_{\alpha,\beta}^{\psi,\mu}(1, v)$. Likewise, $\mathbf{Q}_{\alpha,\beta}^{\psi,\mu}(u, v)$ and $\mathbf{Q}_{\alpha,\beta}^{\psi,\mu+2\beta}(u, v)$ will have a common border along $\mathbf{Q}_{\alpha,\beta}^{\psi,\mu}(u, 1)$. The case where $\mu = \frac{\pi}{2} - \beta$, that is, when the patches meet at the pole will be considered separately.

Proposition 3. *Given the approximation* $\mathbf{Q}_{\alpha,\beta}^{\psi,\mu}(u,v)$ *as defined in* (6),

(i) $\mathbf{Q}_{\alpha,\beta}^{\psi,\mu}$ *and* $\mathbf{Q}_{\alpha,\beta}^{\psi+2\alpha,\mu}$ *joins with geometric continuity* GC^k *along the curve* $\mathbf{Q}_{\alpha,\beta}^{\psi,\mu}(1,v)$ *if* $A_{2\alpha}\mathbf{p}_\alpha(u)$ *and* $\mathbf{p}_\alpha(u)$ *meet with geometric continuity* GC^k *at* $u = -1$ *and* $u = 1$, *respectively.*

(ii) *for* $\mu + \beta < \pi/2$ *we have that* $\mathbf{Q}_{\alpha,\beta}^{\psi,\mu}$ *joins* $\mathbf{Q}_{\alpha,\beta}^{\psi,\mu+2\beta}$ *with geometric continuity* GC^k *along the curve* $\mathbf{Q}_{\alpha,\beta}^{\psi,\mu}(u,1)$ *if* $B_{2\beta}\mathbf{q}_\beta(v)$ *and* $\mathbf{q}_\beta(v)$ *meet with geometric continuity* GC^k *at* $v = -1$ *and* $v = 1$, *respectively.*

Proof: With $\mu + \beta < \pi/2$ the proofs of (i) and (ii) are almost identical, so we will only give the proof of (i) here. Suppose now that $A_{2\alpha}\mathbf{p}_\alpha(u)$ and $\mathbf{p}_\alpha(u)$ meet with the desired continuity, that is, there is an allowable C^k reparameterization ϕ, so that $\phi(-1) = 1$ and

$$A_{2\alpha}\frac{\partial^i}{\partial u^i}\mathbf{p}_\alpha(u)|_{u=-1} = \frac{\partial^i}{\partial u^i}\mathbf{p}_\alpha(\phi(u))|_{u=-1}, \qquad i = 0,\ldots,k. \tag{7}$$

A general definition of geometric continuity for surfaces can be found in e.g. Gregory [7]. Without going into too much detail here, it can be shown (see [5]) that it is sufficient to prove that there is a C^k reparameterization $\overline{\phi}$ so that

$$\frac{\partial^i \mathbf{Q}_{\alpha,\beta}^{\psi+2\alpha,\mu}(u,v)}{\partial u^i}\Bigg|_{u=-1} = \frac{\partial^i \mathbf{Q}_{\alpha,\beta}^{\psi,\mu}(\overline{\phi}(u),v)}{\partial u^i}\Bigg|_{u=-1} \tag{8}$$

for $v \in [-1,1]$. The underlying reason for this being a sufficient condition is that the derivatives *along* the edge $\mathbf{Q}_{\alpha,\beta}^{\psi,\mu}(1,v)$ are all identical since the patches are identical along this edge. If in addition the derivatives up to order k are identical *across* the edge — one choice in this case is to pick the parameter direction u — then *all* derivatives up to order k are identical on the edge.

By letting $\overline{\phi} = \phi$ and expanding the right hand side of (8), we get

$$\frac{\partial^i \mathbf{Q}_{\alpha,\beta}^{\psi,\mu}(\phi(u),v)}{\partial u^i}\Bigg|_{u=-1} = \mathrm{diag}(B_\mu\mathbf{q}_\beta(v))A_\psi\frac{\partial^i}{\partial u^i}\,\mathbf{p}_\alpha(\phi(u))|_{u=-1}. \tag{9}$$

The left hand side of (8) expands to

$$\frac{\partial^i \mathbf{Q}_{\alpha,\beta}^{\psi+2\alpha,\mu}(u,v)}{\partial u^i}\Bigg|_{u=-1} = \mathrm{diag}(B_\mu\mathbf{q}_\beta(v))A_{\psi+2\alpha}\frac{\partial^i}{\partial u^i}\,\mathbf{p}_\alpha(u)|_{u=-1}. \tag{10}$$

Observing that the rotation $A_{\psi+2\alpha}$ is equal to the product $A_\psi A_{2\alpha}$ of two rotations, we have, using (7), established the equality (8). ∎

One special case remains, namely the case where two or more patches meet at one of the poles. One of the edges of the domain would then collapse to a single point. However, this does not exclude the possibility of obtaining

some form of geometric continuity at this point. A definition of geometric
continuity for patches meeting at a common vertex is again given by Gregory
[7, p. 365]. The essence of this definition is that we have GC^k at such a vertex
if the patches are GC^k across common borders leading to the vertex, and in
addition if the patches form a common tangent plane at the vertex.

$\mathbf{Q}_{\alpha,\beta}^{\psi,\pi/2-\beta}(u,v)$ is a patch that is rotated so that $\mathbf{Q}_{\alpha,\beta}^{\psi,\pi/2-\beta}(u,1) = \{0,0,r\}$
for some $r > 0$, then $\{\mathbf{Q}_{\alpha,\beta}^{\psi+j2\alpha,\pi/2-\beta}(u,v)\}$, $j = 0,1,\ldots$ form a patch complex
where the patches meet at $\{0,0,r\}$ and neighboring patches have common
borders leading to this point.

Proposition 4. *Suppose $k \geq 1$, $0 < \beta \leq \pi/2$ and $\alpha = \pi/n, n \in \mathbb{N}$. Then*
$\mathbf{Q}_{\alpha,\beta}^{j2\alpha,\pi/2-\beta}(u,v), (u,v) \in [-1,1] \times [-1,1], j = 1,\ldots,n$ *form a patch complex
of geometric continuity GC^k with a common vertex at $\{0,0,r\} \in \mathbb{R}^3$ for some
$r > 0$ if the conditions of Proposition 3 holds and, in addition,*

$$\mathbf{q}_\beta'(1) = l(-\sin\beta, -\sin\beta, \cos\beta), \tag{11}$$

*for some $l > 0$. That is, the planar circle approximation must have a tangent
that is parallel to the tangent of the circle segment at their respective end
points.*

A proof of Proposition 4 can be found in [5].

§6. Two Examples

The first example use a circle approximation taken from [9]. In his paper
Mørken showed that the quadratic approximation below, which has an error
equal to the scaled Chebyshev polynomial of degree 4, is the best approxima-
tion of degree 2. We define $\bar{\mathbf{p}}_\alpha(t) = \sum_{i=0}^2 \mathbf{c}_i B_{i,2}(t)$, where $B_{i,2}(t) = \binom{2}{i}(1-t)^{2-i}(1+t)^i/4$ is the quadratic Bernstein-Bézier basis on $[-1,1]$. The coefficients are given by $\mathbf{c}_0 = r(\cos\alpha, -\sin\alpha)$, $\mathbf{c}_1 = rl(1,0)$ and $\mathbf{c}_2 = r(\cos\alpha, \sin\alpha)$,
where $r = \{2/(2-\sin^4(\alpha/2))\}^{(1/2)}$ and $l = 2 - \sin\alpha$. This is an approximation
to the circle segment limited by the angles $-\alpha$ to α, and the error is given by

$$E_\alpha = |\bar{e}(0)| = \frac{\sin^4(\alpha/2)}{2 - \sin^4(\alpha/2)}. \tag{12}$$

It can be shown that the error is of order α^4. From this we can construct a
GC^0 spherical approximation with the basic patch

$$\mathbf{P}_{\alpha,\beta}(u,v) = \text{diag}\left[(\bar{p}_{1\beta}(v), \bar{p}_{1\beta}(v), \bar{p}_{2\beta}(v))\right] \cdot (\bar{p}_{1\alpha}(u), \bar{p}_{2\alpha}(u), 1)^T. \tag{13}$$

for $(u,v) \in [-1,1] \times [-1,1]$. Referring to Proposition 1, it is easily shown
that the maximum error for this patch is attained at $(u,v) = (0,0)$ and that
in this special case $E_{\alpha,\beta} = E_\alpha + E_\beta + E_\alpha E_\beta$.

In [3], Eisele shows that there exists, under some conditions, a best approximation in the Euclidian norm to spherical segments. Given four points

on the unit sphere on the form $\mathbf{a}_0 = (a_x, a_y, a_z) = (-\sin\theta\cos\psi, \sin\theta\sin\psi)$, $\mathbf{a}_1 = (-a_x, -a_y, a_z)$, $\mathbf{a}_2 = (-a_x, -a_y, a_z)$ and $\mathbf{a}_3 = (a_x, -a_y, a_z)$ which defines a spherical segment centered on the north pole, Eisele gives a Reméz type algorithm that produces a sequence of approximations converging to this best approximation provided θ is small enough. The result is a bi-quadratic tensor product with an Euclidian error function that equioscillates 13 times and a maximum error of approximately 0.00018 with $\theta = \pi/8$ and $\psi = \pi/4$.

Approximating the same spherical segment (but centering the 4 points on $(1, 0, 0)$) with the method described in this paper, using Mørken's quadratic circle approximation described above, we get a maximum Euclidian error of approximately 0.00019. In other words, an approximation that is very close to Eisele's best approximation. (The two approximations are somewhat different, but span a spherical segment of about the same size.)

The second example is based on a cubic circle approximation found in [2]. This approximation is Hermite and has a three times equioscillating error. Going through the same steps as in the example above we get a spherical approximation that is GC^1 between neighboring patches, but also at the poles since the condition in Proposition 4 is satisfied. Letting $\alpha = \beta = \pi/4$, a quarter circle, we get the error function in Figure 2.

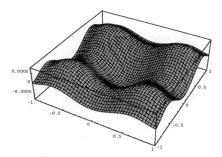

Figure 2: The signed approximation error $e(\mathbf{P}_{\pi/4,\pi/4})$.

§7. Conclusion

All in all, the approximation scheme presented above gives approximations that are both simple to construct and have small errors. However, there is one major handicap in that the approximation degenerates at the poles. This may be a problem when used in applications involving e.g. ray-tracing since the surface normal is not defined at the poles.

We can handle this problem in at least two ways. In [1], Berger and Kallay state that a rectangular domain may be a good tool for modeling a triangular patch, degenerate edge and all. In their paper they give a formula for the unit normal on such an edge which is a continuous extension of the formula for the unit normal away from the edge.

In our approximation scheme the unit normal at the poles is easily found if the approximation is GC^1 or better. Because of the conditions in Proposition

4 we have a unique surface normal at the pole, parallel to the z-axis.

Some may still be unhappy with such an approach, where a point on the surface requires special treatment, however simple. Another way to deal with this problem is to convert the degenerate quadrilateral patch to a non-degenerate triangular one. This is obtained through a reparameterization that can be found in [5]. Note that this is not the standard conversion from a tensor product to *two* triangular patches found in [6], but a conversion that is unique to degenerate quadrilaterals.

References

1. Berger, D. N. and M. Kallay, Computing the unit normal on a degenerate edge, Computer Aided Design **7** (1992), 395–396.
2. Dokken, T., M. Dæhlen, T. Lyche and K. Mørken, Good approximation of circles by curvature-continuous Bézier curves, Comp. Aided Geom. Design **7** (1990), 33–41.
3. Eisele, E. F., Best approximations of some parametric surfaces by Bézier surfaces, in *Mathematics of Surfaces V*, R. Fisher (ed.), Oxford Univ. Press, to appear.
4. Farin, G., B. Piper and A. J. Worsey, The octant of a sphere as a non-degenerate triangular Bézier patch, Comp. Aided Geom. Design **4** (1987), 329–332.
5. Glærum, S., Piecewise polynomial approximations of some simple geometric shapes, PhD thesis, Univ. of Oslo, 1994.
6. Goldman, R. N. and D. J. Filip, Conversion from Bézier rectangles to Bézier triangles, Computer Aided Design **1** (1987), 25–27.
7. Gregory, J. A., Geometric continuity, in *Mathematical Methods in Computer Aided Geometric Design*, T. Lyche and L. L. Schumaker (eds.), Academic Press, Boston, 1989, 353–371.
8. Herron, G., Smooth closed surfaces with discrete triangular interpolants, Comp. Aided Geom. Design **2** (1985), 297–306.
9. Mørken, K., Best approximation of circle segments by quadratic Bézier curves, in *Curves and Surfaces*, P. J. Laurent, A Le Méhauté and L. L. Schumaker (eds.), Academic Press, Boston, 1991, 331–336.
10. Piegl, L., The sphere as a rational Bézier surface, Comp. Aided Geom. Design **3** (1986), 45–52.

Sigurd Glærum
Universitetet i Oslo, Institutt for Informatikk
Box 1080 Blindern
N-0316 Oslo, NORWAY
sigurdg@ ifi.uio.no

Non-polynomial Polar Forms

Dan Gonsor and Marian Neamtu

Abstract. We begin by defining the polar form for a special type of function, namely a trigonometric polynomial, in order to illustrate the similarities between trigonometric polar forms and polynomial polar forms. After deriving properties and developing some results concerning trigonometric polar forms, we consider the generalization to functions that are elements of certain null spaces of constant coefficient differential operators.

§1. Introduction

The concept of a *polar form* or *blossom*, while known for quite some time in an algebraic context, has been introduced into the spline theory by de Casteljau and independently by Ramshaw (see [6], for a comprehensive introduction). Polar forms have proven to be a convenient mathematical tool for describing (piecewise) polynomial functions and for analyzing various spline algorithms such as recurrence relations and knot insertion [3,6,8].

A generalization of polar forms to non-polynomial functions has been given in [5]. There, a geometric approach to polar forms has been developed, whereas our generalization is based on the fact that polynomials form a translation invariant space. Since translation invariant spaces are the null spaces of constant coefficient differential operators, it follows that the polar forms defined here are multivariate exponential polynomials. Hence, these polar forms are *not* multi-affine functions in the classical sense. Nevertheless, they have a similar structure to polynomial polar forms, and therefore could be useful in a context of a general non-polynomial spline theory.

We begin our presentation by considering polar forms for trigonometric functions in Section 2. This will motivate our approach in the more general situation described in Section 3.

Curves and Surfaces in Geometric Design
P. J. Laurent, A. Le Méhauté, and L. L. Schumaker (eds.), pp. 193–200.
Copyright ⓒ 1994 by A K PETERS, Wellesley, MA
ISBN 1-56881-039-3.

§2. Trigonometric Polar Forms

In this section we will define trigonometric polar forms and discuss some of their properties. First we recall some basic definitions and notations. We will let T_n denote the following space of *trigonometric polynomials of order $n + 1$*, $(n \geq 0)$:

$$T_n := \begin{cases} \text{span} \{1, \lfloor 2x \rfloor, \lceil 2x \rceil, \lfloor 4x \rfloor, \lceil 4x \rceil, \ldots, \lfloor nx \rfloor, \lceil nx \rceil\}, & n \text{ even} \\ \text{span} \{\lfloor x \rfloor, \lceil x \rceil, \lfloor 3x \rfloor, \lceil 3x \rceil, \ldots, \lfloor nx \rfloor, \lceil nx \rceil\}, & n \text{ odd}, \end{cases}$$

where, for the sake of shortening the notation, we define

$$\lfloor x \rfloor := \sin x, \quad \lceil x \rceil := \cos x.$$

As is well known [7], the space T_n can be identified with the kernel of the differential operator D_n defined by

$$D_n := \begin{cases} D(D^2 + 2^2)(D^2 + 4^2) \cdots (D^2 + n^2), & n \text{ even} \\ (D^2 + 1^2)(D^2 + 3^2) \cdots (D^2 + n^2), & n \text{ odd}, \end{cases}$$

where $D := d/dx$. Another equivalent way of defining the space T_n is

$$T_n = \text{span} \left\{ \binom{n}{k} \lfloor x \rfloor^{n-k} \lceil x \rceil^k \right\}_{k=0}^n,$$

which justifies the term *degree* used in the context of trigonometric polynomials for the number n. It should be remarked that in the definition of the space T_n we have deviated slightly from the convention used in the literature on trigonometric splines. It is common to define spaces of trigonometric polynomials using trigonometric functions with halved arguments. For example, the space T_2 is most often taken to be the space, span$\{1, \sin x, \cos x\}$ rather than the above defined, span$\{1, \sin 2x, \cos 2x\}$. However, the convention followed in this paper reflects the results in [1], suggesting that the definition of the space T_n given here is more natural.

We are ready to introduce the polar form of a trigonometric function.

Theorem 1. *For every $F \in T_n, n \geq 0$, there exists a unique function $f(x_1, \ldots, x_n)$ of n variables, $x_1, \ldots, x_n \in \mathbb{R}$, called the trigonometric polar form of F, which satisfies the following properties:*

(a) f *is symmetric with respect to* $x_1, \ldots, x_n$,
(b) f *is equal to F on the diagonal i.e.,* $f(x, \ldots, x) = F(x)$, *for all $x \in \mathbb{R}$,*
(c) *for all $m \geq 1$ and all real numbers $y, y_1, \ldots, y_m$, the function f satisfies in each variable the relation*

$$f(\ldots, y, \ldots) = \sum_{j=1}^m \lambda_j f(\ldots, y_j, \ldots), \tag{7}$$

whenever the numbers $\lambda_1, \ldots, \lambda_m$ are chosen so that

$$\sum_{j=1}^{m} \lambda_j \lfloor y_j \rfloor = \lfloor y \rfloor, \quad \text{and} \quad \sum_{j=1}^{m} \lambda_j \lceil y_j \rceil = \lceil y \rceil. \tag{8}$$

Proof: Let $\alpha := (\alpha_1, \ldots, \alpha_n)$ be a multi-index, where $\alpha_i \in \{0, 1\}, i = 1, \ldots, n$, and $1 - \alpha := (1 - \alpha_1, \ldots, 1 - \alpha_n)$. Moreover, for $X := (x_1, \ldots, x_n)$, we write $\lfloor X \rfloor^{\alpha} := \lfloor x_1 \rfloor^{\alpha_1} \cdots \lfloor x_n \rfloor^{\alpha_n}$. First observe that the function

$$s_k(x_1, \ldots, x_n) := \sum_{|\alpha| = k} \lfloor X \rfloor^{1-\alpha} \lceil X \rceil^{\alpha},$$

is a polar form of the function $\binom{n}{k} \lfloor x \rfloor^{n-k} \lceil x \rceil^k \in \mathcal{T}_n$. This is easily proved by verifying all three defining properties (a)–(c). While the proofs of the first two properties are straightforward, the third property follows from the fact that the functions $\lfloor X \rfloor^{1-\alpha} \lceil X \rceil^{\alpha}$ satisfy relation (7). This becomes clear when one observes that these functions are each a product of one of the univariate functions $\lfloor x_i \rfloor$ or $\lceil x_i \rceil$, and both of these functions trivially satisfy (7) under assumption (8).

Next, since the functions $\binom{n}{k} \lfloor x \rfloor^{n-k} \lceil x \rceil^k, k = 0, \ldots, n$, form a basis for $\mathcal{T}_n$, we conclude that the function

$$f(x_1, \ldots, x_n) = \sum_{k=0}^{n} c_k s_k(x_1, \ldots, x_n), \quad c_0, \ldots, c_n \in \mathbb{R}, \tag{9}$$

is a polar form of $F \in \mathcal{T}_n$, given by

$$F(x) = \sum_{k=0}^{n} c_k s_k(x, \ldots, x) = \sum_{k=0}^{n} c_k \lfloor x \rfloor^{n-k} \lceil x \rceil^k.$$

For the uniqueness of representation (9), it suffices to notice that the n-variate functions $s_k(x_1, \ldots, x_n), k = 0, \ldots, n$, are linearly independent since they are linearly independent on their diagonal $x_1 = \cdots = x_n$. ∎

Remark 1. On account of the symmetry of f, the above recursion (7) could be called a *trigonometric multi-affineness* of f.

Remark 2. It is possible to relate polynomial polar forms to trigonometric polar forms. Let $F \in \mathcal{T}_n, n \geq 0$, and let P be such that

$$P\left(\frac{\lfloor x \rfloor}{\lceil x \rceil}\right) = \frac{F(x)}{\lceil x \rceil^n}, \quad \lceil x \rceil \neq 0.$$

Then $P \in \Pi_n$, Π_n the space of (algebraic) polynomials of order $n + 1$, and so it has a polar form, say p. The function

$$f(x_1, \ldots, x_n) := \lceil x_1 \rceil \ldots \lceil x_n \rceil p\left(\frac{\lfloor x_1 \rfloor}{\lceil x_1 \rceil}, \ldots, \frac{\lfloor x_n \rfloor}{\lceil x_n \rceil}\right)$$

coincides with the trigonometric polar form of F, for all $x_1, \ldots, x_n$, such that $\lceil x_1 \rceil \ldots \lceil x_n \rceil \neq 0$.

Remark 3. Let $\sigma_k(x_1, \ldots, x_n)$ be the k-th elementary symmetric polynomial of n arguments. Recall that $\sigma_k(x_1, \ldots, x_n)$ is the (polynomial) polar form of the monomial $\binom{n}{k} x^k$. Therefore, by Remark 2, we have

$$s_k(x_1, \ldots, x_n) = \lceil x_1 \rceil \cdots \lceil x_n \rceil \sigma_k \left(\frac{\lfloor x_1 \rfloor}{\lceil x_1 \rceil}, \ldots, \frac{\lfloor x_n \rfloor}{\lceil x_n \rceil} \right),$$

and hence $s_k(x_1, \ldots, x_n)$ can be viewed as an *elementary symmetric trigono-metric polynomial*. It follows from the proof of Theorem 1 that the trigono-metric polar form of any trigonometric polynomial is a linear combination of the elementary symmetric trigonometric polynomials.

As a consequence of Theorem 1, we have the next three-term recurrence relation for polar forms.

Corollary 1. *Let f be the trigonometric polar form of $F \in \mathcal{T}_n$. Then*

$$f(\ldots, y, \ldots) = \frac{\lfloor y_2 - y \rfloor}{\lfloor y_2 - y_1 \rfloor} f(\ldots, y_1, \ldots) + \frac{\lfloor y - y_1 \rfloor}{\lfloor y_2 - y_1 \rfloor} f(\ldots, y_2, \ldots), \qquad (10)$$

for all $y, y_1, y_2 \in \mathbb{R}$, such that $\lfloor y_2 - y_1 \rfloor \neq 0$.

Proof: Setting $m = 2$, the two equations in (8) represent a linear system for the unknowns λ_1 and λ_2, which by elementary algebra leads to

$$\lambda_1 = \frac{\lfloor y_2 - y \rfloor}{\lfloor y_2 - y_1 \rfloor}, \quad \lambda_2 = \frac{\lfloor y - y_1 \rfloor}{\lfloor y_2 - y_1 \rfloor}. \quad \blacksquare$$

Remark 4. Relation (10) could be considered as an alternative to (7). In-deed, it is not difficult to show that if a function f satisfies (10) for all ad-missible choices of y, y_1, y_2, then it also satisfies the more general recursion (7).

Next, we present an analog of a property of polynomial polar forms [6].

Theorem 2. *(Polar Interpolation) Let $a_0, a_1 \in \mathbb{R}$, $0 < |a_1 - a_0| < \pi$. For $k = 0, \ldots, n$, let $c_k \in \mathbb{R}$ and $t_k := (\overbrace{a_0, \ldots, a_0}^{n-k}, \overbrace{a_1, \ldots, a_1}^{k})$. Then there exists a unique trigonometric polynomial F of degree n whose polar form f satisfies $f(t_k) = c_k, k = 0, \ldots, n$.*

Proof: Let us first assume that $|a_1 - a_0| \neq \pi/2$. We define the functions b_0 and b_1 by

$$b_0(x) := \frac{\lfloor a_1 - x \rfloor}{\lfloor a_1 - a_0 \rfloor} \quad \text{and} \quad b_1(x) := \frac{\lfloor x - a_0 \rfloor}{\lfloor a_1 - a_0 \rfloor}. \qquad (11)$$

Clearly,

$$b_j(a_k) = \delta_{jk}, \quad j, k = 0, 1. \tag{12}$$

Next, let

$$F(x) := \sum_{k=0}^{n} c_k B_k^n(x), \tag{13}$$

where $B_k^n(x) := \binom{n}{k} b_0^{n-k}(x) b_1^k(x)$. It is not difficult to prove that the functions $B_k^n, k = 0, \ldots, n$, are linearly independent. This can be done e.g., by induction on n. It is clear from (12) that the assertion is true for $n = 1$ since in this case, $t_k = a_k, k = 0, 1$. For $n > 1$, let $\sum_{k=0}^{n} d_i B_k^n(x) = 0$, for all $x \in \mathbb{R}$. In particular, this equality must hold for $x = a_0$. Therefore, by (12), $d_0 = 0$. However, the remaining sum is now a product of the function b_1 with a linear combination of functions $B_k^{n-1}, k = 0, \ldots, n - 1$, which are linearly independent by the induction hypothesis. Therefore, the remaining coefficients d_k, for $k = 1, \ldots, n$ must be zero. Hence, the B_k^n are linearly independent and thus the representation of F in (13) is indeed unique.

Next, let f be the polar form of F. We show that the function f satisfies the interpolation conditions $f(t_k) = c_k$. Observe that for an arbitrary but fixed number x, the value $F(x) = f(x, \ldots, x)$ can be expressed as a linear combination of the values $f(t_k)$. For example, applying (10) to the first argument of f leads to

$$f(x, \ldots, x) = b_0(x) f(a_0, x, \ldots, x) + b_1(x) f(a_1, x, \ldots, x).$$

In general, we obtain

$$F(x) = f(x, \ldots, x) = \sum_{k=0}^{n} f(t_k) B_k^n(x), \tag{14}$$

which, in combination with (13), gives the desired result. The uniqueness of the function F satisfying the interpolation conditions follows from representation (14), since $c_k = 0, k = 0, \ldots, n$, clearly forces F to be the zero function.

The proof for the remaining case $|a_1 - a_0| = \pi/2$ is almost identical with the above proof except that now the definitions of the functions b_0 and b_1 have to be modified. Assuming, without loss of generality that $a_1 > a_0$, we can set

$$b_0(x) := \lfloor a_1 - x \rfloor \quad \text{and} \quad b_1(x) := \lfloor x - a_0 \rfloor = \lceil a_1 - x \rceil. \quad \blacksquare \tag{15}$$

Corollary 2. *The functionals* $\mu_j : \mathcal{T}_n \to \mathbb{R}, j = 0, \ldots, n$, *defined by*

$$\mu_j F := f(t_j), \quad F \in \mathcal{T}_n, \quad f - \text{the polar form of } F,$$

form a dual basis for $\{B_k^n\}_{k=0}^n$ *i.e.,*

$$\mu_j B_k^n = \delta_{jk}, \quad j, k = 0, \ldots, n.$$

Remark 5. The above results suggest that the trigonometric functions B_k^n associated with the interval $[a_0, a_1]$ may be viewed as analogs of classical Bernstein polynomials. These trigonometric Bernstein polynomials have been studied in greater detail in [1]. In that paper they have been coined *circular Bernstein polynomials* since the functions b_0 and b_1 defined by (11) and (15) can be considered as circular analogs of barycentric coordinates. From Corollary 2 it follows that, as in the polynomial case, the coefficients $c_k, k = 0, \ldots, n$ of a trigonometric polynomial of the form (13) can be obtained by evaluating the trigonometric polar form f of F at the points t_k.

§3. Polar Forms for Certain Translation Invariant Spaces

In this section we briefly describe how the results of Section 2 can be carried over to a larger class of functions. Let us first observe that the equations (8) are equivalent to

$$\sum_{j=1}^{m} \lambda_j \lfloor y_j - t \rfloor = \lfloor y - t \rfloor, \tag{16}$$

which must hold true for all $t \in \mathbb{R}$. This equation suggests the following generalization of (8). Let d be a real-valued function. In accordance with (16), we require that every $y \in \mathbb{R}$ can be associated with numbers $\lambda_1, \ldots, \lambda_m$ such that

$$\sum_{j=1}^{m} \lambda_j d(y_j - t) = d(y - t), \tag{17}$$

for all $t \in \mathbb{R}$, provided equation (17) is solvable. In particular, we require that it be solvable for $m = 2$, whenever the two functions $d(y_j - \cdot), j = 1, 2$ are linearly independent. This requirement imposes strong restrictions on the function d. To explain this, it will be convenient to introduce the space $\mathcal{D} := \text{span}\{d(\cdot - t), t \in \mathbb{R}\}$, i.e., the linear span of all translates of the function d. Setting $m = 2$ and keeping y, y_1, y_2 fixed, equation (17) implies that every function from $\mathcal{D}$ can be expressed as a linear combination of two fixed functions. Hence, the dimension of $\mathcal{D}$ can be at most two. Another restriction on $\mathcal{D}$ is that it must be a translation invariant space, that is, such that $d \in \mathcal{D}$ implies $d(\cdot - t) \in \mathcal{D}$, for all $t \in \mathbb{R}$. The translation invariance is clearly a consequence of the definition of $\mathcal{D}$. The case, where the dimension of $\mathcal{D}$ is one, is trivial since the only translation invariant one dimensional space is the space of constant functions. In the remainder of this section we will only consider the case where the dimension of $\mathcal{D}$ is two. The space $\mathcal{D}$ is completely characterized by the following

Proposition 1. *A two dimensional space of continuous real-valued functions is translation invariant if and only if it is the null space of a linear second order constant coefficient differential operator.*

Proof: The crux of the proof is in treating the elements of the space $\mathcal{D}$ under consideration as distributions. It is easily seen that a two dimensional translation invariant space $\mathcal{D}$ of distributions is also invariant under differentiation *i.e.*, $f \in \mathcal{D}$ implies $f' \in \mathcal{D}$. Next, let $f \in \mathcal{D}$ such that f and f' are linearly independent. The existence of such an f can be proved as follows. Suppose, on the contrary, that there is no such function. Let f_1, f_2 be two linearly independent elements of $\mathcal{D}$ and let a_1, a_2 be real numbers such that $f_1' = a_1 f_1, f_2' = a_2 f_2$. Moreover, let $f \in \mathcal{D}$ and let a be such that $f' = af$. Thus, $f = b_1 f_1 + b_2 f_2$ for some real coefficients b_1, b_2. By combining these equalities and by the linear independence of f_1, f_2 it follows that $a = a_1 = a_2$. Therefore, since f was arbitrary, $\mathcal{D}$ is the solution space of the equation $f' - af = 0$, which is a one dimensional space. This contradicts the assumption that $\mathcal{D}$ is two dimensional. Therefore, let f be such that f and f' are independent *i.e.*, such that they span $\mathcal{D}$. Since f'' is an element of $\mathcal{D}$, there exist two coefficients a, b such that $f'' = af' + bf$. Thus, f solves a constant coefficient differential homogeneous equation. However, then f' must also solve the same equation which can be easily seen by differentiating both sides of this equation. Since f and f' are independent it follows that all elements of $\mathcal{D}$ must solve this equation. To finish the proof, it is sufficient to realize that the null space of a second order constant coefficient differential operator is translation invariant. ∎

Remark 6. The assertion of Proposition 1 also follows from [2, Thm. 1.3 (a)]. In fact, the assumption of continuity of the functions in $\mathcal{D}$ is unnecessarily strong. It is sufficient to assume that $\mathcal{D}$ is a space of distributions.

In the sequel, let $\mathcal{D}$ be the null space of a second order constant coefficient differential operator and let $\mathcal{D}_n := \text{span}\{d^n, d \in \mathcal{D}\}$. In particular, $\mathcal{D}_0$ is the space of constant functions. We are ready to define a $\mathcal{D}$-polar form of a function $F \in \mathcal{D}_n$.

Theorem 3. *For every* $F \in \mathcal{D}_n$, $n \geq 0$, *there exists a unique function* $f(x_1, \ldots, x_n)$ *of n variables, called a $\mathcal{D}$-polar form of F, satisfying the following properties:*
(a) *f is symmetric with respect to $x_1, \ldots, x_n$,*
(b) *f is equal to F on the diagonal i.e., $F(x) = f(x, \ldots, x)$, for all $x \in \mathbb{R}$,*
(c) *for all $m \geq 1$ and all real numbers $y, y_1, \ldots, y_m$, the function f is $\mathcal{D}$-affine i.e., it satisfies in each variable the relation*

$$f(\ldots, y, \ldots) = \sum_{j=1}^{m} \lambda_j f(\ldots, y_i, \ldots), \tag{20}$$

whenever the coefficients $\lambda_1, \ldots, \lambda_m$ are chosen such that

$$\sum_{j=1}^{m} \lambda_j d(y_j) = d(y), \tag{8}$$

for all $d \in \mathcal{D}$.

Proof: The proof can be done along the same lines as the proof of Theorem 1. Here, the functions $\lfloor x \rfloor$ and $\lceil x \rceil$ should be replaced by two arbitrary linearly independent functions from $\mathcal{D}$. ■

Remark 7. The space $\mathcal{D}_n$ arises in a different context also in [4]. There, spline spaces are considered whose elements can be locally identified with functions from $\mathcal{D}_n$. The associated B-splines turn out to satisfy recurrence relations which are similar to the classical ones for polynomial B-splines.

References

1. Alfeld, P., M. Neamtu, and L. L. Schumaker, Bernstein-Bézier polynomials on circles, spheres, and sphere-like surfaces, preprint, 1993.
2. Ben-Artzi, A., and A. Ron, Translates of exponential box splines and their related spaces, Trans. Amer. Math. Soc. **309** (1988), 683–710.
3. Goldman, R. N., Blossoming and knot insertion algorithms for B–spline curves, Computer-Aided Geom. Design **7** (1989), 69–81.
4. Li, Y., On the recurrence relations for B-splines defined by certain L-splines, J. Approx. Th. **43** (1985), 359–369.
5. Pottmann, H., The geometry of Tchebycheffian splines, Computer-Aided Geom. Design **10** (1993), 181–210.
6. Ramshaw, L., Blossoms are polar forms, Computer-Aided Geom. Design **6** (1989), 323-358.
7. Schumaker, L. L., *Spline Functions: Basic Theory*, Interscience, New York, 1981.
8. Seidel, H.–P., Knot insertion from a blossoming point of view, Computer-Aided Geom. Design **5** (1988), 81–86.

Dan E. Gonsor
Kent State University
Department of Mathematics and Computer Science
Kent, OH 44242
dgonsor@mcs.kent.edu

Marian Neamtu
Vanderbilt University
Department of Mathematics
Nashville, TN 32740
neamtu@athena.cas.vanderbilt.edu

Curvature of Rational Quadratic Splines

T. N. T. Goodman

Abstract. The monotonicity properties of the curvature of a conic arc are studied. Such an arc can have monotone curvature if and only if the tangent turns through an angle of at most $\frac{\pi}{2}$. It is further shown that if two conic arcs are joined with continuous tangent direction and curvature, then the resulting curve can have monotone curvature if and only if the tangent turns through at most π. Finally an algorithm is given for minimising the number of local extrema of curvature for a sequence of conic arcs with prescribed join-points and tangents there.

§1. Introduction

We shall consider a conic arc in Bézier form, i.e., a rational quadratic curve

$$r(t) = \frac{A(1-t)^2 + Bw2t(1-t) + Ct^2}{(1-t)^2 + w2t(1-t) + t^2}, \qquad 0 \le t \le 1, \qquad (1.1)$$

where A, B, C are distinct points in $\mathbb{R}^2$ called the *Bézier points* and $w > 0$ is the *weight*.

It has been suggested that the 'fairness' of a curve is related to the monotonicity of its curvature, see [2] and the references therein, and with this in mind we shall study the local extrema of the curvature of (1.1) in Section 2, extending the work of [2] for a polynomial quadratic. After this work was completed, we were informed that essentially the same results had been obtained by Frey and Field in the report [1], and so we limit Section 2 to little more than a summary of results.

Labelling the angles α and β as in Figure 1, it is shown in Section 2 that if $\alpha + \beta \le \frac{\pi}{2}$, then there is always a choice of weight w for which the curve (1.1) has monotone curvature. For $\alpha + \beta > \frac{\pi}{2}$, the conic arc cannot have monotone curvature. Therefore in Section 3 we consider construction of two conic arcs which join with continuous tangent direction and curvature, and which interpolate end-points A and C with tangent directions $B - A$ and $C - B$ respectively. For $\alpha + \beta \le \pi$, we give necessary and sufficient conditions

Curves and Surfaces in Geometric Design
P. J. Laurent, A. Le Méhauté, and L. L. Schumaker (eds.), pp. 201–208.

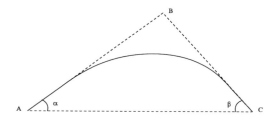

Fig. 1. Bézier points for a conic arc.

on the position of the join and the direction of the tangent there in order to achieve monotone curvature, and we show that these conditions can always be satisfied. (This problem was suggested to me by R. Ait Haddou.)

Finally, in Section 4, we consider a sequence of conic arcs in $\mathbb{R}^2$. Given the end-points and join-points and the tangent directions there, we give an algorithm for constructing a curve with the minimum number of local extrema of curvature.

§2. One Conic Arc

In Figures 2-6 we illustrate the local extrema of curvature for (1.1), $0 < t < 1$, for fixed A, C and w. Without loss of generality we may assume that B lies above the line through A and C. If B lies in a region marked 0, then (1.1) has monotone curvature. If B lies in a region marked m or M, then there is a local minimum or maximum of curvature respectively, while if it lies in a region marked 2, then there are both a local minimum and a local maximum. The solid curves denote boundaries between regions and in all cases are semi-circles of diameter $\frac{|C-A|}{2w^2}$.

We remark that the proof, which is different from that in [1], involves transforming the conic segment to a standard representation in which it is clear where the local extrema of curvature lie.

From Figures 2-6 we can easily deduce the following, where α and β are as in Figure 1.

Theorem 1. *If $\alpha + \beta > \frac{\pi}{2}$, then the curve (1.1) does not have monotone curvature for any $w > 0$. Suppose that $\alpha + \beta \leq \frac{\pi}{2}$, $\alpha \leq \beta$, $w > 0$. Then (1.1) has monotone curvature if and only if*

$$\frac{\cos \alpha \sin(\alpha + \beta)}{2 \sin \beta} \leq w^2 \leq \frac{\cos \beta \sin(\alpha + \beta)}{2 \sin \alpha}. \tag{2.1}$$

Moreover if $w^2 < \frac{\cos \alpha \sin(\alpha+\beta)}{2 \sin \beta}$, then (1.1) has a local minimum of curvature for $0 < t < 1$ while if $w^2 > \frac{\cos \beta \sin(\alpha+\beta)}{2 \sin \alpha}$, it has a local maximum.

Fig. 2. $w > 1$, hyperbola.

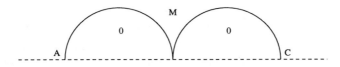

Fig. 3. $w = 1$, parabola.

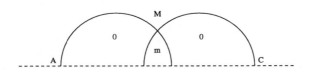

Fig. 4. $\frac{1}{\sqrt{2}} < w < 1$, ellipse.

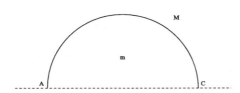

Fig. 5. $w = \frac{1}{\sqrt{2}}$, ellipse.

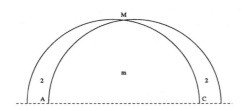

Fig. 6. $0 < w < \frac{1}{\sqrt{2}}$, ellipse.

Clearly if $\beta \leq \alpha$, then α and β are interchanged in (2.1). We note that the interval (2.1) for w reduces to a single value if and only if either $\alpha = \beta$ (when $w = \cos \alpha$ and the conic is a circle) or $\alpha + \beta = \frac{\pi}{2}$ (when $w = \frac{1}{\sqrt{2}}$).

A simple calculation shows that the curvature at A is

$$K_1 = \frac{\sin \alpha \sin^2(\alpha + \beta)}{2w^2 |C - A| \sin^2 \beta}, \tag{2.2}$$

and that at C is

$$K_2 = \frac{\sin \beta \sin^2(\alpha + \beta)}{2w^2 |C - A| \sin^2 \alpha}. \tag{2.3}$$

Thus if $\alpha + \beta \leq \frac{\pi}{2}$, $\alpha \leq \beta$ and (2.1) is satisfied, then the minimum and maximum values of the curvature are given by (2.2) and (2.3) respectively. We note that $\frac{K_1}{K_2}$ is independent of w. Thus in choosing a fair curve it seems natural to choose w to minimise K_1 and K_2, i.e., $w^2 = \frac{\cos \beta \sin(\alpha + \beta)}{2 \sin \alpha}$.

§3. Two Conic Arcs

Consider two conic arcs with Bézier points A, B, C and weight w_1, and with Bézier points C, D, E and weight w_2, respectively. We assume that they have the same tangent direction at C and so B, C, D are collinear. This is illustrated in Figure 7, where we have labelled angles $\alpha, \beta, \alpha_1, \beta_1, \alpha_2, \beta_2$. We note that

$$0 < \alpha_1 < \alpha, 0 < \beta_2 < \beta, \beta_1 > 0, \alpha_2 > 0, \alpha_1 + \beta_1 + \alpha_2 + \beta_2 = \alpha + \beta. \tag{3.1}$$

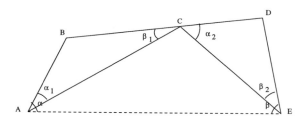

Fig. 7. Bézier points for two conic arcs.

We shall derive conditions for the curve comprising the two arcs from A to E to have continuous non-decreasing curvature. From Theorem 1, this is possible for each arc separately if and only if

$$\alpha_1 \leq \beta_1, \qquad \alpha_2 \leq \beta_2, \tag{3.2}$$

$$\alpha_1 + \beta_1 \leq \frac{\pi}{2}, \qquad \alpha_2 + \beta_2 \leq \frac{\pi}{2}. \tag{3.3}$$

We denote the curvature at C by K. Then by (2.3) and (2.1) we have

$$\frac{\tan \beta_1 \sin(\alpha_1 + \beta_1)}{\sin \alpha_1 |C - A|} \leq K \leq \frac{\sin^2 \beta_1 \sin(\alpha_1 + \beta_1)}{\sin^2 \alpha_1 \cos \alpha_1 |C - A|},$$

while by (2.2) and (2.1) we have

$$\frac{\sin^2 \alpha_2 \sin(\alpha_2 + \beta_2)}{\sin^2 \beta_2 \cos \beta_2 |E - C|} \leq K \leq \frac{\tan \alpha_2 \sin(\alpha_2 + \beta_2)}{\sin \beta_2 |E - C|}.$$

We can find K satisfying these inequalities if and only if

$$\frac{\tan \beta_1 \sin(\alpha_1 + \beta_1)}{\sin \alpha_1 |C - A|} \leq \frac{\tan \alpha_2 \sin(\alpha_2 + \beta_2)}{\sin \beta_2 |E - C|}$$

and

$$\frac{\sin^2 \alpha_2 \sin(\alpha_2 + \beta_2)}{\sin^2 \beta_2 \cos \beta_2 |E - C|} \leq \frac{\sin^2 \beta_1 \sin(\alpha_1 + \beta_1)}{\sin^2 \alpha_1 \cos \alpha_1 |C - A|}.$$

Applying the sine rule, these become

$$\frac{\tan \beta_1 \sin(\alpha_1 + \beta_1) \sin(\alpha - \alpha_1)}{\sin \alpha_1} \leq \frac{\tan \alpha_2 \sin(\alpha_2 + \beta_2) \sin(\beta - \beta_2)}{\sin \beta_2}, \quad (3.4)$$

$$\frac{\sin^2 \alpha_2 \sin(\alpha_2 + \beta_2) \sin(\beta - \beta_2)}{\sin^2 \beta_2 \cos \beta_2} \leq \frac{\sin^2 \beta_1 \sin(\alpha_1 + \beta_1) \sin(\alpha - \alpha_1)}{\sin^2 \alpha_1 \cos \alpha_1}. \quad (3.5)$$

We have thus proved

Theorem 2. *We can construct conic arcs with Bézier points A, B, C and C, D, E (as in Figure 7) which join to form a curve with continuous tangent direction and continuous non-decreasing curvature if and only if (3.1)–(3.5) are satisfied.*

From (3.1) and (3.3) we must have $\alpha + \beta \leq \pi$. We now assume $\alpha + \beta \leq \pi$ and show that, for $\alpha \leq \beta$, we can choose $\alpha_1, \beta_1, \alpha_2, \beta_2$ satisfying (3.1)–(3.5). For $\alpha \geq \beta$ we can similarly construct a curve with non-increasing curvature. First note that (3.4) and (3.5) are equivalent to

$$\frac{\sin 2\alpha_1}{\sin 2\beta_2} \leq \frac{\sin(\alpha - \alpha_1) \sin^2 \beta_1 \sin(\alpha_1 + \beta_1) \sin \beta_2}{\sin(\beta - \beta_2) \sin^2 \alpha_2 \sin(\alpha_2 + \beta_2) \sin \alpha_1} \leq \frac{\sin 2\beta_1}{\sin 2\alpha_2}. \quad (3.6)$$

Now (3.2) and (3.3) imply that

$$\sin 2\alpha_i \leq \sin 2\beta_i, \qquad i = 1, 2,$$

and so

$$\frac{\sin 2\alpha_1}{\sin 2\beta_2} \leq \frac{\sin 2\beta_1}{\sin 2\alpha_2}. \quad (3.7)$$

A crucial observation is that it is sufficient to find some choice of $\alpha_1, \beta_1, \alpha_2, \beta_2$ satisfying (3.1)–(3.4), and some other choice satisfying (3.1)–(3.3) and (3.5), for then by (3.6), (3.7) and continuity, there must be some choice of $\alpha_1, \beta_1, \alpha_2, \beta_2$ which satisfy (3.1)–(3.5).

We first satisfy (3.1)–(3.3) and (3.5). Consider

$$\alpha_1 = \frac{\alpha}{2}, \qquad \beta_2 = \frac{\beta}{2}, \qquad \beta_1 = \frac{\alpha}{2} + \theta, \qquad \alpha_2 = \frac{\beta}{2} - \theta, \qquad (3.8)$$

where

$$0 \le \theta < \frac{\beta}{2}, \qquad \beta - \frac{\pi}{2} \le \theta \le \frac{\pi}{2} - \alpha. \qquad (3.9)$$

Clearly (3.1)–(3.3) are satisfied. Inequality (3.5) becomes

$$\sin^2(\frac{\beta}{2} - \theta)\sin(\beta - \theta)\sin\alpha \le \sin^2(\frac{\alpha}{2} + \theta)\sin(\alpha + \theta)\sin\beta. \qquad (3.10)$$

First suppose that $\alpha + \frac{\beta}{2} \le \frac{\pi}{2}$. As $\theta \to \frac{\beta}{2}^-$, then (3.9) and (3.10) are satisfied. Now suppose that $\alpha + \frac{\beta}{2} > \frac{\pi}{2}$ and put $\theta = \frac{\pi}{2} - \alpha$. Then (3.9) is satisfied and (3.10) becomes

$$\cos^2(\alpha + \frac{\beta}{2})|\cos(\alpha + \beta)|\sin\alpha \le \cos^2\frac{\alpha}{2}\sin\beta. \qquad (3.11)$$

Since $\alpha + \beta \le \pi$, $\alpha \le \beta$, we have $\frac{\pi}{2} - \frac{\alpha}{2} \ge \alpha + \frac{\beta}{2} - \frac{\pi}{2}$ and so $\cos^2\frac{\alpha}{2} \ge \cos^2(\alpha + \frac{\beta}{2})$. Since $\sin\alpha \le \sin\beta$, (3.11) is satisfied.

Thus we have satisfied (3.1)–(3.3) and (3.5). Note that if $\beta \le \frac{\pi}{2}$, then $\theta = 0$ satisfies (3.9), while (3.4) becomes $\sin^2\frac{\alpha}{2} \le \sin^2\frac{\beta}{2}$ which also holds. Thus for $\beta \le \frac{\pi}{2}$ we can always find a solution to (3.1)–(3.5) of form (3.8) for some θ satisfying (3.9). However this is not true in general for $\beta > \frac{\pi}{2}$ and so we consider other ways of satisfying (3.1)–(3.4).

First suppose that $\alpha \le \frac{\pi}{4}$. Then as $\alpha_1 \to \alpha^-$, (3.4) is satisfied for any $\beta_1, \alpha_2, \beta_2$ satisfying (3.1)–(3.3). Next suppose that $\alpha > \frac{\pi}{4}$. In this case we choose

$$\alpha_1 = \beta_1 = \frac{\pi}{4}, \qquad \alpha_2 = \beta_2 = \frac{\alpha + \beta}{2} - \frac{\pi}{4}.$$

It is easily checked that (3.1)–(3.3) are satisfied. Inequality (3.4) becomes

$$\sqrt{2}\sin(\alpha - \frac{\pi}{4}) \le \frac{\sin(\alpha + \beta - \frac{\pi}{2})\sin(\frac{\beta - \alpha}{2} + \frac{\pi}{4})}{\cos(\frac{\alpha + \beta}{2} - \frac{\pi}{4})}$$

$$= 2\sin(\frac{\alpha + \beta}{2} - \frac{\pi}{4})\sin(\frac{\beta - \alpha}{2} + \frac{\pi}{4}),$$

i.e.,

$$\sin\alpha - \cos\alpha \le \cos(\frac{\pi}{2} - \alpha) - \cos\beta,$$

i.e.,

$$\cos\beta \le \cos\alpha,$$

which is true since $0 < \alpha \le \beta < \pi$.

§4. A Sequence of Conic Arcs

For $n \geq 1$, we fix points $A_0, \ldots, A_n, B_1, \ldots, B_n$ in $\mathbb{R}^2$, where for $i = 1, \ldots,$ $n-1$, A_i lies on the line segment joining B_i and B_{i+1}, and $B_i \neq A_i \neq B_{i+1}$. We consider a curve r comprising n conic arcs r_i, $i = 1, \ldots, n$, with Bézier points A_{i-1}, B_i, A_i and weight $w_i > 0$. Thus r interpolates the points $A_0, \ldots, A_n$ and at $A_i, i = 1, \ldots, n-1$, has tangent in the direction of $B_{i+1} - B_i$. We shall assume that for $i = 1, \ldots, n$, the angle between $B_i - A_{i-1}$ and $A_i - B_i$ is at most $\frac{\pi}{2}$ so that, by Theorem 1, there is a choice of w_i so that r_i has monotone curvature. For $i = 1, \ldots, n$ we shall denote by K_i and L_i the curvature of r_i at A_{i-1} and A_i respectively.

Let $C_1, \ldots, C_m$ denote the signed curvature at any m points in order along the curve r other than $A_1, \ldots, A_{n-1}$ (where the curvature may be discontinuous). Let $s(C_1, \ldots, C_m)$ denote the number of strict sign changes in the sequence $C_2 - C_1, \ldots, C_m - C_{m-1}$ and denote by $E(r)$ the maximum value of $s(C_1, \ldots, C_m)$ over all choices of $C_1, \ldots, C_m$, all m. We shall construct an algorithm for choosing the weights $w_1, \ldots, w_n$ so that $E(r)$ has its minimum value. Firstly we prove

Theorem 3. *For curves r as above, the minimum value of $E(r)$ is attained for a curve r for which $r_1, \ldots, r_n$ have monotone curvature.*

Proof: Suppose that for some i, $1 \leq i \leq n$, r_i does not have monotone curvature. We shall show that w_i can be modified so that r_i is changed to a curve with monotone curvature without increasing $E(r)$. We may assume that r_i has positive curvature and $K_i \leq L_i$. We suppose that r_i has a local maximum of curvature for $0 < t < 1$, the case of a local minimum following similarly. Recalling Theorem 1, we decrease w_i, and hence increase K_i and L_i, to ensure that r_i has increasing curvature.

If $L_{i-1} \leq K_i$, then increasing K_i does not change the contribution to $E(r)$ from r_{i-1}. If the curvature of r_{i-1} is decreasing near A_{i-1}, then the neighbourhood of A_{i-1} will contribute 1 to $E(r)$, regardless of the value of K_i. If $L_{i-1} > K_i$ and the curvature of r_{i-1} is increasing near A_{i-1}, then the neighbourhood of A_{i-1} will originally contribute 2 to $E(r)$ and so this contribution cannot be increased by changing r_i. So in all cases the contribution to $E(r)$ from r_{i-1} is not increased by modifying r_i.

If the curvature of r_{i+1} is increasing near A_i, then the neighbourhood of A_i will originally contribute 1 to $E(r)$ and so changing r_i can increase this contribution by at most 1. If the curvature of r_{i+1} is decreasing near A_i, then after modification the neighbourhood of A_i will contribute 1 to $E(r)$, and so this contribution can have been increased by at most 1. So in all cases the contribution to $E(r)$ from r_{i+1} is increased by at most 1 by changing r_i. Since modifying r_i has removed a local maximum of curvature, the value of $E(r)$ is not increased. ∎

Thus to minimise $E(r)$ we may assume that each weight w_i lies in the interval $I_i = [a_i, b_i]$ for which r_i has monotone curvature. Note that for any arc r_i the sign of the curvature and whether the curvature is constant, increasing or decreasing is independent of the choice of w_i in I_i. Therefore we can divide the curve into maximal sections on which the sign and nature of monotonicity of the curvature on arcs r_i are constant. Since the contributions to $E(r)$ by these sections are independent, it is sufficient to minimise $E(r)$ over each section separately. Without loss of generality we may therefore assume that all arcs $r_1, \ldots, r_n$ have positive, increasing curvature.

We now choose $w_1, \ldots, w_n$ recursively as follows. Let $w_1 = b_1$. Assume that we have chosen $w_1, \ldots, w_k$ where $1 \le k \le n-1$. If we can choose w_{k+1} in I_{k+1} so that $K_{k+1} \ge L_k$, then let w_{k+1} be the largest such value. Otherwise let $w_{k+1} = b_{k+1}$.

We now show that this algorithm does indeed minimise $E(r)$. Suppose that for this choice of weights, $L_i > K_{i+1}$ if and only if i has values $i_1 < \cdots < i_m$. Then $E(r) = 2m$. Now take any other choice of weights $\hat{w}_i$ in I_i, $i = 1, \ldots, n$, with corresponding arcs $\hat{r}_i$ with curvatures $\hat{K}_i$ and $\hat{L}_i$ at A_{i-1} and A_i, and $\hat{L}_i > \hat{K}_{i+1}$ if and only if i has values $\hat{i}_1 < \cdots < \hat{i}_{\hat{m}}$. We must show that $m \le \hat{m}$. We do this by proving by induction that for $j = 1, \ldots, m$, there is a value $\hat{i}_j$ with $\hat{i}_j \le i_j$. If $\hat{i}_1 \ge i_1$, then $\hat{L}_{i_1} \ge L_{i_1}$. We cannot choose w_{i_1+1} so that so that $K_{i_1+1} \ge L_{i_1}$ and hence we cannot have $\hat{K}_{i_1+1} \ge \hat{L}_{i_1}$, and so $\hat{i}_1 = i_1$. Thus $\hat{i}_1 \le i_1$. Now assume that $\hat{i}_k \le i_k$ for some k, $1 \le k \le m - 1$. Suppose that $\hat{i}_{k+1} \ge i_{k+1}$. Now $w_{i_k+1} = b_{i_k+1}$ and so $K_{i_k+1} \le \hat{K}_{i_k+1}$. Thus $\hat{L}_{i_{k+1}} \ge L_{i_{k+1}}$. As before we cannot have $\hat{K}_{i_{k+1}+1} \ge \hat{L}_{i_{k+1}}$ and so $\hat{i}_{k+1} = i_{k+1}$. Thus $\hat{i}_{k+1} \le i_{k+1}$ and the inductive step is complete.

Finally we remark that for the case of a closed curve r, there are analogues of Theorem 3 and the above algorithm, but space does not permit their inclusion here.

References

1. Frey, W. H., and D. A. Field, Designing Bézier conic segments with monotone curvature, research report GMR-7845, General Motors Research Laboratories, Warren, Michigan, 1991.
2. Sapidis, N. S., and W. H. Frey, Controlling the curvature of a quadratic Bézier curve, Comp. Aided Geom. Design **9** (1992), 85-91.

T. N. T. Goodman
Dept. of Mathematics and Computer Science
The University
Dundee DD1 4HN, SCOTLAND
tgoodman@mcs.dundee.ac.uk

B-Spline Knot-Line Elimination and Bézier Continuity Conditions

Raúl Gormaz

Abstract. Referring to the work of Dahmen-Micchelli-Seidel [3], where a new scheme for multivariate B-spline is proposed, we present some insight into its geometrical nature. Their scheme provides automatically C^{k-1} continuity with degree k piecewise polynomials. The cost to be paid is a complex structure of the underlying triangulation. We propose an analogy between those B-splines and triangular Bézier patches. The affine conditions between the poles of two adjacents Bezier patches for C^r continuity are replaced by similar conditions between the poles of the B-splines. This time the conditions do not produce additional continuity, but some knot-lines are eliminated.

§1. Introduction

Simplex splines were introduced by C. de Boor [4] in 1976. They are now seen as a fruitful generalization of the univariate B-splines introduced by H.B. Curry and I.J. Schoenberg [1] in 1966. A good survey on simplex splines and some of its extension is given in [6].

Many authors have contributed to prove that simplex splines considered as single functions exhibit the same behaviour as theirs univariate analogs. But if we think of B-splines as "base"-splines, the work is far from being accomplished. Among the different schemes proposed since 1982, the one presented in [3] is the first to combine simplicity and good properties.

The basic idea can be stated as follows. Given a triangulation of $\mathbb{R}^2$, consider each vertex as a knot having multiplicity $k+1$, where k is the degree of the piecewise polynomial scheme. If a C^{k-1} continuous scheme is needed, we have to pull apart the repeated knots. In this way we get a triangulation in which every vertex is replaced by a "cloud" of knots around it. In fact, the $k+1$ knots being different is not sufficient to assure C^{k-1} continuity. Some non-colinearity between some of those knots is also necessary (see (H')).

Once the clouds of knots are fixed, we pick some knot subsets. We recall that $k+3$ knots determines a unique simplex spline of degree k. This enables

Curves and Surfaces in Geometric Design
P. J. Laurent, A. Le Méhauté, and L. L. Schumaker (eds.), pp. 209–216.
Copyright © 1994 by A K PETERS, Wellesley, MA
ISBN 1-56881-039-3.

us to associate to each knot subset a simplex spline, obtaining in this way a
set of "base"-splines or B-splines. The problem is then reduced to finding a
rule for selecting knot subsets in such a way that the spline space generated
by the B-splines possesses nice properties.

To continue along this line, we need some notation. Let $\Delta^I := [t_0^I, t_1^I, t_2^I]$
be the triangle "I" of our triangulation $\mathcal{T}$ of $\mathbb{R}^2$. To each vertex t_i^I we associate
a list of k distinct additional knots

$$t_i^I := t_{i,0}^I, t_{i,1}^I, \ldots, t_{i,k}^I.$$

The simple rule proposed in [3] consisted in producing a subset V_β^I, where
$\beta = (\beta_0, \beta_1, \beta_2)$ are three nonnegative integers, as follows :

$$V_\beta^I := \{t_{0,0}^I, t_{0,1}^I, \ldots, t_{0,\beta_0}^I, t_{1,0}^I, t_{1,1}^I, \ldots, t_{1,\beta_1}^I, t_{2,0}^I, t_{2,1}^I, \ldots, t_{2,\beta_2}^I\}. \tag{1}$$

If we are supposed to produce degree k simplex splines, we must impose
that

$$|\beta| := \beta_0 + \beta_1 + \beta_2 = k. \tag{2}$$

The number of possible β that satisfies (2) is exactly the dimension of
$\Pi^k(\mathbb{R}^2)$, the space of polynomials of (total) degree less than or equal to k.
In fact, $\Pi^k(\mathbb{R}^2)$ is included in the spline space generated by the B-splines.
Moreover, Seidel [10] proved that this spline space contains $\Pi_{k-1}^k(\mathcal{T})$, that
is, all the degree k piecewise polynomials based on the triangulation $\mathcal{T}$ and
which are C^{k-1} continuous.

In Section 2 we recall the main results of Dahmen-Micchelli-Seidel [3], and
Seidel [10]. In Section 3 we present our result about the knot-line elimination
and some illustrations.

§2. B-patches and B-splines

Let us recall the concept of blossom (or polar form) of a polynomial function:

Definition 1. *Let E, Q be vector (or affine) spaces and let $F : E \longrightarrow Q$ be
a degree k polynomial function. The unique function $f : E^k \longrightarrow Q$, k-affine,
symmetric and such that $f(u, \ldots, u) = F(u)$, $\forall u \in E$, is called the blossom
(or polar form) of F.*

L. Ramshaw [8] proved the existence and uniqueness of such an object.
He expressed most of the known properties of piecewise polynomial functions
in a simpler way using the blossom f instead of F itself. H.-P. Seidel ([3,9])
extended the de Casteljau algorithm and produced a generalization of the
Bézier triangular patches, the B-patches. Following the same notation as in
Section 1, we add the following new notation:

$$\Delta_\beta^I := [t_{0,\beta_0}^I, t_{1,\beta_1}^I, t_{2,\beta_2}^I] \quad \text{and}$$
$$X_\beta^I := (t_{0,0}^I, t_{0,1}^I, \ldots, t_{0,\beta_0-1}^I, t_{1,0}^I, \ldots, t_{1,\beta_1-1}^I, t_{2,0}^I, \ldots, t_{2,\beta_2-1}^I) \in (\mathbb{R}^2)^{|\beta|}.$$

If Δ_β^I is non-degenerate, it is possible to define the barycentric coordinates of $u \in \mathbb{R}^2$ with respect to this triangle :

$$u = \sum_{i=0}^{2} \lambda_{\beta,i}^I(u)\, t_{i,\beta_i}^I \quad \text{and} \quad \sum_{i=0}^{2} \lambda_{\beta,i}^I(u) = 1.$$

The generalized algorithm computes $F(u)$ starting from the values $f(X_\beta^I)$, $|\beta| = k$. These values are called the *poles* of F. We define

$$X_\beta^I u^\nu := (X_\beta^I, \underbrace{u, \ldots, u}_{\nu \text{ times}}) \in (\mathbb{R}^2)^{|\beta|+\nu}$$

Putting $C_\beta^\nu(u) := f(X_\beta^I u^\nu)$, with $|\beta| = k - \nu$, the algorithm uses the k-affinity of f to get the following recurrence relation :

$$C_\beta^0(u) := f(X_\beta^I), \quad |\beta| = k$$

$$C_\beta^{\nu+1}(u) := \sum_{i=0}^{2} \lambda_{\beta,i}^I(u)\, C_{\beta+e^i}^\nu(u), \quad |\beta| = k - |\nu| - 1,$$

where e^i denotes the canonical basis vector. It is clear that $C_0^k(u) = F(u)$.

If the algorithm is applied to the values

$$f_\gamma = \begin{cases} 1 & \text{if } \gamma = \beta \\ 0 & \text{if } \gamma \neq \beta, \end{cases}$$

a particular degree k polynomial is obtained, which we will denote by $B_\beta^I(\cdot)$. From the linearity of the algorithm, we obtain

$$F(u) = \sum_{|\beta|=k} f(X_\beta^I)\, B_\beta^I(u). \tag{3}$$

It is possible to relate all this with simplex splines. Assume that

$$\Omega^I := int\Big(\bigcap_{|\beta| \le k} \Delta_\beta^I \Big) \neq \emptyset. \tag{H}$$

This assumption implies in particular that all the triangles Δ_β^I are non-degenerate. The generalized de Casteljau algorithm is then meaningful.

The following results were stated in [3] and [10]. A recent and direct proof is found in [7].

Proposition 2. *The simplex spline $M(\cdot|V_\beta^I)$ based over the knot set V_β^I and restricted to Ω^I is proportional to $B_\beta^I(\cdot)$.*

Definition 3. *The normalized B-spline, $\mathcal{N}_\beta^I(\cdot)$, is the scaled simplex spline based over the knot set V_β^I such that $\mathcal{N}_\beta^I(u) = B_\beta^I(u)$, $\forall u \in \Omega^I$.*

The main result is the following

Proposition 4. *Let S be a degree k piecewise polynomial function over the triangulation $\mathcal{T}$ which is C^{k-1} continuous. Then*

$$S(u) = \sum_I \sum_{|\beta|=k} \mathcal{N}_\beta^I(u)\, f^I(X_\beta^I), \qquad \forall u \in \mathbb{R}^2,$$

where f^I is the blossom of F^I, the restriction of S to the triangle Δ^I.

Finally, applying Proposition 4 to $S \equiv 1$ we get

Corollary 5.

$$\sum_I \sum_{|\beta|=k} \mathcal{N}_\beta^I(u) \equiv 1.$$

That is, the normalized B-splines form a partition of unity. This property enables us to consider poles as real points.

Remark. Seidel's result [10] is more general than Proposition 4. He presents the s-dimensional case and combines multiplicity greater than 1 with lower continuity for S.

§3. Knot-line Elimination

A simplex spline $M(\cdot|K)$, based on a knot set $K = \{x_0, \ldots, x_n\}$, with $n = k+2$, is a piecewise polynomial function of degree k. It is known that if K is in general position (no line contains three points of K), $M(\cdot|K)$ is C^{k-1} continuous. Moreover, the discontinuities are located exactly along the segments $[x_i, x_j]$. We call those segments *knot-lines*. This implies that a spline of the form

$$S(u) = \sum_I \sum_{|\beta|=k} \mathcal{N}_\beta^I(u)\, P_\beta^I, \tag{4}$$

where P_β^I are arbitrary points in Q called *poles of the spline S*, may have discontinuities (of the kth derivative) at most along the corresponding knot-lines of the B-splines $\mathcal{N}_\beta^I$. But for appropriate choices of the poles P_β^I, some of the discontinuities may perhaps cancel each other.

Let us concentrate on two adjacent triangles, say Δ^I and Δ^J. Suppose, for the simplicity of the notation, that the common edge is $[t_0^I = t_0^J, t_1^I = t_1^J]$. On each of the sets Ω^I and Ω^J the spline is a single polynomial, say F^I and F^J. They satisfy

$$F^I(u) = \sum_{|\beta|=k} B_\beta^I(u)\, P_\beta^I, \tag{5}$$

and similarily for F^J. It is clear from (3) that $P_\beta^I = f^I(X_\beta^I)$.

For $|\beta| = k$, let us denote by a_β^I the affine function such that

$$a_\beta^I(t_{0,\beta_0}^I) = a_\beta^I(t_{1,\beta_1}^I) = 0 \ , \quad a_\beta^I(t_{2,\beta_2}^I) = 1 \tag{6}$$

The straight line spanned by t^I_{0,β_0} and t^I_{1,β_1} is

$$L_\beta := \{u \in \mathbb{R}^2 \,|\, a^I_\beta(u) = 0\}.$$

We assume that

$$\text{all the lines } L_\beta \text{ are distinct.} \qquad (H')$$

Those lines induce a partition over $\mathbb{R}^2$, $\mathcal{T}_L$. The following result (see [7], Lemma 7) is our main tool to find out the particular choices for the poles in order to eliminate some knot-lines.

Proposition 6. *There exist a unique degree k piecewise polynomial function S over the partition $\mathcal{T}_L$ such that*

1) *S is C^{k-1} continuous.*
2) *$S(u) = F^I(u) \,\forall u \in \Omega^I$ and $S(u) = F^J(u) \,\forall u \in \Omega^J$.*

Moreover, the function S has the following form :

$$S(u) = F^J(u) + \sum_{|\beta|=k} \mu_\beta \, (a^I_\beta(u))^k_+, \qquad (7)$$

where the real coefficients μ_β satisfy the relations :

$$\sum_{\substack{|\beta|=k \\ \beta_0 \geq \gamma_0 \,,\, \beta_1 \geq \gamma_1}} \mu_\beta \, \varphi_\beta(X^I_\gamma) = f(X^I_\gamma) \qquad (8)$$

for all γ, $|\gamma| = k$, where f is the blossom of $F = F^I - F^J$ and

$$\varphi_\beta(u_1, \ldots, u_k) := \prod_{j=1}^{k} a^I_\beta(u_j)$$

is the blossom of

$$\Phi_\beta(u) := a_\beta(u)^k.$$

It is easily verified that (7) defines a degree k piecewise polynomial function over $\mathcal{T}_L$ which is C^{k-1} and such that

$$S(u) = F^J(u) \quad \forall u \in \Omega^J.$$

Then, the proof (see [7] for more details) is reduced to checking that

$$S(u) = F^I(u) \quad \forall u \in \Omega^I,$$

and that this solution is unique. But this last condition is equivalent to

$$F^I = F^J + \sum_{|\beta|=k} \mu_\beta \, \Phi_\beta.$$

The relation (8) express the fact that the poles of $F = F^I - F^J$ are the same as those of $\sum \mu_\beta \, \Phi_\beta$. The scope of the sum is reduced because of the zeros of φ_β (see (6)).

From this relation, the coefficients μ_γ can be computed recursively as

$$\mu_\gamma = \frac{f(X_\gamma^I) - \sum_{\beta \in R_\gamma} \mu_\beta \, \varphi_\beta(X_\gamma^I)}{\varphi_\gamma(X_\gamma^I)}, \tag{9}$$

where

$$R_\gamma = \{ \beta \in \mathbb{Z}_+^3 \mid |\beta| = k, \beta_0 \geq \gamma_0, \beta_1 \geq \gamma_1, \beta \neq \gamma \}.$$

A knot-line L_β will be a discontinuity line for the spline S given by (4) if the coefficient μ_β of the proposition is different from zero.

Proposition 7. *The spline S given by the expression (4) has no discontinuity of its kth derivative along the lines*

$$[t_{0,\beta_0}^I, t_{1,\beta_1}^I], \qquad \forall \beta \; |\beta| = k, \; \beta_2 \leq r$$

iff

$$P_\beta^I = f^J(X_\beta^I), \qquad \forall \beta \; |\beta| = k, \; \beta_2 \leq r. \tag{10}$$

Proof: From the recurrence relation (9), the coefficients μ_β are zero iff

$$f^I(X_\beta^I) = 0, \qquad \forall \beta, \; |\beta| = k, \; \beta_2 \leq r.$$

But $f = f^I - f^J$, then

$$f^I(X_\beta^I) = f^J(X_\beta^I), \qquad \forall \beta \; |\beta| = k, \; \beta_2 \leq r,$$

and the result follows from the fact that $P_\beta^I = f^I(X_\beta^I)$. ∎

Remark. The relations (10) correspond to affine relations between the poles of F^I and F^J. Moreover, if the cloud of knots corresponding to each vertex is collapsed over its original vertex, the relations (10) become exactly the affine relations for C^r continuity between the Bézier points of the two triangular Bézier patches F^I and F^J.

We finish this presentation with three figures showing the effect of the relations (10). The subjacent triangulation is made of 6 triangles with a common vertex in the center point. On the left we present the poles and on the right, the resulting surface. All the three figures are quadratic splines, hence all are C^1.

Figure 1 shows the situation of no relation between two adjacent patches. Visually it seems to be a discontinuous situation. Figure 2 corresponds to the relation (10) with $r = 0$. The condition is reduced in this case to the identification of the three border poles between adjacent triangles. Visually, a C^0 effect is obtained. Figure 3 shows the effect of (10) with $r = 1$ just for one common edge between two triangles. In this case, one of the crease lines has been eliminated.

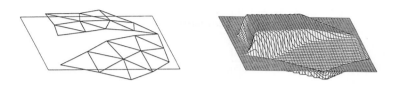

Figure 1. No conditions between poles of adjacents patches.

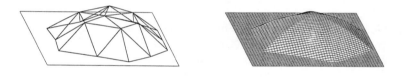

Figure 2. Conditions between poles of adjacent patches for $r = 0$.

Figure 3. Conditions between poles of two adjacent patches for $r = 1$.

Acknowledgements. This work was developed while the author was staying at the LMC-IMAG laboratory, Grenoble-France. Many fruitful discussions were held with Prof. P.-J. Laurent to whom I am very indebted.

References

1. Curry, H.B., Schoenberg I.J., On Pólya frequency functions IV: The fundamental spline functions and their limits, J. d'Analyse Math. **17** (1966), 71–107.
2. Dahmen, W., Micchelli C.A., On the linear independence of multivariate B-splines II : Complete configurations, Math. Comp. **41** (1983), 143–162.
3. Dahmen, W., Micchelli C.A., Seidel H.P., Blossoming Begets B-splines built better by B-patches, Math. Comp. **59** (1992), 97–115.
4. De Boor, C., Splines as linear combinations of B-splines, in *Approximation Theory 2*, G. G. Lorentz, C. K. Chui and L. L.Schumaker (eds.), Academic Press, New York, 1976, 1–47.
5. Fong, P., Seidel H.P. An implementation of triangular B-spline surfaces over arbitrary triangulations, Computer-Aided Geom. Design **10** (1993), 267–275.
6. Goodman, T.N.T. Polyhedral splines, in: *Computation of curves and surfaces*, W. Dahmen et al. (eds), Kluwer Acad. Publis., 1990, 347–382.
7. Gormaz, R., Laurent P.J., Some results on blossoming and multivariate B-splines, in *Multivariate Approximation: From CAGD to Wavelets*, K. Jetter and F. Utreras (eds.), Word Scientific, Singapore, 1993.
8. Ramshaw, L., Blossoming : A connect-the-dots approach to splines, DEC System Research Center Report 19, June, 1987.
9. Seidel, H.P., Polar forms and triangular B-splines surfaces, in *Blossoming : The new polar form approach to spline curves and surfaces*, SIG-GRAPH'91, 1991, Course notes #26, 8.1–8.52.
10. Seidel, H.P., Representing piecewise polynomials as linear combinations of multivariate B-splines, in *Curves and Surfaces*, T. Lyche and L. L. Schumaker (eds.), Academic Press, 1992, 559–566.

Raúl Gormaz
Universidad de Chile
Facultad de Ciencias Físicas y Matemáticas
Departamento de Ingeniería Matemática
Casilla 170/3   Correo 3   Santiago, CHILE

rgormaz@uchcecvm.cec.uchile.cl

Applications of Constrained Polynomials to Curve and Surface Approximation

R. J. Goult

Abstract. This paper describes the application of families of constrained Chebyshev polynomials and constrained orthogonal polynomials to parametric curve and surface approximation problems. The first method described is a generalisation of the classical Chebyshev economisation method and is applied to reduce the degree of parametric polynomial curves and surfaces, with selected boundary constraints. The second method uses constrained orthogonal polynomials to obtain polynomial approximations to any form of parametric curve or surface.

§1. Introduction

The development of the methods described here was initiated at Cranfield University as a participant in the Esprit CAD*I project. The work was motivated by requirements to communicate curve and surface data between heterogeneous CAD systems with a minimum loss of information. Two distinct types of problem were addressed, degree reduction and more general approximation.

The constrained Chebyshev polynomials were developed to make it possible to reduce the degree of a parametric polynomial curve or surface whilst maintaining continuity conditions at the boundary. The maintenance of continuity is particularly important when subdivision of the curve or surface is necessary. Important objectives in the development were the production of an optimal solution and the ability to predict error bounds.

The constrained orthogonal polynomials make it possible to obtain a parametric polynomial approximation of chosen degree to any smooth parametric curve or surface. Continuity constraints can be applied at the boundaries and a least squares error bound is provided at negligible computational cost. Whereas the application of the constrained Chebyshev method is restricted to single segment polynomial curves or single patch bi-parametric

Curves and Surfaces in Geometric Design
P. J. Laurent, A. Le Méhauté, and L. L. Schumaker (eds.), pp. 217–224.
Copyright Ⓒ 1994 by A K PETERS, Wellesley, MA
ISBN 1-56881-039-3.

polynomial surfaces, the orthogonal method can be a wide range of problems including rational curves and surfaces, procedurally defined surfaces and multi-patch surfaces.

The two methods are described briefly in Sections 2 and 3 and some test results are presented in Section 3 of this paper.

§2. Constrained Chebyshev Economisation

In the classical Chebyshev economisation method a function $f(t)$ defined over the range $[-1 \le t \le 1]$

$$f(t) = a_n t^n + a_{n-1} t^{n-1} + \ldots + a_0 \tag{1}$$

is reduced to a polynomial of degree at most $n-1$ by subtracting $a_n 2^{1-n} T_n(t)$, where $T_n(t)$ is the Chebyshev polynomial of degree n. The equi-oscillatory properties of the Chebyshev polynomials ensure that the resulting lower degree approximating polynomial is the one of degree at most $n-1$ which minimises the absolute deviation in the interval $[-1, 1]$. The absolute deviation in this interval is $2^{1-n}|a_n|$ and this is attained precisely n times. Although the Chebyshev economisation method is optimal, it has the important disadvantage that the approximating function deviates from the original polynomial function at the end points of the interval, this makes the method inappropriate for our applications. As a generalisation of this method, Lachance [2] and Sherar [3] developed the constrained Chebyshev polynomials, and have applied them to curve and surface degree reduction problems.

For convenience in CAD applications, the monic constrained Chebyshev polynomials of degree n are defined over the shifted range $[0 \le t \le 1]$. They are constrained to have roots at $t = 0$ and $t = 1$, and have equi-oscillatory properties. If the constrained Chebyshev polynomial has a root of multiplicity k at $t = 0$, and a root of multiplicity m at $t = 1$, then the result of subtracting any multiple of this polynomial from $f(t)$ is a function which matches $f(t)$ and its first $k - 1$ derivatives at $t = 0$, and matches $f(t)$ and its first $m - 1$ derivatives at $t = 1$. Constrained Chebyshev polynomials of various degrees have been defined for k, m in the range $[1 \ldots 4]$; for practical application the polynomials with symmetric end conditions with $k = m = 1, 2$, or 3 have been found to be most useful. Such a polynomial is denoted $C_n^k(t)$ in this paper. A modified Remes algorithm has been used to compute $C_n^k(t)$ for $k = 1, 2, 3$ and values of n up to 15.

$C_n^k(t)$ is equi-oscillatory over the interval $[0, 1]$ and has a bound E_n^k which depends upon n and k. For $k = 0$ the bound is equal to the Chebyshev bound 2^{1-n} and is slightly larger for positive values of k. The result of subtracting $a_n C_n^k(t)$ from $f(t)$, as defined in (1), is a polynomial of degree at most $n - 1$ which is identical in value to $f(t)$ and to its first $k - 1$ derivatives at $t = 0$ and at $t = 1$. Over the range $[0 < t < 1]$ the maximum absolute deviation from $f(t)$ is $|a_n| E_n^k$.

The constrained Chebyshev polynomials can be applied to problems of degree reduction of parametric polynomial curves as follows. Let

$$\mathbf{r}(t) = \sum_{i=0}^{n} \mathbf{a_i} t^i$$

be a parametric polynomial curve of degree n with vector coefficients, defined over the range $[0 \leq t \leq 1]$. The approximating curve of degree at most $n-1$ is then defined as

$$\mathbf{s}(t) = \mathbf{r}(t) - \mathbf{a_n} C_n^k(t). \tag{2}$$

$\mathbf{s}(t)$ is the optimum polynomial of degree $n-1$ or less, which minimises $\max_{0 \leq t \leq 1} |\mathbf{r}(t) - \mathbf{s}(t)|$, and maintains continuity of the curve and it first $k-1$ derivatives. The maximum deviation is $|\mathbf{a_n}| E_n^k$. Whilst this is the optimum solution from the point of view of approximating both the curve and its parametrisation, it may be possible to find a curve of degree $n-1$ which is geometrically closer to the original curve but has a greater deviation between points with the same parameter value. For a single degree reduction step the absolute error bound given above is attained, when the method is used iteratively over a number of degree reduction steps the cumulative bound produced is a conservative one. The iterative process proceeds by determining the vector coefficient of highest degree in $\mathbf{s}(t)$ and subtracting this multiple of $C_n^k(t)$.

For surface degree reduction the process is generalised as follows, let

$$\mathbf{R}(u, v) = \sum_{i=0}^{m} \sum_{j=0}^{n} \mathbf{A_{ij}} u^i v^j \tag{3}$$

be a bi-parametric polynomial surface defined over the ranges $[0 \leq u \leq 1]$, $[0 \leq v \leq 1]$. Equation (3) can be re-written as a polynomial in v as

$$\mathbf{R}(u, v) = \sum_{j=0}^{n} \mathbf{P_j}(u) v^j, \tag{3a}$$

where $\mathbf{P_j}(u)$ is a polynomial in u with vector coefficients. A degree reduction step in v is then defined by

$$\mathbf{S}(u, v) = \mathbf{R}(u, v) - \mathbf{P_m}(u) C_m^k(v). \tag{4}$$

A similar method can be used to define a degree reduction step in u, a slightly more complicated formula given in [3] is used to define a simultaneous degree reduction step in both variables. Using the surface degree reduction technique gives identical results on the boundary curves, for example $\mathbf{R}(u, 1)$, as the curve degree reduction technique described in the previous section. A consequence of this is that when two surface patches share a common boundary curve then the same property will remain valid after degree reduction. The resulting degree reduced surface matches exactly the original surface, and the appropriate number of derivatives, at the four corner points. Error bounds for the surface degree reduction are computed, in a similar manner to the curve errors, from E_m^k and the norm of $\mathbf{P_m}(u)$, these bounds are conservative for a multi-step degree reduction process.

The paragraphs above describe the degree reduction process for a single patch surface. If the errors after this process are too large the surface must be subdivided at selected parameter values and the process repeated after computing the polynomial coefficients to take account of the revised parametric limits for each subpatch. In practice it has been difficult to implement an optimal subdivision strategy which minimises the number of subpatches.

§3. Approximation using Orthogonal Polynomials

Families $(u_1(t), u_2(t), u_N(t))$ of constrained orthogonal polynomials are defined to have roots of the required multiplicity at t=0 and t=1 and to satisfy the orthogonality and normalisation condition $\int_0^1 u_i(t)u_j(t)dt = \delta_{ij}$. In order to constrain the first $k-1$ derivatives at $t=0$ and the first $m-1$ derivatives at $t=1$ the polynomials can be defined recursively by defining

$$f_1(t) = t^k(t-1)^m,$$

and for any n,

$$a_n^2 = \int_0^1 (f_n(t))^2 dt, \qquad u_n(t) = f_n(t)/a_n$$

$$b_n = \int_0^1 t(u_n(t))^2 dt, \, f_{n+1}(t) = tu_n(t) - b_n u_n(t) - a_n u_{n-1}(t). \qquad (5)$$

As with the Chebyshev polynomials the must useful orthogonal polynomials for geometric applications are those with symmetric constraints at the end points, and orthogonal polynomials up to degree 23 have been generated for $k = m = 1, 2$ or 3. The integrals in equations (5) can be evaluated analytically and, although explicit expressions for the polynomials can be obtained, the most convenient computational algorithm is to store the coefficients a_n and b_n and use (5) to evaluate the polynomials as required.

The orthogonal polynomials can be applied to obtain the parametric polynomial curve $\mathbf{s}(t)$ which minimises the integral

$$E^2 = \int_0^1 (\mathbf{r}(t) - \mathbf{s}(t))^2 dt, \qquad (6)$$

where $\mathbf{r}(t)$ is the parametric curve to be approximated. It should be noted that $\mathbf{r}(t)$ can now be any parametric curve, not necessarily a polynomial. The approximation process commences with the definition of a first approximation $\mathbf{s}_0(t)$ which can be any polynomial curve of the target degree, or less, and which matches $\mathbf{r}(t)$ and its first $k-1$ derivatives at the end points. For the currently implemented version of the software a Bézier curve of the target degree is used to define $\mathbf{s_0}(t)$. For this Bézier curve the control points at and near the ends are fully defined by the constraints, the inner control points, if any, are arbitrarily chosen to follow the shape of the original curve. An error function $\mathbf{r_e}(t)$ is then defined as

$$\mathbf{r_e}(t) = \mathbf{r}(t) - \mathbf{s_0}(t). \tag{7}$$

This vector valued function has the properties that the function itself and the first $k-1$ derivatives vanish at $t=0$ and at $t=1$, as such it can be approximated by the constrained orthogonal polynomials defined by (5). The approximation is

$$\mathbf{c_i} = \int_0^1 (\mathbf{r_e}(t)u_i(t))dt,$$

$$\mathbf{s_1}(t) = \sum_{i=1}^N \mathbf{c_i}u_i(t). \tag{8}$$

The mean square error in the approximation (8) for $\mathbf{r_0}(t)$ is then given by the formula

$$E^2 = \int_0^1 (\mathbf{r_e}(t))^2 dt - \sum_{i=1}^N (\mathbf{c_i})^2. \tag{9}$$

The final approximation to $\mathbf{r}(t)$ is then given by

$$\mathbf{s}(t) = \mathbf{s_0}(t) + s_1(t),$$

with the mean square error given by (9). $\mathbf{s}(t)$ will then coincide with $\mathbf{r}(t)$ at the end points (C^{k-1} continuity), and is optimal in the least squares sense defined by (6).

Surface approximation uses a similar method to approximate $\mathbf{R}(u,v)$, by first obtaining an initial approximation $\mathbf{S_0}(u,v)$ which has the required behaviour around the boundaries. In order to compute $\mathbf{S_0}(u,v)$ it is necessary to first use the curve approximation technique described above to approximate the boundary curves of $\mathbf{S}$, and, if derivative continuity is required, the cross boundary partial derivative functions. The results of these curve approximations are then used to define the control points around the boundaries of a Bézier surface $\mathbf{S_0}(u,v)$ of the required degree which coincides with $\mathbf{R}(u,v)$ at the corner points $(u=0, u=1), (v=0, v=1)$. An "error surface" $\mathbf{R_e}(u,v)$ is then defined as

$$\mathbf{R_e}(u,v) = \mathbf{R}(u,v) - \mathbf{S_0}(u,v). \tag{10}$$

This surface can then be approximated by the constrained orthogonal polynomials, the resulting approximation being given by the formulae:

$$\mathbf{c}_{ij} = \int_0^1 \int_0^1 (\mathbf{R_e}(u,v)u_i(u)u_j(v))dudv$$

$$\mathbf{S}_1(u, v) = \sum_{i=1}^{N} \sum_{j=1}^{M} \mathbf{c}_{ij} u_i(u) u_j(v). \tag{11}$$

As in the curve case the final approximation is given by

$$\mathbf{S}(u, v) = \mathbf{S}_0(u, v) + \mathbf{S}_1(u, v),$$

with mean square error

$$E^2 = \int_0^1 \int_0^1 (\mathbf{R}_e(t))^2 dt - \sum_{i=1}^{N} \sum_{j=1}^{M} (\mathbf{c}_{ij})^2 du dv. \tag{12}$$

The method outlined above, and more comprehensively described in [1], can be used to approximate any parametric surface for which the integrals defined in (11) can be evaluated. In practice these integrals are evaluated by multi-point Gaussian quadrature, which must be done carefully due to the oscillatory nature of the basis functions involved. If the error estimate provided by (12) indicates that the approximations not sufficiently precise the surface must be subdivided.

§4. Test Results.

In its final form the data for the tests was supplied as Bézier control points. Only the first example, of a single patch surface of degree 9 was suitable for the degree reduction software. In order to process this surface it was necessary to convert the input data to explicit polynomial coefficient format. The results from the degree reduction were then converted back to Bézier format. Table 1 gives the results obtained. The errors quoted here are the error bounds from the degree reduction process, this provides an upper bound for the actual errors which is generally not attained. In all cases the CPU times apply to a VAX 8600 series computer in a clustered time-sharing environment. The level of continuity defines both the level of agreement with the original surface at the 4 corners and the minimum patch to patch continuity after subdivision of the surface. In principle the other test surfaces could have been degree reduced on a patch by patch basis, but this did not seem to be a worthwhile endeavour. For the orthogonal polynomial method, the version of the software used provided automatic first derivative continuity with the first basis function as $f_1(t) = t^2(t-1)^2$. This implied that in seeking a degree 3 approximation no orthogonal polynomials were actually used. In this exceptional case equation (12) can still be used to compute the errors in the initial approximation and initiate the subdivision strategy. The method is used as described for the degree 5 approximants, using 2 orthonormal basis functions only. The method generally works better when a higher degree approximation is required. The output is obtained by directly converting the orthogonal basis functions to Bézier format. In order to evaluate the integrals defined in (11) it was necessary to write a general evaluator to evaluate points and partial derivatives at any point on a Bézier surface. Tables 2 and 3 show the results obtained. Those for Brode are in line with expectations showing errors (in this case root mean square errors) which are of a similar order of magnitude to those in Table 1. For the test data defining SEITE 1 it was not possible to obtain a really close approximation and it is suspected that there may have been some corruption in the data used by the surface evaluator. The results quoted in Table 3 should give a reliable indication of the CPU time but the error estimates may not be accurate. Due to a restriction on the available disk space it was not possible to compute sample results for the final test case.

Table 1: Chebyshev Degree Reduction						
surface: BRODE ; degree: 9×9 ; segments: 1×1						
degree	segm.	minimal contin.	compr. factor	error toler.	meas. error	CPU
3×3	2×2	C^0 , C^0	1.56	0.1	0.099	4.6 sec
	4×4	C^1 , C^1	0.39	0.1	0.035	17.4 sec
5×5	2×3	C^1 , C^1	0.46	0.01	0.005	6.9 sec
	4×3	C^1 , C^1	0.23	0.002	0.0014	13.8 sec
7×7	2×1	C^0 , C^0	0.78	0.1	0.08	2.4 sec

Table 2: Constrained Orthogonal Approximation						
surface: BRODE ; degree: 9×9 ; segments: 1×1						
degree	segm.	minimal contin.	compr. factor	error toler.	meas. error	CPU
3×3	1×1	C^1 , C^1	6.25	1.0	0.5	3.1 sec
	2×2	C^1 , C^1	1.56	0.5	0.25	12.1 sec
	5×5	C^1 , C^1	0.25	0.01	0.009	12.1 sec
5×5	1×1	C^1 , C^1	2.78	0.1	0.03	6.9 sec
	2×2	C^1 , C^1	0.69	0.01	0.005	15.2 sec

Table 3: Constrained Orthogonal Approximation						
surface: SEITE 1 ; degree: 3×3 ; segments: 9×7						
degree	segm.	minimal contin.	compr. factor	error toler.	meas. error	CPU
3×3	3×5	C^1 , C^1	4.2	5.0	5.1	18.9 sec
5×5	2×2	C^1 , C^1	7.0	10.0	9.8	8.4 sec
	4×4	C^1 , C^1	1.75	5.0	4.3	32.8 sec
11×11	2×2	C^1 , C^1	1.75	2.0	1.6	22.3 sec

Acknowledgements. I would like to thank Peter Sherar of Cranfield University for his assistance with the computational tests.

References

1. Goult, R. J., Parametric Curve and Surface Approximation, in *The Mathematics of Surfaces III*, D. C. Handscomb (ed.), Oxford University Press, 1989.
2. Lachance, M. A., E. B. Saff and R. S. Varga, Bounds on incomplete polynomials vanishing at both end points of an interval, in *Constructive Approaches to Mathematical Models*, Academic Press, New York, 1979.
3. Sherar, P. A., Constrained Chebyshev Economisation for Parametric Curves and Surfaces. Cranfield M.Sc. thesis, 1987.

R. J. Goult
LMR Systems
33 Filgrave
Newport Pagnell
Bucks
MK 16 9ET, ENGLAND

Semi-regular B-spline Surfaces: Generalized Lofting by B-splines

J. Gravesen

Abstract. A generalization of B-spline surfaces is defined. These surfaces need only ordinary B-spline algorithms for evaluation, they allow in a natural way singularities in the interior, and can to some extent be used to interpolate between B-spline curves with different knot vectors.

§1. Introduction

In technical applications, it is often necessary to design surfaces with knuckle lines (kinks), see [4]. We propose a generalization of B-spline surfaces, which allow such knuckle lines in a natural and well behaved way.

We will consider a parameterized surface $r(u, v)$ as the locus of the curve $r_v : u \mapsto r(u, v)$ which is moved and distorted by varying the parameter v. An example is a tensor-product B-spline surface of degree (n, m) with knot vectors $\mathbf{u} = u_0, u_1, \ldots$ and $\mathbf{v} = v_0, v_1, \ldots$ and control points $\mathbf{d}_{ij}$,

$$r(u, v) = \sum_{i,j} \mathbf{d}_{ij} N_i^n(\mathbf{u}; u) N_j^m(\mathbf{v}; v),$$

where the functions $N_i^n(\mathbf{u}; u)$ are the B-splines of degree n on the knot vector $\mathbf{u}$. The surface can be written as the locus of a B-spline curve on the knot vector $\mathbf{u}$,

$$r(u, v) = r_v(u) = \sum_i \mathbf{d}_i(v) N_i^n(\mathbf{u}; u),$$

where the control points are defined by $\mathbf{d}_i(v) = \sum_j \mathbf{d}_{ij} N_j^m(\mathbf{v}; v)$, i.e., the curve is distorted and moved by varying the control points along B-spline curves with knot vector $\mathbf{v}$.

The idea is now also to vary the knot vector $\mathbf{u}$ of the curve $r_v(u)$, i.e., we will consider surfaces which are the locus of a B-spline curve,

$$r(u, v) = r_v(u) = \sum_i \mathbf{d}_i(v) N_i^n(\mathbf{u}(v); u),$$

Curves and Surfaces in Geometric Design
P. J. Laurent, A. Le Méhauté, and L. L. Schumaker (eds.), pp. 225–232.

where both the control points $\mathbf{d}_i(v)$ and the knot vector $\mathbf{u}(v)$ are B-spline curves on the knot vector $\mathbf{v}$, see definition 2.

If we look in the parameter plane, then the knot vectors $\mathbf{u}$ and $\mathbf{v}$ of a tensor-product B-spline surface form a regular grid, while the knot vectors $\mathbf{u}(v)$ and $\mathbf{v}$ of the new type of surface form a semi-regular grid, see Figure 1b; thus we call these surfaces *semi-regular B-spline surfaces*.

The B-spline curve $r_v(u)$ can be smooth for some values of v and have *knuckle points (kinks)* for other values of v. For the latter to happen we just have to allow some of the knots in the knot vector $\mathbf{u}(v)$ to be equal for certain values of v. The transition from a smooth curve to a curve with a knuckle point is always well behaved, see Theorem 1. This is in contrast to the situation for a tensor-product B-spline surface where the control points have to satisfy certain conditions in order to avoid wriggles in the transition zone, see [2], pp. 261–263, and [4].

As the knot vector of the curve $r_v(u)$ varies, a semi-regular surface can be used to interpolate between curves with different knot vectors. We will not discuss this aspect here, but just point out that there are problems, for example, if we interpolate between knot vectors $\mathbf{u}_j$ with a B-spline curve $\mathbf{u}(v)$ such that $\mathbf{u}(v_j) = \mathbf{u}_j$, then $\mathbf{u}(v)$ need not be a knot vector for each v.

The rest of the paper is organized as follows. In § 2 we give the precise definition of a semi-regular B-spline surface, illustrate it with an example and indicate how to develop methods to evaluate $r(u,v)$, $\frac{\partial r}{\partial u}(u,v)$, $\frac{\partial r}{\partial v}(u,v)$, etc. In § 3 we prove Theorem 1 which precisely describes the transition from a smooth curve to a curve with a knuckle point. In § 4 we present conclusions and final remarks.

§2. Definitions and Algorithms

If $\mathbf{u} = u_0, \ldots, u_{N+n-1}$ is a knot vector (*i.e.*, $u_0 \leq \cdots \leq u_{N+n-1}$), then we denote the B-splines of degree n with knot vector $\mathbf{u}$ by $N_i^n(\mathbf{u};u)$, $i = 0, \ldots, N$. The B-splines can be defined by the Mansfield-de Boor-Cox recursion:

$$N_i^0(\mathbf{u};u) = \begin{cases} 1 & \text{if } u \in [u_{i-1}, u_i[, \\ 0 & \text{if } u \notin [u_{i-1}, u_i[, \end{cases}$$

$$\delta_i = u_{i+k} - u_i, \quad \alpha_i = \frac{u - u_{i-1}}{\delta_{i-1}}, \quad \beta_i = \frac{u_{i+k} - u}{\delta_i}, \qquad (1)$$

$$N_i^k(\mathbf{u};u) = \alpha_i N_i^{k-1}(\mathbf{u};u) + \beta_i N_{i+1}^{k-1}(\mathbf{u};u),$$

see [1] or [2]. We need knot vectors $\mathbf{u}(v)$ which are B-spline curves, *i.e.*, where each knot is of the form

$$u_i(v) = \sum_j u_{ij} N_j^m(\mathbf{v};v).$$

In order to have a knot vector, we must have $u_0(v) \leq u_1(v) \leq \cdots$. A sufficient condition is $u_{0j} \leq u_{1j} \leq \cdots$ all j, *i.e.*, each sequence $\mathbf{u}_j = u_{0j}, u_{1j}, \ldots$ is a knot vector. Hence we make the following definitions.

Definition 1. *A knot net is a sequence* $\mathbf{u}_1, \ldots, \mathbf{u}_M$ *of knot vectors* $\mathbf{u}_j = u_{0j}, \ldots, u_{(N+n-1)j}$.

Definition 2. *A semi-regular B-spline surface of degree* n, m *with knot vector* $\mathbf{v} = v_0, \ldots, v_{M+m-1}$, *knot net* $(u_{ij})_{i=0,\ldots,N+n-1,j=0,\ldots,M}$ *and control net* $(\mathbf{d}_{ij})_{i=0,\ldots,N,j=0,\ldots,M}$, *is the locus of the B-spline curve*

$$r(u,v) = r_v(u) = \sum_{i=0}^{N} \mathbf{d}_i(v) N_i^n(\mathbf{u}(v); u),$$

where the control points $\mathbf{d}_i(v)$ *and the knot vector* $\mathbf{u}(v) = u_0(v), u_1(v), \ldots$ *are B-spline curves on the knot vector* $\mathbf{v}$

$$\mathbf{d}_i(v) = \sum_{j=0}^{M} \mathbf{d}_{ij} N_j^m(\mathbf{v}; v), \qquad i = 0, \ldots, N$$

$$u_i(v) = \sum_{j=0}^{M} u_{ij} N_j^m(\mathbf{v}; v), \qquad i = 0, \ldots, N+n-1.$$

We illustrate this type of surface with an example where the curve $r_v(u)$ is C^2 for $v \in [0, \frac{4}{5}[$ and has a knuckle point at $u = \frac{1}{2}$ for $v \in [\frac{4}{5}, 1]$. In order to concentrate on the effect of a variable knot vector, the control polygon will be essentially independent of v. To make the transition from a smooth curve to a singular curve as difficult as possible, we let the control polygon oscillate as much as possible.

Example. We consider a semi-regular B-spline surface of degree $3, 3$ with knot vector $\mathbf{v}$, knot net $\mathbf{u}_0, \ldots, \mathbf{u}_7$ and control net $\mathbf{d}_{ij} = (x_{ij}, y_{ij}, z_{ij})$, where

$$\mathbf{v} = 0, 0, 0, \tfrac{1}{5}, \tfrac{2}{5}, \tfrac{3}{5}, \tfrac{4}{5}, 1, 1, 1$$

$$\mathbf{u}_j = \begin{cases} 0, 0, 0, \tfrac{1}{4}, \tfrac{1}{2}, \tfrac{3}{4}, 1, 1, 1 & \text{for } j = 0, 1, 2, 3 \\ 0, 0, 0, \tfrac{1}{2}, \tfrac{1}{2}, \tfrac{1}{2}, 1, 1, 1 & \text{for } j = 4, 5, 6, 7 \end{cases}$$

$$x_{ij} = \tfrac{i}{6}, \quad y_{ij} = \tfrac{j}{7}, \quad z_{ij} = \begin{cases} 0 & \text{for } i = 0, 2, 4, 6 \\ \tfrac{3}{4} & \text{for } j = 0, 1, 2, 3 \text{ and } i = 1, 3, 5 \\ 1 & \text{for } j = 4, 5, 6, 7 \text{ and } i = 1, 5 \\ \tfrac{1}{2} & \text{for } j = 4, 5, 6, 7 \text{ and } i = 3 \end{cases}$$

see Figure 1a.

In Figure 1b the knot vector $\mathbf{u}(v)$ and the control points $\mathbf{d}_i(v)$ are plotted. Notice that $\mathbf{u}(v) = 0, 0, 0, \tfrac{1}{2}, \tfrac{1}{2}, \tfrac{1}{2}, 1, 1, 1$ for $v \in [\frac{4}{5}, 1]$, thus the surface has a knuckle line for $u = \frac{1}{2}$ and $v \in [\frac{4}{5}, 1]$. The final surface is plotted in Figure 1c, together with a blowup around the point $(\frac{1}{2}, \frac{4}{5})$, where the knuckle line appears. Notice the well behaved transition from a C^2 surface to a C^0 surface.

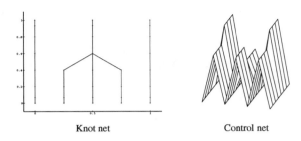

Knot net   Control net

Fig. 1a. The data needed for a semi-regular B-spline surface.

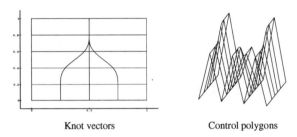

Knot vectors   Control polygons

Fig. 1b. The data for the curves $r_v(u)$.

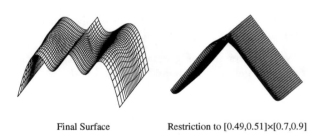

Final Surface   Restriction to [0.49,0.51]×[0.7,0.9]

Fig. 1c. The final surface and a blowup at $(\frac{1}{2}, \frac{2}{5})$.

As long as we only want 0-order information about the surface (*i.e.*, points), we only need to evaluate B-splines curves. Given a (u, v) we first evaluate the B-splines curves $u_i(v)$ and $\mathbf{d}_i(v)$ on the knot vector $\mathbf{v}$, and then evaluate the B-spline curve $r_v(u)$ on the knot vector $\mathbf{u}(v)$. If we want higher order information such as area or curvature, we need to find derivatives. We have

$$\frac{\partial^k r}{\partial u^k}(u, v) = \frac{d^k r_v}{du^k}(u),$$

for which well-known formulas exists, see [1] and [2]. So we only miss the derivatives in the v direction. This boils down to differentiating a B-spline (curve) with respect to the knots. We will only do it for the first derivative, as the higher order derivatives can be handled by analogous means. We have

$$\frac{\partial r}{\partial v}(u, v) = \sum_i \left(\frac{d\mathbf{d}_i}{dv}(v) N_i^n(\mathbf{u}(v); u) + \mathbf{d}_i(v) \frac{\partial N_i^n(\mathbf{u}(v); u)}{\partial v} \right).$$

Using the above mentioned formulas we can find

$$\dot{u}(v) = \frac{du_i}{dv}(v), \qquad \dot{\mathbf{d}}_i(v) = \frac{d\mathbf{d}_i}{dv}(v),$$

and by differentiating (1) we get

Proposition 1. *If the knot vector* $\mathbf{u} = \mathbf{u}(v)$ *depends on a parameter* v, *and* $u_i = u_i(v)$, *and* $\dot{u}_i = u_i'(v)$, *then the derivative* $\frac{\partial N_i^n(\mathbf{u};u)}{\partial v}$ *can be determined by the following recursion formulas*

$$\frac{\partial N_i^0(\mathbf{u}; u)}{\partial v} = 0,$$

$$\dot{\delta}_i = \dot{u}_{i+k} - \dot{u}_i, \quad \dot{\alpha}_i = \frac{-\dot{u}_{i-1} - \alpha_i \dot{\delta}_{i-1}}{\delta_{i-1}}, \quad \dot{\beta}_i = \frac{\dot{u}_{i+k} - \beta_i \dot{\delta}_i}{\delta_i},$$

$$\frac{\partial N_i^k(\mathbf{u}; u)}{\partial v} = \alpha_i \frac{\partial N_i^{k-1}(\mathbf{u}, u)}{\partial v} + \beta_i \frac{\partial N_{i+1}^{k-1}(\mathbf{u}, u)}{\partial v}$$

$$+ \dot{\alpha}_i N_i^{k-1}(\mathbf{u}, u) + \dot{\beta}_i N_{i+1}^{k-1}(\mathbf{u}, u). \tag{2}$$

Another approach to the evaluation of B-splines is to use knot insertion. We want to insert a knot u in the knot vector $\mathbf{u}(v)$, and in order to be able to speak of derivatives we have to insert the knot in $\mathbf{u}(v)$ for all v in some small open set. In this situation it might happen that u crosses some of the knots in $\mathbf{u}(v)$. Then the new knot vector and control polygon will have singularities, (they cancel in the end, but this makes it difficult to speak of derivatives). So we will assume this does not happen (to first order).

Proposition 2. *Let* $r_v(u)$ *be a B-spline curve of degree* n *with knot vector* $\mathbf{u}(v)$ *and control points* $\mathbf{d}_i(v)$ *and insert the knot* $u = u(v)$. *Let*

$$\dot{u}_i = u_i'(v), \quad \dot{\mathbf{d}}_i = \mathbf{d}_i'(v), \quad \dot{u} = u'(v),$$

and denote the new control points $\widehat{\mathbf{d}}_i$. *If* $u = u_k \Rightarrow \dot{u} = \dot{u}_k$, *then for each* i *such that* $u_{i-1} < u < u_{i+n-1}$, *we have*

$$\delta_i = u_{i+n} - u_{i-1}, \qquad \alpha_i = \frac{u_{i+n} - u}{\delta_i} \tag{3}$$

$$\widehat{\mathbf{d}}_i = \alpha_i \mathbf{d}_{i-1} + (1 - \alpha_i)\mathbf{d}_i$$

and

$$\dot{\delta}_i = \dot{u}_{i+n} - \dot{u}_{i-1}, \qquad \dot{\alpha}_i = \frac{\dot{u}_{i+n} - \dot{u} - \alpha_i \dot{\delta}_i}{\delta_i} \tag{4}$$

$$\widehat{\mathbf{d}}_i' = \alpha_i \dot{\mathbf{d}}_{i-1} + (1 - \alpha_i)\dot{\mathbf{d}}_i + \dot{\alpha}_i(\mathbf{d}_{i-1} - \mathbf{d}_i)$$

Proof: The formulas in (3) just constitute the knot insertion algorithm, see [1] and [2]. Eqs. (4) are obtained by differentiation of (3). ∎

§3. The Shape of a Collapsing Segment

The main motivation for this generalization of B-spline surfaces is the possibility to have knuckle lines. If the curve $r_v(u)$ has degree n, then we can accomplish this by letting n knots meet, see Figure 1b, where 3 knots meet at $v = \frac{4}{5}$. When n knots meet, the segment of the curve defined for u between these n knots collapses to a point. We will investigate the transition from the smooth part of the surface to the part with the knuckle line by determining the control polygon for the collapsing segment. If the n knots have contact of order m at the point where they come together then (to order m) the curve segment has a control polygon which forms a wedge, see Theorem 1 and Figure 2.

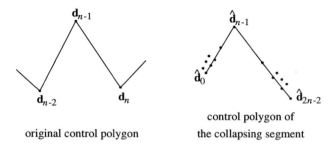

original control polygon   control polygon of
                                  the collapsing segment

Fig. 2. The control polygon for the B-spline and the collapsing segment.

Theorem 1. *Let the knots of a B-spline curve $r_v(u)$ satisfy*

$$u_i(v) = \widehat{u} + v^m \dot{u}_i + O(v^{m+1}), \quad \text{for } i = n - 1, \dots, 2n - 2.$$

If $\widehat{\mathbf{d}}_i$ denotes the control points of the segment, defined on $[u_{n-1}, u_{2n-2}]$, then

$$\Delta\widehat{\mathbf{d}}_i = \begin{cases} v^m \dfrac{\dot{u}_{n+1} - \dot{u}_{n-1}}{\widehat{u} - u_{i-1}} \Delta\mathbf{d}_{n-2} + O(v^{m+1}) & \text{for } i = 0, \dots, n - 2 \\[2ex] v^m \dfrac{\dot{u}_{2n-2} - \dot{u}_i}{u_{2n-1} - \widehat{u}} \Delta\mathbf{d}_{n-1} + O(v^{m+1}) & \text{for } i = n - 1, \dots, 2n - 3 \end{cases}$$

Proof: If we insert the knots u_{n-1} and u_{n-2} to full multiplicity, then we obtain the control polygon for the collapsing segment and this control polygon determines the shape of the segment. The knot insertion is described by the diagram,

$$\begin{array}{ccccccc} \mathbf{d}_0 & \cdots & \mathbf{d}_{n-1} & & \cdots & & \mathbf{d}_{2n-2} \\ \mathbf{d}_1^1 & \cdots & \mathbf{d}_{n-1}^1 & \mathbf{d}_n^1 & \cdots & \mathbf{d}_{2n-2}^1 & \\ & \ddots & \cdots & & \ddots & \cdots & \\ & \mathbf{d}_{n-1}^{n-1} & & & \mathbf{d}_{2n-2}^{n-1} & & \end{array}$$

If we let $\mathbf{d}_i^0 = \mathbf{d}_i$, the control points of the collapsing segment are given by

$$\widehat{\mathbf{d}}_i = \begin{cases} \mathbf{d}_{n-1}^{n-1-i} & \text{for } i = 0, \dots, n - 1, \\ \mathbf{d}_i^{i-n+1} & \text{for } i = n, \dots, 2n - 2. \end{cases}$$

For $k = 1, \ldots, n-1$ and $i = k, \ldots, n-1$, we have

$$\mathbf{d}_i^k = \frac{u_{n-k+i} - u_{n-1}}{u_{n-k+i} - u_{i-1}} \mathbf{d}_{i-1}^{k-1} + \frac{u_{n-1} - u_{i-1}}{u_{n-k+i} - u_{i-1}} \mathbf{d}_i^{k-1}$$

$$= \mathbf{d}_i^{k-1} + \frac{u_{n-k+i} - u_{n-1}}{u_{n-k+i} - u_{i-1}} (\mathbf{d}_{i-1}^{k-1} - \mathbf{d}_i^{k-1}).$$

Inserting $u_j = \widehat{u} + v^m \dot{u}_j + O(v^{m+1})$, for $j = n - k - 1, n - 1$ yields

$$\mathbf{d}_i^k = \mathbf{d}_i^{k-1} - v^m \frac{\dot{u}_{n-k+i} - \dot{u}_{n-1}}{\widehat{u} - u_{i-1}} \Delta \mathbf{d}_{i-1}^{k-1} + O(v^{m+1}),$$

where Δ denotes the forward difference. In particular

$$\Delta \mathbf{d}_{i-1}^k = \Delta \mathbf{d}_{i-1}^{k-1} + O(v^m),$$

so induction on k yields

$$\mathbf{d}_i^k = \mathbf{d}_i^{k-1} - v^m \frac{\dot{u}_{n-k+i} - \dot{u}_{n-1}}{\widehat{u} - u_{i-1}} \Delta \mathbf{d}_{i-1} + O(v^{m+1}).$$

When $i = n - 1$, we have for $k = 1, \ldots, n-1$,

$$\widehat{\mathbf{d}}_{n-1-k} = \mathbf{d}_{n-1}^k = \mathbf{d}_{n-1}^{k-1} - v^m \frac{\dot{u}_{2n-k-1} - \dot{u}_{n-1}}{\widehat{u} - u_{n-2}} \Delta \mathbf{d}_{n-2} + O(v^{m+1}),$$

or

$$\Delta \widehat{\mathbf{d}}_{n-1-k} = \mathbf{d}_{n-1}^{k-1} - \mathbf{d}_{n-1}^k = v^m \frac{\dot{u}_{2n-k-1} - \dot{u}_{n-1}}{\widehat{u} - u_{n-2}} \Delta \mathbf{d}_{n-2} + O(v^{m+1}). \quad (5)$$

We can similarly show that

$$\Delta \widehat{\mathbf{d}}_{n+k-2} = \mathbf{d}_{k+n-1}^k - \mathbf{d}_{k+n-2}^{k-1} = v^m \frac{\dot{u}_{2n-2} - \dot{u}_{k+n-2}}{u_{2n-1} - \widehat{u}} \Delta \mathbf{d}_{n-1} + O(v^{m+1}), \quad (6)$$

for $k = 1, \ldots, n-1$. Together (5) and (6) establish the theorem. ∎

§4. Conclusion.

In [3] Hoitsma and Lee suggest another generalization of B-spline surfaces. They consider surfaces of the form

$$r(u, v) = \sum \mathbf{d}_{i,j}(u, v) N_i^{n_j}(\mathbf{u}_j; u) N_j^{m_i}(\mathbf{v}_i; v),$$

where $\mathbf{d}_{i,j}(u, v)$ in [3] are assumed constant. The method presented here has the advantages of being conceptual and computational simple. Compared to an ordinary tensor-product B-spline surface with the same control polygon the amount of data needed to define a semi-regular B-spline surface is roughly only 4/3 times as large.

The main feature is the natural ability to model surfaces with knuckle lines (kinks). Another feature is the possibility to interpolate between B-spline curves with different knot vectors, but as mentioned in the introduction this is not straightforward.

References

1. de Boor C. *A Practical Guide to Splines*, Springer-Verlag, New York, 1978.
2. Farin G. *Curves and Surfaces for Computer Aided Geometric Design, A Practical Guide*, Second Edition, Academic Press, New York, 1990.
3. Hoitsma H. H., and Lee M., Generalized Rational B-spline Surfaces, in *NURBS for Curve and Surface Design*, G. Farin (ed.), SIAM, Philadelphia, 1991, 87–102.
4. Jensen J. J., and Baatrup J., Transformation of Body Planes to a B-spline Surface, in *Computer Applications in the Automation of Shipyard Operation and Ship Design, VI*, D. Lin, Z. Wang and C. Kuo (eds.), Procedings of the Sixth IFIP International Conference, Shanghai, 1988, 3–11.

Jens Gravesen
Mathematical Institute
The Technical University of Denmark
Building 303
DK-2800 Lyngby, DENMARK
gravesen@mat.dth.dk

On Best Convex Interpolation of Curves

Christoph Henninger and Karl Scherer

Abstract. The following problem is considered: let $P_0, \ldots, P_N$ be data points in $\mathbb{R}^2$ such that the interpolating polygon is convex. Then determine a curve $F(t) : t \in [a, b] \to \mathbb{R}^2$ such that the functional $J(F) := \int_a^b |F''(t)|^2 dt$ is minimized under the constraints $F(t_i) = P_i$, $i = 0, \ldots, N$, and $\kappa(F) = F' \otimes F'' \leq 0$. Here $|\cdot|$ denotes the Euclidean and $\otimes$ the cross product. The nodes satisfy $a = t_0 < t_1 < \cdots < t_n = b$. It is shown that the above problem does always have a solution. In each segment (t_i, t_{i+1}) every (regular) solution is C^2-continuous and piecewise cubic with at most two extra knots. Criteria are given which allow to determine their number and positions exactly. Based on this characterization an algorithm for determining the solution is constructed.

§1. Introduction

In recent years the classical results on best interpolation of data have been extended to problems with constraints such as the monotonicity, convexity or positivity of the interpolants. There exists already a rich literature on this subject, we refer e.g. to [10].

However, only scalar-valued functions or data (the non-parametric case) seem to be considered so far. Here we consider convex data in the plane and want to construct a "best" interpolating curve which is convex. The precise formulation is as follows: suppose we are given points

$$P_0, \ldots, P_N \in \mathbb{R}^2, \quad P_i \neq P_{i+1}, \tag{1}$$

satisfying the "discrete convexity condition"

$$D_{i-1} \otimes D_i \leq 0, \quad 1 \leq i \leq N - 1, \quad D_i := P_{i+1} - P_i. \tag{2}$$

Here, $a \otimes b$ denotes the cross product $a_1 b_2 - a_2 b_1$ of two vectors $a = (a_1, a_2)$, $b = (b_1, b_2) \in \mathbb{R}^2$. Geometrically, condition (2) says that the interpolating polygon is convex. We consider smooth curves $F(t) = (f_1(t), f_2(t)) : [a, b] \to \mathbb{R}^2$

Curves and Surfaces in Geometric Design
P. J. Laurent, A. Le Méhauté, and L. L. Schumaker (eds.), pp. 233–240.
Copyright ⓒ 1994 by A K PETERS, Wellesley, MA
ISBN 1-56881-039-3.

with components $f_j(t) \in L_2^2(a, b)$, i.e., f_j is continuously differentiable, f_j' is absolutely continuous and $f_j'' \in L_2(a, b)$, and define for a fixed set of nodes

$$\underline{t} : a = t_0 < t_1 < \cdots < t_N = b \tag{3}$$

the class of smooth convex interpolating curves by

$$Z(\underline{t}) := \{F(t) = (f_1(t), f_2(t)) : f_j \in L_2^2(a, b), F(t_i) = P_i, 0 \leq i \leq N,$$
$$\kappa(F)(t) := F'(t) \otimes F''(t) \leq 0 \text{ a.e.}\}. \tag{4}$$

The aim is then to find an optimal curve F^* in $Z(\underline{t})$ in the sense that

$$J(F^*) = \inf_{F \in Z(\underline{t})} \{J(F) := \int_a^b (F''(t), F''(t))dt\}. \tag{P}$$

Here $(\cdot, \cdot)$ denotes the scalar product in $\mathbb{R}^2$. We call a solution of this problem a *best convex interpolant of the data* $(P_i)_{i=0}^N$ with respect to the sequence $\underline{t}$ of nodes. Due to the nature of the "curvature constraint" in (4) it is a non-convex problem, i.e., $Z(\underline{t})$ is in general not a convex set. This contrasts to the corresponding scalar case where problem (P) simplifies to a convex problem since then the constraint in (4) $f''(t) \leq 0$ becomes a linear one. (In fact in the literature mentioned above only convex problems are considered.)

Before stating an existence theorem, we remark that in order to admit variable nodes, problem (P) can be extended to

$$\inf_{\underline{t}} \inf_{F \in Z(\underline{t})} \{J(F) := \int_a^b (F''(t), F''(t))dt\}. \tag{5}$$

This model for an optimal convex interpolating curve also takes care of the influence of parametrization. It suggests to introduce reparametrization of the curves $F(t)$ in (5) with respect to arclength. One can show

Lemma 1. *The infimum in Problem (P) has a value less than or equal to*

$$\inf_{\underline{t}} \inf_{F \in Z(\underline{t})} \{\int_a^b K(F)^2|F'(t)|dt : L(F) := \int_a^b |F'(t)|dt = b - a\}, \tag{6}$$

where $|F(t)|^2 = (F(t), F(t))$ and $K(F) = (F' \otimes F'')/|F'|^3$ denotes the curvature of F.

In (6) the variational principle is more geometrically oriented, and the curvature is minimized with prescribed arclength. Such problems have been considered in the unconstrained case by many authors. We refer to the monograph of Fisher and Jerome [2], and to Malcolm [7]. In view of the equivalence in Lemma 1, it is not surprising that the method in [4] can be extended to an existence proof for the constrained problem (6).

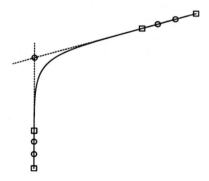

Fig. 1. $P_0 = (0,0), P_1 = (1,1), M_0 = (0,2.4), M_1 = (3.4,1), t_0 = 0, t_1 = 1.$

Theorem 1. *There exists a solution of problem* (5).

In the case of fixed nodes $\underline{t}$ one can easily see that the set $Z(\underline{t})$ is always non-void. This implies

Corollary 1. *There exists a solution of problem* (P) *for any fixed* $\underline{t}$.

For a proof of these results as well as of Lemma 1, we refer to [4]. We concentrate now on problem (P) for fixed knots since the corresponding problem in (6) does not allow a nice reformulation as in the non-parametric case (which would result in a convex problem, cf. [10]).

§2. The Local Problem

The local problem considers only interpolation at two points. The motivation for this comes from the idea to construct a solution of (P) by composition of two-point interpolating curves. However C^1-smoothness must be preserved when doing this. So we prescribe for the local problem the derivatives at the end points as well. We fix points P_0, P_1 and derivatives M_0, M_1 at the end points of an interval $[t_0, t_1]$ and try to find a solution of

$$\inf_{F \in Z(t_0,t_1)} \{J_0(F) := \int_{t_0}^{t_1} (F''(t), F''(t))dt : F'(t_i) = M_i, i = 0, 1\}. \qquad (P_{loc})$$

Here we used the notation $Z(t_0, t_1)$ for $Z(\underline{t})$ since $\underline{t}$ consists of the points t_0, t_1. Furthermore we make the assumption that (see Figure 1)

$$M_0 \otimes D, \ D \otimes M_1 \leq 0, \quad M_0 \otimes D = 0 \iff D \otimes M_1 = 0, \qquad (A)$$

where $D = P_1 - P_0$. This is no restriction since the data D, M_i comes from a global solution which is convex.

Problem (P_{loc}) can be solved completely. In the following we briefly review the basic ideas, for details refer to [4]. First, methods of control theory will be applied (for spline problems this has been done previously by Mangasarian–Schumaker [8] and Opfer–Oberle [9]) in order to obtain a characterization of the solution. To this end, problem (P_{loc}) is reformulated in the form

$$\inf \int_{t_0}^{t_1} L(t, X, U) \tag{7a}$$

$$X' = G(t, X, U) = (g_1(t, X, U), \ldots, g_4(t, X, U)), \tag{7b}$$

$$\phi(t, X, U) \leq 0 \text{ a.e. in } [t_0, t_1], \tag{7c}$$

$$X(t_i) = (P_i, M_i), \quad i = 0, 1, \tag{7d}$$

where $X(t) = (x_1(t), x_2(t), x_3(t), x_4(t))$ are the so-called *state variables* and $U(t) = (u_1(t), u_2(t))$ the *control variables*. They are defined in our case by

$$x_j(t) = f_j(t), \quad x_{j+2} = f_j'(t), \quad u_j(t) = f_j''(t), \quad j = 1, 2. \tag{8}$$

Then we find from the definition of (P_{loc}) that

$$L(t, X, U) := (U, U) = (F'', F''), \tag{9a}$$

$$g_j(t, X, U) := x_{j+2}, \quad g_{j+2}(t, X, U) := u_j, \quad j = 1, 2, \tag{9b}$$

$$\phi(t, X, U) := x_3 u_2 - x_4 u_1 = F' \otimes F''. \tag{9c}$$

Using well-known results of control theory (see Hestenes [5, Chapter 7]) one now obtains the following necessary conditions for a solution X^*, U^* of (7):

Theorem 2. *Let* $X^*(t), U^*(t)$ *be a regular solution of (7), i.e.,*

$$rank \left(\frac{\partial \phi}{\partial u_j} \right)_{j=1,2} = rank\, (x_3(t), x_4(t)) = 1, \quad \forall t \in [t_0, t_1].$$

Then there exist multipliers $\lambda \geq 0, \Pi(t) = (\pi_1(t), \ldots, \pi_4(t))$ *and* $\mu(t)$ *such that the Hamilton function*

$$H(t, X, U, \Pi, \mu, \lambda) = \sum_{i=1}^{4} \pi_i(t) g_i(t, X, U) - \lambda L(t, X, U) - \mu(t)\phi(t, X, U)$$

$$= \pi_1 x_3 + \pi_2 x_4 + \pi_3 u_1 + \pi_4 u_2 - \lambda(u_1^2 + u_2^2)$$

$$- \mu(x_3 u_2 - x_4 u_1)$$

satisfies the following conditions:

(1) $\mu(t)$ *is piecewise continuous on* $[t_0, t_1]$ *and continuous at each point of continuity of* U^* *(i.e.,* $F^{*''}$*). Moreover,* $\mu(t) \geq 0$ *with* $\mu(t) = 0$ *at each value of* t *for which* $\phi(t, X^*, U^*) < 0$.

(2) The multipliers $\pi_i(t)$ are continuous and have piecewise continuous derivatives, and the function $H(t, X^*(t), U^*(t), \Pi(t), \mu(t), \lambda)$ is continuous on $[t_0, t_1]$.

(3) The functions $X^*(t), U^*(t), \Pi(t), \mu(t)$ satisfy the Euler-Lagrange equations

$$x_i' = H_{\pi_i}, \quad \pi_i' = -H_{x_i}, \quad 0 = H_{u_i}, \quad \frac{d}{dt}H = H_t.$$

We remark that the case $\lambda = 0$ corresponds to the degenerate case $U \equiv 0$, i.e., $M_0 = M_1$ and M_0 such that $P_1 - P_0 = (t_1 - t_0)M_0$ and hence $F^*(t) = P_0 + (t - t_0)M_0$. In the case $\lambda > 0$ one can always achieve $\lambda = 1$ by scaling.

In view of (8), a regular solution has the property $|F'(t)| > 0 \ \forall t \in [t_0, t_1]$. Theorem 1 does not guarantee this for a solution so that the above characterization may not be applicable. Nevertheless, we will work with the necessary conditions above and later see that they determine uniquely the solution of (P_{loc}) if it is regular. Numerical experiments indicate that indeed (P_{loc}) may have a non-regular solution (derived from the above conditions) for choices of extremely large M_0, M_1. On the other hand, the experiments have shown that in the neighborhood of a solution of the global problem (P) the corresponding solutions of (P_{loc}) are in fact regular.

Although the Hamilton function satisfies still other properties (see [5, p.346ff]) the above properties (1)-(3) contain enough information to describe a solution exactly. At first observe that the first set of Euler-Lagrange equations is just the state equation (7b) which is nothing but a reformulation of the definition (8). The next two sets yield the four equations

$$\pi_1' = 0 = \pi_2', \quad \pi_3' = -\pi_1 + \mu f_2'', \quad \pi_4' = -\pi_2 - \mu f_1'', \tag{10}$$

and the two equations

$$-2f_1'' + \mu f_2'' + \pi_3 = 0 = -2f_2'' - \mu f_1'' + \pi_4. \tag{11}$$

Let us first consider the unconstrained case where the curvature constraint is never active and thus $\mu \equiv 0$. Then we see from (10) that π_3, π_4 must be linear functions and hence $F(t) = (f_1(t), f_2(t))$ a cubic polynomial curve by (11). But then the boundary conditions (7d) determine the solution of the unconstrained local problem uniquely by the Hermite interpolation problem for cubic polynomial curves with the data of (7d). We call this curve a *free solution* or *unconstrained solution*. One can show more generally ([4])

Theorem 3. *Each component of a (regular) solution F^* of (P_{loc}) must be a piecewise cubic function. Moreover, on each (open) segment the curvature is either identically zero (active constraint) or always < 0.*

Hence a solution F^* consists of pieces of cubic arcs and straight lines, parametrized by cubic polynomials, which alternate with each other. For the proof of this latter property one has to prove in advance

Lemma 2. *A (regular) solution F^* of (P_{loc}) is everywhere curvature continuous.*

Concerning the proof we remark that properties (1),(2) of Theorem 2 are essential. One can sharpen Lemma 2 to

Lemma 3. *$F^{*\prime\prime}$ is continuous on $[t_0, t_1]$.*

As a corollary, we obtain the fact that a (regular) solution of (P_{loc}) must be a C^2-cubic spline curve. The question now is how many pieces such a solution can contain. From Theorem 3 it is immediate that there are countably many. To settle this question, the following results are proved in [4]:

Lemma 4. *Let assumption (A) be satified. If the free solution of (P_{loc}) satisfies the curvature constraint at the boundary points t_0, t_1, it does so everywhere in $[t_0, t_1]$, i.e., it is the solution of (P_{loc}) itself.*

We remark that the proof uses the representation in Bézier form. This seems not to be noticed in the literature though the connection between convexity of the curve and the underlying control polygon is well known (see e.g. [3]). A further auxiliary result is

Lemma 5. *Suppose assumption (A) is satisfied. Then a regular subarc consisting of two free arcs connected by a straight line cannot be optimal.*

If the free solution of (P_{loc}) does not satisfy the curvature constraint at a boundary point a (constrained) solution of (P_{loc}) must start at this point with a cubic polynomial piece having curvature identically zero.

Based on these lemmas one can finally conclude

Theorem 4. *A (regular) solution F^* of (P_{loc}) can consist of at most three cubic polynomial curves connected to a C^2-curve.*

Hence, there only remain four possibilities:
(1) the solution is equal to the free solution, hence, a cubic polynomial curve on all of $[t_0, t_1]$,
(2) the solution begins with a straight line and continues with a cubic arc,
(3) the solution begins with a cubic arc and continues with a straight line,
(4) the solution begins at both end points with straight lines which are connected by a cubic arc.

One can show then that a solution of one of the above types always exists and is unique.

§4. The Global Problem

In order to find a solution of (P) we solve on each interval $[t_i, t_{i+1}]$ problem (P_{loc}) for a fixed choice of derivatives $\underline{M} := (M_i)_{i=0}^{N}$, $M_i \in \mathbb{R}^2$, and minimize then with respect to the M_i. This procedure is justified by

Theorem 5. *Let*

$$Z_i(\underline{t}, \underline{M}) = \{F : f_j \in L_2^2(t_i, t_{i+1}), \ F(t_j) = P_j, \ F'(t_j) = M_j, \ j = i, i+1,$$
$$F' \otimes F'' \leq 0\}$$

be the set of feasible curves on $[t_i, t_{i+1}]$. *Then there holds*

$$\min_{F \in Z(\underline{t})} \int_a^b (F'', F'') = \min_{\underline{M}} \sum_{i=0}^{N-1} \min_{F \in Z_i(\underline{t}, \underline{M})} \int_{t_i}^{t_{i+1}} (F'', F'').$$

The proof of this theorem is not difficult. We refer to [4] as well as for an analogous result for problem (6).

Next we describe how the solution of (P_{loc}) is computed with the help of Theorem 4. We do this for the most general case (4) of Theorem 4. The principal observation is that the Bézier polygon for the middle arc must have control points $Q_1, \ldots, Q_4 \in \mathbb{R}^2$ where $Q_2 = Q_3 =: Q^*$ is obtained as the intersection of the straight lines passing through P_0 in direction M_0 and through P_1 in direction M_1. This determines the points Q_1, Q_4 as well as the control points of the other two Bézier polygons by the boundary conditions and the continuity conditions (up to C^2) once the additional two knots τ_1, τ_2 in (t_0, t_1) have been fixed. Figure 1 illustrates this. Indicated are the Bézier points of the three cubic patches, where circles are the interior Bézier points.

Of course the integral $\int_{t_i}^{t_{i+1}} (F'', F'')$ must still be minimized with respect to τ_1, τ_2 in order to obtain the (regular) solution of (P_{loc}). In [4] this problem has been reduced to the determination of the roots of a 5-th degree polynomial. For details as well as corresponding formulae for the cases (2),(3) of Theorem 4 we refer to [4].

To solve the global problem (P) we apply a quasi-Newton method for minimization of J with respect to $\underline{M}$. To ensure feasibility the inequality constraints (A) have to be maintained. Since the D_i are fixed these constraints become linear ones. On the other hand the partial derivatives with respect to the M_i cannot be computed explicitly (at least in case (4) of Theorem 4) so they have to be approximated by difference quotients. Hence, we employ the method described by Bräuninger [1], which has been developed for problems of such a type. Further attention had to be paid to the case of colinear data where in view of (A) a reduction of the number of variables was necessary. As starting values for M_i in most cases we chose $M_i = (\bar{D}_i + \bar{D}_{i+1})/2$, $\bar{D}_i = D_i/|D_i|$. For all tested examples, the resulting algorithm needed only a few iterations until convergence.

§5. Final Remarks

The preceding discussion shows that the variational principle of problem (P) leads to a concrete numerical method for constructing convex interpolating curves which are optimal in a well defined sense. Other methods considered so far for this purpose do not seem to have such a property. In particular the local problem (P_{loc}) has been solved completely except that a criterion excluding the existence of a non-regular solution is lacking. As mentioned above this can be expected only for data in a neighborhood of a solution of (P). For a positive answer one would need more information about (P), in particular a uniqueness proof would be desirable.

The same question is of interest also for problem (5) with variable knots because of its connection to the more geometric problem (6). Another direction for further research is to extend problem (P) to data with "inflection points". Then the constraint has to be formulated such that a solution does not induce artificial inflection points. In the non-parametric case this has been pursued in [6]. Here in the parametric case one might profit from the almost complete result of Theorem 4 for problem (P_{loc}).

References

1. Bräuninger, J., A quasi-Newton method for minimization under linear constraints without evaluation of any derivatives, Computing **21** (1979), 127–141.
2. Fisher, S. D., and J. W. Jerome, *Minimum Norm Extremals in Function Spaces*, Lecture Notes in Mathematics 479, Springer, Berlin, 1975.
3. Goodman, T. N. T., and K. Unsworth, Shape preserving interpolation by curvature continuous parametric curves, Computer-Aided Geom. Design **5** (1988), 323–340.
4. Henninger, C., Beste formerhaltende parametrische Interpolation, dissertation, Universität Bonn, 1994.
5. Hestenes, M., *Calculus of Variations and Optimal Control Theory*, Wiley, New York, 1966.
6. Irvine, L. D., S. P. Marin, and P. W. Smith, Constrained interpolation and smoothing, Constr. Approx. **2** (1986), 129–151.
7. Malcolm, M. A., On the computation of nonlinear spline functions, SIAM J. Numer. Anal. **14** (1977), 254–282.
8. Mangasarian, O. L., and L. L. Schumaker, Splines via optimal control, in *Approximations with Special Emphasis on Spline Functions*, I. J. Schoenberg (ed.), Academic Press, New York, 1969, 119–155.
9. Opfer, G., and H. J. Oberle, The derivation of cubic splines with obstacles by methods of optimization and optimal control theory, Numer. Math. **52** (1988), 17–32.
10. Ward, J. D., Some constrained approximation problems, in *Approximation Theory V*, C. K. Chui, L. L. Schumaker, and J. L. Ward (eds.), Academic Press, New York, 1986, 211–229.

Christoph Henninger, Karl Scherer
Institut für Angewandte Mathematik
Wegelerstr. 6
D – 53115 Bonn, GERMANY
UNM11C@IBM.rhrz.uni-bonn.de

Approximate Conversion and Data Compression of Integral and Rational B-spline Surfaces

Josef Hoschek and Franz-Josef Schneider

Abstract. Conversion methods are required for data exchange. We introduce approximation and conversion algorithms for integral or rational B-spline surfaces which use least squares with discrete L_2-norm and parameter correction. The method can be extended to trimmed surfaces.

§1. Introduction

Most computer aided design systems for free-form curve and surface modelling use parametric representations. Nowadays, rational representations are also used because of their larger degree of freedom, and the fact that they permit exact representation of conics. Nevertheless, the representation schemes used within these systems differ a lot with regard to the types of polynomial bases and the maximum polynomial degrees provided. Bernstein-Bézier, B-Spline, and monomial basis functions are frequently used in various systems. Polynomial degrees vary between 3 and about 20. Conversion from one polynomial base to another can be achieved by direct matrix multiplication whenever the number and degrees of polynomial terms in both representations are equal, or the degree of the polynomials should be elevated [9], [5].

If two systems do not allow the same maximum polynomial degree, then approximate conversions of high degree into low order functions (reducing combined with splitting spline segments) and perhaps vice versa (elevating and merging spline segments) are inevitable. This causes approximation errors which must be minimized [4,10,15,1,12].

None of these methods change the parametrization. In the development [7,8] we have used for integral surfaces a suitable parameter correction so that the error vectors, which are to be minimized, are (approximately) perpendicular to the approximating surface. Parameter correction is a very effective tool to reduce dramatically the number of surface patches. Thus the algorithms developed below can also be used for data compression, i.e. the reduction of the number of coefficients in the B-spline representation.

Curves and Surfaces in Geometric Design
P. J. Laurent, A. Le Méhauté, and L. L. Schumaker (eds.), pp. 241–250.
Copyright © 1994 by A K PETERS, Wellesley, MA
ISBN 1-56881-039-3.

§2. Norm, Segments and Points

We assume that the given surfaces $\mathbf{Y}(u,v)$ are sets of (integral or rational) Bézier-patches or B-spline patches. The set may have a global parametrization over a rectangle in the parametric domain. We shall approximate this set of surfaces by integral or rational tensor product B-splines using least square methods with discrete L_2-norm, therefore we must first determine a suitable knot sequence and a suitable set of points on the given set.

To get the number of segments we proceed heuristically: As we have observed a cubic curve approximates easily a curve with one minimum of curvature, a quartic curve approximates easily a curve with two minima or in general a curve of order k can easily approximate $k-3$ minima of curvature. Therefore we discretize the set of surfaces in the two parameter directions u and v for instance with $N \times M$ equidistant points on the boundary curves, where N and M are suitable chosen. The parameter values of these points may be $u_i = const.$ or $v_k = const.$. On each line u_i (or v_k resp.) we determine the minima of the curvature κ with help of an additional (finer) discretization of the line u_i (or v_k). For instance, on the parametric line u_i, these points may have the parametric value ν_j. We have a minimum of curvature if

$$\kappa(\nu_{j-1}) - \kappa(\nu_j) < 0 \qquad \text{and} \qquad \kappa(\nu_{j+1}) - \kappa(\nu_j) > 0. \tag{1}$$

Analogously, we discuss the lines $v_k = const.$. The maximal number of minima in the u-direction may be U; the maximal number of minima in the v-direction may be V. Because of our heuristic model, we need in the u-direction for a generic B-spline curve of order k (degree $k-1$, (and analogously in the v-direction)

$$S_u = int \left[\frac{U}{k-3} \right] + 1 \tag{2}$$

segments or a knot sequence with boundary values of multiplicity k and inner values of multiplicity 1. The maximal summation index n_u (resp. n_v) in the corresponding B-spline representation for simple knots is $n_u = S_u + k - 2$. If multiplicity l is required for the interior knots, we get $n_u = (S_u - 1) \cdot l + k - 1$. Our implementation starts with these values (n_u, n_v) and a suitable chosen order k. If the required error tolerance cannot be fullfilled, we can proceed with alternative strategies:

a) fix k (continuity C^{k-2}) and introduce a further segment with $S_u^* = S_u + 1$,

b) change the summation index to $n_u^* = n_u + 1$ and determine an appropriate order k^* out of a suitable chosen interval $k_0 \leq k^* \leq k_1$ of the order k, while the continuity class C^r (r arbitrarily) in u-direction (resp. v) is constant. This approach leads to a minimal data volume within the chosen interval of the order of the surface. In our implementation we have chosen $k_0 = 4, k_1 = 8$ and $r = 2$.

For the conversion of the test examples in Sect. 5 we have used a knot sequence with (in general) uniformly distributed knots. The first and the last interval in the knot sequence has only half the length of the other intervals to attain a higher approximation quality of the boundary regions of the surfaces. With this trick we can transmit more information from the given surface to the approximation surface, and more easily maintain (within a given error tolerance of the normal vectors) the quality of the continuity between neighboring patches.

As we are using the discrete L_2-norm, we must choose a suitable set of points on the given structure. We shall determine the number of points with the following strategy:

1) choose a mean distance δL of the points on the boundary curves with respect to the shape of the curves.

2) subdivide the boundary curves in n_u (or n_v) equidistant intervals in u-direction (v-direction).

3) join corresponding points of the subdivided boundaries by lines. The parameter values of the intersection points of these lines determine the points $P(u_i, v_j)$ nearly equidistant on the given surface [8].

4) make the approximation process faster, by reducing the interior points with the following strategy: the interior points of the odd rows and columns will be cancelled but only if the row or the column is not a neighbouring row or column to the boundary curves. Thus the boundary areas contain more information of the given surface.

§3. Approximation with NUBS- or NURBS-Surfaces

We will now approximate the set of points constructed in the last section by a NUBS- or NURBS-surface. First we will generate the boundary curves then the interior of the surface.

The parameter values of the points on the boundary curves will be determined by centripetal parametrization. Thus we get for the points P_i the parameter values $t_i (i = 0(1)N)$ with respect to a knot sequence T constructed according in Sect. 2. The required curve may have the parametric representation

$$\mathbf{X}(t) = \frac{\sum_{i=0}^{n} \beta_i \, \mathbf{d}_i \, N_{ik}(t)}{\sum_{i=0}^{n} \beta_i \, N_{ik}(t)} \tag{4}$$

with the control points $\mathbf{d}_i$ and the weights β_i. For $\beta_i = 1$ we get the integral case. The index n is determined as in Sect. 2. For digitized points an estimate of the summation index n is developed in [13].

If we use the Gaussian least square method for the approximation process, we have to minimize the total error sum

$$D = \sum_{i=0}^{N} (\mathbf{P}_i - \mathbf{X}(t_i))^2. \tag{5}$$

The minimization problem (5) is nonlinear, therefore in [Schn92] a linearization was introduced which works very effectively: First we introduce new vectors with $(m = 1(1)3)$

$$\mathbf{P}^m = (P_0^m, P_1^m, ..., P_N^m)^T, \quad \mathbf{d}^m = (d_0^m, d_1^m, ..., d_l^m)^T,$$
$$\boldsymbol{\beta} = (\beta_0, \beta_1, ..., \beta_l)^T. \tag{6a}$$

Additionally we will use the abbreviations for the rational basis functions

$$\Psi_{ij} = \frac{\beta_j N_{jk}(t_i)}{\sum_{q=0}^{l} \beta_q N_{qk}(t_i)} \qquad (i = 0(1)N, \; j = 0(1)l), \tag{6b}$$

and denote the corresponding matrix by $\Psi(\boldsymbol{\beta}) = \{\Psi_{ij}\}$. With these abbreviations the total error sum (5) can be written as

$$\mathbf{d} = \sum_{m=1}^{3} |\mathbf{P}^m - \Psi(\boldsymbol{\beta})\,\mathbf{D}^m|^2. \tag{7}$$

Now we introduce the Moore-Penrose- or pseudo-inverse [3]

$$\mathbf{D}^m := \Psi^+(\boldsymbol{\beta})\,\mathbf{P}^m, \tag{8}$$

and (7) becomes

$$\mathbf{d} = \sum_{m=1}^{3} |(\mathbf{I} - \mathbf{P}_\Psi)\,\mathbf{P}^m|^2 \tag{8a}$$

with the matrix $\mathbf{P}_\Psi := \Psi \cdot \Psi^+ =: \{\Theta_{ij}\}$. In the integral case the pseudo-inverse is the ordinary inverse matrix and we can minimize (7) or (8a) directly with methods like the Householder transformation.

For the rational case, the operator $\mathbf{P}_\Psi$ is linearized at a position $\mathbf{P}_0$ with the help of a Taylor expansion. Thus the operator $\mathbf{P}_\Psi$ is transformed into

$$\mathbf{P}_\Psi \approx \mathbf{P}_\Psi(\boldsymbol{\beta}^0) + D[\mathbf{P}_\Psi(\boldsymbol{\beta}^0)]\,(\boldsymbol{\beta} - \boldsymbol{\beta}^0)^T =: \mathbf{P}_\Psi^L, \tag{9}$$

with

$$D[\mathbf{P}_\Psi(\boldsymbol{\beta})] = \frac{\partial\,\Theta_{ij}}{\partial\,\beta_m} \qquad (m = 1(1)l).$$

After this linearization, we have to minimize the linear error functional (instead of (8a))

$$d^L = \sum_{m=1}^{3} |(\mathbf{I} - \mathbf{P}_\Psi^L)\,\mathbf{P}^m|^2.$$

This step is the crucial point of the method: even though the pseudo-inverse is unknown up to now, it was proved in [6] (see e.g. [13]) that the first derivative can be evaluated from the well-known matrix $\Psi(\beta)$. A necessary condition for the minimum is

$$Dd^L = -2 \sum_{m+1}^{3} (\mathbf{I} - \mathbf{P}_\Psi^L) \, \mathbf{P}^m \, D[\mathbf{P}_\Psi^L] \, \mathbf{P}^m = 0. \tag{10}$$

If we introduce corresponding abbreviations, we can write (10) as

$$\mathbf{M}(\beta^0) \, (\beta - \beta^0)^T = \mathbf{R}(\beta^0), \tag{11}$$

where the matrix $\mathbf{M}$ is regular and symmetric. (11) is now solved iteratively with a damped Newton algorithm (damping factor δ) with the recursion formula

$$\beta^{(k+1)} = \beta^{(k)} + \delta \, \mathbf{b}(\beta^{(k)}) \qquad \text{with} \qquad \mathbf{b}(\beta^{(k)}) = \mathbf{M}^{-1} \, (\beta^{(k)}) \, \mathbf{R}(\beta^{(k)}). \tag{12}$$

To maintain the convex hull property for B-spline curves, it is necessary that the weights are always positive. Therefore, in (5) the weights are additionally mapped from $z_j \mapsto \beta_j$ by

$$\beta_j := \epsilon + \frac{\pi}{2} + \arctan(z_j)$$

with $\epsilon > 0$, $z_j \in \mathbb{R}$ as a new variable. Thus, β_j can only move between ϵ and $\epsilon + \pi$.

The approximation procedure can be divided into two parts: we start with equal values for all weights $\beta^0 = \{\epsilon + \frac{\pi}{2}\}$ as mean value of the permissible interval and approximate the given point set with an integral B-spline curve by solving (5) with least square methods. To reduce the error tolerance we additionally change the parametrization with help of parameter correction methods introduced in [Hos92]. After this first step, we solve equation (11) iteratively on the basis of the control points determined in the first step and the weights as unknowns. We use the damped Newton algorithm (12) combined with parameter correction. Thus we get new weights; the control points follow from (8). Now we start again with these new weights, hold them constant and use the least square solution of (5). This procedure is repeated iteratively until a required error tolerance holds. Otherwise, the number of segments or the index n in (4) will be enlarged.

After determining the four boundary curves, we approximate the interior of the required surface. The points $\mathbf{P}_s \, (s = 0(1)M)$ may have the parameter values (u_s, v_s) as determined in Sect. 2. Now we have to reduce the total error sum

$$d = \sum_{j=0}^{M} \delta_j^2 = \sum_{j=0}^{M} (\mathbf{X}(u_j, v_j) - \mathbf{P}_j)^2, \tag{13}$$

where $\mathbf{X}(u, v)$ is the given set of Bézier- or B-Spline surfaces. Analogously to (6a) and (6b), we introduce vectors for the components of the given points, the control points and the weights and proceed analogously to the approximation of curves. We also use parameter correction as described in [9,14].

The approximation procedures stop as soon as the maximal distance d_i between the given points and the approximation surface is less than a prescribed error tolerance, otherwise the procedure in Sect. 2 starts. – An approach to NURBS-approximation with nonlinear optimization techniques is introduced in [12].

§4. Conversion of Trimming Curves

We assume that the curves C_i on the given trimmed surface $\mathbf{Y}$ are described in the parametric domain by nonuniform rational B-spline representations

$$C_i(t) = \begin{pmatrix} u_i(t) \\ v_i(t) \end{pmatrix} = \frac{\sum_{j=0}^{p_i} \beta_{ij} d_{ij} N_{jk_i}(t)}{\sum_{j=0}^{p_i} \beta_{ij} N_{jk_i}(t)}$$

with k_i as order of the curves C_i.

We shall convert these given curves into a set of integral B-spline curves over a uniform knot vector and a number of segments determined analogously to Sect. 1. The de Boor points $\mathbf{d}_{ij}^*$ of these curves C_i^* are unknown. The approximation process runs through following steps:

1) First we choose points (u_l, v_l) $(l = 1(1)s)$ equidistant on C_i in the parametric domain $E_1(u, v)$. These points are mapped onto the surface $\mathbf{Y}$, thus we get points $\{\mathbf{P}_i\}$ on the curve $\mathbf{X}(C_i)$.

2) Then we project $\{\mathbf{P}_i\}$ onto the approximation surface $\mathbf{X}$ with help of the surface normals of $\mathbf{X}$. For this projection we use the parameter correction as introduced above. The foot points of the perpendiculars lead to the points $\mathbf{P}_i^*$ on $\mathbf{Y}$ with the parameter values $(\bar{u}_l, \bar{v}_l)$, $l = 1(1)s$. They may have the (chordal) parameter values $\bar{t}_l$.

3) The points $(\bar{u}_l, \bar{v}_l)$ in the parametric domain $E_2(\bar{u}, \bar{v})$ will be approximated with the B-spline curve C^* by minimizing the error vectors

$$\delta_l = \begin{pmatrix} \bar{u}_l \\ \bar{v}_l \end{pmatrix} - C^*(\bar{t}_l^*).$$

Again we use parameter correction during the approximation process [8].

§5. Test Examples

The minisymposium on spline conversion methods at Chamonix included presentations by Eck (Darmstadt), Goult (Cambridge), Hoschek/Schneider (Darmstadt), Lyche/Dokken (Oslo), and Patrikalakis/Wolter (Boston). To compare these different methods on spline conversion we proposed three test examples (bench marks):

1) the surface BRODE, a 9×9 Bézier patch which spans about $100 \times 140\ mm$,

2) the surface SEITE1, a bicubic set of 9×7 Bézier patches which span about $500 \times 2200\ mm$ (continuity C^0),

3) the surface SURFB, a biquintic set of 17×63 Bézier patches which span about $450 \times 1800\ mm$.

Each participant of the minisymposium was asked to convert these surfaces to a 3×3 or a 5×5 Bézier- or B-spline surface with a maximal error tolerance of 0.1 and $0.01\ mm$.

While the first example was developed in [2], the other two examples are parts of a car body panel (courtesy of a BMW CAD Development Division). Figures 1–3 show the shapes of the surfaces using a grid of reflection lines (courtesy of Mental Images, Berlin). We do not present the corresponding figures showing the converted surfaces since they look virtually the same.

To measure the efficiency of the conversion methods, we introduce a compression factor k. When converting to integral Bézier patches,

$$k = \frac{\text{number of Bézier points of the given patches}}{\text{number of Bézier points of the converted patches}},$$

while converting to integral B-spline surfaces

$$k = \frac{3\,(\text{number of Bézier points of the given patches})}{3\,(\text{number of B-spline control points}) + \text{knots in the knot vectors}}.$$

In the rational case we have to use the factor 4 instead the factor 3 in the denominator.

§6. Results

In our approach we have used the discrete L_2-norm. The error distance is only controlled at discrete points $\{P_i\}$ of the given Bézier surface patches $Y(u, v)$ (For the choice of points, see Sect. 2). We measure the maximal distances with respect to these points $\{Pi\}$ during the parameter correction. The calculations for SURFB were done with our implementation by Stefan Augustiniack from a cad development division of BMW, Munich. Instead of a 5×5 approximation of SURFB, only the optimal solutions with respect to the chosen error tolerance and the degree restriction $3 \leq degree \leq 5$ (see Sect. 2) was determined to get a lower data volume. The table for SEITE1 also shows an optimal case. The CPU-time contains the calculation of the segmentation.

Besides the integral conversion we have also determined rational approximations. Because of restrictions of the workspace it would be necessary to subdivide the given patch structure additionally. In these cases we could only maintain the C^0-continuity, and therefore we have removed these examples. Unfortunately, we must observe that for the rational approximation the CPU-time increases considerably because we are using the components during the approximation process.

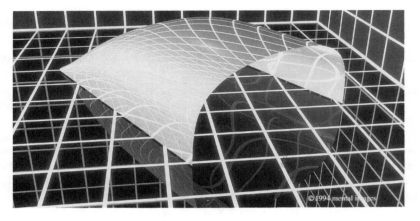

Fig. 1. The BRODE surface.

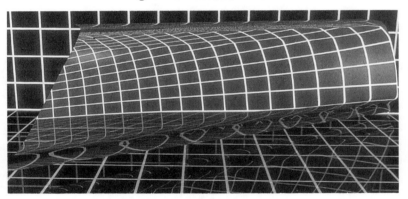

Fig. 2. The SEITE surface.

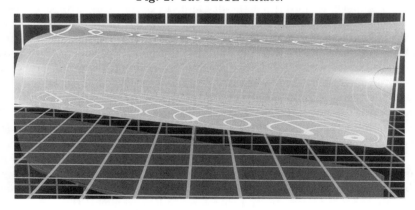

Fig. 3. The SURFB surface.

surface BRODE (s.a.[2]): degree 9×9, 1 segment, points: 20 on each boundary, 121 in interior						
integral approximation						
degree	segm.	minimal contin.	compr. factor	error toler.	meas. error	CPU HP 9000/715
3×3	3×2	C^2, C^2	2.7	0.01	0.009	4.5 sec
	2×2	C^2, C^2	3.2	0.1	0.05	4 sec
5×5	2×1	C^2, C^2	1.5	0.01	0.009	4 sec
	1×1	C^5, C^5	2.2	0.1	0.02	3.5 sec
rational approximation						
3×3	2×2	C^2, C^2	2.5	0.01	0.009	280 sec
	2×1	C^2, C^2	3.0	0.1	0.095	90 sec
5×5	1×1	C^5, C^5	1.7	0.1	0.008	68 sec

surface SEITE1: degree 3×3, segments 9×7 points: 50 on each boundary, 676 in interior						
integral approximation						
degree	segm.	minimal contin.	compr. factor	error toler.	meas. error	CPU HP 9000/715
3×3	11×19	C^2, C^2	3.1	0.01	0.009	52 sec
	5×18	C^2, C^2	5.6	0.1	0.06	50 sec
3×5	4×5	C^2, C^2	6.5	0.1	0.06	77 sec
5×5	4×8	C^2, C^2	2.3	0.01	0.09	85 sec
	2×5	C^2, C^2	5.7	0.1	0.07	60 sec
rational approximation						
3×3	4×14	C^2, C^2	5.9	0.1	0.008	426 sec
5×5	1×13	C^5, C^5	6.4	0.1	0.09	631 sec

surface SURFB: degree 5×5, segments 17×63 points: 50 on each boundary, 676 in interior						
integral approximation						
degree	segm.	minimal contin.	compr. factor	error toler.	meas. error	CPU Indigo
3×3	23×24	C^2, C^2	53.4	0.01	0.002	46 sec
	16×22	C^2, C^2	78.4	0.1	0.07	78 sec
5×3	8×23	C^2, C^2	53.3	0.01	0.009	50 sec
4×4	8×11	C^2, C^2	85.8	0.1	0.02	106 sec

Acknowledgements. The authors thank the Deutsche Forschungsgemeinschaft and the Volkswagen-Stiftung for their financial support.

References

1. Bardis, L., Patrikalakis, N. M, Approximate conversion of rational splines. Computer Aided Geometric Design **6** (1989) 189-204.
2. Brode, J, Konvertieren von Polynomen in CAGD. Diplomarbeit, TU Braunschweig 1990.
3. Campbell, S. L., and Meyer, C. D, Generalized Inverses of Linear Transformations. Pittmann 1979.
4. Dannenberg, L., and Nowacki, H, Approximate conversion of surface representations with polynomial bases. CAGD **2** (1985), 123–132.
5. Farin, G, *Curves and Surfaces for Computer Aided Geometric Design. A Practical Guide.* 3. ed., Academic Press, 1992.
6. Golub, G. H., and Pereyra, V, The differentiation of pseudo-inverses and nonlinear least square problems whose variables seperate. SIAM Numerical Analysis **10** (1973), 413–432.
7. Hoschek, J., Schneider, F.-J., and Wassum, P, Optimal approximate conversion of spline surfaces. CAGD **6** (1989), 293–306.
8. Hoschek, J., and Schneider, F.-J, Approximate spline conversion for integral and for rational Bézier and B-spline surfaces – spline approximation of offset-surfaces. in: Barnhill, R.E. (ed.): Geometric Modelling SIAM Publication (1991).
9. Hoschek, J., and Lasser, D, *Fundamentals of Computer Aided Geometric Design.* A K Peters, 1993.
10. Lachance, M. A, Chebyshev economization for parametric surfaces. Computer Aided Geometric Design **5** (1988), 195–208.
11. Ma, W., and Kruth, J. P, Mathematical modelling of free-form curve and surfaces from discrete points with NURBS, in *Curves and Surfaces in Geometric Design*, P.-J. Laurent, A. Le Méhauté, and L. L. Schumaker (eds.), A K Peters, Wellesley, 1994, 319–326.
12. Patrikalakis, N. M, Approximate conversion of rational splines. Computer Aided Geometric Design **6** (1989), 155–166.
13. Schneider, F.-J, Interpolation, Approximation und Konvertierung mit rationalen B-Spline-Kurven und Flächen. Dissertation, Darmstadt 1992.
14. Wassum, P, Bedingungen und Konstruktionen zur Geometrischen Stetigkeit und Anwendungen auf approximative Basistransformationen. Dissertation, Darmstadt 1991.
15. Watkins, M. A., Worsey, A. J, Degree reduction of Bézier curves. Computer-aided design **20** (1988), 398–405.

Prof. Dr. J. Hoschek, Dr. F.-J. Schneider
University of Technology, Dept. of Mathematics
Schlossgartenstr. 7, 64289 Darmstadt, GERMANY
hoschek@mathematik.th-darmstadt.de

A Geometrical Approach
to Interpolation on Quadric Surfaces

B. Jüttler and R. Dietz

Abstract. The paper presents a very powerful construction for rational curves and surfaces on quadrics. Based on a result of number theory, a generalization of the stereographic projection is introduced in the case of the unit sphere. With the help of this map, interpolation with spherical rational curves is shown to be a linear problem. The existence of quadratic triangular and biquadratic tensor–product Bézier patches on the sphere interpolating given boundaries is discussed. The final section outlines the extension to arbitrary non–degenerate quadric surfaces.

§1. Introduction

Quadric surfaces (like ellipsoids, hyperboloids of one or two sheets, etc.) traditionally play an important role in several industrial applications. In order to include them into computer-aided design systems, a mathematical description of curve segments and surface patches on quadrics is required. Rational parametric representations of curves and surfaces (*e.g.*, NURBS-curves and -surfaces) support the exact description of conic sections and quadric surfaces.

In recent years, several authors have developed different constructions for rational curves and surfaces on quadrics [1,8,9,12,14], etc. Most of these papers are based on the use of the classical *stereographic projection.*

In the case of the unit sphere U, the stereographic projection σ connects the points of the equator plane P $(z = 0)$ with those of the sphere. The north pole $\mathbf{z}$ of the sphere is chosen as the centre of projection. The line connecting the north pole with an arbitrary point $\mathbf{p}$ of the equator plane intersects the sphere in $\mathbf{z}$ and in a second intersection point $\sigma(\mathbf{p})$ (see Fig. 1).

The map $\mathbf{p} \mapsto \sigma(\mathbf{p})$ from the equator plane P to the unit sphere U is called the *stereographic projection* with centre $\mathbf{z}$. It will be described with help of homogeneous coordinates. These coordinates are defined by the relation

$$1 : x : y : z = x_0 : x_1 : x_2 : x_3 \tag{1}$$

Curves and Surfaces in Geometric Design
P. J. Laurent, A. Le Méhauté, and L. L. Schumaker (eds.), pp. 251–258.
Copyright © 1994 by A K PETERS, Wellesley, MA
ISBN 1-56881-039-3.

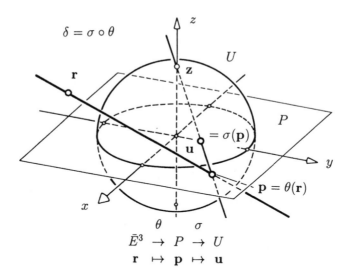

Fig. 1. The generalized stereographic projection.

(see [2]), and the bold letter $\mathbf{x} = (x_0\ x_1\ x_2\ x_3)^\top$ denotes the homogeneous coordinates of a point in three-dimensional space.

The image of a point $\mathbf{p} = (p_0\ p_1\ p_2\ 0)^\top$ on the equator plane P under the stereographic projection σ is the point

$$\sigma(\mathbf{p}) = \begin{pmatrix} p_0^2 + p_1^2 + p_2^2 \\ 2p_0 p_1 \\ 2p_0 p_2 \\ p_1^2 + p_2^2 - p_0^2 \end{pmatrix} \tag{2}$$

on the sphere. The image of a rational curve (resp. surface) on the equator plane under σ is a rational curve (resp. surface) on the sphere. But the stereographic projection has an important disadvantage: *it does not yield all irreducible rational curves and surfaces on quadrics.* (A rational curve resp. surface is said to be irreducible, if its homogeneous coordinates do not have any common linear factors, *i.e.*, if gcd $(x_0, x_1, x_2, x_3) = 1$ holds). For instance, it is impossible to construct all biquadratic rational Bézier surface patches as the images of bilinear patches under a stereographic projection. (Some counterexamples have been found by Geise and Langbecker [9], Boehm and Hansford [1], and Fink [8].)

The present paper deals with a generalization of the stereographic projection. Based on a result of number theory, the generalized stereographic projection has been developed by Dietz et al. [5,6]. This paper discusses some geometrical aspects of the method.

At first, the generalized stereographic projection is introduced in the case of the unit sphere as a representative of the class of oval quadric surfaces (ellipsoids, hyperboloids of two sheets, etc.). Then, the method is applied to interpolation with spherical rational curves and to the construction of rational surface patches on the sphere. The final section outlines the extension of the results to the hyperbolic paraboloid as a representative of the class of doubly–ruled quadric surfaces (hyperbolic paraboloids and hyperboloids of one sheet).

§2. The Generalized Stereographic Projection

Using homogeneous coordinates (1), the unit sphere U is given by the implicit equation

$$u_0^2 = u_1^2 + u_2^2 + u_3^2. \tag{3}$$

The following considerations are based on an algebraic approach to rational curves and surfaces on the sphere as introduced in [11]. The homogeneous coordinates of a rational curve (resp. of a rational tensor–product surface) on the sphere have to satisfy equation (3). Thus, this curve resp. this surface can be considered as a solution of the diophantine equation (3) in the ring of polynomials $\mathbb{R}[t]$ (resp. in the ring of bivariate polynomials $\mathbb{R}[u, v]$).

Already in 1868, V. A. Lebesgue found the representation formula

$$\begin{pmatrix} u_0 \\ u_1 \\ u_2 \\ u_3 \end{pmatrix} = \begin{pmatrix} r_0^2 + r_1^2 + r_2^2 + r_3^2 \\ 2r_0r_1 - 2r_2r_3 \\ 2r_1r_3 + 2r_0r_2 \\ r_1^2 + r_2^2 - r_0^2 - r_3^2 \end{pmatrix} =: \delta(\mathbf{r}) \tag{4}$$

for *all irreducible solutions* of equation (3) in the ring of integers [3]. This formula can be directly generalized to arbitrary polynomial rings [5].

Of course, a geometric interpretation of the representation formula (4) would be very helpful for the construction of rational curves and surfaces on quadrics. So we define a map:

Definition 1. *The map* $\mathbf{r} \in \bar{E}^3 \mapsto \delta(\mathbf{r}) \in U$ *from the three-dimensional space (which is projectively completed by adding points at infinity) to the unit sphere* U *(see (4)) is called the generalized stereographic projection.*

Now, the algebraic properties of the representation formula (4) can be formulated geometrically: *any irreducible spherical rational curve of polynomial degree $2n$ (resp. spherical rational tensor–product surface of degree $(2m, 2n)$) can be constructed as the image of a spatial rational curve of polynomial degree n (resp. of a spatial rational tensor–product surface of degree (m, n)) under the generalized stereographic projection δ.*

For $r_3 = 0$, the representation formula (4) yields exactly the classical stereographic projection (2). Thus we have the following

Proposition 2. *The restriction of the generalized stereographic projection δ : $\bar{E}^3 \to U$ to the equator plane $r_3 = 0$ of the unit sphere U is the stereographic projection $\sigma : P \to U$.*

Of course, the properties of the classical stereographic projection σ are well known. For example, this map preserves circles, *i.e.*, the image of a circle or a line on the equator plane under the stereographic projection is a circle on the sphere. In order to discuss the generalized stereographic projection, this map is decomposed into the classical stereographic projection and an auxiliary map θ.

Theorem 3. *The generalized stereographic projection* $\delta : \bar{E}^3 \to U$ *is the composition of the hyperbolic projection* $\theta : \mathbf{r} \in \bar{E}^3 \mapsto \theta(\mathbf{r}) \in P$, *where*

$$\theta(\mathbf{r}) = \begin{pmatrix} r_0^2 + r_3^2 \\ r_0 r_1 - r_2 r_3 \\ r_1 r_3 + r_0 r_2 \\ 0 \end{pmatrix}, \tag{5}$$

with the stereographic projection $\sigma : P \to U$, *i.e.,* $\delta(\mathbf{r}) = \sigma(\theta(\mathbf{r}))$.

The proof results from straightforward calculations. Figure 1 shows the generalized stereographic projection and its decomposition. At first, the points of the three-dimensional space are mapped to the equator plane P by the hyperbolic projection, and then they are mapped to the unit sphere U with help of the stereographic projection.

The discussion of the hyperbolic projection starts with the inverse image of a point.

Proposition 4. *The inverse image of a point* $\mathbf{p} = (\, p_0 \; p_1 \; p_2 \; 0\,)^\mathsf{T}$ *of the equator plane* P *under the hyperbolic projection* θ *(see (5)) is the line*

$$\lambda \begin{pmatrix} p_0 \\ p_1 \\ p_2 \\ 0 \end{pmatrix} + \mu \begin{pmatrix} 0 \\ p_2 \\ -p_1 \\ p_0 \end{pmatrix} \quad (\lambda, \mu \in \mathbb{R}) \tag{6}$$

in three-dimensional space. This line intersects the equator plane in its image under θ, *so it will be called a projecting line of the hyperbolic projection. Resulting from Theorem 3, the inverse image of a point* $\mathbf{u} = (\, u_0 \; u_1 \; u_2 \; u_3\,)^\mathsf{T}$ *under the generalized stereographic projection* δ *is the line (6), where* $\mathbf{p} = \sigma^{-1}(\mathbf{u}) = (\, (u_0 - u_3) \; u_1 \; u_2 \; 0\,)^\mathsf{T}$.

The projecting lines (6) of the hyperbolic projection are located on hyperboloids of revolution around the z–axis. They form an *elliptic linear congruence of lines* [10] (or a *net of lines*), i.e., they pass through two distinct focal lines, which are both conjugate–complex and at infinity. The hyperbolic projection θ (see (5)) is a special net projection, i.e., a projection with respect to a net of lines.

The net projection has been introduced by Tuschel in 1911 in order to develop a constructive geometry for helices [13]. A second approach to the net projection has been discovered by Wunderlich in 1936: it can be considered as a non–Euclidean parallel projection [15].

The next proposition summarizes some properties of the hyperbolic projection (cf. [7]) and the resulting properties of the generalized stereographic projection.

Proposition 5. *The image of an arbitrary non–projecting line under the hyperbolic projection is a circle or a line on the equator plane P. Its image under the generalized stereographic projection is a circle on the sphere (because the stereographic projection preserves circles).*

The inverse image of a circle on the sphere under the generalized stereographic projection is a ruled surface formed by projecting lines. This ruled surface proves to be either a hyperbolic paraboloid or a hyperboloid of one sheet.

Any plane in three-dimensional space contains exactly one projecting line (6) of the hyperbolic projection. Thus, the generalized stereographic projection maps the lines of a fixed plane to the circles through one fixed point. This point corresponds to the projecting line contained in the given plane.

The special role of the circles in the equator plane P results from the fact that the two focal lines of the elliptic linear congruence of lines intersect the equator plane in the two circular points at infinity.

The next section applies the generalized stereographic projection to the construction of an interpolating spherical rational curve.

§3. Interpolation with Spherical Rational Curves

Let $m+1$ points $\mathbf{p}_i$ on the unit sphere U with parameters $t_i \in \mathbb{R}$ $(i = 0, .., m)$ be given. These points are to be interpolated by a spherical rational curve $\mathbf{x}(t)$. Such a curve can be constructed with help of the following *algorithm:*

1) *Find the inverse images of the given points $\mathbf{p}_i$ under the generalized stereographic projection (cf. Proposition 4). These inverse images are certain projecting lines in three-dimensional space.*

2) *Construct a spatial rational curve $\mathbf{y}(t)$ which passes through the inverse images of the given points. The point $\mathbf{y}(t_i)$ has to be located on the inverse image of the given point $\mathbf{p}_i$. The spatial curve $\mathbf{y}(t)$ can be found by solving a linear system of equations.*

3) *Apply the generalized stereographic projection δ to the spatial curve $\mathbf{y}(t)$. Its image is the required interpolating curve $\mathbf{x}(t)$ on the unit sphere.*

Note that the interpolating spherical curve is found by solving a linear system of equations. Thus, interpolation with rational curves on quadrics proves to be a *linear* problem. The details of the method and some properties of the obtained solution will be discussed in [4].

§4. Rational Surface Patches on the Sphere

This section briefly discusses the construction of quadratic triangular and biquadratic tensor–product surface patches on the sphere. Any spherical quadratic triangular patch

$$\mathbf{x}(u, v, w) = \sum_{i+j+k=2} B^2_{i,j,k}(u, v, w)\, \mathbf{c}_{i,j,k} \quad (u, v, w \geq 0;\ u + v + w = 1) \quad (7)$$

(where the $B_{i,j,k}^n(u,v,w) = \frac{n!}{i!j!k!}u^i v^j w^k$ denote the trivariate Bernstein polynomials and the $\mathbf{c}_{i,j,k} \in \mathbb{R}^4$ are the homogeneous control points) can be
constructed as the image of a *linear* triangular patch under the generalized
stereographic projection. The boundaries of this linear patch are three lines
in three-dimensional space, and these three lines are contained in a plane. Resulting from the third part of Proposition 5, the three boundary circles of the
quadratic triangular patch (7) (which are the images of the three boundary
lines of the linear patch under δ) must intersect in one point.

Theorem 6. *If the quadratic triangular surface patch (7) is part of the
sphere, then its three boundaries intersect in one point. Conversely, if three
spherical circles satisfy this condition, then they can be interpolated by a
spherical quadratic triangular patch.*

This theorem has been discovered by Sederberg and Anderson by discussing
Steiner surface patches on quadrics [12].

Now, the method is applied to biquadratic tensor–product surface patches
on the sphere. Any spherical biquadratic patch

$$\mathbf{y}(u,v) = \sum_{i=0}^{2}\sum_{j=0}^{2} B_i^2(u)B_j^2(v)\,\mathbf{b}_{i,j} \quad ((u,v) \in [0,1] \times [0,1]) \tag{8}$$

(where the $B_i^n(t) = \binom{n}{i}t^i(1-t)^{n-i}$ denote the Bernstein polynomials and
the $\mathbf{b}_{i,j} \in \mathbb{R}^4$ are the homogeneous control points) can be constructed as the
image of a *bilinear* patch under the generalized stereographic projection. The
four boundary circles of the spherical patch (8) intersect in the four corner
points $\mathbf{p}_i$ and in four second intersection points $\mathbf{q}_i$ ($i=1,..,4$), see Fig. 2.

In [5], a condition analogous to that of Theorem 6 has been derived:

Theorem 7. *If the biquadratic patch (8) is part of the sphere, then the four
points $\mathbf{p}_1, \mathbf{q}_2, \mathbf{p}_3, \mathbf{q}_4$ (or equivalently $\mathbf{q}_1, \mathbf{p}_2, \mathbf{q}_3, \mathbf{p}_4$) are located on one circle
(see Fig. 2). Conversely, if four spherical circles satisfy this condition, then
they can be interpolated by a spherical biquadratic patch.*

Using the generalized stereographic projection, this theorem can be proved
directly (see [5]). The four control points of the bilinear patch (which is
the preimage of the biquadratic patch) span two planes, and the circle from
Theorem 7 is the image of the line, in which the two planes intersect.

The equivalence of the existence of the two circles in Theorem 7 is known
as *Miquel's Theorem* in the foundations of geometry.

§5. Extension to Other Quadric Surfaces

This section outlines the discussion of the hyperbolic paraboloid

$$h_0 h_3 = h_1 h_2 \tag{9}$$

(a) (b)

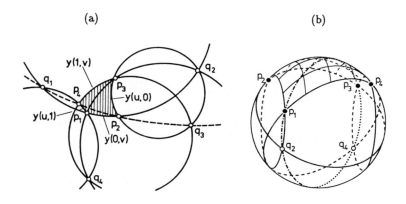

Fig. 2. The boundaries of the spherical biquadratic patch
(a) Scheme of the boundaries, (b) The spherical patch.

as a representative of doubly–ruled quadric surfaces. In [5], the representation
formula

$$h_0 = r_0 r_3 \qquad h_1 = r_1 r_3$$
$$h_2 = r_0 r_2 \qquad h_3 = r_1 r_2 \tag{10}$$

has been given which is analogous to that of the unit sphere (2).

Again, this formula is considered as a generalized stereographic projection. This map proves to be the composition of a net projection (with respect
to a hyperbolic linear congruence of lines) with a stereographic projection.
Rational curves and surfaces on the hyperbolic paraboloid can be constructed
similarly to the case of the unit sphere, see [6].

Any non–degenerated oval (resp. doubly–ruled quadric) surface is the
image of the unit sphere (resp. of the hyperbolic paraboloid) under an appropriate projective map. (For instance, this map can be constructed with help
of a principal axes transformation, see [6].) Thus, the methods and results of
this paper can be directly generalized to arbitrary non–degenerated quadric
surfaces.

References

1. Boehm, W., and D. Hansford, Bézier patches on quadrics, in *NURBS
   for Curve and Surface Design*, G. Farin (ed.), SIAM, Philadelphia, 1991,
   1–14.
2. Coxeter, H. S. M., *Projective Geometry*, Blaisdell, New York, 1964.

3. Dickson, L. E., *History of the Theory of Numbers, Vol. II*, Chelsea, New York, 1952.

4. Dietz, R., Rationale Bézier–Kurven und Bézier–Flächenstücke auf Quadriken, Dissertation, Technische Hochschule Darmstadt, 1994, to appear.

5. Dietz, R., J. Hoschek and B. Jüttler, An algebraic approach to curves and surfaces on the sphere and on other quadrics, Computer-Aided Geom. Design **10** (1993), 211-229.

6. Dietz, R., J. Hoschek and B. Jüttler, Rational patches on quadric surfaces, submitted to Comput. Aided Design.

7. Eckhart, L., *Konstruktive Abbildungsverfahren*, Springer, Wien, 1926.

8. Fink, U., Biquadratische Bézier-Flächenstücke auf Quadriken, Dissertation, Universität Stuttgart, 1992.

9. Geise, G., and U. Langbecker, Finite quadric segments with four conic boundary curves, Computer-Aided Geom. Design **7** (1990), 141–150.

10. Hlavatý, V., *Differential Line Geometry*, Noordhoff, Groningen–Holland, 1953.

11. Hoschek, J., and G. Seemann, Spherical splines, Math. Modelling and Numer. Anal. **26** (1992), 1–22.

12. Sederberg, T., and D. Anderson, Steiner surface patches, IEEE Comp. Graphics Appl. **5** (1985), 23–26.

13. Tuschel, L., Über die Schraubliniengeometrie und deren konstruktive Verwendung, Sitzungsber. Kais. Akad. Wiss. Wien **120** (1911), 233–254.

14. Warren, J., and S. Lodha, A Bézier representation for quadric surface patches, Comput. Aided Design **22** (1990), 574–579.

15. Wunderlich, W., Darstellende Geometrie nichteuklidischer Schraubflächen, Monatsh. Math. Phys. **44** (1936), 249–279.

Bert Jüttler and Roland Dietz
Technische Hochschule Darmstadt
Fachbereich Mathematik
AG Differentialgeometrie und Kinematik
Schloßgartenstraße 7
D – 64289 Darmstadt, GERMANY

juettler@mathematik.th-darmstadt.de
rdietz@mathematik.th-darmstadt.de

Finding Shortest Paths on Surfaces

R. Kimmel, A. Amir, and A. M. Bruckstein

Abstract. This paper presents a new algorithm for determining minimal length paths between two points or regions on a three dimensional surface. The numerical implementation is based on finding equal distance contours from a given point or area. These contours are calculated as zero sets of a bivariate function designed to evolve so as to track the equal distance curves on the given surface. The algorithm produces all minimal length paths between the source and the destination areas on the surface given as height values on a rectangular grid. Complexity and accuracy are governed by the grid resolution and the distance step size in the iterative scheme.

§1. Introduction

Finding paths of minimal length between two areas on a three dimensional surface is of great importance in many fields such as computer-aided neuroanatomy, robotic motion planning (autonomous vehicle navigation), geophysics, terrain navigation, etc. Paths of minimal Euclidean distance between two points on a graph-surface will be referred to as *minimal geodesics* in this paper.

A variational approach to finding a geodesic path on a given surface was presented by Beck et al. in [1]. In other cases it is natural to approximate a surface with planar polygonal patches. Some algorithms that solve this *discrete geodesic problem* were presented recently [10,20]. Schwartz et al. used such an algorithm as a preliminary step in solving the *mapmaker problem*, which is the problem of the gradient–descent surface flattening of a polyhedral surface [15]. Kiryati and Székely [9] considered *voxel* representations of three–dimensional surfaces and used 3D–length estimators to derive an efficient algorithm for the minimal distance geodesic problem.

In this paper we introduce a different way of dealing with the problem of finding the minimal geodesics. The surface is given as height samples on a rectangular grid. As a first step, a distance map from the source area is calculated. The distance map is computed via *equal distance curve* propagation on the surface. Equal distance curves are calculated as the zero sets of a bivariate function evolving in time. This formulation of curve evolution processes

Curves and Surfaces in Geometric Design
P. J. Laurent, A. Le Méhauté, and L. L. Schumaker (eds.), pp. 259–268.

is due to Osher and Sethian, [11]. It overcomes some topological changes
and numerical problems encountered in direct implementations of curve evolutions using parametric representations. The implicit representation of the
evolving curve produces a stable and accurate numerical scheme for tracing
shock waves in fluid dynamics. This formulation has subsequently found applications in doing planar shape analysis in computer vision [4,5], in solving
the shape from shading problem in computer vision [8], in simulating crystal
growth [18], offsetting shapes in computer-aided design [6], in implementing
continuous–scale morphology operations on a digitized picture [13], and more.

The relation between minimal paths, geodesics and equal distance contours may be found in elementary differential geometry textbooks, *e.g.*, [3].
Geodesics are locally shortest paths in the sense that any perturbation of a
geodesic curve will increase its length. The minimal length paths between
two points are geodesics connecting those points. A simple way of determining minimal geodesics is by constructing a so–called *geodesic polar coordinate
system* on the surface around the source area. Using such a coordinate system
readily provides the *geodesic circle map*, or the map of equal distance contours
on the surface.

In the next Section an analytic model for the equal distance contour evolution is discussed. In Section 3, a numerical implementation of the analytic
propagation is presented. It is based on ideas of Osher and Sethian [11,17].
The results of the numerical algorithm are demonstrated in several examples
in Section 4. We conclude with a discussion of some possible extensions of
the algorithm and comment on its complexity in Section 5.

§2. The Analytic Model

In this section we determine a differential equation describing the propagation
of equal distance contours on a smooth surface, from a point or a source region
on the surface. This equation describes the evolution of a planar curve, a curve
that is the projection of the three dimensional equal distance contour (a $3D$
curve) on the (x, y)–plane.

2.1. Equal Distance Contours

We first define the equal distance contour. Given a source area $S \in \mathbb{R}^3$ on a
surface $\mathcal{S} \in \mathbb{R}^3$ that can be described by a function $z(x, y) : \mathbb{R}^2 \rightarrow \mathbb{R}$, let the
$3D$ equal distance contour of distance t from S be defined as

$$\{\mathbf{p} \in \mathcal{S} : \ d_z(\mathbf{p}, S) = t\} = \alpha(*, t),$$

where $d_z(\mathbf{p}, S)$ is the *minimal Euclidean distance* determined by the the shortest paths from a point $\mathbf{p}$ to an area S on surface $\mathcal{S}$.

We proved [7] that the $3D$ parametric representations of $\alpha(*, t)$ on $\mathcal{S}$ can
be obtained by the equal distance contour propagation: Let $\alpha(u, t)$ be a $3D$
curve propagating on the surface $\mathcal{S} \subset \mathbb{R}^3$, where u is the parameter and t is
the propagation time. Then

Lemma 1. *The equal distance contour evolution is given by*

$$\frac{\partial}{\partial t}\alpha = N \times \vec{\mathbf{t}}^{\alpha} \qquad given \qquad \alpha(0), \tag{1}$$

where N is the surface normal and $\vec{\mathbf{t}}^{\alpha}$ is the tangent to the contour α.

The traces of constant parameter along the curve evolving according to (1) are geodesics, and these geodesics are locally shortest paths:

Lemma 2. *Define the curve $\beta(t) = \alpha(u,t)|_{u=u_0}$. Then, for any u_0, the curve $\beta(t)$ is a geodesic.*

See [7] for the proof. Lemma 1 provides the evolution equation of the equal distance contour. Starting from the boundary of the source area $\alpha(0) = \{(x,y,z(x,y))|(x,y,z(x,y)) \in \partial S\}$, it is possible to find the equal distance contour for any desired distance d, by using the evolution equation to calculate $\alpha(u,t)|_{t=d}$. This propagation may be used to build the distance map for each point on the surface.

Implementing the three dimensional curve evolution is quite a complicated task. We are therefore interested in considering the projection of the $3D$ curve on the (x,y)–plane, $C(t) = \{(x,y)|(x,y,z(x,y)) \in \alpha(t)\} \equiv \pi \circ \alpha(t)$.

The trace of the propagating planar curve may be determined merely by its normal velocity [14], while the tangential component affects only the internal parameterization of the evolving curve. Let us consider the projection of the above evolution on the (x,y)–plane. The knowledge of how this projected contour behaves allows us to construct a simple, accurate and stable numerical algorithm that can be used to produce these equal distance contours.

In [7] we calculate the planar normal component of the projected velocity of the evolving equal distance contour, $V_N = \langle \vec{\mathbf{n}}, \pi \circ (N \times \vec{\mathbf{t}}^{\alpha}) \rangle$. Using this velocity we construct a differential equation describing the projected equal distance contour evolution of the form

$$\frac{\partial}{\partial t}C = V_N \vec{\mathbf{n}} \qquad given \qquad C(0) = \{(x,y) : (x,y,z(x,y)) \in \partial S\} \equiv \partial\tilde{S}, \tag{2}$$

where $\vec{\mathbf{n}}$ is the planar normal direction, and V_N depends on the surface gradient ($p = \frac{\partial z}{\partial x}$ and $q = \frac{\partial z}{\partial y}$) and $\vec{\mathbf{n}}$.

It is possible to construct a direct numerical scheme form the above curve evolution equation, but implementing this direct formulation involves some difficulties which are discussed in detail in Section 3.

2.2. Finding The Minimal Path

The procedure that calculates the equal distance contours allows us to build a Euclidean distance map on the surface, from a given area. Assuming we have reliable distance map procedure in hand, we can construct a simple procedure that finds the minimal path from a source area S to a destination area D (where $S, D \in \mathcal{S}$).

Define $\mathcal{M}_A$ as the distance map of area A as

$$\mathcal{M}_A(x,y) = d_z((x,y,z(x,y)), A).$$

We readily have the following result:

Lemma 3. *All minimal paths between S and D on S are given by the set $G \subset S$,*

$$G = \{(x, y, z(x, y)) : \mathcal{M}_S(x, y) + \mathcal{M}_D(x, y) = g_m\} \tag{3}$$

where $g_m \equiv \min_{(x,y)}(\mathcal{M}_S + \mathcal{M}_D)$ is the global minimum of the sum of the source and destination distance maps.

See [7] for the proof. We also have the following result.

Corollary 1. *All paths between S and D which are defined by G (equation (3)), are minimal geodesics.*

In the next section a numerical scheme based on the level set representation of the evolving planar curve is presented. Note that the shortest paths are minimal value level sets of the function $\mathcal{M}_S + \mathcal{M}_D$. This observation will later be used on to find the minimal geodesics.

§3. The Numerical Approximation

When implementing curve evolution equations such as (2) on a digital computer, a number of problems must be solved.

- Topological changes: Topological changes may occur while the curve evolves, *i.e.* the curve may change its topology from one connected curve to two separate evolving curves, or, two curves may merge into one, see Figure 1b.

- Stability and accuracy: In [11,17] some numerical problems which characterize a direct formulation of (2) are described. The problems are caused due to a time varying coordinate system (u, t) of the direct representation (where u is the parameterization, and t is the time).

- Singularities: Even an initial smooth curve can develop curvature singularities (see Figure 1a). The question is how to continue the evolution after singularities appear. The natural way is to choose the solution which agrees with the *Huygens principle* [16]. Viewing the curve as the front of a burning flame, this solution states that *once a particle is burnt, it cannot be re-ignited* [2]. It can also be proved that from all the *weak* solutions of (2) part the singularities, the one derived from the Huygens principle is unique, and can be obtained by a constraint denoted as the *entropy condition* [16].

Sethian and Osher [11,17] proposed an algorithm for curve and surface evolution that elegantly solves these problems. As a first step in constructing the algorithm, the curve is embedded in a higher dimensional function. Then, evolution equations for the implicit representation of the curve are solved using numerical techniques derived from hyperbolic conservation laws [19].

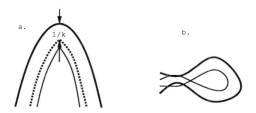

Fig. 1. Problems in curve evolution implementation.

3.1. The Eulerian Formulation

Let the curve $C(t)$ be represented by the zero level set of a smooth Lipschitz continuous function $\phi : \mathbb{R}^2 \times [0, T) \to \mathbb{R}$, so that ϕ is negative in the interior and positive in the exterior of the zero level set $\phi = 0$. Consider the zero level set defined by $\{C(t) \in \mathbb{R}^2 : \phi(C, t) = 0\}$. We have to find the evolution rule of ϕ, so that the evolving curve $C(t)$ can be represented by the evolving zero level set $\phi(C(t), t) = 0$. Using the chain rule we get $\nabla \phi(C, t) \cdot C_t + \phi_t(C, t) = 0$. Note that for any level set the planar normal can be written as $\vec{n} = \frac{\nabla \phi}{\|\nabla \phi\|}$. Using this relation in conjunction with the condition equation (2) we obtain

$$\frac{\partial}{\partial t} \phi = -V_N \|\nabla \phi\|, \tag{4}$$

where the curve $C(t)$ is obtained as the zero level set of ϕ. This procedure is known as the Eulerian formulation [17].

This formulation of planar curve evolution processes frees us from the need to take care of the possible topological changes in the propagating curve. In Figure 2 we can see that embedding the curve in a higher dimensional function automatically solves a number of topological problems. The evolving curve is obtained as a level set of the evolving surface, which remains continuous (and connected) even when topological changes occur. The numerical implementation of (4) is based on *monotone* and *conservative* numerical algorithms, derived from hyperbolic conservation laws and the Hamilton–Jacobi "type" equations of the derivatives, (for details see [11]). For some normal velocities these numerical schemes automatically enforce the *entropy condition*, a condition equivalent to *Huygens principle*, see [16].

Inserting the normal component of the velocity derived in [7] into equation (4), we get

$$\frac{\partial}{\partial t} \phi = \sqrt{\frac{1}{1 + p^2 + q^2} \left((1 + q^2)\phi_x^2 + (1 + p^2)\phi_y^2 - 2pq\phi_x\phi_y \right)} \tag{5}$$

This equation describes the propagation rule for the surface ϕ. For numerical implementation see [7] which is based on [11,17,19].

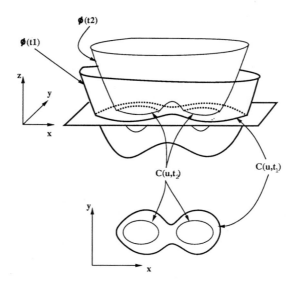

Fig. 2. The Eulerian formulation solves topological problems. See Text.

3.2. Finding Minimal Geodesics

Having $\mathcal{M}_S$ and $\mathcal{M}_D$ on the grid, the minimal geodesic may be found by using Lemma 3 in a simple way. Recalling that $g_m = \min(\mathcal{M}_S + \mathcal{M}_D)$, the projection of the minimal geodesic, G, on to the (x, y)–plane is $\tilde{G} = \{(x, y) | (\mathcal{M}_S(x, y) + \mathcal{M}_D(x, y)) = g_m\}$. The desired minimal geodesics are therefor achieved by finding the minimal level set $\mathcal{M}_S + \mathcal{M}_D = g_m$.

Finding the minimal geodesics is only a simple example of a wide variety of possibilities for using the calculated distance map on the surface.

§4. Examples and Results

Let us demonstrate the performance of the algorithm by applying it to a number of synthetic surfaces and finding the paths of minimal length. The synthetic surfaces are given on a mesh of 256×256 points.

Two simple surfaces are considered. In the first two cases the source and destination areas are points located near the upper right and lower left corners. In the first example, Figure 3a, the source and destination are on opposite sides of the "mountain". Figure 3b shows the sum of the distance map obtained from the source point and the distance map form the destination point. We search for the minimal level set of that sum, shown as black smooth curve in Figure 3c. Here the gray lines represent the equal distance contour from each point, and the pixel chain line in the middle of the minimal level

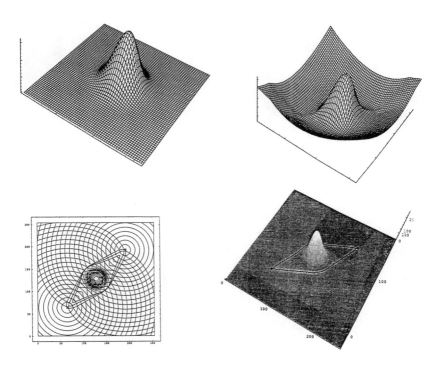

Fig. 3. A simple "mountain" surface. See text for more details.

set is the desired path obtained by a simple thinning algorithm. The thinning algorithm make use of the distance values of the pixels in the interior of the minimal level set. In Figure 3d the two minimal geodesics are drawn as black lines on the surface.

In the next example two cuts are made in the "egg-box" surface, see Figure 4a. Infinite walls in the sum of the distance maps in Figure 4b represent impenetrable walls created by the cuts.

The last example presents the possibility to find the minimal paths between two initial areas. In Figure 5, a circle and a square areas are located on opposite sides of the mountain. The algorithm finds the minimal paths, two in this symmetric case, which start from different locations on the boundary of the initial areas.

§5. Concluding Remarks

We have described a numerical method for calculating a distance map from a given area on a graph surface, so that topological problems in the propagated

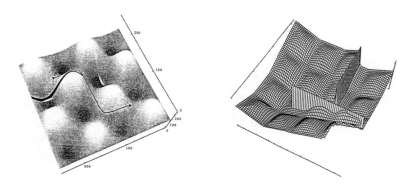

Fig. 4. Introducing obstacles in the "egg–box" surface. See text.

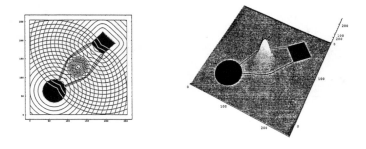

Fig. 5. Finding the minimal path between two areas.

equal distance contours are inherently avoided. An algorithm for finding the minimal geodesics between two areas on the surface based on the distance mapping was constructed. The algorithm works on a grid; therefore it is easy to implement the algorithm in parallel using each mesh point as a small calculating device which communicates with its four close neighbors. In each iteration we need to calculate the values of $\phi(x, y, t)$ in those grid points close to the current contour and the rest of the grid points serve as sign holders. This can be exploited to reduce calculation effort. When not considering any possible redundancy, the calculation effort is of order $O(\frac{L}{\Delta t}m \cdot n)$, where L is the length of the minimal geodesic and $m \cdot n$ is the number of grid points.

Acknowledgements. We would like to thank Dr. N. Kiryati for introducing to us the problem of finding the shortest path, and to Dr. Guillermo Sapiro, Prof. A. Guez and Mr. D. Shaked for the many suggestions and discussions on this topic.

References

1. Beck, J. M., R. T. Farouki, and J. K. Hinds, Surface analysis methods, IEEE CG&A (1986), 18–37.
2. Blum H., Biological shape and visual science, J. Theor. Biology **38** (1973), 205–287.
3. Do Carmo M. P., *Differential Geometry of Curves and Surfaces*, Prentice–Hall, Inc., New Jersey, 1976.
4. Kimia, B. B., A. Tannenbaum, and S. W. Zucker, Toward a computational theory of shape: an overview, in *Lecture Notes in Computer Science* **427**, Springer–Verlag, New York, 1990, 402–407.
5. Kimia, B. B., A. Tannenbaum, and S. W. Zucker, On the evolution of curves via a function of curvature, I: the classical case, J. of Math. Analysis and Applications **163** (1992), 438–458.
6. Kimmel, R., and A. M. Bruckstein, Shape offsets via level sets, *CAD* **25** (1993), 154–162.
7. Kimmel, R., A. Amir, and A. M. Bruckstein, Finding shortest paths on graph surfaces, CIS Report #9301, Center for Intelligent Systems, Technion, Israel, January, 1993.
8. Kimmel, R., and A. M. Bruckstein, Shape from shading via level sets, CIS Report #9209, Center for Intelligent Systems, Technion, Israel, June, 1992.
9. Kiryati, N., and Székely, G., Estimating shortest paths and minimal distances on digitized three dimensional surfaces, Pattern Recog. **26**, 11 (1993), 1623–1637.
10. Mitchel, J. S. B., D. M. Mount, and C. H. Papadimitriou, The discrete geodesic problem, SIAM J. Comput. **16**, 4 (1987), 647–668.
11. Osher, S. J., and J. A. Sethian, Fronts propagating with curvature dependent speed: algorithms based on Hamilton–Jacobi formulations, J. of Comp. Phys. **79** (1988), 12–49.
12. Osher, S. J., and L. I. Rudin, Feature–oriented image enhancement using shock filters, SIAM J. Numer. Anal. **27**, 4 (1990), 919–940.
13. Sapiro, G., R. Kimmel, D. Shaked, and A. M. Bruckstein, Implementing continuous–scale morphology, CIS Report #9208, Center for Intelligent Systems, Technion, Israel, June, 1992.
14. Sapiro, G., and A. Tannenbaum, Affine invariant scale–space, (to appear) Int. J. of Comput. Vision
15. Schwartz, E. L., A. Shaw, and E. Wolfson, A numerical solution to the generalized mapmaker's problem: flattening non convex polyhedral surfaces, IEEE Trans. on PAMI **11**, 9 (1989), 1005–1008.
16. Sethian, J. A., Curvature and the evolution of fronts, Commun. in Math. Phys. **101** (1985), 487–499.

17. Sethian, J. A., A review of recent numerical algorithms for hypersurfaces moving with curvature dependent speed, J. of Diff. Geom. **33** (1989), 131–161.
18. Sethian, J. A., and J. Strain, Crystal growth and dendritic solidification, J. of Computational Physics, **98** (1992), 231–253.
19. Sod, G. A., *Numerical Methods in Fluid Dynamics*, Cambridge Univ. Press, 1985.
20. Wolfson, E., and E. L. Schwartz, Computing minimal distances on polyhedral surfaces, IEEE Trans. on PAMI **11** (1989), 1001–1005.

Ron Kimmel
Dept. of Electrical Engineering
Technion – Israel Institute of Technology
Haifa 32000, ISRAEL
ron@ techunix.technion.ac.il

Arnon Amir
Dept. of Computer Science
Technion – Israel Institute of Technology
Haifa 32000, ISRAEL
arnon@ csc.cs.technion.ac.il

Alfred M. Bruckstein
Dept. of Computer Science
Technion – Israel Institute of Technology
Haifa 32000, ISRAEL
freddy@ cs.technion.ac.il

Polygonalization of Algebraic Surfaces

R. Klein

Abstract. An algorithm is presented for the polygonalization of real algebraic surfaces, which is based on an adaptive finite subdivision of an enclosing tetrahedron and a Bernstein-Bézier-representation of the polynomials. Using the Bernstein-Bézier-representation of the polynomial, an estimate of the curvature of the surface in a tetrahedral region can be achieved. The curvature bound is used to eliminate small triangles superfluously produced by the algorithm, for example in the neighbourhood of singular points.

§1. Introduction

A real algebraic surface is given by an implicit equation of the form $f(x, y, z) = 0$, where $f(x, y, z)$ is a polynomial. In recent years there has been a growing interest in the use of algebraic surfaces for geometric modeling. Examples are blending between algebraic surfaces [9,12,13] or schemes for modeling free-form objects directly with algebraic surfaces [2,3,4,14].

We consider the problem of computing piecewise linear approximations of real algebraic surfaces. Modern graphics hardware accepts such polygonal approximations and accurately render the complicated surfaces using sophisticated lighting and shading models. Similar, structured linear approximations of surfaces are required for finite element approaches to solve systems of partial differential equations. The problem of constructing a polygonal approximation, especially for finite element meshes, is complicated by the need for a correct topology of the mesh even in the presence of singularities and multiple sheets of the real algebraic surface.

An early polygonization algorithm for algebraic surfaces is described in [6]. It samples $f(x, y, z)$ over a three-dimensional rectangular grid of points and linearly interpolates polygons in regions where the function values change signs. A similar approach based on a tetrahedral subdivision of space was first described by Tindle [16]. Bloomenthal [5] used an adaptive marching cube technique involving an adaptively refined sample grid to generate more polygons in regions of greater surface complexity. These techniques work well in

Curves and Surfaces in Geometric Design
P. J. Laurent, A. Le Méhauté, and L. L. Schumaker (eds.), pp. 269–275.
Copyright © 1994 by A K PETERS, Wellesley, MA
ISBN 1-56881-039-3.

many cases but due to the sampling they may miss singularities and small components of the surface. Arnon [1] is addressing the singularity problem by applying the cylindrical algebraic decomposition algorithm to obtain topologically correct polygonalizations. Hall and Warren [8] presented an algorithm based on a tetrahedral subdivision in conjunction with a Bernstein-Bézier-representation. The Bernstein-Bézier-representation is used to exclude regions which cannot contain parts of a surface. Therefore, the algorithm is able to detect also small components which may not be detected by a simple sampling of the defining polynomial on grid points. This algorithm does not take into account singularities. In 1991 we developed an algorithm based on sampling in conjunction with numerical curve-tracing [11]. This algorithm produces more regular triangulations and is able to detect singularities of order 1 and to include them correctly into the triangulation. But due to the sampling it may miss small components of the surface. Therefore in the algorithm presented here, we also use a Bernstein-Bézier-representation to avoid such cases.

§2. Recursive Subdivision of Space

In the algorithm a starting tetrahedron is recursively subdivided until a subdivision criterion stops the recursion. The tetrahedra created in the subdivision process build up a set T of subtetrahedra, such that any two $t, t' \in T$ intersect either in a common vertex, in a common edge or in a common face. To maintain such a partition of the starting tetrahedron, the adaptive subdivision algorithm decides independently for each tetrahedron whether it should be subdivided or not. It defers the processing of tetrahedra not yet recursively subdivided until it has considered all tetrahedra. It then makes a second pass through the list of unsubdivided tetrahedra and subdivides these in dependence on the subdivision of its neighbour-tetrahedra.

The subdivision technique used by Hall and Warren in [8] first subdivides a regular tetrahedra into four regular tetrahedra and one octahedron by slicing off each corner of the original tetrahedron. The remaining regular octahedron is further split into eight similar tetrahedra by inserting a vertex at the center of the octahedron and by the projection of edges to each of its corners. The eight tetrahedra are called *cubic tetrahedra*, since such a tetrahedron may also be formed by slicing a corner off a cube. For this cubic tetrahedra a second subdivision scheme is used subdividing such a tetrahedron into one regular tetrahedron and three pairs of cubic tetrahedra. These different subdivision schemes have also to be taken into account for the subdivision of the tetrahedra adjacent to a subdivided tetrahedron during the subdivision process.

We use the following subdivision scheme to simplify this process: After slicing off each corner of the original tetrahedron the remaining octahedron is subdivided into four similar subtetrahedra as shown in Figure 1. These subtetrahedra are no longer regular tetrahedra since one of their edges always is $\sqrt{2}$-times longer than the other ones. Using in the next subdivision step the same scheme again, such a tetrahedron subdivides into two regular tetrahedra and six tetrahedra of the same type. Thus, the whole space can be filled

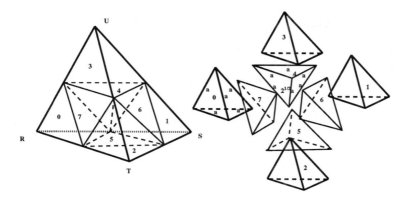

Fig. 1. Subdivision of a tetrahedron.

making use of only these two types of tetrahedra. The advantage of this scheme is that all tetrahedra can be subdivided in the same manner.

§3. Subdivision Criteria

To get a subdivision criteria which

a) terminates subdivision if the tetrahedron does not contain the zero contour of the defining function f and

b) terminates subdivision if the zero-contour of f lie within some user-defined tolerance of linear approximation

we use a Bernstein-Bézier-representation for polynomial functions. A polynomial of degree n can be written in the form

$$f(r, s, t, u) = \sum_{i+j+k+l=n} b_{i,j,k,l} B^n_{ijkl}(r, s, t, u),$$

where r, s, t, u are the barycentric coordinates of the point $(x, y, z) \in \mathbb{R}^3$ with respect to the domain tetrahedron $\Delta(R, S, T, U)$, and $b_{i,j,k,l} \in \mathbb{R}$ are the Bézier-ordinates, n is the degree of f and $B^n_{ijkl}(r, s, t, u) = \frac{n!}{i!j!k!l!} r^i s^j t^k u^l$ are the trivariate Bernstein-polynomials. If all Bézier-ordinates $b_{i,j,k,l}$, $i + j + k + l = n$ have the same sign, then because of the convex-hull property of the Bernstein-Bézier-polynomials, f is also single signed over the entire tetrahedron. Therefore, in the algorithm the Bernstein-Bézier-ordinates are computed and their signs are checked. If all the signs are the same, the tetrahedron does not contain the zero contour and further recursive subdivision is stopped.

The linear approximation error can be controlled in bounding the zero contour $f = 0$ into a pair of parallel planes defined by the values of f at the vertices R, S, T, V of the tetrahedron [8]. This values are $b_{n000}, b_{0n00}, b_{00n0}, b_{000n}$, respectively, and they define uniquely a interpolating linear function l. Now f is rescaled in such a way that the linear function $l(r, s, t, u)$ measures the distance from the point (r, s, t, u) to $l(r, s, t, u)$ and two parallel planes are defined in Bernstein-Bézier-representations by the Bézier-ordinates

$$c_{ijkl} = (i/n)b_{n000} + (j/n)b_{0n00} + (k/n)b_{00n0} + (l/n)b_{000n} - \epsilon$$

$$d_{ijkl} = (i/n)b_{n000} + (j/n)b_{0n00} + (k/n)b_{00n0} + (l/n)b_{000n} + \epsilon.$$

If $c_{ijkl} \leq b_{ijkl} \leq d_{ijkl}$ for all $i, j, k, l \geq 0$, $i + j + k + l = n$, then the zero-contour $f = 0$ lies between the planes $l - \epsilon$ and $l + \epsilon$, and an approximating triangle is computed intersecting the corresponding edges of the tetrahedron with the surface itself.

The resulting surface mesh might be very irregular, even when the surface is relatively flat. This can happen because a surface passes near a vertex of the original triangulation space, the positions of two different parts of the surface are close to each other or a tetrahedron contains a singular point. In the third case the approximation error only decreases with the diameter of the tetrahedron, see e.g. Figure 2 for the 2D-case. To improve the resulting mesh by removing small unwanted triangles, in [8] a heuristic method is used: If a vertex of the original triangulation space lies near the surface (typically 5 percent of the shortest edge of the tetrahedron), the vertex is "snapped" to the surface. In this way, small triangles of the original mesh are eliminated. This will not work if small triangles are produced due to singularities in a tetrahedron or in the case that the positions of different parts of the surface are close to each other. In these cases the generation of extremely small triangles can hardly be avoided. Actually, this problem appears in the algorithm independently of the curvature of the surface. Therefore, in addition to the subdivision criteria mentioned above, we also keep track of two additional polynomials in Bernstein-Bézier-representation. Using these two polynomials we are able to control the maximum curvature of the zero contour and the possible presence of singularities in the tetrahedron. In a postprocessing step the bounds on the curvature in each of these small tetrahedra are used to reduce the number of triangles in the resulting mesh by removing vertices. This decimation algorithm works as follows: Multiple passes are made over all vertices in the mesh. Each vertex is candidate for removal. We consider a submesh consisting of all triangles that use one of these vertices and compute for this submesh a common bound c_{max} of the curvature and its maximum diameter d_{max}. The error of a linear approximation of a local triangulation of this submesh after removing the vertex depends on c_{max} and d_{max}. If this value is smaller than the user-defined error the vertex can be removed.

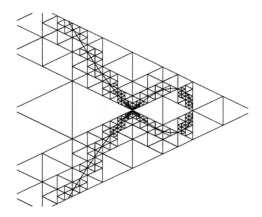

Fig. 2. Singularities cause small triangles to be created.

Curvature Estimation.

For a regular point p on the zero contour of the polynomial f the normal curvature $\kappa_v^n(p, v)$ in direction of v is given by

$$\kappa_v^n(p) = \frac{-v^t \cdot H_{|p} \cdot v}{||N_{|p}||},$$

where $N_{|p} := \nabla f(p)$ and $H_{|p}$ is the Hessian of f in p.

Let Δ denote the reference tetrahedron. Using the above formula the maximum curvature of the zero-contour can be estimated:

$$\kappa_{max}^n = sup_{p \in \Delta} sup_v |\kappa_v(p)| = sup_{p \in \Delta} sup_v \frac{||v \cdot H_{|p} \cdot v||}{||N_{|p}||}$$

$$\leq sup_{p \in \Delta} \frac{||H_{|p}||}{||N_{|p}||} \leq \frac{sup_{p \in \Delta} ||H_{|p}||}{inf_{p \in \Delta} ||N_{|p}||}$$

To get an estimation of the supremum and infimum over the whole tetrahedron in the above formula, we define two polynomials g and h

$$g := \frac{\partial f}{\partial x^2} + \frac{\partial f}{\partial y^2} + \frac{\partial f}{\partial z^2}$$

and

$$h := \frac{\partial^2 f}{\partial^2 x^2} + \frac{\partial^2 f}{\partial^2 y^2} + \frac{\partial^2 f}{\partial^2 z^2} + 2\frac{\partial^2 f}{\partial x \partial y} + 2\frac{\partial^2 f}{\partial x \partial z^2} + 2\frac{\partial^2 f}{\partial y \partial z^2}$$

of degree $2 \cdot (n - 1)$ and $2 \cdot (n - 2)$, respectively. These polynomials are represented in a Bernstein-Bézier-form with Bézier-ordinates $c_{i,j,k,l}$, $i + j + k + l = 2(n - 1)$ and $d_{i,j,k,l}$, $i + j + k + l = 2(n - 2)$. Using these polynomials we get

$$
\begin{aligned}
\kappa_{max}^n &\leq \frac{\sup_{p \in \Delta} \|H_{|p}\|}{\inf_{p \in \Delta} \|N_{|p}\|} \\
&= \frac{\sup_{p \in \Delta} \sqrt{h(p)}}{\inf_{p \in \Delta} \sqrt{g(p)}} \\
&\leq \frac{\sqrt{max_{i+j+k+l=2(n-2)}(d_{i,j,k,l})}}{\sqrt{min_{i+j+k+l=2(n-1)}(c_{i,j,k,l})}}.
\end{aligned}
$$

Singularities

If all Bézier-ordinates $c_{i,j,k,l}$, $i + j + k + l = 2(n - 1)$ of the polynomial g are strictly positive, we are sure that no singularities are contained in the corresponding tetrahedron. If all the Bézier-ordinates $d_{i,j,k,l}$, $i + j + k + l = 2(n - 2)$ of the polynomial h are positive, the Hessian of f is regular in every point p of the tetrahedron and we try to find the singular point using a simple Newton method. If a singularity is found in the tetrahedron, it is not difficult to include it into the approximating mesh of triangles.

To compute also higher order singularities, we are working on an algorithm which performs the subdivision for the polynomials $\frac{\partial f}{\partial x}$, $\frac{\partial f}{\partial y}$ and $\frac{\partial f}{\partial z}$ and uses the convex hull property of these polynomials to detect singularities. In future work the algorithm will be based on interval arithmetic operations in order to deal with the problem of finite precision arithmetic that arises with the computation of zero-contours and singularities.

References

1. D. S. Arnon and J. Rauen, On the display of cell decompositions of algebraic surfaces, Technical report, Xerox Corporation, Palo Alto Research Center, 1988.
2. C. L. Bajaj, The emergence of algebraic curves and surfaces in geometric design, Technical Report CSD-TR-92-056, Computer Science Department, Purdue University, West Lafayette, IN 47907, 1992.
3. C. L. Bajaj, Surface fitting with implicit algebraic surface patches, in *Topics in Surface Modeling*, H. Hagen (ed.), SIAM Publications, 1992, 23–52.
4. C. L. Bajaj and I. Ihm, Smoothing polyhedra using implicit algebraic splines, in *Computer Graphics (SIGGRAPH '92 Proceedings)*, volume 26, E. E. Catmull (ed.), 1992, 79–88.
5. J. Bloomenthal, Polygonization of implicit surfaces, Computer Aided Geometric Design 5 (1988), 341–356.
6. C. B. Bradshaw, Surfaces of functions of three variables, Master's thesis, Department of Civil Engineering, Brigham Young University, 1982.

7. L. H. de Figueiredo, J. de Miranda Gomes, D. Terzopoulos, and L. Velho, Physically-based methods for polygonization of implicit surfaces, in *Proceedings of Graphics Interface '92*, 1992, 250–257.

8. M. Hall and J. Warren, Adaptive polygonalization of implicitly defined surfaces, IEEE Computer Graphics and Applications **10** (6) (1990), 33–42.

9. C. Hoffmann and J. Hopcroft, The potential method for blending surfaces and corners, in *Geometric Modeling: Algorithms and New Trends*, G. Farin (ed.), SIAM, Philadelphia, 1987, 347–365.

10. C. Hoffmann and J. Hopcroft, Automatic surface generation in computer aided design, The Visual Computer **1** (2) (1985), 92–100.

11. R. Klein, Visualisierung algebraischer Flächen, Technical report, Universität Tübingen, WSI/GRIS, 1992.

12. A. E. Middleditch and K. H. Sears, Blend surfaces for set theoretic volume modeling systems, in *Computer Graphics (SIGGRAPH '85 Proceedings)*, volume 19, B. A. Barsky, editor, 1985, 161–170.

13. A. Rockwood and J. Owen, Blending surfaces in solid modeling, in *Geometric Modeling: Algorithms and New Trends*, G. Farin (ed.), SIAM, Philadelphia, 1987, 367–383.

14. T. Sederberg, Piecewise algebraic surface patches, Computer Aided Geometric Design **2** (1985), 53–60.

15. H. P. Seidel, A general subdivision theorem for Bézier triangles, in *Mathematical Methods in Computer Aided Geometric Design*, T. Lyche and L. Schumaker (eds.), Academic Press, New York, 1989,, 573–582.

16. G. L. Tindle, Tetrahedral triangulation, in *The Mathematics of Surfaces II*, R.R. Martin (ed.), Clarendon Press, Oxford, 1987, 387–394.

17. L. Velho, Adaptive polygonization of implicit surfaces using simplicial decomposition and boundary constraints, in *Eurographics '90*, C. E. Vandoni and D. A. Duce (eds.), North-Holland, 1990, 125–136.

Reinhard Klein
Universität Tübingen, WSI/GRIS
Auf der Morgenstelle 10
D-72076 Tübingen, GERMANY
reinhard@ gris.informatik.uni-tuebingen.de

A Knot Removal Strategy for Spline Curves

J. C. Koua Brou

Abstract. This paper gives a strategy for removing one or several knots from the B-spline representation of a curve without perturbing it more than a given tolerance.

§1. Introduction

The problem of reducing the number of data in the representation of a function or a curve is not new. Some knot removal techniques have been already published [4,5]. Lafranche and Le Méhauté [4] proposed an approach using a Bézier approximation of a function in $\mathbb{R}^2$, while Lyche and Morken [5] considered the problem using a B-spline representation. We give here another approach using also this representation.

Let k and n be positive integers such as $n \geq k \geq 2$. Let $\mathbf{t} = (t_i)_{i=1}^{n+k}$ be a real sequence called a *knots vector*. We restrict ourself in this paper to the cases such as $t_1 \leq \cdots \leq t_k$, $t_i < t_{i+1}$ for $i = k+1$ to n, and $t_{n+1} \leq \cdots \leq t_{n+k}$. Given $(C_i)_{i=1}^n$ a sequence of n vectors, $C_i \in \mathbb{R}^d$ with $d = 2$ or 3, let

$$S(x) = \sum_{i=1}^{n} C_i B_{i,k,\mathbf{t}}(x)$$

be a $\mathbb{R}^d$ spline curve of order k defined on $\mathbf{t}$; where $B_{i,k,\mathbf{t}}(x)$ are the normalized B-splines functions associated with $\mathbf{t}$.

If z is a real number such as $t_l < z < t_{l+1}$, and denote $\tilde{\mathbf{t}} = \mathbf{t} \cup \{z\}$, we can express S in the basis associated with $\tilde{\mathbf{t}}$ by using the Boehm's knot insertion algorithm (see [1]):

$$S(x) = \sum_{i=1}^{n+1} \tilde{C}_i B_{i,k,\tilde{\mathbf{t}}}(x)$$

with

$$\tilde{C}_i = \begin{cases} C_i & i \leq l - k + 1 \\ \alpha_i C_i + (1 - \alpha_i) C_{i-1} & l - k + 2 \leq i \leq l \\ C_{i-1} & l + 1 \leq i \leq n + 1 \end{cases} \tag{1}$$

Curves and Surfaces in Geometric Design
P. J. Laurent, A. Le Méhauté, and L. L. Schumaker (eds.), pp. 277–284.

And

$$\alpha_i = \frac{z - t_i}{t_{i+k-1} - t_i}.$$

§2. Norms used

Given a knot vector $\mathbf{t} = (t_i)_{i=1}^{n+k}$ and a curve f of $\mathbb{R}^d$, $d = 2$ or 3, such that

$$f(x) = \sum_{j=1}^n C_j B_{j,k,\mathbf{t}}(x); \quad C_j = (C_{1j}, \ldots, C_{dj})^\top \in \mathbb{R}^d. \tag{2}$$

Let $f = (f_1, \ldots, f_d)^\top$, where for $i = 1$ to d,

$$f_i(x) = \sum_{j=1}^n C_{ij} B_{j,k,\mathbf{t}}(x).$$

Let

$$\|f\|_{L^\infty(\mathbb{R}^d)} = \|f\|_{L^\infty,d} = \max_{1 \le i \le d} \|f_i\|_{L^\infty(\mathbb{R})}.$$

Suppose now, given D a $d \times n$ matrix, and denote $D = (D_1, \ldots, D_n)$ with $D_j \in \mathbb{R}^d$ the jth column vector of D. Let

$$\|D\|_{l^\infty,d,n} = \max_{1 \le i \le d} \max_{1 \le j \le n} |D_{ij}|.$$

Lemma. $\| \cdot \|_{l^\infty,d,n}$ *is a norm.*

If $S_{k,\mathbf{t}}$ denotes the space of all spline curves of order k defined on $\mathbf{t}$, let f be as in (2); and if $C = (C_1, \ldots, C_n)$ let

$$\|f\|_{S_{k,\mathbf{t}}} = \|C\|_{l^\infty,d,n}.$$

Proposition. $\| \cdot \|_{L^\infty,d}$ *is equivalent to* $\| \cdot \|_{S_{k,\mathbf{t}}}$.

Thus we set $\| \cdot \|_\top = \| \cdot \|_{L^\infty,d}$ and $\| \cdot \|_* = \| \cdot \|_{S_{k,\mathbf{t}}}$ hereafter.

§3. The principle

Let $\mathbf{t} = (t_i)_{i=1}^{m+k}$ be a knot vector as in the introduction. Assume, f is a spline curve defined on $\mathbf{t}$ which has the form

$$f(x) = \sum_{i=1}^m C_i B_{i,k,\mathbf{t}}(x), \qquad C_i \in \mathbb{R}^d.$$

Suppose we have to remove a knot $z = t_j$ from $\mathbf{t}$; where z is an interior knot of $\mathbf{t}$, $(k + 1 \le j \le m)$. Let $\mu = \mathbf{t} \setminus \{z\}$. We can make a reasonable hypothesis:

Hypothesis. f stems from a B-spline curve g^j of order k defined on μ by inserting the knot t_j.

Thus, g^j has the form

$$g^j(x) = \sum_{i=1}^{m-1} A_i B_{i,k,\mu}(x).$$

Let us reinsert z into μ; note that $\mu_{j-1} < z < \mu_j$; so using (1) with $l = j - 1$ we have

$$g^j(x) = \sum_{i=1}^{m} \tilde{A}_i B_{i,k,\mathbf{t}}(x)$$

where as in (1) the coefficients $\tilde{A}_i$ are given by the relation

$$\tilde{A}_i = \begin{cases} A_i & i \leq j - k \\ \beta_i A_i + (1 - \beta_i) A_{i-1} & j - k + 1 \leq i \leq j - 1 \\ A_{i-1} & j \leq i \leq m \end{cases} \tag{3}$$

with

$$\beta_i = \frac{z - \mu_i}{\mu_{i+k-1} - \mu_i}.$$

The coefficients of f are known, and we want to determine those of g^j. According to the hypothesis, f comes from g^j, so we can expect the coefficients of g^j in (3) to be equal to C_i for all i. Thus, we have

$$C_i = \tilde{A}_i = \begin{cases} A_i & i \leq j - k \\ \beta_i A_i + (1 - \beta_i) A_{i-1} & j - k + 1 \leq i \leq j - 1 \\ A_{i-1} & j \leq i \leq m \end{cases}. \tag{4}$$

The immediate result is

$$A_i = \begin{cases} C_i & i \leq j - k \\ C_{i+1} & j - 1 \leq i \leq m - 1. \end{cases}$$

The other coefficients of g^j can be determined by using one of the two algorithms given below:

A.1) "forward"

Let $A_{j-k} = C_{j-k}$

For $i = j - k + 1$ to $j - 2$ do

$\quad \beta_i = \frac{z - \mu_i}{\mu_{i+k+1} - \mu_i}$

$\quad A_i = [C_i - (1 - \beta_i) A_{i-1}]/\beta_i$

endfor

A.2) "backward"

Let $A_{j-1} = C_j$

For $i = j - 1$ to $j - k + 2$ step -1 do

$\quad \beta_i = \frac{z - \mu_i}{\mu_{i+k+1} - \mu_i}$

$\quad A_{i-1} = [C_i - \beta_i A_{i-1}]/(1 - \beta_i)$

endfor

Unfortunately, there are two drawbacks. Using the coefficients computed by "forward" and reinserting the knot z into μ leads to

$$\tilde{A}_i = C_i \qquad \text{for all} \quad i \quad \text{but} \quad i = j - 1$$

and "backward" gives

$$\tilde{A}_i = C_i \qquad \text{for all} \quad i \quad \text{but} \quad i = j - k + 1.$$

Therefore, the condition (4) cannot be satisfied entirely. And, if we call g_1^j the curve determined using "forward" and g_2^j the other one, we cannot ensure $g_1^j = g_2^j$; so we have to choose one of these curves to be g^j.

Let us recall that, in fact, we want to approximate f; so if $\| \, . \, \|_*$ is a certain norm to be defined, it seems reasonable to define g^j as follows

$$g^j = \begin{cases} g_1^j & \text{if } \|f - g_1^j\|_* \leq \|f - g_2^j\|_* \\ g_2^j & \text{else.} \end{cases}$$

Hereafter, we can assign the number $\omega_j = \|f - g^j\|_*$ to t_j. Let us note that if $\omega_j = 0$ then $f = g^j$, this means that $z = t_j$ is not necessary for the representation of f. In fact, the bigger ω_j is, the more important is z for the representation of f. So we say that ω_j, called a *weight* associated with t_j, quantifies the significance of z in the representation of f.

Let ε, called the *tolerance*, be a positive real number given. The knot t_j could be removed if $\omega_j \leq \varepsilon$.

§4. The strategy

The strategy can be divided into three main parts which are:
* Computation of weights.
* Selection of knots to be removed.
* Determination (or reconstruction) of the approximated curve.

Computation of weights.

First, we have to assign a weight to each interior knot t_j, $k + 1 \leq j \leq m$. Let V be an integer array called *the decision array*, such that

$$DA(j) = \begin{cases} 1 & \text{if "forward" is to be used} \\ 2 & \text{if "backward" is to be used.} \end{cases}$$

Hereafter, we use the algorithm described below to reach our goal.

Computing weights

For $j = k + 1$ to n do
   $z = t_j$
   $\tilde{\mathbf{t}} = \mathbf{t} \setminus \{z\}$
   For $i = 1$ to $j - k$ do
      $A_i = C_i$
   endfor
   For $i = j - 1$ to $n - 1$ do
      $A_i = C_{i+1}$
   endfor
   For $i = j - k + 1$ to $j - 2$ do
      $\beta_i = \frac{z - \tilde{t}_i}{\tilde{t}_{i+k-1} - \tilde{t}_i}$
      $A_i = [C_i - (1 - \beta_i)A_{i-1}]/\beta_i$
   endfor
   $\omega_j^m = \|A - C\|_T = \|A_{j-1} - C_{j-1}\|_{l^\infty}$
   For $i = j - 1$ to $j - k + 2$ step -1 do
      $\beta_i = \frac{z - \tilde{t}_i}{\tilde{t}_{i+k-1} - \tilde{t}_i}$
      $A_{i-1} = [C_i - \beta_i A_{i-1}]/(1 - \beta_i)$
   endfor
   $\omega_j^d = \|A - C\|_* = \|A_{j-k+1} - C_{j-k+1}\|_{l^\infty}$
   If $(\omega_j^m \le \omega_j^d)$ then
      $\omega_j = \omega_j^m$
      $DA(j) = 1$
   else
      $\omega_j = \omega_j^m$
      $DA(j) = 2$
   endif
endfor

Remark. We have $\beta_i \ne 0$ and $\beta_i \ne 1$ for i such that $i = j - k + 1$ to $j - 1$. This is due to the construction of the knot vector $\mathbf{t}$.

Selection of knots to be removed.

Now we attempt to remove several knots at the same time. First of all, we look for all the knots with a weight less than ε; and we group them in a vector V of p elements. We then need to introduce a *vicinity constraint* when two knots, say t_{j_1} and t_{j_2}, are to be removed together : $|j_1 - j_2| > k$. Unless we compute a new weight associated with t_{j_1} and t_{j_2}, corresponding to the removal of these two knots together, which is, of course, always possible, we have to impose such a vicinity constraint which allows us to pick two knots without perturbing the weights calculated before. This constraint is obtained by taking into account the supports (which contain a knot t_j) of all B-splines and the process of determining the weight associated with this knot (the coefficients used).

By taking into account the vicinity constraint, and using one of the selection mode enumerated below, we get a vector U of length q which contains the knots to remove at the same time:

* A binary search.

* A random search.

* A forward approach, consisting of ranking the knots by subscripts in increasing order before removing them.

* A backward approach which is the same as the approach above with knots ranked in decreasing order.

Determination of the approximated curve.

At this stage, the array of decision DA is already filled. Assume U is the array of the subscript of the q knots to be removed from $\mathbf{t}$. Following the algorithm below we can reconstruct the curve g which approximates f:

Reconstruction algorithm

> (Initiation)
> $A^0 = C$
> $\tau^{(0)} = \mathbf{t}$
> (Process)
> For $i = 1$ to q do
> $j = U(i)$
> $\tau^{(i)} = \tau^{(i-1)} \setminus \{t_j\}$
>   if $DA(j) = 1$ then
>   use "forward" to compute $A^{(i)}$
>   else
>   use "backward" to compute $A^{(i)}$
>   endif
> endfor
> $A = A^{(q)}$
> $\tau = \tau^{(q)}$

where $A = (A_i)_{i=1}^{n-q}$ denotes the B-spline coefficients of g, and τ a knot vector of length $n - q + k$ on which g is defined.

§5. Numerical results

Suppose we are given $m = 501$ points $(x_i, y_i)_{i=1}^{m}$ in $\mathbb{R}^2$ defining a discrete representation of a Lissajous curve:

$$\left. \begin{array}{l} x_i = 2\sin(2\xi_i) \\ y_i = 3\sin(3\xi_i) \end{array} \right\} \qquad i = 1, \ldots, m,$$

where

$$\xi_i = \frac{2\pi(i-1)}{m-1}, \qquad i = 1, \ldots, m.$$

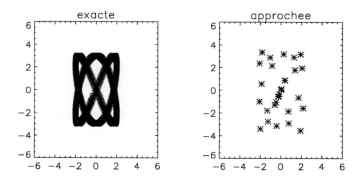

Fig. 1. Discrete curves.

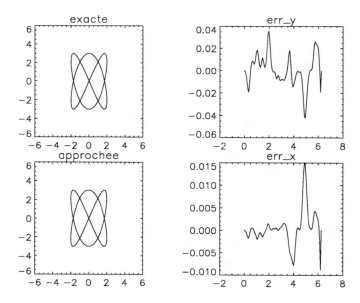

Fig. 2. Continuous curves.

We can determine a cubic B-spline curve f which interpolates these values, and which is defined on $\mathbf{t} = (t_i)_{i=1}^{m+k}$, with $k = 4$, $t_1 = \cdots = t_k = \xi_1$, $t_i = \xi_{i-2}$ for $i = k+1$ to m, and $t_{m+1} = \cdots = t_{m+k} = \xi_m$. Using a tolerance $\varepsilon = 0.05$ and this curve f as starting point, the strategy apply on f gives after some iterations, a curve g of the same order defined on a knot sequence τ of 33 elements. The discrete initial curve denoted here "exacte" which is given by the plotting of the B-spline coefficients of f, and the discrete final curve named "approchee" given by the coefficients of g are shown in Figure 1. We also give (see Fig. 2) the continuous initial curve ("exacte"), the final continuous curve ("approchee"), the error in the x-component ("err_x") and the error in the y-component ("err_y"). Some other examples can been found in [3].

§6. Remarks

We note that our strategy works independently of the curve component treating the curve in its whole entity, in contrast to the strategy of Lyche (see [5]). It is different from the strategy given in [2], although both start from the same idea. Notice also that it is nondestructive as in [4].

Acknowledgements. I would like to thank Alain J. Y. Le Méhauté for his useful discussions and comments.

References

1. Boehm, W., Inserting new knots into splines curves, Comput. Aided Design **12** (1980), 199–201.
2. Eck, M., and Handenfeld, J., Knot Removal for B-spline Curves, preprint.
3. Koua, Brou J. C., Réduction de Bases de Données et Approximation de Fonctions, Thèse INSA de Rennes, Octobre 1993.
4. Le Méhauté, A. Lafranche, Y., A knot removal strategy for scattered Data in $\mathbb{R}^2$, in *Mathematical Methods in Computer Aided Geometric Design*, T. Lyche and L. L. Schumaker (eds), 419–426
5. Lyche, T., and Mørken, K., Knot removal for parametric B-spline curves and surfaces, Computer-Aided Geom. Design 4 (1987), 217–230.

Jean Claude Koua Brou
Département Mathématiques et Systèmes de Communication
Ecole Nationale des Télécommunications de Bretagne
B. P. 832, 29285 Brest cedex FRANCE
Koua_brou@Gti.Enst-Bretagne.fr

Computation of Curvatures Related
to Surface-Surface Blending

I. Lef

Abstract. To ensure accurate construction of fillets/chamfers between two parametric surfaces, one has to know curvature values (in the object space as well as in parameter spaces) of the spine and contact curves. The paper shows how to compute them for various blend types.

§1. Introduction

Throughout this paper, bold letters will denote vector quantities. Dot, cross, and triple products will be denoted as $(\mathbf{a}, \mathbf{b})$, $\mathbf{a} \times \mathbf{b}$, and $(\mathbf{a}, \mathbf{b}, \mathbf{c})$, respectively, and $\|\mathbf{a}\|$ will denote the norm of $\mathbf{a}$.

Blend (fillet or chamfer) constructions between two parametric surfaces usually employ tracing algorithms. Accuracy requirements call for careful choice of the step size at every iteration of the algorithm. To determine the step size, we have to know exact curvature values for contact and spine curves in the object space as well as in the parameter spaces of both surfaces.

Formulas for curvatures of the intersection curve between two surfaces were developed in [1,3]. They will serve as the basis for further derivations. Let $\mathbf{r}^{(1)}(u, v)$ and $\mathbf{r}^{(2)}(p, q)$ be two regular parametric surfaces defined in parameter spaces (u, v) and (p, q), respectively. The intersection between $\mathbf{r}^{(1)}$ and $\mathbf{r}^{(2)}$ is a curve $\mathbf{\Gamma}$ that can be expressed through the parameter spaces of the surfaces:

$$\mathbf{\Gamma}(t) = \mathbf{r}^{(1)}(u(t), v(t)) = \mathbf{r}^{(2)}(p(t), q(t)),$$

where t is a parameter common to all three representations.

Consider the *kernels*

$$\omega_1 = \dot{u}\ddot{v} - \ddot{u}\dot{v}$$

$$\omega_2 = \dot{p}\ddot{q} - \ddot{p}\dot{q}.$$

Curves and Surfaces in Geometric Design
P. J. Laurent, A. Le Méhauté, and L. L. Schumaker (eds.), pp. 285–292.
Copyright © 1994 by A K PETERS, Wellesley, MA
ISBN 1-56881-039-3.

Once the kernels are known, the parametric curvatures κ_1 and κ_2 can be found as

$$\kappa_1 = \frac{\omega_1}{(\dot{u}^2 + \dot{v}^2)^{3/2}}$$
$$\kappa_2 = \frac{\omega_2}{(\dot{p}^2 + \dot{q}^2)^{3/2}}.$$

(1)

Let $\mathbf{N}_1$ and $\mathbf{N}_2$ be the *non-normalized* surface normals

$$\mathbf{N}_1 = \mathbf{r}_u^{(1)} \times \mathbf{r}_v^{(1)}$$
$$\mathbf{N}_2 = \mathbf{r}_p^{(2)} \times \mathbf{r}_q^{(2)},$$

$\mathbf{A}_1$ and $\mathbf{A}_2$ the quadratic forms

$$\mathbf{A}_1 = \mathbf{r}_{uu}^{(1)}\dot{u}^2 + 2\mathbf{r}_{uv}^{(1)}\dot{u}\dot{v} + \mathbf{r}_{vv}^{(1)}\dot{v}^2$$
$$\mathbf{A}_2 = \mathbf{r}_{pp}^{(2)}\dot{p}^2 + 2\mathbf{r}_{pq}^{(2)}\dot{p}\dot{q} + \mathbf{r}_{qq}^{(2)}\dot{q}^2,$$

$\dot{\mathbf{\Gamma}}$ the tangent vector to the intersection curve in the object space, and D the triple product

$$D = 2(\mathbf{N}_1, \dot{\mathbf{\Gamma}}, \mathbf{N}_2).$$

The following two theorems were proved in [1].

Theorem 1. *For transversal intersection points, the kernels can be found as*

$$\omega_i = \frac{2(\mathbf{A}_{3-i} - \mathbf{A}_i, \mathbf{N}_{3-i})}{D}, \qquad i = 1, 2.$$

Theorem 2. *The object space curvature κ can be found as*

$$\kappa = \frac{\|\mathbf{A}_i \times \dot{\mathbf{\Gamma}} + \omega_i \mathbf{N}_i\|}{\|\dot{\mathbf{\Gamma}}(t)\|^3} \qquad i = 1, 2.$$

(2)

If $\dot{\gamma}_1 = (\dot{u}, \dot{v})$ and $\dot{\gamma}_2 = (\dot{p}, \dot{q})$ are tangent vectors to the intersection curve in corresponding parameter spaces, then, substituting (1) into (2), we have

$$\kappa = \frac{\|\mathbf{A}_i \times \dot{\mathbf{\Gamma}} + \kappa_i \|\dot{\gamma}_i\|^3 \mathbf{N}_i\|}{\|\dot{\mathbf{\Gamma}}(t)\|^3}, \qquad i = 1, 2.$$

(3)

This can be also viewed as a formula for the 3D curvature of a surface curve (not necessarily an intersection curve), expressed through its 2D curvature, 2D tangent vector, and local properties of the surface.

§2. Standard Blends

This kind of blend is usually described by the *rolling ball* model: a ball of a constant radius R is rolling along both surfaces staying in contact with them. The trajectory of its center is called the *spine curve*, while the trajectories of its contact points are called the *contact curves*. Once the contact curves are found, their corresponding points can be connected, to form a blend surface, by line segments (chamfering, G_0 blend continuity), circular arcs (filleting, G_1 continuity), splines (general blending, G_2 continuity).

The spine curve can be found as the intersection of two offset surfaces, $\mathbf{S}_1$ and $\mathbf{S}_2$:

$$
\begin{cases}
\mathbf{S}^{(1)}(u, v) = \mathbf{r}^{(1)}(u, v) + R\,\mathbf{n}^{(1)}(u, v) \\
\mathbf{S}^{(2)}(p, q) = \mathbf{r}^{(2)}(p, q) + R\,\mathbf{n}^{(2)}(p, q).
\end{cases}
\tag{4}
$$

Theorems 1 and 2, when applied to the offset surfaces, allow us to find parametric curvatures of the spine curve in the object space and in the parameter spaces of the offset surfaces.

Since each original surface and its offset share the parametrization, and the spine curve is the image of the contact curve under the mapping (4), 2D properties of the contact curve are the same as those of the spine curve. In particular, 2D curvature of the contact curve is the same as that of the spine curve. Applying (3) to each pre-image of the intersection curve, we obtain the 3D curvature of the contact curves.

§3. Variable Radius Blends

This type of blend can be described by a modified rolling ball model, namely, with the ball radius changing in the process to meet specified restrictions. To proceed with curvature computations, we will use a concept of *Voronoi surface*.

Definition 1. *The Voronoi surface* $\mathbf{V}$ *of surfaces* $\mathbf{r}^{(1)}$ *and* $\mathbf{r}^{(2)}$ *is the locus of points equidistant from* $\mathbf{r}^{(1)}$ *and* $\mathbf{r}^{(2)}$.

The spine curve of the blend between two surfaces belongs to their Voronoi surface. We want to represent the Voronoi surface as a variable offset of either $\mathbf{r}^{(1)}$ or $\mathbf{r}^{(2)}$, thus $\mathbf{V}^{(1)}$ and $\mathbf{V}^{(2)}$ both describing, geometrically, the same surface:

$$
\begin{bmatrix}
\mathbf{V}^{(1)}(u, v) = \mathbf{r}^{(1)}(u, v) + \mathcal{R}^{(1)}(u, v)\,\mathbf{n}^{(1)}(u, v) \\
\mathbf{V}^{(2)}(p, q) = \mathbf{r}^{(2)}(p, q) + \mathcal{R}^{(2)}(p, q)\,\mathbf{n}^{(2)}(p, q),
\end{bmatrix}
\tag{5}
$$

where $\mathcal{R}^{(1)}(u, v)$ and $\mathcal{R}^{(2)}(p, q)$ are appropriate scalar functions on (u, v) and (p, q), respectively. We also want the mappings (5) to be local homeomorphisms. Such a representation is not always possible. There are cases when almost entire Voronoi surface cannot be represented as the image of any mapping of type (5), and cannot be represented by homeomorphisms of type (5) anywhere. On the other hand, points on $\mathbf{r}^{(1)}$ or $\mathbf{r}^{(2)}$ where (5) is not a homeomorphism, are not suitable for hosting points of contact curves under the rolling ball model. Therefore, we assume that, at every point of interest on $\mathbf{r}^{(1)}$ or $\mathbf{r}^{(2)}$, there exists $\epsilon > 0$ such that

$$1 - 2HR + KR^2 \geq \epsilon, \tag{6}$$

where H and K are the mean and Gaussian curvatures, respectively, of the original surface, and R is the local value of the corresponding offset function. The above inequality assures that (5) are homemorphisms and that the normals of the Voronoi surfaces do not flip. For simplicity, we will also assume that the offset directions are the same as the directions of normals for both surfaces.

Mappings (5) now define two *local* parametrizations of the relevant parts of the Voronoi surface, inherited from parametrizations of $\mathbf{r}^{(1)}$ and $\mathbf{r}^{(2)}$. Our goal is to relate one parametrization to the other, and to find *local* representations of the functions $\mathcal{R}^{(i)}$. (Another approach to constructing Voronoi surfaces, the global one, is described in [2].)

Two points, $\mathbf{r}^{(1)}(u, v)$ and $\mathbf{r}^{(2)}(p, q)$, define a point on the Voronoi surface iff there exist (equal) scalars $\mathcal{R}^{(1)}(u, v)$ and $\mathcal{R}^{(2)}(p, q)$ such that

$$\mathbf{r}^{(1)}(u, v) + \mathcal{R}^{(1)}(u, v)\,\mathbf{n}^{(1)} = \mathbf{r}^{(2)}(p, q) + \mathcal{R}^{(2)}(p, q)\,\mathbf{n}^{(2)}. \tag{7}$$

Let $\boldsymbol{\Delta}$ be the difference vector $\boldsymbol{\Delta} = \mathbf{r}^{(2)}(p, q) - \mathbf{r}^{(1)}(u, v)$. Then the offset values, for a matching pair of points, can be expressed as

$$\begin{cases} \mathcal{R}^{(1)}(u, v) = \dfrac{(\boldsymbol{\Delta}, \boldsymbol{\Delta})}{2(\boldsymbol{\Delta}, \mathbf{n}^{(1)})} \\[3mm] \mathcal{R}^{(2)}(p, q) = -\dfrac{(\boldsymbol{\Delta}, \boldsymbol{\Delta})}{2(\boldsymbol{\Delta}, \mathbf{n}^{(2)})}. \end{cases} \tag{8}$$

The negative sign reflects asymmetry of $\boldsymbol{\Delta}$ with respect to $\mathbf{r}^{(1)}$ and $\mathbf{r}^{(2)}$.

To find the connection between two Voronoi surface parametrizations, we substitute (8) into (7) and differentiate the result with respect to u. We obtain a linear vector equation with two unknowns $\frac{\partial p}{\partial u}$ and $\frac{\partial q}{\partial u}$, given as

$$\mathbf{a}\frac{\partial p}{\partial u} + \mathbf{b}\frac{\partial q}{\partial u} = \mathbf{c}. \tag{9}$$

Vectors $\mathbf{a}, \mathbf{b}, \mathbf{c}$ are functions of $\boldsymbol{\Delta}$ and partial derivatives of $\mathbf{r}^{(1)}$ and $\mathbf{r}^{(2)}$. Taking dot products of (9) with $\mathbf{a}$ and $\mathbf{b}$, we obtain a system of two linear equations with unknowns $\frac{\partial p}{\partial u}$ and $\frac{\partial q}{\partial u}$.

Theorem 3. *The linear system of equations (9) is not degenerate.*

Proof: One can show by straightforward calculations that the absolute value $\mathcal{D}$ of the system determinant is

$$\mathcal{D} = \frac{\epsilon\sqrt{EF - G^2}\sqrt{1 - \cos\theta}}{\sqrt{2}},$$

with E, F, G being the first fundamental form coefficients of $\mathbf{r}^{(2)}$, θ being the angle between the normals $\mathbf{n}^{(1)}$ and $\mathbf{n}^{(2)}$, and ϵ from (6). The determinant $\mathcal{D}$ is 0 only when $\theta = 0$. This corresponds to an infinitely remote point on the Voronoi surface. ∎

Solving the system and differentiating (9) once more with respect to u, we arrive at the linear system for $\frac{\partial^2 p}{\partial u^2}$ and $\frac{\partial^2 q}{\partial u^2}$, with the same determinant. Proceeding this way, we find all the derivatives:

$$\frac{\partial p}{\partial u}, \frac{\partial q}{\partial u}, \frac{\partial p}{\partial v}, \quad \cdots \quad , \frac{\partial^2 u}{\partial p^2}, \quad \cdots \quad , \frac{\partial^2 p}{\partial q^2}, \quad \cdots \quad , etc. \tag{10}$$

Thus, both *local* parametric representations of the Voronoi surface are completely known.

3.1. Locally Defined Blends

Definition 2. *A blend is called locally defined if its current radius depends only on the local properties of the surfaces.*

For example, the radius of a standard *angular* fillet is defined as

$$\mathcal{R}_a = R\Big(1 - (\mathbf{n}^{(1)}, \mathbf{n}^{(2)})^2\Big).$$

The radius $\mathcal{R}_a$ depends only on the angle between two normals (and, of course, on the constant R). $\mathcal{R}_a$ is equal to R when the angle is $\pi/2$. The values of the radius, as well as its first derivatives, are 0 when the angle is 0 or π. This periodic property makes $\mathcal{R}_a$ suitable for filleting surfaces where the angle can approach 0 or π.

Equating $\mathcal{R}^{(i)}$ for either surface with $\mathcal{R}_a$, we obtain an implicit equation of the spine curve in the corresponding parameter space. Assuming the regularity of the curve (which is true in most practical cases), we can compute 2D curvature of, and tangent vector to the curve at any given point, provided that for each parametrization, we choose appropriate transformation values from (10) depending on the parameter space we work with. Again, the 2D curvature of the contact curve is the same as that of the spine curve (in the corresponding parameter space of $\mathbf{V}$), and applying formula (3) we obtain the 3D curvature of the contact curve.

3.2. Globally Defined Blends

Definition 3. *A blend is called globally defined if its radius depends on global properties of the surfaces.*

A common example is the *linear* fillet whose radius changes linearly between start and end values. The main problem with globally defined blends is how to build a geometric representation of the radius function domain. A frequent choice is the intersection curve of the blended surfaces. Its advantage is in the relative ease of obtaining it *before* the beginning of blend construction. However, mapping a current point of the spine curve onto this curve to get the current parameter value and, subsequently, the radius value, usually involves heuristics.

The natural choice for the geometric representation of the domain, is the spine curve of the *future* blend. The mapping becomes trivial but another problem, that of computing the current parameter value, can be solved only approximately since the spine curve length is not known until the blend is built.

Let us start with the (u, v) parametrization of the Voronoi surface. Let $(u(s), v(s))$ be the representation of the spine curve in the parameter space of $\mathbf{V}^{(1)}$ and let parameter s be the arc-length of the spine curve in the object space. Let the desired blend radius function, $\Re(t)$, be a scalar function with $t \in [0, 1]$.

Suppose we have found a point, (u, v), on the Voronoi surface, where the offset value $\mathcal{R}^{(1)}(u, v)$ is equal to the value $\Re(t_0)$ of the input radius function at the current parameter t_0. Our immediate goal is to find the step direction, $(\dot{u}, \dot{v})$, to match the first derivative of the offset function, $\mathcal{R}^{(1)}(u(s), v(s))$, along the spine curve, with the first derivative of the input radius function, $\Re'(t_0)$, at t_0. If s_0 is the length of the already built part of the spine curve, then the estimated length of the spine curve is $l = s_0/t_0$. We consider the system of equations:

$$\begin{cases} l\,\Re'(t_0) = \mathcal{R}_u^{(1)}\dot{u} + \mathcal{R}_v^{(1)}\dot{v} \\ \qquad 1 = E\dot{u}^2 + 2F\dot{u}\dot{v} + G\dot{v}^2. \end{cases} \tag{11}$$

The first equation in (11) does the matching of the derivatives, the second one states that tangent vector to the curve is of unit length in 3D, representing the assumption that the spine curve parametrization is arc-length in 3D. Quantities E, F, G are the coefficients of the first fundamental form of the Voronoi surface $\mathbf{V}^{(1)}$.

This system of equations can be reduced to one quadratic equation. The solutions of (11) represent two (possibly) different step directions. One has to choose a direction that is consistent with the direction of the spine curve. In case of no solution, the blend construction is either impossible or the value of t_0 is not consistent with the arc position. (However, in the absence of an exact solution of (11), the minimum norm solution might be worth considering for practical implementations of tracing algorithms. We will not discuss the details of the global algorithm here.)

Our next goal is to find the vector of the second derivative, $(\ddot{u}, \ddot{v})$. Here is the linear system to determine it:

$$\begin{cases} l^2\,\Re''(t_0) = \mathcal{R}_u^1\ddot{u} + \mathcal{R}_v^1\ddot{v} + \mathcal{R}_{uu}^1\dot{u}^2 + 2\mathcal{R}_{uv}^1\dot{u}\dot{v} + \mathcal{R}_{vv}^1\dot{v}^2 \\ \qquad 0 = \left(\mathbf{T}^{(1)},\ \mathbf{V}_u^{(1)}\ddot{u} + \mathbf{V}_v^{(1)}\ddot{v} + \mathbf{V}_{uu}^{(1)}\dot{u}^2 + 2\mathbf{V}_{uv}^{(1)}\dot{u}\dot{v} + \mathbf{V}_{vv}^{(1)}\dot{v}^2\right), \end{cases} \tag{12}$$

where $\mathbf{T}^{(1)} = \mathbf{V}_u^{(1)}\dot{u} + \mathbf{V}_v^{(1)}\dot{v}$ is the tangent vector to the spine curve in 3D.

The first equation matches the second derivative $\Re''(t_0)$ of the input radius function with the second derivative $\frac{d^2}{ds^2}\mathcal{R}^{(1)}(u(s), v(s))$ of the offset function. The second equation states that the second derivative of the spine curve is orthogonal in 3D to its first derivative. This is also the effect of arc-length parametrization of the spine curve. (As before, if the system (12) is inconsistent, one might consider the minimum norm solution.)

Once (12) is solved, we can compute parametric and object space curvatures of the spine curve. Again, the parametric curvature of the contact curve is the same as that of the spine curve. Knowing the parametric tangent direction $(\dot{u}, \dot{v})$, we can use (3) to find the 3D curvature of the contact curve. Similar computations can be done for the second parametrization (p, q) of the Voronoi surface.

§4. Non-isosceles Blends

This kind of blend arises when offset values are different for the two surfaces. Non-isosceles chamfers are very common in design practice. While isosceles fillets make use of circular arcs, non-isosceles fillets can use elliptical arcs.

The approach to computation of curvatures for non-isosceles blends is the same, only the Voronoi surface is defined slightly more generally.

Definition 4. *Let $\alpha \in [0, \pi]$. The α-Voronoi surface $\mathbf{V}^\alpha$ of surfaces $\mathbf{r}^{(1)}$ and $\mathbf{r}^{(2)}$ is the locus of points for which the ratio of distances D_1 and D_2 from corresponding surfaces is*

$$D_1 : D_2 = \cos\alpha : \sin\alpha$$

The radius functions, for a matching pair of points, can be expressed as:

$$\begin{cases} \mathcal{R}^{(1)}(u, v) = \dfrac{(\boldsymbol{\Delta}, \boldsymbol{\Delta})\cos\alpha}{(\boldsymbol{\Delta}, \mathbf{n}^{(1)})\cos\alpha - (\boldsymbol{\Delta}, \mathbf{n}^{(2)})\sin\alpha} \\[4mm] \mathcal{R}^{(2)}(p, q) = \dfrac{(\boldsymbol{\Delta}, \boldsymbol{\Delta})\sin\alpha}{(\boldsymbol{\Delta}, \mathbf{n}^{(1)})\cos\alpha - (\boldsymbol{\Delta}, \mathbf{n}^{(2)})\sin\alpha}. \end{cases}$$

We can now proceed with curvature computations in exactly the way described previously.

§5. Varying Non-isosceles Blends

The next generalization is to vary the index α, defining it either locally or globally. This will allow arc center points to travel from one α-Voronoi surface to another, thus leading to more general types of blend surfaces. The curvature computation algorithm described above can be applied to these blends as well.

References

1. Belyaev, A., and I. Lef, Curvature computations for surface-surface inter-section curves, in *Proceedings of the Schlumberger Software Conference*, Austin, TX, 1990, 1.037–1.041.
2. Chandru, V., and C. M. Hoffman, Variable radius blending using Dupin Cyclides, in *Geometric Modeling for Product Engineering*, M. J. Wozny, J. U. Turner, and K. Preiss (eds.), North-Holland, New York, 1990, 39–57.
3. Lef, I., Curvature computations at points of tangency on surface-surface intersection curves, in *Proceedings of the Schlumberger Software Confer-ence*, Austin, TX, 1991, 2.125–2.131.

Isaac Lef
Applicon, Inc.
Billerica, MA 01821
lef@ billerica.applicon.slb.com

Least-Squares Optimization of Thread Surfaces

M. Léger, J.-M. Morvan, and M. Thibaut

Abstract. In geology, faults may sometimes be considered as slipping surfaces between two approximately rigid blocks. Geometrically, we define these surfaces as follows: they are preserved by at least one single-parameter family of moves, and we call them *threads* by analogy with a screw and a nut. Besides, we need to optimize our subsurface models for oil exploration purposes. Therefore, we propose a method that optimizes a surface with respect to several least-squares criteria related to curvature, proximity to known points, and the "thread property".

§1. Introduction

Geophysics aims to determine and image a subsurface by using seismic data and geologic information. Inversion [5] is a popular technique to achieve this goal, but it requires a least-squares formulation of geological knowledge [2]. We focus our attention on the shape of faults. Sometimes, faults separate blocks which may be considered rigid. We formulate in least-squares terms this geological knowledge: "a fault surface separates two approximately rigid blocks" and we optimize the surface with respect to that criterion. We also constrain the surface to be smooth and to be close to given points.

In the first section, we derive the geometrical consequences of the rigid block hypothesis which leads us to introduce the concept of *thread* surface. We briefly review inversion in the second section. Next, we describe the physical objective functions related to following data: the *thread* property, the proximity to given points and the smoothness of the surface. In the fourth section, we add three nonphysical criteria to make the problem well-posed. Finally, we present numerical results and conclusions.

§2. Rigid Blocks and Thread Surfaces

We consider a fault as a surface, *i.e.*, we neglect the thickness of the possible gauge zone. We represent the surface using a parameterization $\Phi(u, v)$ (u and v are curvilinear coordinates) because many faults are almost vertical and therefore $z(x, y)$ representations are awkward. We discretize the parameterization by using B-spline tensor products.

Curves and Surfaces in Geometric Design
P. J. Laurent, A. Le Méhauté, and L. L. Schumaker (eds.), pp. 293–300.

2.1. The rigid block approximation

In some geologic circumstances, strains in the blocks separated by a fault are
weak. In these cases, the rigid block approximation is valid. Besides, blocks
usually remain in contact during faulting, and consequently the fault surface
remains the same during the move. We call *threads* the surfaces that have
this property, by analogy with a screw and a nut.

Definition 1. *A surface S is a thread if and only if there exists a single
parameter family of moves that preserves surface S.*

Surfaces of revolution and cylinders are typical examples of threads. On the
contrary, an egg-box is not a thread.

2.2. A characteristic property of threads

We review the definition of a twistor.

Definition 2. *A twistor $\mathcal{T}$ is a vector field such that, for any M,*

$$\mathcal{T} = \begin{bmatrix} T(\vec{O}) \\ \vec{\Omega} \end{bmatrix}_O = \begin{bmatrix} T(\vec{O}) + \vec{\Omega} \wedge \vec{OM} \\ \vec{\Omega} \end{bmatrix}_M.$$

According to an idea of C.-M. Marle, Definition 1 is equivalent to

Definition 3. *A surface S is a thread if and only if there exists a nonzero
twistor $\mathcal{T}$ such that vector $T(P)$ belongs to the tangent plane $T_P(S)$ to S at
any point P of S.*

In the general case (Figure 1), the field lines of a twistor are helices with the
same axis and the same pitch. If $\Omega = 0$, the vector field is constant, the
thread is a cylinder and the relative move between the blocks is a translation.
If $T(M)$ is everywhere orthogonal to Ω, the thread is a surface of revolution,
and the relative move is a rotation. A thread may be considered as a single
parameter family of field lines of a nonzero twistor. Geologists call *striae* the
lines made on one block by the bumps of the other. We call *computed striae*
the *field lines* of the (projection of the) twistor on a (quasi-) thread surface.

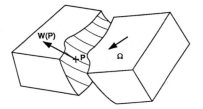

Fig. 1. Two rigid blocks.

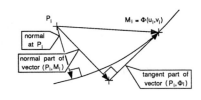

Figure 2. Proximity to a given point.

§3. Inversion

We have chosen the inversion approach because it can take various kinds of data into account, which is our problem since we wish to constrain a surface to be smooth, to be close to given points and to approach a thread.

3.1. The direct and inverse problems

The laws of physics, represented by a map f, enable us to compute predicted observations d as a function of a known model m, $f(m) = d$. This is called the *direct problem.* In geophysics, the situation differs since we wish to determine an earth model m from known observed data d, $f^{-1}(d) = m$. This is called the *inverse problem.* Unfortunately, the map f is generally not invertible, and the model parameters cannot be computed by using that equation. Therefore, we look for a model $\tilde{m}$ such that computed data $\tilde{d} = f(\tilde{m})$ are as close as possible to observed data d.

3.2. The objective function

We need to make clear what "as close as possible" means. To do that, we define an objective function Q which measures the discrepancy between computed and observed data. If we assume that the data space is a vector space, a norm on the residual vector $d - \tilde{d}$ will define objective function Q. We choose an L_2-norm for convenience $Q(m) = \frac{1}{2}\|d - \tilde{d}\|^2$. The Gauss-Newton method gives the solution model by minimization of objective function Q.

§4. The Physical Data

In our problem, the data space is the product of three data spaces, one for each kind of information. We use geometric terms to translate the geological problem and we obtain a geometric objective function depending on surface S and twistor $\mathcal{T}$. Next, according to the functional viewpoint, we introduce a parameterization Φ by a change of variable.

4.1. The thread criterion

According to the above definition, threads are the only surfaces that zero the geometric objective function

$$Q_T(S, \mathcal{T}) = \frac{1}{2} \int_S \langle T(M), N(M) \rangle^2 dS,$$

where $N(M)$ is the vector normal to the surface at M and $T(M)$ is the value of twistor $\mathcal{T}$ at M. Here, the data space is the space of functions $\langle T(M), N(M) \rangle$ defined on S. In this case, the "observed" dataset is the zero function in this space. Note that the twistor is *a priori* unknown, therefore it is an auxiliary unknown.

4.2. The given points

Sometimes, wells cross faults and hence we know that intersection points P_i belong to the thread surface. Since well trajectories are not perfectly known, the location of points P_i is not exactly known. We introduce the geometric objective function $Q_P(S) = \sum_i \mathcal{D}(P_i, S)$, where $\mathcal{D}(P_i, S)$ is the Euclidean distance between surface S and point P_i.

This distance is a complicated function of parameterization Φ. Therefore, we approximate it by the normal component of vector $\vec{M_i P_i}$ where $M_i = \Phi(m_i)$ is a point of S specified by its curvilinear coordinates $m_i = (u_i, v_i) \in U$ (Figure 2). This approximation is satisfactory if the modulus of $\vec{M_i P_i}$ is much smaller than the principal radii of curvature of the surface. Then, the objective function is $\tilde{Q}_P^\perp(\Phi) = \frac{1}{2} \sum_i \|\langle \vec{P_i M_i}, N(M) \rangle\|^2$.

4.3. The curvature criterion

We smooth the surface by minimizing the geometric objective function

$$Q_C(S) = \frac{1}{2} \int_S (\|h\|^2 \, dS) = \frac{1}{2} \int_U (\lambda_1^2 + \lambda_2^2) \circ \Phi \, |d\Phi| \, dU,$$

where h is the norm of the second fundamental form of a surface. Note that $h^2 = \lambda_1^2 + \lambda_2^2$, where λ_1 and λ_2 are the principal curvatures. As in the case of the thread criterion, the "observed" data are zero. A curvature-based criterion is more satisfactory than a second derivative-based criterion because the former is intrinsic, $i.e.$, insensitive to the parameterization, whereas the latter is not. The "smoothing effect" of a second derivative-based criterion is sensitive to the first derivative of the parameterization, and then may be anisotropic or heterogeneous.

4.4. The physical objective function

Finally, we state our physical problem as the minimization of overall geometric objective function Q_φ which is the weighted sum of the above objective functions $Q_\varphi(S, \mathcal{T}) = W_T . Q_T + W_P . Q_P^\perp + W_C . Q_C$, where W_T, W_P, W_C are weights that represent the confidence we have in each kind of data (the higher the confidence, the higher the weights).

From a functional viewpoint, the physical objective function is written

$$\tilde{Q}_\varphi(\Phi, \mathcal{T}) = W_T . \tilde{Q}_T + W_P . \tilde{Q}_P^\perp + W_C . \tilde{Q}_C.$$

§5. The Three Causes for Indetermination

Since there is an infinity of parameterizations describing the same surface, the minimization of $\tilde{Q}_\varphi(\Phi, \mathcal{T})$ is always an ill-posed problem, even if the minimization of $Q_\varphi(S, \mathcal{T})$ is a well-posed one. We call that the *canonical indetermination*.

Besides, the null twistor zeroes Q_T for any given surface, hence, the surface that optimizes $W_P.Q_P^\perp + W_C.Q_C$ will always optimize Q_φ. This is not acceptable because every surface would be a thread in such a case. To avoid this situation, and to match the definition of a thread, we normalize the twistor.

Moreover, the given point criterion controls only the normal component of vectors $P_i \vec{M}_i$, and thus we need to introduce the tangential component of these vectors.

5.1. The additional criterion

In order to solve the canonical indetermination problem, we introduce an additional criterion $\tilde{Q}_A$, which automatically selects one particular parameterization in the set of those that describe the optimal surface. This criterion which is added to $\tilde{Q}_\varphi$ should not modify the optimal surface (condition 1) and should be sufficiently constraining to make the problem well-posed (condition 2).

Let us consider the space $\mathcal{P}$ of parameterizations, *i.e.*, the set of $\mathcal{C}^2$-diffeomorphisms from the curvilinear coordinate domain U to $\mathbb{R}^3$. We define $\mathcal{S}$ as the set of surfaces described by parameterizations in $\mathcal{P}$. We define a map s from $\mathcal{P}$ to $\mathcal{S}$ which associates the surface S with a parameterization Φ describing S. We define an equivalence relationship $\sim$ on $\mathcal{P}$ as follows $\Phi_1 \sim \Phi_2 \iff s(\Phi_1) = s(\Phi_2)$. Hence, we have $\mathcal{S} = \mathcal{P}/\sim$.

Now, we introduce an additional objective function $\tilde{Q}_a$ which meets condition 2, and we will later modify it to make it compatible with condition 1. Generally, there is no intersection between the set of the critical points of $\tilde{Q}_a$ and the set of the critical points of $\tilde{Q}_\varphi$. In other words, the physical solution changes if we optimize $\tilde{Q}_a + \tilde{Q}_\varphi$. We need to modify $\tilde{Q}_a$ to meet condition 1. Therefore, at any point Φ of $\mathcal{P}$, we project the gradient of $\tilde{Q}_a$ on the space $T_\Phi \mathcal{S}$ tangent to the local leaf $S = s(\Phi)$. It can be shown [3] that there exists, locally, a function $\tilde{Q}_A$ on $\mathcal{P}$ whose gradient is this projected gradient.

In practice, we choose to define $\tilde{Q}_a$ as a second derivative L_2-seminorm by analogy with the curvature criterion.

5.2. The twistor normalization

To normalize the twistor, we define the objective function

$$\tilde{Q}_N(\Phi, \mathcal{T}) = \frac{1}{2} \| \int_U \|T(M)\|^2 \circ \Phi |d\Phi| \, dU - \int_U |d\Phi| \, dU \|^2.$$

Clearly, optimizing this objective function makes the RMS value of twistor $\mathcal{T}$ on surface S close to 1.

5.3. The "tangential distance"

To control tangential perturbations of the parameterizations, we define the objective function

$$\tilde{Q}_P^{\parallel}(\Phi) = \sum_i \|\vec{M_i P_i}\|^2 - \langle \vec{M_i P_i}, N(M_i)\rangle^2.$$

5.4. The overall objective function

The overall objective function $\tilde{Q}$ is the sum of physical objective function $\tilde{Q}_\varphi$ and nonphysical objective function $W_A\tilde{Q}_A + W_N\tilde{Q}_N + W_P\tilde{Q}_P^{\parallel}$.

§6. Numerical Examples

We now describe several numerical experiments. First, we solve an approximation problem, *i.e.*, we optimize the curvature and proximity to data point criteria, but not the thread criterion. Then, we optimize the surface with respect to a known twistor. Finally, we solve the inverse problem with both the twistor and the surface unknown.

6.1. Approximation problem

Figure 3 gives information about 25 given points P_i defined by their Cartesian coordinates (x_i, y_i, z_i). They are displayed by crosses if $z_i = 0$ and diamonds if $z_i = 1$. Each point P_i is associated with a point $M_i = \Phi(m_i)$ lying on the surface, the curvilinear coordinates of which are $m_i = (u_i, v_i)$. We used these data points in all numerical experiments.

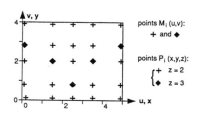
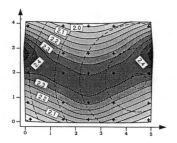

Fig. 3. Given points.　　　**Figure 4.** Approximation surface.

To solve the approximation problem, we simply choose a piece of the plane as the initial model. Figure 4 shows the result. This surface is smooth, and as close as possible to data points. Indeed, the RMS distance between data points and the surface is 0.008 instead of 0.2 in the initial model. Note that it is clearly not a thread because of the two bumps.

6.2. Unknown surface and fixed twistor

Figure 5 shows the result of the optimization of the surface with respect to all the criteria (except the normalization criterion) under the constraint of a fixed twistor. We used a translation twistor such that $T(M) = (1, 0, 0)$ everywhere. The surface becomes almost a cylinder, and the computed striae (dashed lines in Figures 4 – 6) are almost straight lines. This surface is smooth, and almost as close to the given points as the surface displayed in Figure 4 which was the initial surface. Dashed lines in Figure 4 suggest the initial twistor.

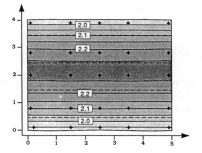 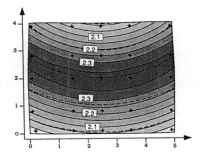

Fig. 5. Unknown surface and fixed twistor.   **Fig. 6.** Unknown surface and twistor.

6.3. Unknown surface and twistor

Figure 6 shows the result of the complete inverse problem, *i.e.*, the twistor and the surface are both unknown. After 12 iterations, convergence is achieved (the gradient is divided by 500,000). This surface is clearly a surface of revolution (and hence a thread). For this twistor, the reduction elements are $T(O) = (1.32, -0.50, 0.0091)$ and $\Omega = (-0.0003, 0.0033, 0.20)$, and the axis Δ is defined parametrically by $\Delta = \{M / \ \vec{OM} = \vec{OA} + \mu\vec{V} / \ \mu \in \mathcal{R}\}$ with $A = (-2.20, -6.6, -0.104)$ and $\vec{V} = (-0.0079, 0.082, 5.0)$. The pitch of the thread is 0.023 (the theoretical value is 0 for surfaces of revolution) and the thread turns to the right. The RMS value of the (twistor-tangent plane) angle is 0.15 degree, and 3.5 degrees for Figure 4. We conclude that we found an optimal twistor which is almost tangent to the surface.

To sum up, a slight downgrade of the proximity and curvature criteria is compatible with a great improvement of the thread criterion.

§7. Conclusions

Our results demonstrate that it is possible to constrain the shape of faults with more elaborate geological knowledge than only smoothness and proximity to data points. The least-squares "thread criterion", which derives from the rigid block approximation, improves predictions about the shape of faults, especially if only a few wells cross it, because a simple approximation may give inaccurate results in this case. The "thread criterion" yields the direction of striae as a by-product. The method is flexible with respect to the available knowledge about the twistor which represents the relative move between the two blocks. Moreover, if a surface is represented parametrically, its smoothing by the minimization of principal curvatures leads to more satisfying results than by the minimization of second derivatives because curvature is an intrinsic quantity. Besides, the "additional cirterion method" solves the canonical indermination problem due to the multiplicity of the parametrizations for one surface. In the future, we plan to test our method with field data. Moreover, striae measurements could be used to formulate another criterion, if available.

Acknowledgments. We would like to thank L. Nguyen Luc for his valuable help with the figures. We also thank J. Brac who implemented the spline routines, and C.-M. Marle for his advice concering the thread criterion.

References

1. Apprato, D., Spline fitting along a set of curves, Mathematical Modelling and Numerical Analysis **25** (1991), 193–212.
2. Léger, M., Morvan, J-M., Rakotoarisoa, H., Inversion of 3D structure geometry using geologic least-squares criteria, submitted to Geophysical Journal International.
3. Rakotoarisoa, H., Modélisation géométrique et optimisation de structures géologiques 3D, Thesis, 1992, (in French).
4. Spivak, M., *A Comprehensive Introduction to Differential Geometry*, second edition, 5 volumes: Publish or Perish, Inc. Houston, 1979.
5. Tarantola, A., *Inverse Problem Theory*, Paris, 1987.

Michel Léger
Institut Français du Pétrole
1-4 avenue de Bois Préau
92500 Rueil Malmaison, FRANCE
legerm@irsun1.ifp.fr

A Multivariate Generalization of
the de Boor-Fix Formula

Suresh Lodha and Ron Goldman

Abstract. Duality between B-bases and L-bases is established by extending the de Boor-Fix formula for the dual functionals from curves to surfaces. The de Boor-Fix formula is then used to demonstrate the duality between certain recurrence diagrams for B-bases and L-bases and to derive the Cavaretta-Micchelli identity. Surprisingly, the multivariate generalization of the de Boor-Fix formula is simpler than the standard univariate formula. The relationship between the multivariate and the univariate forms of the formula is clarified.

§1. Introduction

B-patches were introduced by Seidel [8] and shown to agree with multivariate B-splines on a certain region [6]. The basis functions for the B-patches, referred to as B-bases, are known to be local multivariate generalizations of B-splines. Lineal bases, or L-bases, that is, multivariate polynomial bases formed from products of linear polynomials, were studied by Cavaretta and Micchelli and shown to be dual to B-bases using a multivariate polynomial identity [2]. After briefly reviewing some essential properties of B-bases and L-bases, we establish this duality by generalizing the de Boor-Fix dual functionals [4] from curves to surfaces. This formula for the dual functionals is then used to demonstrate the duality between certain recurrence diagrams for B-bases and L-bases and to derive the Cavaretta-Micchelli identity. Our work easily generalizes to higher dimensions. Nevertheless, for the sake of simplicity, the results are presented and derived here only for surfaces.

Throughout this paper, we shall adopt the following notation. A multi-index α is a 3-tuple of non-negative integers. If $\alpha = (\alpha_1, \alpha_2, \alpha_3)$, then $|\alpha| = \alpha_1 + \alpha_2 + \alpha_3$ and $\alpha! = \alpha_1! \alpha_2! \alpha_3!$. Other multi-indices will be denoted by β and γ. A unit multi-index e_k is a 3-tuple with 1 in k-th position and 0 everywhere else. Scalar indices will be denoted by i, j, k, l. Finally, given a homogeneous polynomial $f(x, y, z)$, $D^\alpha f$ denotes $\frac{\partial^{|\alpha|} f}{\partial x^{\alpha_1} \partial y^{\alpha_2} \partial z^{\alpha_3}}$.

Curves and Surfaces in Geometric Design
P. J. Laurent, A. Le Méhauté, and L. L. Schumaker (eds.), pp. 301–310.
Copyright © 1994 by A K PETERS, Wellesley, MA
ISBN 1-56881-039-3.

301

§2. B-bases

A collection $\mathcal{U}$ of 3 sets $\{\mathbf{u}_{1,j}\}$, $\{\mathbf{u}_{2,j}\}$, $\{\mathbf{u}_{3,j}\}$, $j = 1, \cdots, n$ of vectors in $\mathbb{R}^3$ is called a *knot-net* of vectors if $(\mathbf{u}_{1,\alpha_1+1}, \mathbf{u}_{2,\alpha_2+1}, \mathbf{u}_{3,\alpha_3+1})$ are linearly independent vectors in $\mathbb{R}^3$ for $0 \le |\alpha| \le n - 1$. One can write any vector $\mathbf{u}$ in $\mathbb{R}^3$ in terms of the basis $(\mathbf{u}_{1,\alpha_1+1}, \mathbf{u}_{2,\alpha_2+1}, \mathbf{u}_{3,\alpha_3+1})$ so that

$$\mathbf{u} = \sum_{k=1}^{3} h_{k,\alpha}(\mathbf{u})\mathbf{u}_{k,\alpha_k+1}.$$

Notice that $h_{k,\alpha}$ are trivariate linear homogeneous polynomials. Given coefficients $C_\alpha \in \mathbb{R}^m$ for $|\alpha| = n$, a B-patch of degree n over the knot-net $\mathcal{U}$ is a homogeneous trivariate polynomial $B : \mathbb{R}^3 \to \mathbb{R}^m$ defined by the following recurrence. The initial conditions for the recurrence are given by setting: $C_\alpha^0(\mathbf{u}) = C_\alpha$ for $|\alpha| = n$. The recurrence is defined for $|\alpha| = n - l$, $l = 1, \cdots, n$ by

$$C_\alpha^l(\mathbf{u}) = \sum_{k=1}^{3} h_{k,\alpha}(\mathbf{u})C_{\alpha+e_k}^{l-1}(\mathbf{u}). \tag{1}$$

The B-patch is then defined as $B(\mathbf{u}) = C_0^n(\mathbf{u})$. This algorithm is known as the *up recurrence*; it generalizes the de Boor evaluation algorithm for B-spline curves [3]. A *B-basis* $\{b_\alpha^n, |\alpha| = n\}$ is a collection of $\binom{n+2}{2}$ homogeneous trivariate polynomials from $\mathbb{R}^3$ to $\mathbb{R}$ defined by choosing the constants $C_\beta \in R$ as follows:

$$C_\beta = \begin{cases} 1, & \text{if } \beta = \alpha \\ 0, & \text{otherwise.} \end{cases}$$

It has been shown that $\{b_\alpha^n, |\alpha| = n\}$ is, in fact, a basis for the space of homogeneous polynomials in $\mathbb{R}^3$ [2]. Moreover, an arbitrary B-patch of degree n can be represented in terms of a B-basis as follows:

$$B(\mathbf{u}) = \sum_{|\alpha|=n} C_\alpha b_\alpha^n(\mathbf{u}).$$

The up recurrence also generates a recurrence relation between B-basis polynomials of different degrees by "reversing" the recurrence [1]. If $\{b_\alpha^k\}$ are the B-basis functions of degree k over the knot-net of vectors $\{(\mathbf{u}_{1,j}, \mathbf{u}_{2,j}, \mathbf{u}_{3,j}), j = 1, \cdots, k\}$, then the following recurrence is obtained for $l = 1, \cdots, n$, where $b_0^0 = 1$:

$$b_\alpha^l(\mathbf{u}) = \sum_{k=1}^{3} h_{k,\alpha-e_k}(\mathbf{u})b_{\alpha-e_k}^{l-1}(\mathbf{u}). \tag{2}$$

This *down recurrence* is a generalization of the Cox-de Boor-Mansfield recurrence for the univariate B-spline basis functions.

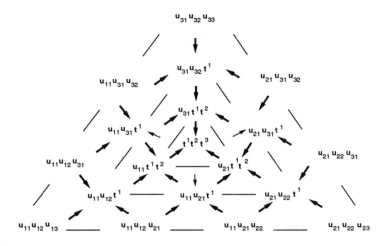

Fig. 1. Recurrence diagram for B-bases.

2.1. Recurrence Diagram

We now describe a schematic recurrence diagram, which presents clearly the role the independence condition on the vectors $(\mathbf{u}_{1,\alpha_1+1}, \mathbf{u}_{2,\alpha_2+1}, \mathbf{u}_{3,\alpha_3+1})$ plays in the definition of a B-patch. The recurrence diagram for the cubic case is shown in Figure 1. Clusters of three vectors from the knot-net lie at the base of the pyramid. Let $\mathbf{t}^i$, $i = 1, 2, 3$, be three arbitrary vectors. Recall that the linear independence condition on the vectors $(\mathbf{u}_{1,\alpha_1+1}, \mathbf{u}_{2,\alpha_2+1}, \mathbf{u}_{3,\alpha_3+1})$ implies that any vector $\mathbf{t}^i$ can always be expressed as a linear combination of $(\mathbf{u}_{1,\alpha_1+1}, \mathbf{u}_{2,\alpha_2+1}, \mathbf{u}_{3,\alpha_3+1})$ that is,

$$\mathbf{t}^i = \sum_{k=1}^{3} h_{k,\alpha}^i \mathbf{u}_{k,\alpha_k+1}.$$

In the recurrence diagram the vector $\mathbf{t}^i$ is introduced at level i. The constants $h_{k,\alpha}^i$ that should appear along the edges at level i are not shown to avoid cluttering the figure. For every node of the pyramid, there are three clusters of vectors one level below from which the cluster of vectors at the node is computed. The three clusters have all common factors except for precisely one factor. Thus using this recurrence one can represent any cluster of vectors at the apex as a linear combination of clusters from the knot-net of a B-patch.

2.2. Blossoming Property

Let $B_f(\mathbf{u}_1, \cdots, \mathbf{u}_n)$ denote the blossom [8] of the function f evaluated at the vectors $\mathbf{u}_1, \cdots, \mathbf{u}_n$. The schematic recurrence diagram of the previous section can also be viewed as the blossoming recurrence for a B-patch simply

by interpreting each cluster of vectors as the value of the blossom of the
B-patch evaluated at that cluster [8]. Given a knot-net of vectors $\mathbf{u}_{i,j}$, for
$1 \leq i \leq 3, 1 \leq j \leq n$, and any $\alpha = (\alpha_1, \alpha_2, \alpha_3)$ with $|\alpha| = n$, we shall
make the following judicious choice of the vectors: $\mathbf{u}_i = \mathbf{u}_{1,i}$ for $i = 1, \cdots, \alpha_1$,
$\mathbf{u}_i = \mathbf{u}_{2,i-\alpha_1}$ for $i = \alpha_1 + 1, \cdots, \alpha_1 + \alpha_2$, and $\mathbf{u}_i = \mathbf{u}_{3,i-\alpha_1-\alpha_2}$ for $i =$
$\alpha_1 + \alpha_2 + 1, \cdots, n$. Denote this cluster of vectors by $\mathcal{U}_\alpha$, and the blossom
of a function f evaluated at this cluster by $B_f(\mathcal{U}_\alpha)$. The following lemma
describes the blossoming property of B-bases [8].

Lemma 1.

$$B_{b_\beta^n}(\mathcal{U}_\alpha) = \delta_{\alpha\beta}. \tag{3}$$

Corollary 1.

$$B_{B(\mathbf{u})}(\mathcal{U}_\alpha) = \mathcal{C}_\alpha. \tag{4}$$

§3. L-bases

A collection $\mathcal{L}$ of 3 sets $\{L_{1,j}\}$, $\{L_{2,j}\}$, $\{L_{3,j}\}$, $j = 1, \cdots, n$ of linear homogeneous polynomials in three variables is called a *knot-net* of polynomials if
$(L_{1,\alpha_1+1}, L_{2,\alpha_2+1}, L_{3,\alpha_3+1})$ are linearly independent polynomials for $0 \leq |\alpha| \leq$
$n - 1$. An *L-basis* $\{l_\alpha^n, |\alpha| = n\}$ is a collection of $\binom{n+2}{2}$ trivariate polynomials
from $\mathbb{R}^3$ to $\mathbb{R}$ defined as follows:

$$l_\alpha^n = \prod_{i=1}^{\alpha_1} L_{1i} \prod_{j=1}^{\alpha_2} L_{2j} \prod_{k=1}^{\alpha_3} L_{3k}. \tag{5}$$

Observe that each L-basis polynomial function l_α^n is formed by taking the
product of n linear polynomials. We prove in the next subsection using a
recurrence diagram that $\{l_\alpha^n, |\alpha| = n\}$ is, in fact, a basis for the space of
homogeneous degree n polynomials in $\mathbb{R}^3$.

3.1. Recurrence Diagram

We now describe a recurrence diagram, which presents clearly the role the
independence condition on the polynomials $(L_{1,\alpha_1+1}, L_{2,\alpha_2+1}, L_{3,\alpha_3+1})$ plays
in the definition of an L-basis. The recurrence diagram for the cubic case
is shown in Figure 2. The L-basis functions lie at the base of the tetrahedron. Let M^1, M^2, M^3 be three arbitrary linear homogeneous polynomials. Observe that the linear independence condition on the polynomials
$(L_{1,\alpha_1+1}, L_{2,\alpha_2+1}, L_{3,\alpha_3+1})$ implies that any linear polynomial M^i can always
be expressed as a linear combination of $(L_{1,\alpha_1+1}, L_{2,\alpha_2+1}, L_{3,\alpha_3+1})$, that is,

$$M^i = \sum_{k=1}^{3} g_{k,\alpha}^i L_{k,\alpha_k+1}.$$

For every node of the tetrahedron, there are three polynomials one level below
from which the polynomial at the node is computed. The three polynomials

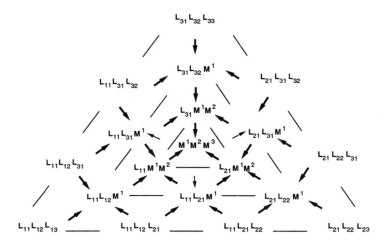

Fig. 2. Recurrence diagram for L-bases.

are lineal polynomials with all common factors except precisely one. In the recurrence diagram the polynomial M^i is introduced at level i. The multipliers $g_{k,\alpha}^i$ that should appear along the edges at level i are not shown in Figure 2 to avoid cluttering the diagram. Thus using this recurrence one can write any lineal polynomial, that is, any product of linear polynomials as a linear combination of L-basis functions. In particular, the polynomials $x^i y^j z^k$ are lineal polynomials and hence can be represented in terms of L-basis functions. Moreover, since any polynomial can be represented as a sum of lineal polynomials (for example using Taylor expansion) it follows that any polynomial can be written as a linear combination of L-basis functions. Thus these basis functions span the space $H^n(\mathbb{R}^3)$ of homogeneous polynomials of degree n in $\mathbb{R}^3$. Moreover, since the number of basis functions $\binom{n+2}{2}$ agrees with the dimension of $H^n(\mathbb{R}^3)$, the L-basis functions do indeed form a basis for $H^n(\mathbb{R}^3)$.

3.2. Symmetry Property

In this section, certain derivatives of L-basis functions are related to symmetric functions. This relationship is needed later to establish the duality between L-bases and B-bases.

Let $\mathbf{u}_i = (a_i, b_i, c_i), i = 1, \cdots, n$ be n vectors in $\mathbb{R}^3$. A symmetric function $S_\alpha(\mathbf{u}_1, \cdots, \mathbf{u}_n)$ of order α with $|\alpha| = n$ is defined as follows where the sum is taken over all possible permutations $(i_1, \cdots, i_{\alpha_1}, j_1, \cdots, j_{\alpha_2}, k_1, \cdots, k_{\alpha_3})$ of the integers $1, \cdots, n$:

$$S_\alpha(\mathbf{u}_1, \cdots, \mathbf{u}_n) = \sum a_{i_1} \cdots a_{i_{\alpha_1}} b_{j_1} \cdots b_{j_{\alpha_2}} c_{k_1} \cdots c_{k_{\alpha_3}},$$

Symmetric functions are closely related to blossoms. In fact, it follows immediately from the definition that

$$B_{x^{\alpha_1} y^{\alpha_2} z^{\alpha_3}}(\mathbf{u}_1, \cdots, \mathbf{u}_n) = \frac{1}{n!} S_\alpha(\mathbf{u}_1, \cdots, \mathbf{u}_n). \tag{6}$$

We shall apply this formula below in the proof of Theorem 1.

Now let $L_{i,j} = a_{ij}x + b_{ij}y + c_{ij}z$ be a knot-net of polynomials. Then $\mathbf{u}_{ij} = (a_{ij}, b_{ij}, c_{ij})$ is a knot-net of vectors. Define $\mathcal{U}_\alpha$ as in Section 2.2. Then the following lemma is an easy consequence of (5):

Lemma 2.

$$D^\alpha l_\alpha^n = S_\alpha(\mathcal{U}_\alpha). \tag{7}$$

§4. Duality

We now establish the duality between L-bases and B-bases by generalizing the de Boor-Fix formula for the dual functionals from curves to surfaces. This duality has been established by Cavaretta and Micchelli [2] using a multinomial identity, which can be viewed as a generalization of Marsden's identity for univariate B-splines. Here we take a different approach to duality. We then derive the Cavaretta-Micchelli identity as a simple consequence of our approach.

We introduce the following bracket operator. Given any two homogeneous polynomials $f, g : \mathbb{R}^3 \to \mathbb{R}$, define

$$[f, g](\mathbf{u}) = \frac{1}{n!} \sum_{|\alpha|=n} \frac{D^\alpha f(\mathbf{u}) * D^\alpha g(\mathbf{u})}{\alpha!}.$$

Note the bracket operator depends upon n, and therefore strictly speaking, the notation $[f, g]_n$ is more appropriate. However, we shall suppress the subscript n. This bracket operator has three important properties:

1) $[f, g]$ is linear in each parameter f and g; that is $[af_1 + bf_2, g] = a[f_1, g] + b[f_2, g]$, and $[f, ag_1 + bg_2] = a[f, g_1] + b[f, g_2]$. This property follows immediately from the linearity of differentiation.

2) $[f, g]$ is a constant independent of $\mathbf{u}$ if both f and g are homogeneous polynomials of degree n. This follows from observing that any partial derivative of order n of any homogeneous polynomial of degree n always yields a constant.

3) $[(a_1x + a_2y + a_3z)^n, g(\mathbf{u})] = g(a_1, a_2, a_3)$, if g is a polynomial of degree n or less. To see this, let $g(\mathbf{u}) = \sum_{|\beta|=n} g_\beta x^{\beta_1} y^{\beta_2} z^{\beta_3}$. Then, $D^\alpha g(\mathbf{u}) = \alpha! g_\alpha$. Moreover, $D^\alpha (a_1x + a_2y + a_3z)^n = n! a_1^{\alpha_1} a_2^{\alpha_2} a_3^{\alpha_3}$. Combining these two relations yields the desired result.

4.1. A Multivariate Generalization of the de Boor-Fix Formula

Given a knot-net of vectors $\mathbf{u}_{i,j}$ in $\mathbb{R}^3$, consider the knot-net of linear homogeneous polynomials $L_{i,j}$ defined by the following correspondence:

$$(a, b, c) \leftrightarrow (ax + by + cz).$$

Let l_α^n be the L-basis functions defined by the knot-net $L_{i,j}$, and let b_β^n be the B-basis functions defined by the knot-net $\mathbf{u}_{i,j}$.

Theorem 1.

$$[l_\alpha^n, b_\beta^n] = \delta_{\alpha\beta}.$$

Proof: The proof follows from the homogenization of Taylor's formula, the symmetry property of L-bases and the blossoming property of B-bases.

Taylor's formula for a polynomial function $g(x, y)$ of degree n in two variables states that:

$$g(x, y) = \sum_{\alpha_1 + \alpha_2 \leq n; \alpha_1, \alpha_2 \geq 0} \frac{g_{\alpha_1 \alpha_2} x^{\alpha_1} y^{\alpha_2}}{\alpha_1! \alpha_2!},$$

where $g_{ij} = \frac{\partial^{i+j} g}{\partial x^i \partial y^j}$, evaluated at $(0,0)$. Let $f(x, y, z)$ denote the homogeneous polynomial obtained by homogenizing $g(x, y)$. Then $D^\alpha f = \alpha! g_{\alpha_1, \alpha_2}$. Therefore, the homogenized Taylor's formula for a homogeneous polynomial of degree n in three variables is :

$$f(x, y, z) = \sum_{|\alpha|=n} \frac{D^\alpha f * x^{\alpha_1} y^{\alpha_2} z^{\alpha_3}}{\alpha!}.$$

Blossoming both sides of this equation evaluating at $\mathbf{u}_1, \cdots, \mathbf{u}_n$, and applying (6), we obtain:

$$B_f(\mathbf{u}_1, \cdots, \mathbf{u}_n) = \frac{1}{n!} \sum_{|\alpha|=n} \frac{D^\alpha f * S_\alpha(\mathbf{u}_1, \cdots, \mathbf{u}_n)}{\alpha!}. \tag{8}$$

Choosing the function f to be the B-basis function b_β^n and the cluster $(\mathbf{u}_1, \cdots, \mathbf{u}_n)$ to be $\mathcal{U}_\alpha$ as defined in Section 2.2, we observe that

1) $B_{b_\beta^n}(\mathcal{U}_\alpha) = \delta_{\alpha\beta}$ by Lemma 1.
2) $D^\alpha l_\alpha^n = S_\alpha(\mathcal{U}_\alpha)$ by Lemma 2.

Substituting these formulas into (8) gives the desired result. ∎

Corollary 2.

$$[l_\alpha^n, B(\mathbf{u})] = C_\alpha.$$

Corollary 3. *Cavaretta-Micchelli Identity: [2]*

$$\sum_{|\alpha|=n} l_\alpha^n(x, y, z) b_\alpha^n(a, b, c) = (ax + by + cz)^n.$$

Proof: Since the functions $b_\alpha^n(a, b, c)$ form a basis, one can write $(ax + by + cz)^n$ as a linear combination of $b_\alpha^n(a, b, c)$. Therefore there exists functions $c_\alpha(x, y, z)$ so that

$$(ax + by + cz)^n = \sum_{|\alpha|=n} c_\alpha(x, y, z) b_\alpha^n(a, b, c).$$

Bracketing both sides with l_α^n and applying Corollary 2, we obtain:

$$[l_\alpha^n, (ax + by + cz)^n] = c_\alpha(x, y, z).$$

But by the third important property of the bracket operator

$$[l_\alpha^n, (ax + by + cz)^n] = l_\alpha^n(x, y, z). \quad \blacksquare$$

The duality between L-bases and the B-bases is derived in Theorem 1 for the homogeneous case where a knot-net of polynomials in $\mathbb{R}^3$ is dual to a knot-net of vectors in $\mathbb{R}^3$ via the correspondence: $ax + by + cz \leftrightarrow (a, b, c)$. We now discuss the recurrence diagram duality, which is made explicit by the generalized de Boor-Fix formula. We make the simple but critical observation that the recurrence diagrams of Figures 1 and 2 have the same coefficients along the edges, if the knot-net for these diagrams are dual to each other and the vectors $\mathbf{t}^i$ introduced at level i of Figure 1 are dual to the polynomials M^i introduced at level i of Figure 2. This follows from the fact that the coefficients of a vector $\mathbf{t}^i$ with respect to a linearly independent basis $\mathbf{u}_1, \mathbf{u}_2, \mathbf{u}_3$ are the same as the coefficients of the dual linear polynomial M^i with respect to the dual linearly independent set of polynomials L_1, L_2, L_3, because these two representations are dual interpretations of the same set of algebraic equations.

In essence, the schematic recurrences in Figures 1 and 2 are the same diagrams with dual interpretations:

1) Nodes **uvw** in Figure 1 are interpreted as blossom values $B_{B(\mathbf{u})}(\mathbf{u}, \mathbf{v}, \mathbf{w})$.

2) Nodes LMN in Figure 2 are interpreted as bracket values $[LMN, B(\mathbf{u})]$.

By Corollaries 1 and 2, it follows that under the knot-net correspondence $[L_{1i} L_{2j} L_{3k}, B(\mathbf{u})] = B_{B(\mathbf{u})}(\mathbf{u}_{1i} \mathbf{u}_{2j} \mathbf{u}_{3k})$. Therefore the bases of the tetrahedra are identical. Since the coefficients along the edges are identical, these two figures are dual representations of the same recurrence.

§5. Relationship to the Univariate Case

The duality between B-splines and Polya polynomials in the univariate case has been established by using the following bracket operator for any two homogeneous polynomials $f, g : \mathbb{R}^2 \to \mathbb{R}$ of degree n:

$$\{f, g\}(\mathbf{u}) = \sum_{k=0}^{n} \frac{(-1)^{n-k}}{n!} D^{(k,0)} f(\mathbf{u}) D^{(n-k,0)} g(\mathbf{u}), \tag{9}$$

where $\mathbf{u} = (t, s) \in \mathbb{R}^2$. If B_α and P_β are the homogeneous B-spline bases and the corresponding Polya bases of degree n, then the well-known de Boor-Fix formula can be stated as $\{B_\alpha, P_\beta\} = \delta_{\alpha\beta}$ [1]. The correspondence between the B-spline bases and the Polya bases is established by defining the Polya polynomials to be the homogeneous polynomial of degree n, whose roots are the knots of the corresponding B-spline basis functions with the normalization that the coefficient of t^n is $(-1)^n$. For example, in the cubic case, the homogeneous Polya polynomials associated with the knots $\{t_1, \cdots, t_6\}$ of the cubic B-spline basis functions are $(t_1 s - t)(t_2 s - t)(t_3 s - t)$, $(t_2 s - t)(t_3 s - t)(t_4 s - t)$, $(t_3 s - t)(t_4 s - t)(t_5 s - t)$, and $(t_4 s - t)(t_5 s - t)(t_6 s - t)$. Now suppose that we define modified Polya polynomials P'_β to be the homogeneous polynomial of degree n whose roots are negative reciprocals of the knots of the B-Spline basis functions with the normalization that the coefficient of s^n is 1. For example, in the cubic case, these modified Polya polynomials are $(tt_1 + s)(tt_2 + s)(tt_3 + s)$, $(tt_2 + s)(tt_3 + s)(tt_4 + s)$, $(tt_3 + s)(tt_4 + s)(tt_5 + s)$, and $(tt_4 + s)(tt_5 + s)(tt_6 + s)$. It is then easy to see that $[B_\alpha, P'_\beta] = \delta_{\alpha\beta}$ with the bracket operator as defined in Section 4 and the alternating sum of the bracket operator as defined in (9) is not needed. At a computational level, one observes that the univariate de Boor-Fix formula can be simplified by using the modified Polya polynomials, which are obtained by the transformation $t \to (\frac{-1}{t})$. More importantly, it is in this simplified form that the de Boor-Fix formula generalizes to higher dimensions. The correspondence between the knots and the roots, which is used to establish the de Boor-Fix formula in the univariate case is not available in higher dimensions. Instead the correspondence between the vectors (a, b, c) and the polynomials $ax + by + cz$ must be used to establish this duality in higher dimensions.

§6. Conclusions and Future Work

The de Boor-Fix formula for the dual functionals plays a key role in deriving many important properties of B-splines [5]. In particular, in the univariate setting many change of basis algorithms between B-spline bases and Polya bases can be derived using the de Boor-Fix formula [7]. These change of basis algorithms are very important in computer aided geometric design applications. This work has generalized the de Boor-Fix formula from curves to surfaces and higher dimensions, and this paves the way to study and explore some important properties of B-bases, L-bases and their relationships.

Acknowledgments. This work was partially supported by National Science Foundation grants CCR-9309738 and CCR-9113239. This research was also supported in part by faculty research funds granted by the University of California, Santa Cruz.

References

1. Barry, P. and R. Goldman, Algorithms for progressive curves: Extending B-spline and blossoming techniques to the monomial, power, and Newton dual bases, in *Knot Insertion and Deletion Algorithms for B-Spline Curves and Surfaces*, R. Goldman and T. Lyche (eds.), Siam, 1993, 11–64.
2. Cavaretta, A. S. and C. A. Micchelli, Pyramid patches provide potential polynomial paradigms, in *Mathematical Methods in CAGD II*, T. Lyche and L. L. Schumaker (eds.), Academic Press, 1992, 69–100.
3. de Boor, C., On calculating with B-splines, Journal of Approximation Theory **6** (1972), 50–62.
4. de Boor, C. and G. Fix, Spline approximation by quasi-interpolants, Journal of Approximation Theory **8** (1973), 19–45.
5. de Boor, C. and K. Hollig, B-splines without divided differences, in *Geometric Modeling: Algorithms and New Trends*, G. Farin (ed.), Siam, 1987, 21–27.
6. Dahmen, W., C. A. Micchelli, and H. P. Seidel, Blossoming begets B-spline bases built better by B-patches, Mathematics of Computation **59** (1992), 97–115.
7. Goldman, R. and P. Barry, Wonderful triangle: A simple, unified, algorithmic approach to change of basis procedures in computer aided geometric design, in *Mathematical Methods in CAGD II*, T. Lyche and L. L. Schumaker (eds.), Academic Press, 1992, 297–320.
8. Seidel, H. P., Symmetric recursive algorithms for surfaces: B-patches and the de Boor algorithm for polynomials over triangles, Constructive Approximation **7** (1991), 259–279.

Suresh Lodha
Computer and Information Sciences
University of California
Santa Cruz, CA 95064
lodha@cse.ucsc.edu

Ron Goldman
Department of Computer Science
Rice University
Houston, TX 77251-1892
rng@cs.rice.edu

A Metric for Parametric Approximation

T. Lyche and K. Mørken

Abstract. We define a metric on the set of parametric curves that is more convenient for computational purposes than the Hausdorff metric. We then give a definition of approximation rate for parametric approximation schemes in terms of this metric, and present a simple family of odd degree parametric polynomial approximations to circle segments with approximation order twice the degree of the polynomial.

§1. Introduction

Most traditional methods for parametric curve approximation are based on approximation of functions. A parametric curve is approximated by applying standard approximation schemes for functions to the component functions of the curve. Recently, there has been increasing interest in what we call *parametric* methods, where the geometric properties of the curve are utilized for approximation purposes. One advantage of such schemes is that for polynomial approximants of fixed degree, they often provide higher approximation rates than the traditional schemes; see [1] for an early result in this direction and [3] for a good bibliography.

In order to achieve such high approximation orders, it is of fundamental importance that the error is measured suitably. In Section 2, we define a metric on the set of parametric curves which will provide us with a family of error measures suitable for our purposes. In particular, this metric leads to a natural definition of (parametric) approximation order, see [3] for a similar definition.

In earlier papers, we have studied approximation of circle segments by quadratic and cubic polynomial curves which are fourth and sixth order accurate, see [4,5]. In Section 3, we generalize this to higher degrees.

Curves and Surfaces in Geometric Design   311
P. J. Laurent, A. Le Méhauté, and L. L. Schumaker (eds.), pp. 311–318.

§2. A Metric for Parametric Curves

Let $\mathbf{f} = (f_1, f_2)$ be a parametric curve in $\mathbb{R}^2$. A typical way of computing an approximation $\mathbf{p} = (p_1, p_2)$ to $\mathbf{f}$ is to let p_1 and p_2 be approximations to f_1 and f_2, respectively. The obvious advantage of this approach is that we can make use of the extensive theoretical and practical knowledge accumulated for approximation of functions. We can for instance let p_1 and p_2 be the least squares approximations to f_1 and f_2, or we can determine p_1 and p_2 by interpolating f_1 and f_2 at suitable points. In Figure 1 we have approximated the half circle $\mathbf{f}(t) = (\cos t, \sin t)$ for $t \in [0, \pi]$ by letting p_1 and p_2 be the cubic polynomials that interpolate $\cos t$ and $\sin t$ and their first derivatives at the end points $t = 0$ and $t = \pi$. This interpolant is shown with short dashes. The curve with longer dashes is a cubic interpolant to a reparametrization of the half circle; it interpolates $\mathbf{f}(\phi(t))$ (with $\phi(0) = 0$, $\phi(\pi) = \pi$ and $\phi'(t) > 0$) and its first *two* derivatives at $t = 0$ and $t = \pi$. The lines from the approximations to the half circle connect points with the same parameter value.

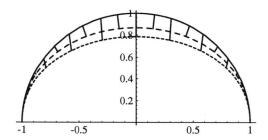

Fig. 1. Two cubic Hermite interpolants approximating a half circle.

Figure 1 illustrates the fact that traditional approximation methods for curves often lead to inefficient parametrizations, which again means that we get an overestimate of the error when we compute it by comparing the values at common values of the parameter. By allowing the curve to be reparametrized before approximation, we can both get better approximations and easily compute better error estimates.

In this section we will use the idea of reparametrization to define a suitable metric on the set of parametric curves. But let us first define precisely what we mean by parametric curves. The material to follow in the next three definitions is quite standard and is included only for the convenience of the reader.

Definition 1. *A regular parametric representation* $\mathbf{f}$ *in* $\mathbb{R}^d$ *is a mapping of a real interval* $I = [a, b]$ *into* $\mathbb{R}^d$ *such that* $\mathbf{f}'(t) \neq 0$ *for all* $t \in I$.

To generate reparametrizations of $\mathbf{f}$, we roughly compose it with increasing, real functions. To be more precise, let $I_1 = [a, b]$ and $I_2 = [c, d]$ be two real intervals. The set A_{I_1,I_2} is defined by

$$A_{I_1,I_2} = \{\phi \in C^1(I_1) \mid \phi(I_1) = I_2 \text{ and } \phi'(t) > 0 \text{ for } t \in I_1\},$$

the set of increasing functions on I_1 with continuous derivative that map I_1 onto I_2. In the following, symbols like I, I_1, I_2 will always denote real intervals. When I_1 and I_2 are irrelevant or obvious from the context we will often abbreviate A_{I_1,I_2} to A. A function $\phi \in A_{I_1,I_2}$ is often called an allowable parameter change from I_1 to I_2.

Definition 2. *A regular reparametrization of a parametric representation* $\mathbf{f} : I_1 \mapsto \mathbb{R}^d$ *is a regular parametric representation* $\mathbf{g} : I_2 \mapsto \mathbb{R}^d$ *such that* $\mathbf{g}(\phi(t)) = \mathbf{f}(t)$ *for some function* $\phi \in A_{I_1,I_2}$.

Definition 3. *A regular parametric curve is the equivalence class of all reparametrizations of a parametric representation* $\mathbf{f}$, *i.e., the set* $\{\mathbf{f} \circ \phi \mid \phi \in A\}$.

If $\mathbf{f}$ is defined on I_1, then $\phi \in A$ in Definition 3 means $\phi \in \cup_{I_2} A_{I_1,I_2}$, where the union is over all subintervals I_2 of $\mathbb{R}$. Observe however that there is no essential loss in only letting ϕ vary in A_{I_1,I_1}.

In line with common practice, we will usually identify a particular parametric representation with the curve it represents.

In order to discuss errors and approximation rates for curves, we need some way to measure the distance between two curves. An easy way to do this when the two curves $\mathbf{f}$ and $\mathbf{g}$ are parametrized over the same interval is to compare $\mathbf{g}(t)$ with $\mathbf{f}(t)$ and compute some norm of the difference between the two. As indicated above, this is not in general a particularly good approach.

We already have a metric for measuring the distance between two general sets, the Hausdorff metric. If we specialize this to curves, we find that the distance between $\mathbf{f}$ and $\mathbf{g}$ (parametrized on the intervals I_1 and I_2 respectively) is given by

$$d_H(\mathbf{f}, \mathbf{g}) = \max\left\{\max_{t \in I_1} \min_{s \in I_2} \|\mathbf{f}(t) - \mathbf{g}(s)\|, \quad \max_{s \in I_2} \min_{t \in I_1} \|\mathbf{f}(t) - \mathbf{g}(s)\|\right\},$$

with $\|\cdot\|$ some vector norm in $\mathbb{R}^d$. This metric is however not very appropriate for dealing with curves, and it is complicated to compute, see [2].

Let us consider the expression $\max_{t \in I_1} \min_{s \in I_2} \|\mathbf{f}(t) - \mathbf{g}(s)\|$ in more detail. Fix t, and let $\phi(t)$ denote an s for which the minimum is attained; then we have

$$\max_{t \in I_1} \min_{s \in I_2} \|\mathbf{f}(t) - \mathbf{g}(s)\| = \max_{t \in I_1} \|\mathbf{f}(t) - \mathbf{g}(\phi(t))\|. \tag{1}$$

The function $\phi(t)$ is in general neither monotone nor continuous even if $\mathbf{f}$ and $\mathbf{g}$ are well behaved. However, keeping in mind the definition of a parametric curve, the right hand side of (1) suggests the following metric.

Definition 4. *Let* **f** *and* **g** *be two parametric curves defined on the intervals* I_1 *and* I_2. *The* parametric distance *between* **f** *and* **g** *is defined by*

$$d_P(\mathbf{f}, \mathbf{g}) = d_P(\mathbf{f}, \mathbf{g})_{I_1} = \inf_{\phi \in A_{I_1, I_2}} \max_{t \in I_1} \|\mathbf{f}(t) - \mathbf{g}(\phi(t))\|.$$

At a first glance this definition looks unsymmetric. But it is not too hard to see that we also have

$$d_P(\mathbf{f}, \mathbf{g})_{I_1} = \inf_{\psi \in A_{I_2, I_1}} \max_{s \in I_2} \|\mathbf{f}(\psi(s)) - \mathbf{g}(s)\| = d_P(\mathbf{g}, \mathbf{f})_{I_2}. \tag{2}$$

From this it is clear that

$$d_P(\mathbf{f}, \mathbf{g}) \geq d_H(\mathbf{f}, \mathbf{g}).$$

Let us make sure that d_P really is a metric for parametric curves.

Proposition 5. *The function* d_P *is a metric on the set of parametric curves in* $\mathbb{R}^d$.

Proof: Observe first that d_P is independent of the particular parametric representations we have picked for **f** and **g**. If $\mathbf{f}_1 = \mathbf{f} \circ \phi$ and $\mathbf{g}_1 = \mathbf{g} \circ \psi$ are two allowable reparametrizations of **f** and **g**, we have

$$d_P(\mathbf{f}_1, \mathbf{g}_1) = d_P(\mathbf{f} \circ \phi, \mathbf{g} \circ \psi) = d_P(\mathbf{f}, \mathbf{g} \circ \psi \circ \phi^{-1}) = d_P(\mathbf{f}, \mathbf{g}).$$

The inverse ϕ^{-1} of ϕ exists since $\phi'(t) > 0$ for all t. The last equality follows since $\psi \circ \phi^{-1} \circ \xi$ generates all allowable parameter changes when ξ varies over A.

Let us next check the axioms for metric spaces. Nonnegativity of d_P is trivial. The symmetry $d_P(\mathbf{f}, \mathbf{g}) = d_P(\mathbf{g}, \mathbf{f})$ follows from (2). If **g** is a reparametrization of **f**, we clearly have $d_P(\mathbf{f}, \mathbf{g}) = 0$. To prove the opposite, suppose that **f** and **g** are two given curves with $d_P(\mathbf{f}, \mathbf{g}) = 0$. Then there is a sequence of mappings $\phi_n \in A$ such that

$$\lim_{n \to \infty} \max_t \|\mathbf{f}(t) - \mathbf{g}(\phi_n(t))\| = 0. \tag{3}$$

Then there must be a ϕ such that $\mathbf{f} = \mathbf{g} \circ \phi$. A priori, we know nothing about this ϕ; we will first show that it is unique if we require it to be continuous.

Let $\mathbf{g}(s) = \mathbf{f}(\phi(t))$ be a point that does not intersect any other points of **g**. To any given $\epsilon > 0$ we can then find a $\delta_1 > 0$ such that if $\|\mathbf{g}(s) - \mathbf{g}(s_1)\| < \delta_1$, then we must have $|s - s_1| < \epsilon$, where $s_1 = \phi(t_1)$. But $\mathbf{g} \circ \phi = \mathbf{f}$ is continuous, so there is some $\delta > 0$ such that if $|t - t_1| < \delta$, then $\|\mathbf{g}(\phi(t)) - \mathbf{g}(\phi(t_1))\| < \delta_1$. From this we conclude that if $|t - t_1| < \delta$, then $|\phi(t) - \phi(t_1)| < \epsilon$, in other words ϕ is continuous at t.

If **g** is not simple, e.g., it has one loop so that $\mathbf{g}(s_1) = \mathbf{g}(s_2)$ for $s_1 < s_2$, we conclude from the above that ϕ must be continuous in each open subinterval where it is simple. Since $\mathbf{g} \circ \phi = \mathbf{f}$, there must also be t_1 and t_2 such that $\mathbf{f}(t_1) = \mathbf{f}(t_2)$ and

$$\phi((a, t_1)) = (a, s_1), \qquad \phi((t_1, t_2)) = (s_1, s_2), \qquad \phi((t_2, b)) = (s_2, b).$$

Since $\mathbf{f}(t_1) = \mathbf{f}(t_2) = \mathbf{g}(s_1) = \mathbf{g}(s_2)$, we see that the convergence in (3) is not sufficiently strong for ϕ to distinguish the values s_1 and s_2. But then we can *choose* ϕ such that $\phi(t_1) = s_1$ and $\phi(t_2) = s_2$ which makes it continuous. The case where $\mathbf{g}$ has several loops is similar.

It remains to prove that ϕ has a continuous derivative. Recall from elementary analysis that if $\lim x_n y_n = z$ and $\lim x_n = x$, then if $x \neq 0$ the sequence (y_n) is also convergent and converges to z/x. Consider now the expression

$$\begin{aligned}
\mathbf{f}'(t) &= \lim_{h \to 0} \frac{\mathbf{g}(\phi(t+h)) - \mathbf{g}(\phi(t))}{h} \\
&= \lim_{h \to 0} \frac{\mathbf{g}(\phi(t+h)) - \mathbf{g}(\phi(t))}{\phi(t+h) - \phi(t)} \cdot \frac{\phi(t+h) - \phi(t)}{h}.
\end{aligned} \tag{4}$$

We know that the first quotient on the right converges to $\mathbf{g}'(\phi(t))$ which we have assumed is nonzero for all t. Therefore we can apply the result quoted above to a nonzero component of (4) to conclude that $\lim_{h \to 0} (\phi(t+h) - \phi(t))/h = \phi'(t)$ exists. From this we also obtain that ϕ' must be continuous since it is the quotient of two continuous functions. We therefore have $\mathbf{f} = \mathbf{g} \circ \phi$ for a unique $\phi \in A$.

To prove the triangle inequality, let $\mathbf{f}, \mathbf{g}$ and $\mathbf{h}$ be parametric curves, fix ψ and ϕ in A and consider the inequality

$$\max_t \|\mathbf{f}(t) - \mathbf{g}(\phi(t))\| \leq \max_t \|\mathbf{f}(t) - \mathbf{h}(\psi(t))\| + \max_t \|\mathbf{h}(\psi(t)) - \mathbf{g}(\phi(t))\|.$$

Since the left hand side is independent of ψ, for any given $\epsilon > 0$, we can find a $\psi_0 \in A$ so that

$$\|\mathbf{f}(t) - \mathbf{g}(\phi(t))\| \leq d_P(\mathbf{f}, \mathbf{h}) + \|\mathbf{h}(\psi_0(t)) - \mathbf{g}(\phi(t))\| + \epsilon.$$

Taking the inf over ϕ, we can find a $\phi_0 \in A$ such that

$$\|\mathbf{f}(t) - \mathbf{g}(\phi_0(t))\| \leq d_P(\mathbf{f}, \mathbf{h}) + d_P(\mathbf{h}, \mathbf{g}) + 2\epsilon.$$

Since $d_P(\mathbf{f}, \mathbf{g})$ is less than or equal to the left hand side and ϵ is arbitrary, the triangle inequality follows. ■

Note that we have

$$d_P(\mathbf{f}, \mathbf{g}) \leq \delta_{\phi, \psi}(\mathbf{f}, \mathbf{g}) := \max_t \|\mathbf{f}(\psi(t)) - \mathbf{g}(\phi(t))\|$$

for arbitrary ϕ and ψ in A. In practice, one therefore often estimates $d_P(\mathbf{f}, \mathbf{g})$ by giving some ψ and ϕ and then using the above inequality.

Our main interest is in approximation of arbitrary curves with polynomial curves. The parametric metric d_P is then a convenient tool for measuring the error in the approximation. Of particular interest in approximation theory is the approximation order of an approximation scheme.

Definition 6. *Let S_h be an approximation scheme that to each parametric curve $\mathbf{f} \in \mathbb{R}^d$ defined on an interval $[a, b]$ and for each $h < b - a$ assigns an approximation $S_h(\mathbf{f})$ to the part of $\mathbf{f}$ defined on $I_h[a, a + h]$. Then S_h is said to have approximation order m if the inequality*

$$d_P(\mathbf{f}, S_h(\mathbf{f}))_{I_h} \leq C h^m,$$

holds for some constant C independent of h.

To establish that a scheme has approximation order m, it will be sufficient to show that

$$\max_t \|\mathbf{f}(\phi(t)) - S_h(\mathbf{f})(t)\| \leq C h^m,$$

for some C independent of ϕ and h, see also [3].

§3. Parametric Approximation of Circle Segments

In two earlier papers, we have studied approximation of circle segments by parametric quadratic and cubic polynomials. In particular, it was shown in [5] that there is a quadratic polynomial curve that approximates a segment of angular width α with error proportional to α^4. In [4] it was shown that if we use cubic polynomials, then there are approximations that give an error which is proportional to α^6. The following result introduces a high order approximation to circle segments for any odd degree n. Its relation to the metric of the previous section is discussed below.

Lemma 7. *Let n be a positive, odd integer and define the two functions $x_n(t)$ and $y_n(t)$ by*

$$x_n(t) = 2 \sum_{i=1}^{(n-1)/2} (-1)^{i-1} t^{2i-1} + (-1)^{(n-1)/2} t^n$$

$$y_n(t) = 1 - 2 \sum_{i=1}^{(n-1)/2} (-1)^{i-1} t^{2i}. \tag{5}$$

Then the relation

$$x_n(t)^2 + y_n(t)^2 = 1 + t^{2n} \tag{6}$$

holds for all t in $\mathbb{R}$.

Proof: Summing the geometric series in (5), we find

$$x_n(t) = x_0(t) - (-1)^{(n-1)/2} t^n y_0(t) \quad \text{and} \quad y_n(t) = y_0(t) + (-1)^{(n-1)/2} t^n x_0(t),$$

where

$$x_0(t) = \frac{2t}{1 + t^2} \quad \text{and} \quad y_0(t) = \frac{1 - t^2}{1 + t^2}$$

are standard rational parametrizations of circle segments so that $x_0(t)^2 + y_0(t)^2 = 1$. Squaring the expressions for $x_n(t)$ and $y_n(t)$ we obtain (6). ■

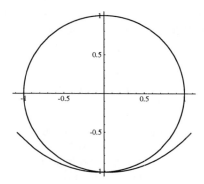

Fig. 2. Polynomial curve approximation to a circle.

As is evident from the lemma, the approximation $(x_n(t), y_n(t))$ approximates a circle segment with approximation order $2n$.

Theorem 8. *Let $\mathbf{r}(s) = (\sin s, \cos s)$ be a parametrization of the circle, and let $\mathbf{p}_n(t) = (x_n(t), y_n(t))$ denote the polynomial curve of Lemma 7. Then $\mathbf{p}_n$ provides a circle approximation with approximation order $2n$ in a neighbourhood of $(0,1)$.*

Proof: The mapping given by $s(t) = \arctan\big(x_n(t)/y_n(t)\big)$ satisfies $s(0) = 0$ and $s'(0) = 1$. Therefore, s maps some small interval $I_h = [-h, h]$ one-to-one and onto some small interval $[-\alpha, \alpha]$. Note that then $\mathbf{p}_n(t)$ and $\mathbf{r}(s)$ lie on the same ray from the origin. We will compare $\mathbf{p}_n$ and $\mathbf{r}$ by comparing $\mathbf{p}_n(t)$ with $\mathbf{r}(s(t))$. Since those two vectors are parallel, we find

$$||\mathbf{p}_n(t) - \mathbf{r}(s(t))|| = \Big| \, ||\mathbf{p}_n(t)|| - ||\mathbf{r}(s(t))|| \, \Big|$$
$$= \Big|(x_n(t)^2 + y_n(t)^2 - 1\Big|\Big/\Big(\sqrt{x_n(t)^2 + y_n(t)^2} + 1\Big)$$
$$\leq \Big|(x_n(t)^2 + y_n(t)^2 - 1\Big|.$$

For $t \in [-h, h]$ we therefore have

$$d_P(\mathbf{r}, \mathbf{p}_n)_{I_h} \leq \max_{t \in I_h} ||\mathbf{r}(s(t)) - \mathbf{p}_n(t)||$$
$$\leq \max_{t \in I_h} |x_n(t)^2 + y_n(t)^2 - 1|$$
$$\leq h^{2n},$$

by Lemma 7. ∎

Since $|x_n(t)^2 + y_n(t)^2 - 1| \leq 2^{-2n}$ for $t \leq 1/2$, it is clear that for such values of t we have better and better approximations to a part of the circle as n increases. One can ask the question about how large a segment of the circle this represents. It can be seen that $x_n(t)$ is increasing and $y_n(t)$ is decreasing for $t \leq 1/2$. Using the formulas for $x_n(t)$ and $y_n(t)$ in the proof of Lemma 7 we also see that for $|t| < 1$

$$\lim_{n \to \infty} \frac{x_n(t)}{y_n(t)} = \frac{2t}{1 - t^2}.$$

For $t = 1/2$ the right hand side of this expression is equal to $4/3$. Thus the curve $(x_n(t), y_n(t))$ will be a good approximation to a segment of a circle of angular width α_n with $\lim_{n \to \infty} \tan \alpha_n = 4/3$.

Finding approximations of even order n is not so simple. For general even degree n there does not appear to be any simple polynomials x_n and y_n such that $x_n(t)^2 + y_n(t)^2 - 1 = \alpha t^{2n}$ for a suitable constant α. A solution for $n = 6$ is shown in Figure 2. Here $x(t) = 2t - (3 - \sqrt{3})t^3 + 2(2 - \sqrt{3})^2 t^5$, $y(t) = 1 - 2t^2 + (4 - 2\sqrt{3})t^4 - (26 - 15\sqrt{3})t^6$, and $x(t)^2 + y(t)^2 - 1 = (1351 - 780\sqrt{3})t^{12} \approx 0.00037 t^{12}$. The curve shown corresponds to $|t| \leq 1.8$ and we have an approximation to a full circle with error 0.02.

Acknowledgements. The first author was supported in part by the Research Council of Norway through project STP28402 and SINTEF SI.

References

1. de Boor, C., K. Höllig, and M. Sabin, High accuracy geometric Hermite interpolation, Computer-Aided Geom. Design **4** (1988), 269–278.
2. Degen, W. L. F., Best approximations of parametric curves by splines, in *Mathematical Methods in Computer Aided Geometric Design II*, T. Lyche and L. L. Schumaker (eds.), Academic Press, Boston, 1992, 171–184.
3. Degen, W. L. F., High accurate rational approximation of parametric curves, Computer-Aided Geom. Design **10** (1993), 293–313.
4. Dokken, T., M. Dæhlen, T. Lyche, and K. Mørken, Good approximation of circles by curvature-continuous Bézier curves, Computer-Aided Geom. Design **7** (1990), 33–41.
5. Mørken, K., Best approximation of circles segments by quadratic Bézier curves, in *Curves and Surfaces*, P. J. Laurent, A. Le Méhauté, and L. L. Schumaker, (eds.), Academic Press, New York, 1991, 331–336.

T. Lyche and K. Mørken
Dept. of Informatics, University of Oslo
P. B. 1080, Blindern
0316 Oslo, NORWAY
tom@ifi.uio.no
knutm@ifi.uio.no

Mathematical Modelling
of Free-Form Curves and Surfaces
from Discrete Points with NURBS

Weiyin Ma and Jean-Pierre Kruth

Abstract. This paper presents theoretical and practical approaches on mathematical modelling of free-form curves and surfaces from discrete points with Non-Uniform Rational B-Splines (NURBS). Interpolation and fitting solutions are studied. Numerical algorithms for both general and feasible solutions are summarized.

§1. Introduction

Given a set of m discrete points

$$\mathcal{M} := \{\overline{\mathbf{P}}_i = [\overline{x}_i, \overline{y}_i, \overline{z}_i]^T\}_1^m \tag{1}$$

representing a free-form curve or surface, we shall construct an interpolation or fitting Non-Uniform Rational B-spline (NURBS) curve or surface from $\mathcal{M}$.

Over the past years, various interpolation schemes have been developed for conics, see [1,5] and references therein. There is also extensive research on using the weights of NURBS as shape modifiers in curve and surface modelling from discrete points. For general NURBS curves and surfaces, some authors have suggested to interpolate or fit the weighted data points $\{[\overline{w}_i\overline{\mathbf{P}}_i, \overline{w}_i]\}_1^m$ in 4D homogeneous space [1,5]. There are, however, no existing algorithms to chose such $\overline{w}_i$'s. This paper presents a two-step approach to the interpolation and fitting of general NURBS curves and surfaces from discrete points. All the *weights of the control points* $\{w_i\}_1^n$ are first identified. Singular Value Decomposition (SVD) technique is applied to achieve the general solution $\{w_i \in \mathbb{R}^1\}_{i=1}^n$ as defined by equation (15). Quadratic programming subject to positive weights can also be applied to find a feasible solution in the positive subspace of $\mathbb{R}^{1+} = \{\mathbb{R}^1(w) : w > 0\}$. The *control points* are then further solved in the least squares sense by applying Householder transformations.

Curves and Surfaces in Geometric Design
P. J. Laurent, A. Le Méhauté, and L. L. Schumaker (eds.), pp. 319–326.
Copyright © 1994 by A K PETERS, Wellesley, MA
ISBN 1-56881-039-3.

§2. NURBS Curves and Surfaces

In 3D Euclidean space, a NURBS curve [1,5] is defined by

$$\mathbf{p}(u) = \frac{\sum_{i=1}^{n} w_i \mathbf{v}_i B_i(u)}{\sum_{i=1}^{n} w_i B_i(u)}, \qquad u \in [t_k, t_{n+1}], \tag{2}$$

where $\{\mathbf{v}_i \in \mathbb{R}^3\}_1^n$ are the control points, $\{w_i\}_1^n$ are the related weights, and $\{B_i(u)\}_1^n$ are normalized B-splines which are defined over a knot sequence $\mathcal{T} := \{t_i\}_1^{n+k}$ with order k (or degree $k-1$). Similarly, a NURBS surface [1,5] for $u \in [t_{uk_u}, t_{un_u+1}]$ and $v \in [t_{vk_v}, t_{vn_v+1}]$ is defined by

$$\mathbf{p}(u,v) = \frac{\sum_{i=1}^{n_u} \sum_{j=1}^{n_v} w_{ij} \mathbf{v}_{ij} B_{ui}(u) B_{vj}(v)}{\sum_{i=1}^{n_u} \sum_{j=1}^{n_v} w_{ij} B_{ui}(u) B_{vj}(v)}, \tag{3}$$

where $\{\{\mathbf{v}_{ij} \in \mathbb{R}^3\}_{j=1}^{n_v}\}_{i=1}^{n_u}$ are the $n = n_u \times n_v$ control points, $\{\{w_{ij}\}_{j=1}^{n_v}\}_{i=1}^{n_u}$ are the related weights, $\{B_{ui}(u)\}_{i=1}^{n_u}$, and $\{B_{vj}(v)\}_{j=1}^{n_v}$ are normalized B-splines which are defined over $\mathcal{T}_u := \{t_{ui}\}_{i=1}^{n_u+k_u}$ and $\mathcal{T}_v := \{t_{vj}\}_{j=1}^{n_v+k_v}$ with orders k_u and k_v respectively.

Writing equations (2) and (3) in matrix form, moving the denominator to the other side, and switching the left and right sides, these two equations have the following uniform representation

$$\begin{cases} \mathbf{b}^T(\cdot) \cdot \mathbf{X} = x(\cdot) \cdot \mathbf{b}^T(\cdot) \cdot \mathbf{w} \\ \mathbf{b}^T(\cdot) \cdot \mathbf{Y} = y(\cdot) \cdot \mathbf{b}^T(\cdot) \cdot \mathbf{w} \\ \mathbf{b}^T(\cdot) \cdot \mathbf{Z} = z(\cdot) \cdot \mathbf{b}^T(\cdot) \cdot \mathbf{w}, \end{cases} \tag{4}$$

where $x(\cdot)$, $y(\cdot)$ and $z(\cdot)$ are coordinates of $\mathbf{p}$, $\mathbf{b}(\cdot) = [B_1(\cdot), B_2(\cdot), ..., B_n(\cdot)]^T$ is the collection of the n basis functions, $\mathbf{X} = [w_1 x_1, w_2 x_2, ..., w_n x_n]^T$, $\mathbf{Y} = [w_1 y_1, w_2 y_2, ..., w_n y_n]^T$, $\mathbf{Z} = [w_1 z_1, w_2 z_2, ..., w_n z_n]^T$ and $\mathbf{w} = [w_1, w_2, ..., w_n]^T$ represent the collection of the coordinates of the control points in 4D homogeneous space, and $(\cdot) = (u)$ for NURBS curves and $(\cdot) = (u, v)$ for NURBS surfaces are *location parameters* locating a point on the curve or surface. For the case of a NURBS surface, $\{B_i(\cdot)\}_1^n = \{\{B_{ui}(u) B_{vj}(v)\}_{j=1}^{n_v}\}_{i=1}^{n_u}$.

§3. Observation Equations

Suppose that the location parameters of the discrete points $\mathcal{M}$, i.e., $\mathcal{U} := \{u_i\}_1^m$ for curve points and $\mathcal{U} := \{u_i\}_1^m$ and $\mathcal{V} := \{v_i\}_1^m$ for surface points, are assigned by some means, see [3] and references therein. Furthermore, suppose that the order, the number of control points, and the complete knot sequence are also known [1,5]. By introducing the discrete points into equation (4), the following observation equations are obtained:

$$\begin{cases} \mathbf{B} \cdot \mathbf{X} = \overline{\mathbf{X}} \cdot \mathbf{B} \cdot \mathbf{w} \\ \mathbf{B} \cdot \mathbf{Y} = \overline{\mathbf{Y}} \cdot \mathbf{B} \cdot \mathbf{w} \\ \mathbf{B} \cdot \mathbf{Z} = \overline{\mathbf{Z}} \cdot \mathbf{B} \cdot \mathbf{w} \end{cases} \tag{5}$$

in which $\overline{\mathbf{X}} = diag\{\overline{x}_1, \overline{x}_2, ..., \overline{x}_m\}$, $\overline{\mathbf{Y}} = diag\{\overline{y}_1, \overline{y}_2, ..., \overline{y}_m\}$ and $\overline{\mathbf{Z}} = diag\{\overline{z}_1, \overline{z}_2, ..., \overline{z}_m\}$ are diagonal matrices,

$$\mathbf{B} = \begin{bmatrix} B_1(\cdot_1) & B_2(\cdot_1) & B_3(\cdot_1) & ... & B_n(\cdot_1) \\ B_1(\cdot_2) & B_2(\cdot_2) & B_3(\cdot_2) & ... & B_n(\cdot_2) \\ \vdots & \vdots & \vdots & & \vdots \\ B_1(\cdot_m) & B_2(\cdot_m) & B_3(\cdot_m) & ... & B_n(\cdot_m) \end{bmatrix}_{m \times n} \tag{6}$$

denotes the observation matrix, $(\cdot_i) = (u_i)$ for NURBS curves and $(\cdot_i) = (u_i, v_i)$ for NURBS surfaces for $i = 1, 2, ..., m$ are the location parameters. In a compact matrix form equation (5) can be further written as

$$\mathbf{A} \cdot \begin{bmatrix} \mathbf{X} \\ \mathbf{Y} \\ \mathbf{Z} \\ \mathbf{w} \end{bmatrix} = \begin{bmatrix} \mathbf{B} & \mathbf{0} & \mathbf{0} & -\overline{\mathbf{X}}\mathbf{B} \\ \mathbf{0} & \mathbf{B} & \mathbf{0} & -\overline{\mathbf{Y}}\mathbf{B} \\ \mathbf{0} & \mathbf{0} & \mathbf{B} & -\overline{\mathbf{Z}}\mathbf{B} \end{bmatrix}_{3m \times 4n} \cdot \begin{bmatrix} \mathbf{X} \\ \mathbf{Y} \\ \mathbf{Z} \\ \mathbf{w} \end{bmatrix}_{4n \times 1} = [\mathbf{0}]_{3m \times 1} . \tag{7}$$

Fitting a curve or surface through $\mathcal{M}$, reduces to solving the overdetermined equations (5) or (7) for the unknowns $\mathbf{X}$, $\mathbf{Y}$, $\mathbf{Z}$ and $\mathbf{w}$.

Numerically, it is difficult to solve (5) or (7) directly because of computer memory, possible negative weights, and numerical instability. Fortunately, by manipulating the block matrix $\mathbf{A}$, it is possible to separate the weights from the control points, and then solve for them separately. First we multiply equation (7) by the transpose of $\mathbf{A}$ to get

$$\begin{bmatrix} \mathbf{B}^T\mathbf{B} & \mathbf{0} & \mathbf{0} & -\mathbf{B}^T\overline{\mathbf{X}}\mathbf{B} \\ \mathbf{0} & \mathbf{B}^T\mathbf{B} & \mathbf{0} & -\mathbf{B}^T\overline{\mathbf{Y}}\mathbf{B} \\ \mathbf{0} & \mathbf{0} & \mathbf{B}^T\mathbf{B} & -\mathbf{B}^T\overline{\mathbf{Z}}\mathbf{B} \\ -\mathbf{B}^T\overline{\mathbf{X}}\mathbf{B} & -\mathbf{B}^T\overline{\mathbf{Y}}\mathbf{B} & -\mathbf{B}^T\overline{\mathbf{Z}}\mathbf{B} & \mathbf{M_0} \end{bmatrix}_{4n \times 4n} \cdot \begin{bmatrix} \mathbf{X} \\ \mathbf{Y} \\ \mathbf{Z} \\ \mathbf{w} \end{bmatrix} = [\mathbf{0}]_{4n \times 1} \tag{8}$$

where $\mathbf{M_0}$ is a $n \times n$ matrix

$$\mathbf{M_0} = \mathbf{B}^T\overline{\mathbf{D}}^2\mathbf{B} = \mathbf{B}^T\overline{\mathbf{X}}^2\mathbf{B} + \mathbf{B}^T\overline{\mathbf{Y}}^2\mathbf{B} + \mathbf{B}^T\overline{\mathbf{Z}}^2\mathbf{B}. \tag{9}$$

Eliminating the first three elements of the fourth row of (8), we have

$$\begin{bmatrix} \mathbf{B}^T\mathbf{B} & \mathbf{0} & \mathbf{0} & -\mathbf{B}^T\overline{\mathbf{X}}\mathbf{B} \\ \mathbf{0} & \mathbf{B}^T\mathbf{B} & \mathbf{0} & -\mathbf{B}^T\overline{\mathbf{Y}}\mathbf{B} \\ \mathbf{0} & \mathbf{0} & \mathbf{B}^T\mathbf{B} & -\mathbf{B}^T\overline{\mathbf{Z}}\mathbf{B} \\ \mathbf{0} & \mathbf{0} & \mathbf{0} & \mathbf{M} \end{bmatrix} \cdot \begin{bmatrix} \mathbf{X} \\ \mathbf{Y} \\ \mathbf{Z} \\ \mathbf{w} \end{bmatrix} = [\mathbf{0}]_{4n \times 1} \tag{10}$$

where $\mathbf{M}$ is a $n \times n$ matrix

$$\mathbf{M} = \mathbf{M_0} - \left[(\mathbf{B}^T\overline{\mathbf{X}}\mathbf{B})(\mathbf{B}^T\mathbf{B})^{-1}(\mathbf{B}^T\overline{\mathbf{X}}\mathbf{B}) + (\mathbf{B}^T\overline{\mathbf{Y}}\mathbf{B})(\mathbf{B}^T\mathbf{B})^{-1}(\mathbf{B}^T\overline{\mathbf{Y}}\mathbf{B}) \right.$$
$$\left. + (\mathbf{B}^T\overline{\mathbf{Z}}\mathbf{B})(\mathbf{B}^T\mathbf{B})^{-1}(\mathbf{B}^T\overline{\mathbf{Z}}\mathbf{B}) \right].$$
$$\tag{11}$$

Thus the weights are separated from the control points as

$$\mathbf{M} \cdot \mathbf{w} = [\mathbf{0}]_{n \times 1} . \tag{12}$$

Now the weights and the control points can be found from (12) and (5), respectively.

§4. Numerical Algorithms

In general, the weights in the real space $\mathbb{R}^1$ can be computed from the singular value decomposition (SVD) of $\mathbf{M}$ [4], in which $\mathbf{M}$ is factorized as

$$\mathbf{M} = \mathbf{QDP}^T, \tag{13}$$

where $\mathbf{Q}$ is a $n \times n$ matrix with orthonormal columns, $\mathbf{P}$ is a $n \times n$ orthogonal matrix, and $\mathbf{D}$ is a $n \times n$ diagonal matrix

$$\mathbf{D} = diag\{d_1, d_2, ..., d_r, d_{r+1}, ..., d_n\} \qquad (d_i \geq d_{i+1}) \tag{14}$$

The diagonal elements of $\mathbf{D}$ are the singular values of $\mathbf{M}$. They are non-negative and are arranged in decreasing order of magnitude. The columns $\{\mathbf{q}_i\}_1^n$ of $\mathbf{Q}$ and $\{\mathbf{p}_i\}_1^n$ of $\mathbf{P}$ are the left and right singular vectors of $\mathbf{M}$, respectively. If the singular values $d_{r+1}, ..., d_n$ are zero or negligible, but d_r is not, then the rank of $\mathbf{M}$ is r and the general solutions of $\mathbf{w}$ are given by

$$\mathbf{w} = \sum_{i=r+1}^{n} \alpha_i \cdot \mathbf{p}_i, \tag{15}$$

where α_i's are arbitrary coefficients. By assigning the $n - r$ free coefficients properly, a set of positive weights can be found. In this case, an interpolation NURBS curve or surface is guaranteed. If such a set of positive weights does not exist in the space spanned by $\{\mathbf{p}_i\}_{r+1}^n$, i.e., $\mathcal{W} = span\{\mathbf{p}_i\}_{r+1}^n$, one can expand $\mathcal{W}$ by including more singular vectors one at a time until a set of positive weights exists. For such a case fitting solutions are found.

An alternative algorithm is also available to find a set of feasible weights in the positive sub-space $\mathbb{R}^{1+} = \{\mathbb{R}^1(w) | w > 0\}$. This leads to a quadratic programming problem (QP) subject to positive weights [4]

$$\min_{\mathbf{w}} F(\mathbf{w}) = 0.5 \cdot \parallel \mathbf{M} \cdot \mathbf{w} \parallel_2^2 \qquad \text{subject to} \qquad \{w_i \geq w_0 \mid i = 1, 2, ..., n\}, \tag{16}$$

where w_0 is a positive constant. For a given set of discrete points and a knot sequence, if the smallest singular value d_n is negligible and all the elements of the corresponding singular vector are positive, this method will converge to the global solution which is the same as that of SVD method, i.e., the scaled last singular vector in $\mathbf{P}$. Otherwise, the algorithm will be constraint dependent, and two or more elements in $\mathbf{w}$ will stick to the boundary of the constraints. In the extreme case all elements of $\mathbf{w}$ will be w_0, which reduces to the case of a B-spline curve or surface. In general, a set of weights which are better than uniform ones can be reached.

When the weights are available, equation (5) can then be used to solve for the control points. Householder Transformations and stepwise refinement are employed to achieve an accurate solution [2,4]. The control points in $\mathbb{R}^3$ can be recovered from the control points in homogeneous space divided by the related weight.

§5. Some Examples and Discussions

In this section δ_{max}, δ_{min}, $\bar{\delta}$ and σ are used as the maximum, minimum, average, and standard deviations measured according to the l_2-norm from the discrete points to the corresponding fitted curve or surface points. We shall also use $\gamma = \parallel \mathbf{M}\mathbf{w}_0 \parallel_2$ as a measure of the rationality of the observation system, where $\mathbf{w}_0$ is the unit vector of $\mathbf{w}$. Under the same conditions, the smaller these parameters, the closer the discrete points are to the fitted curve or surface. The magnitude of these parameters are closely related to the selection of location parameters and knots.

5.1. Fitting with exact parametrization

Fig. 1 illustrates a family of conic sections fitted with exact parametrization from simulated points. In this figure, (b) is the complimentary arc of (a), and (d) is part of the complimentary arc of (c). The original conics are defined with the same control points $\mathbf{v} = \{[1,0], [1,1], [0,1]\}$, and over the same knots $\mathbf{t} = \{0, 0, 0, 1, 1, 1\}$ with $k = 3$ and $n = 3$. The weights for each of the arcs are $\mathbf{w}_a = \{1, 0.5, 1\}$, $\mathbf{w}_b = \{1, -0.5, 1\}$, $\mathbf{w}_c = \{1, 5, 1\}$ and $\mathbf{w}_d = \{1, -5, 1\}$.

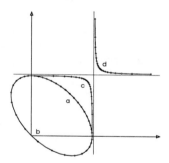

Fig. 1. Fitting with exact parametrization.

The simulated points of (a), (b) and (c) are 20 parametrically equally spaced points, and those of (d) are 20 of the 30 parametrically equally spaced points on the complementary arc of (c). Exact parametrizations are used, i.e., the original location parameters of each point and the original knot sequence are taken as known parameters. The unknowns are the control points and the weights. With both the SVD and QP ($w_0 = 1$) methods, the original control points and the related weights for (a), (b) and (c) are recovered. In Fig. 1a for example, with the SVD decomposition, the three singular values are $\mathbf{D} = \{0.13532 \times 10^0, 0.18047 \times 10^{-1}, 0.94832 \times 10^{-16}\}$ corresponding to the following right-hand singular vectors:

$$\mathbf{P}^T = \begin{bmatrix} 0.23570 & -0.94281 & 0.23570 \\ 0.70711 & 0.00000 & -0.70711 \\ 0.66667 & 0.33333 & 0.66667 \end{bmatrix} .$$

Since $d_2 >> d_3$, the rank of $\mathbf{M}$ is 2, and the general solution of $\mathbf{w}$ is given by

$$\mathbf{w} = \alpha_3 \cdot [0.66667, 0.33333, 0.66667]^T ,$$

where α_3 is an arbitrary coefficient. By setting $\alpha_3 = 1.5$, the original weights are recovered. Applying the Householder Transformation method, the original control points are also recovered. For (d), as only 20 of the 30 simulated points are picked up, the fitting curve is a new interpolation conic arc with control points $\mathbf{v} = \{[1.0417, 1.9616], [1.0434, 1.0434], [1.9616, 1.0417]\}$ and weights $\mathbf{w}_d = \{1.0, 4.6160, 1.0\}$. In general, one can always recover a NURBS curve or surface via fitting with exact parametrization as long as $\mathbf{B}$ is non-singular and sufficient number of well-distributed sample points are used. The minimum number of sample points is $m = n + n/d$ for d-dimensional curves and $m = (n_u + n_u/d)(n_v + n_v/d)$ for d-dimensional surfaces. It turns out that for such a case $rank\{\mathbf{M}\} = n - 1$ and $\mathbf{w} = \mathbf{p}_n$, i.e., $\mathbf{w}$ is the last singular vector of $\mathbf{P}$.

5.2. Practical fitting and interpolation

For practical applications, exact parametrization is not possible, and we have to find a way to parametrize the points to be fitted and select knots for further fitting. Fig. 2 presents some fitting (interpolation) results with different parametrization and fitting methods. All of them are fitted from the same set of 15 discrete points. Chord length method is used to parametrize the discrete points and uniform knots are selected. It is evident that with NURBS, better fitting results are achieved since there are more degree of freedoms.

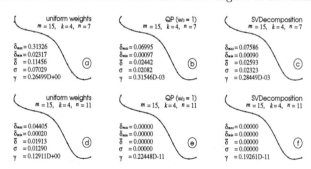

Fig. 2. Practical curve fitting and interpolation.

For case (c), i.e. fitting with the SVD method using 7 control points, the singular values are $\{3.48708, 0.87859, 0.46608, 0.26939, 0.05658, 0.01901, 0.00028\}$ with corresponding singular vectors $\mathbf{P}^T$

$$\begin{bmatrix} \vdots & \vdots & \vdots & \vdots & \vdots & \vdots & \vdots \\ 0.7994 & 0.4775 & 0.3209 & 0.1478 & 0.0826 & 0.0346 & 0.0097 \end{bmatrix}$$

As the smallest singular value $d_7 = 0.00028 \neq 0.0$, the matrix $\mathbf{M}$ is of full rank, and we can only achieve a fitting solution with weights being the corresponding singular vector $\mathbf{p}_7$. Scaling the smallest element of $\mathbf{p}_7$ to 1, we have

$\mathbf{w} = \{82.1757, 49.0867, 32.9855, 15.1924, 8.4868, 3.5552, 1.0000\}$. With the QP $(w_0 = 1)$ method, case (b), similar results are achieved. If the number of control points is further increased to $n = 11$, interpolation solutions are achieved with NURBS, i.e., all deviations are zero. With the SVD method, i.e. case (f), the rank of $\mathbf{M}$ is 8, the singular values are $\{1.1108, ..., 0.0002, 0.3204 \times 10^{-11}, 0.2258 \times 10^{-11}, 0.1524 \times 10^{-11}\}$, and the singular vectors corresponding to the three zero singular values are

$$
\begin{bmatrix}
\vdots & \vdots & \vdots & \vdots & \vdots \\
0.1138 & -0.4335 & -0.5525 & -0.5199 & -0.2969 \\
-0.0517 & 0.1417 & 0.1863 & 0.1933 & 0.1422 \\
0.9748 & 0.2187 & 0.0407 & -0.0006 & -0.0040 \\
\end{bmatrix}
$$

$$
\begin{bmatrix}
\vdots & \vdots & \vdots & \vdots & \vdots & \vdots \\
-0.0931 & 0.0437 & 0.1497 & 0.2082 & 0.1634 & 0.1800 \\
0.0949 & 0.0931 & 0.2082 & 0.4722 & 0.4910 & 0.6000 \\
-0.0012 & 0.0009 & 0.0035 & 0.0072 & 0.0070 & 0.0084 \\
\end{bmatrix}
$$

Setting $\alpha_9 = 0.0$, $\alpha_{10} = 1.0$ and $\alpha_{11} = 1.0$ and then scaling the smallest element of $\mathbf{w}$ to 1 we have a set of positive weights $\mathbf{w} = \{9.8509, 3.8457, 2.4217, 2.0566, 1.4747, 1.0000, 1.0024, 2.2588, 5.1155, 5.3143, 6.4924\}$. With QP $(w_0 = 1)$ method, case (e), we achieve another set of positive weights $\mathbf{w} = \{1.0000, 1.8067, 1.9879, 2.0022, 1.4764, 1.0000, 1.0000, 2.2454, 5.0764, 5.2714, 6.4396\}$ which also interpolates through all the 15 points. One can verify that this set of weights is also a linear combination of the last three singular vectors of $\mathbf{M}$ with $\alpha_9 = 0.1162$, $\alpha_{10} = 10.6713$ and $\alpha_{11} = 1.5783$. In general for practical NURBS curve fitting, $rank\{\mathbf{M}\} \leq min\{n, d(m-n)\}$ for d-dimensional curves. In other words, in order to interpolate m discrete points we need at least $n = (dm + 1)/(d + 1)$ control points.

The case of NURBS surface is similar. Fig. 3 presents some surface fitting examples with different parametrization and fitting methods. All of the cases are fitted from the same set of 218 randomly distributed points. The *Base Surface* method [3] is used to parametrize the discrete points and uniform knots are applied. We discuss case (d) here.

The smallest singular value for case (d) is $d_{25} = 0.1631$ and the corresponding singular vector is $\mathbf{p}_{25} = \{0.1085, 0.0599, 0.0138, -0.0391, -0.1068, 0.0844, 0.0887, 0.0309, -0.0334, -0.1473, 0.0962, 0.1813, 0.1277, 0.1290, 0.0345, 0.0537, 0.0987, 0.2183, 0.2677, 0.1692, 0.0665, 0.2003, 0.3600, 0.4736, 0.5454\}$ with four of its elements being negative. As it is dangerous to use negative weights, one can include one more singular vector in order to achieve a set of positive weights. With $\alpha_{24} = 1.0$ and $\alpha_{25} = 1.0$ and scaling the smallest element of $\mathbf{w}$ to 1, a set of positive weights $\mathbf{w} = \{29.3028, 13.4260, 7.9052, 1.0000, 4.9393, 22.3876, 15.8447, 9.5111, 1.3094, 4.8101, 33.7550, 27.8685, 12.7882, 14.5213, 6.1194, 22.8528, 13.9349, 23.7585, 14.7843, 8.1135, 32.8295, 34.9460, 29.8065, 25.0247, 12.7717\}$ is achieved. From the figure we see that this solution is still better than uniform weights. In general, one can include additional singular vectors one at a time from the smallest singular value upwards in order to achieve a set of positive weights.

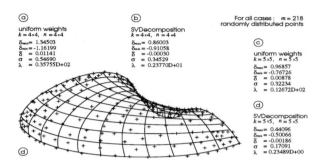

Fig. 3. Practical surface fitting and interpolation.

Acknowledgements. This research is supported in part by the Commission of the European Communities through a Brite-Euram project BE-4527 and by the Katholieke Universiteit Leuven through a doctoral scholarship.

References

1. Farin, G., From conics to NURBS, IEEE Computer Graphics and Applications **12** (1992), 78–86
2. Kruth, J. P., and W. Ma, CAD modelling of sculptured surfaces from digitized data of CMMs, in *Proceedings of the 4th International Symposium on Dimensional Metrology in Production and Quality Control*, Tampere University of Technology, Finland, June 22-25, 1992, 371–387.
3. Ma, W., and J. P. Kruth, Parametrization of randomly measured points for the least squares fitting of B-spline curves and surfaces, Computer-Aided Design, submitted.
4. NAG, *The NAG Fortran Library Manual, Mark 15*, Numerical Algorithms Group Limited, 1991, Chapter E04 and F04.
5. Piegl, L., On NURBS: a survey, IEEE Computer Graphics & Applications **11** (1991), 55–71.

W. Ma and J. P. Kruth
Katholieke Universiteit Leuven
Celestijnenlaan 300B, B-3001 Heverlee, BELGIUM
weiyin@mech.kuleuven.ac.be,   kruth@mech.kuleuven.ac.be

Evaluating Surface Intersections
in Lower Dimensions

Dinesh Manocha, Amitabh Varshney, and Hans Weber

Abstract. We highlight a new algorithm for evaluating the surface intersection curve using a matrix formulation. The projection of the intersection curve is represented as the singular set of a bivariate matrix polynomial. The resulting algorithm for evaluating the intersection curve is based on matrix computations like eigendecomposition and singular value decomposition. Furthermore, at each stage of the algorithm we make use of inverse power iterations to march back to the curve. We also describe the performance of the resulting robust and accurate approach.

§1. Introduction

Evaluating the intersection of parametric and algebraic surfaces is a recurring operation in geometric and solid modeling. Its applications include boundary evaluations, simulation of manufacturing processes, contouring of scattered data, and finite element mesh generations. Surface intersections have been extensively studied in the literature, and the main approaches may be classified as analytic, lattice, marching or subdivision methods. The analytic approach is considered slow in practice due to the algebraic complexity of the intersection curve, and approaches based on subdivision, lattice, and marching methods may not be robust. Furthermore, their accuracy varies with the surface degree, the local surface geometry at the intersection curve, and the angles at which the surfaces intersect. As a result, it is believed that any surface intersection algorithm has to balance three conflicting goals of accuracy, robustness, and efficiency [7,13].

Earlier approaches to surface intersection used the subdivision properties of NURBS surfaces based on a 'divide and conquer' paradigm [13]. However, this approach may be slow and is not guaranteed to be robust in terms of finding all components of the intersection curve. In the last decade 'marching methods' have received a lot of attention for the evaluation of intersection curves [13,1,2,3,9]. Tracing techniques involve the computation of a starting point on each component and locating all the singular points. Given the

Curves and Surfaces in Geometric Design
P. J. Laurent, A. Le Méhauté, and L. L. Schumaker (eds.), pp. 327–334.
Copyright © 1994 by A K PETERS, Wellesley, MA
ISBN 1-56881-039-3.

starting points, these algorithms use marching methods to trace the intersection curve and in the process use robust methods to determine all the branches at singular points. In particular, the tracing algorithms use the algebraic formulation of the intersection problem (e.g. three algebraic equations in four unknowns for NURBS surfaces) and find successive points on the curve with first order approximations and refinement using Newton-Raphson's method. However, it is not clear what a good step size should be at each stage. A small step size makes the overall approach slow, and a large step size may result in convergence problems. It is possible to compute a higher order local approximation at each point on the curve [1], however this involves a great deal of symbolic computation and may not be efficient for high degree intersection curves. Therefore, implementing a robust tracer based on Newton-Raphson's method can be fairly non-trivial and in many cases may not work at all [5]. Other approaches consist of approximating the intersection curve using lattice methods or geometric Hermite approximation [15].

It is possible to represent the intersection curve as an algebraic set in the higher dimensional space spanned by the parameters of two surfaces. Given such a formulation, techniques based on elimination theory can be used to project the intersection curve to a plane curve [14,10,11]. However, the degree of the resulting curve is fairly high (e.g. 108 for the intersection of two bicubic patches) and computation and evaluation based on such a representation can involve efficiency and accuracy problems [8]. However, it is possible to represent the plane curve as a matrix determinant and use matrix computations for evaluation. In this paper, we present a robust algorithm for tracing curves based on a matrix representation. In particular, we represent the intersection curve as the singular set of a bivariate matrix polynomial and use algorithms based on eigendecomposition and singular value decomposition to trace the resulting curve. This involves the use of inverse power iterations for eigenvalue computation and step sizes based on the local geometry of the curve. The main advantage of this approach is in its robustness. Tracing in lower dimension reduces the *geometric complexity* of the curve. Moreover the convergence of power iterations is well understood, and this representation is used in computing the appropriate step sizes. Furthermore, the matrix representation is used for computing the singular points on the intersection curve.

The rest of the paper is organized in the following manner. In §2 we present some background material on surface intersection and review the matrix determinant representation of intersection curves. In §3, we characterize the intersection curve in terms of singular sets of matrix polynomials and characterize the accuracy of various matrix operations performed on these polynomials. §4 describes our tracing algorithm, including choice of step size and the use of power iterations to march along the curve. We also describe how the structure of the matrix formulation can be exploited when performing the inverse power iterations. Finally, in §5 we discuss our implementation and its application to surface intersection and the computation of superbolas for toolpath generation.

§2. Background

The intersection of parametric surfaces results in a high degree algebraic curve. The *algebraic complexity* makes it difficult to compute an exact representation as an algebraic set, and, therefore, most of the current techniques aim at an approximate representation as a piecewise linear curve (obtained by subdivision or tracing methods). However, this representation is not robust.

The topology of the intersection curve can be very complicated. The intersection curve may have more than one component and may have singular points, thereby adding to the *geometric complexity* of the problem. Simple cases like intersection of two cylinders can give rise to a singularity. In this case the intersection curve is an algebraic space curve of degree four. For tensor product bicubic Bézier patches the intersection curve is a space curve of degree 324 in the (x, y, z) space, and degree 108 in the (u, v) domains, and it is simple to come up with cases where the intersection curve has more than one component.

It is well known in algebraic geometry that if one of the surfaces is represented parametrically and the other one implicitly, an implicit representation of the intersection curve can be obtained by substituting the parametric formulation into the implicit representation. Since a Bézier surface is a rational parametric surface, and we wish to compute the intersection of two such surfaces, we need to implicitize the parametric representation of one of the surfaces. It has been shown that if a parameterization has no base points, the resultant of the parametric equations corresponds exactly to the implicit representation. Using Dixon's formulation[4], the resultant corresponds to the determinant of a matrix.

Thus, given two Bézier surfaces $\mathbf{F}(s, t) = (X(s, t), Y(s, t), Z(s, t), W(s, t))$ and $\mathbf{G}(u, v) = (\overline{X}(u, v), \overline{Y}(u, v), \overline{Z}(u, v), \overline{W}(u, v))$, we can implicitize $\mathbf{F}(s, t)$ into an algebraic surface of the form $f(x, y, z, w) = 0$, where

$$
f(x, y, z, w) = \det \begin{pmatrix} f_{11}(x, y, z, w) & \cdots & f_{1n}(x, y, z, w) \\ f_{21}(x, y, z, w) & \cdots & f_{2n}(x, y, z, w) \\ \vdots & \ddots & \vdots \\ f_{n1}(x, y, z, w) & \cdots & f_{nn}(x, y, z, w) \end{pmatrix}.
$$

The algebraic plane curve, birational to the intersection curve, is obtained by substituting the parameterization $\mathbf{G}(u, v)$ into $f(x, y, z, w)$. As a result, the intersection curve is represented as zero set of the determinant of a matrix $M(u, v)$. The corresponding matrix $M(u, v)$ is

$$
f(\overline{X}(u, v), \overline{Y}(u, v), \overline{Z}(u, v), \overline{W}(u, v)) = \begin{pmatrix} g_{11}(u, v) & \cdots & g_{1n}(u, v) \\ g_{21}(u, v) & \cdots & g_{2n}(u, v) \\ \vdots & \ddots & \vdots \\ g_{n1}(u, v) & \cdots & g_{nn}(u, v)) \end{pmatrix},
$$

where

$$g_{ij}(u,v) = f_{ij}(\overline{X}(u,v), \overline{Y}(u,v), \overline{Z}(u,v), \overline{W}(u,v)).$$

Let $D(u,v)$ be the polynomial corresponding to $\det(M(u,v))$. Then the algebraic plane curve corresponds to $D(u,v) = 0$.

§3. Properties of the Matrix Representation

Let the degree of a tensor product Bézier patch $\mathbf{F}(s,t)$ be m_f in t and n_f in s. Similarly, let the degree of a tensor product Bézier patch $\mathbf{G}(u,v)$ be m_g in u and n_g in v. The order of the matrix arising from the implicitization of $\mathbf{F}(s,t)$ by using Dixon's Resultant will be $2m_f n_f$. After substituting the parameterization of $\mathbf{G}(u,v)$ into the implicit representation of $\mathbf{F}(s,t)$, we get a bivariate matrix polynomial $M(u,v)$. Given $v = v_i$, we can write $M(u,v_i)$ as a univariate matrix polynomial of degree m_g in u:

$$M(u,v_i) = M_a u^a + M_{a-1} u^{a-1}(1-u) + \ldots + M_0(1-u)^a$$

where $a = m_g$, and each of the matrices M_j have order $b = 2m_f n_f$. After dividing this formulation by $(1-u)^a$ and re-parameterizing by $w = \frac{u}{1-u}$, we get a matrix polynomial of the form

$$M'(w) = M_a w^a + M_{a-1} w^{a-1} + \ldots + M_0.$$

When M_a is a non-singular and well conditioned matrix, the roots of the matrix polynomial $M'(w)$ are given by the eigenvalues of the companion matrix [11]:

$$C(w) = \begin{pmatrix} \mathbf{0_b} & \mathbf{I_b} & \mathbf{0_b} & \ldots & \mathbf{0_b} \\ \mathbf{0_b} & \mathbf{0_b} & \mathbf{I_b} & \ldots & \mathbf{0_b} \\ \vdots & \vdots & \vdots & \ddots & \vdots \\ \mathbf{0_b} & \mathbf{0_b} & \mathbf{0_b} & \ldots & \mathbf{I_b} \\ -\overline{M}_0 & -\overline{M}_1 & -\overline{M}_2 & \ldots & -\overline{M}_{a-1} \end{pmatrix},$$

where $\mathbf{0_b}$ and $\mathbf{I_b}$ are null and identity matrices respectively, each of order $b = 2m_f n_f$, and $\overline{M}_i = M_a^{-1} M_i, 0 \le i < a$. When M_a is singular or close to being singular, the roots of the matrix polynomial $M'(w)$ are given by the eigenvalues of the generalized eigensystem [11].

We utilize algorithms which rely on computing eigenvalues and eigenvectors of matrices because eigenvalue and eigenvector computations are backwards stable and thus provide improved accuracy. To find all of the roots of the matrix polynomial we can perform a full eigendecomposition on the companion matrix, $C(w)$, using the QR algorithm [6]. More importantly, if we are interested in only one of the roots of the matrix polynomial $M'(w)$, and if we are given a value λ' close to that root, we can perform inverse power iterations to quickly and robustly converge to the actual root λ [6]. [11] also points out that the corresponding eigenvector is of the form

$$[\, \mathbf{V}, \lambda \mathbf{V}, \lambda^2 \mathbf{V}, \ldots, \lambda^{m_g - 1} \mathbf{V} \,],$$

where

$$\mathbf{V} = [\, 1, s, s^2, \ldots, s^{m_f-1}, t, st, \ldots, s^{m_f-1}t^{2n_f-1} \,].$$

Finally, using a method described in [11], we can compute the partial derivatives of $M(u,v)$ at a given point (u_i, v_i). This computation proceeds by a modified Gaussian elimination method, and the accuracy of the partial derivative calculation is approximately the same as that of the standard Gaussian elimination process [12].

For further details on the properties of this matrix representation, the interested reader can refer to [11] as a starting point.

§4. Tracing Algorithm

Our aim has been to develop a tracing algorithm that satisfies to a reasonable extent the conflicting goals of accuracy, efficiency, and robustness. To this end, we have designed an algorithm based on a robust formulation which relies on accurate matrix operations and takes advantage of the special structure of the problem and the matrices involved for efficiency.

The algorithm proceeds as follows:

(1) Given two tensor product Bézier surfaces $\mathbf{F}$ and $\mathbf{G}$, we first generate the matrix representation $M(u,v)$ as described in §2.

(2) Given a starting point (u_i, v_i) on the intersection curve, we find the tangent $(\delta u_i, \delta v_i)$, in the uv space, to the curve at (u_i, v_i).

(3) Given a step size $\sigma(u_i, v_i)$, and assuming $\delta v_i = \max(\delta u_i, \delta v_i)$ [1], we move to (u_i, v_{i+1}), where $v_{i+1} = v_i + \sigma(u_i, v_i)\delta v_i$.

(4) To compute u_{i+1}, we form the companion matrix $C(w)$ (where $w = \frac{u}{1-u}$, as in §3) of the matrix polynomial $M(u, v_{i+1})$, and then perform inverse power iterations on $C(w)$ to find the nearest eigenvalue λ_i to u_i. As discussed in §2, $D(\lambda_i, v_{i+1}) = 0$. Thus, $u_{i+1} = \lambda_i$ and we return to (2), with (u_{i+1}, v_{i+1}) as our starting point.

To find the tangent $(\delta u_i, \delta v_i)$ at (u_i, v_i), we compute $\delta u_i = D^u(u_i, v_i)$ and $\delta v_i = D^v(u_i, v_i)$ as mentioned in §3. The step size $\sigma(u_i, v_i)$ can be adaptively based upon the higher order derivatives of the intersection curve in the uv-space. Higher order derivatives such as D^{uu}, D^{vv}, etc., can be computed in an analogous fashion to the computation of D^u and D^v. Such derivatives can be used to formulate higher order approximants for use as $(\delta u_i, \delta v_i)$. The decision on the number of higher order derivatives to be computed is a trade-off between robustness and efficiency.

Inverse power iterations normally require that we perform a full LU decomposition of the matrix involved, which in this case is the companion matrix of order ab, where $a = m_g$ and $b = 2m_f n_f$. LU decomposition for such a matrix would normally require $\mathcal{O}(a^3 b^3)$ operations. However, by taking advantage of the sparse structure of the matrix, we can reduce this to $\mathcal{O}(ab^3)$

---

[1] The case where $\delta u_i = \max(\delta u_i, \delta v_i)$ proceeds similarly.

operations. If **F** and **G** are tensor product bicubic patches, this provides a speedup by a factor of about 9.

When computing the intersection of two parametric surfaces like **F** and **G**, we are often only interested in the region where $0 \leq u, v, s, t \leq 1$. Since we are tracing in the uv-space, it is obvious when we have left the region of interest for the patch $\mathbf{G}(u, v)$. To find out when whether we still lie in the region $0 \leq s, t \leq 1$, we can use the special structure of the eigenvector corresponding to the eigenvalue λ_i (as noted in §3) to derive (s_{i+1}, t_{i+1}).

The tracing algorithm also provides a simple method by which we can compute the starting points for the open components of the intersection curve. If the patch boundaries are $0 \leq u, v, s, t \leq 1$, then we can find the starting points along the boundary $v = 0$ simply by forming the companion matrix corresponding to $M(u, 0)$ and performing a full eigendecomposition. The real eigenvalues in the interval $[0, 1]$ correspond to starting points for tracing the open components of the intersection curve which intersect $v = 0$. This can be repeated for $u, s, t = 0$ and $u, v, s, t = 1$. The closed loops can be found by tracing paths in the complex space [12].

§5. Implementation and Results

We have implemented the tracing algorithm described in §4 in C on a variety of platforms using double precision calls to the Fortran libraries LAPACK and BLAS. A simple window interface has been set up in which we can observe the tracing progressing in the uv, st, and xyz spaces at the same time. We take two tensor product Bézier patches, calculate Dixon's resultant to implicitize one, and then substitute the other into the resultant to generate $M(u, v)$. We currently only trace the open components for which we obtain starting points by the process described at the end of §4. Our code uses a fixed step size σ and the first order partials of $D(u, v)$ to form the local approximant. We also take full advantage of the sparse structure of the companion matrix when performing the inverse power iterations. To get a rough feeling for the performance of the algorithm, we ran it on an SGI Onyx, which was configured with the default of two processors. We did not parallelize the algorithm. Also, during our test runs the load of the networked machine (excluding our process) was close to zero. We came up with the following average timing figures (obtained using the 'gettimeofday' call) for tracing the intersection curve of two bicubic tensor product Bézier patches:

Implicitization of one surface: 0.089 secs
Full eigendecomposition to find starting points on one edge: 0.149 secs
Tracing one point of a path using inverse power iterations: 0.072 secs

The tracing algorithm is really just a method for tracing a one dimensional algebraic set which is represented in our matrix formulation, so for a sample test of the robustness we set up the matrix so it would trace the degree 20 superbola described in a toolpath generation example in [5]. Field and Field found that conventional tracing and marching techniques failed to properly follow the curve, but when using the matrix formulation our tracing algorithm encountered no problems in generating the superbola.

While implementing this tracing algorithm, we came upon a number of ideas which could be used to improve and augment the tracer.

First, there is an obvious coarse grain parallel structure to the algorithm, and this could easily be exploited on a MIMD or SIMD architecture. Given n starting points, we could distribute one starting point and copies of $M(u, v)$, $F(s, t)$, and $G(u, v)$ to each of a set of n processors, and they could trace out a set of components in parallel.

One of the major reasons for tracing in the lower dimensional space is that we have reduced the *geometric complexity* of the problem. Performing any kind of tracing method in higher dimensional spaces is conceptually much more difficult than performing a similar algorithm in the plane. Furthermore, there has been a fair amount of work done on the resolution of singularities in a plane curve [7], and we can reason much more easily about singularities on a plane than about singularities in higher dimensional spaces. Based on the matrix formulation, we can detect singularities on paths as we trace out the curve and use branching computations to resolve them [12].

Other Applications and Future Work: The algorithm presented above is applicable to all geometric applications related to the computation of one dimensional algebraic sets. These include silhouette computations, offsets and blends, voronoi curves, etc. This algorithm, combined with robust techniques for finding loops and all branches around a singular point, is part of a system being developed for performing CSG operations on spline surfaces [12].

Acknowledgments. Dinesh Manocha is supported in part by a Junior Faculty Award and ARPA Contract #DAEA 18-90-C-0044. Amitabh Varshney is supported under the NIH NCRR Grant #5-P41-RR02170. Hans Weber is supported under ARPA Contract #DAEA 18-90-C-0044.

References

1. Bajaj, C. L., C. M. Hoffmann, J. E. H. Hopcroft, and R. E. Lynch, Tracing surface intersections, Computer Aided Geometric Design **5** (1988), 285–307.
2. Barnhill, R., G. Farin, M. Jordan, and B. Piper, Surface/surface intersection, Computer Aided Geometric Design **4** (1987), 3–16.
3. Barnhill, R. E., and S. N. Kersey, A marching method for parametric surface/surface intersection, Computer Aided Geometric Design **7** (1990), 257–280.
4. Dixon, A. L., The eliminant of three quantics in two independent variables, Proceedings of the London Mathematical Society **6** (1908), 49–69, 209–236.
5. Field, D. A., and R. Field, A new family of curves for industrial applications, Technical Report GMR-7571, General Motors Research Laboratories, 1992.

6. Golub, G. H., and C. F. Van Loan, *Matrix Computations*, John Hopkins Press, Baltimore, 1989.
7. Hoffman, C. M., *Geometric and Solid Modeling*, Morgan Kaufmann, San Mateo, California, 1989.
8. Hoffman, C. M., A dimensionality paradigm for surface interrogations, Computer Aided Geometric Design **7** (1990), 517–532.
9. Kriezis, G. A., P. V. Prakash, and N. M. Patrikalakis, Method for intersecting algebraic surfaces with rational polynomial patches, Computer-Aided Design **22**(10) (1990), 645–654.
10. Manocha, D., Algebraic and numeric techniques for modeling and robotics, dissertation, Computer Science Division, Department of Electrical Engineering and Computer Science, University of California, Berkeley, 1992.
11. Manocha, D., and J. F. Canny, A new approach for surface intersection, International Journal of Computational Geometry and Applications **1**(4) (1991), 491–516.
12. Manocha, D., and S. Krishnan, Robust and efficient surface intersections, manuscript.
13. Pratt, M. J., Surface/surface intersection problems, in *The Mathematics of Surfaces II*, J.A. Gregory (ed.), Claredon Press, Oxford, 1986, 117–142.
14. Sederberg, T. W., Implicit and parametric curves and surfaces, dissertation, Purdue University, 1983.
15. Sederberg, T. W., and T. Nishita, Geometric Hermite approximation of surface patch intersection curves, Computer Aided Geometric Design **8** (1991), 97–114.

Dinesh Manocha
CB #3175, Sitterson Hall
University of North Carolina at Chapel Hill
Chapel Hill, NC 27599-3175
manocha@cs.unc.edu

The Iterative Solution of a Nonlinear Inverse Problem from Industry: Design of Reflectors

A. Neubauer

Abstract. We report on an industrial project which concerns the computer–aided design of surfaces in $I\!R^3$ that can be used as reflectors where the light coming from a point source is reflected in such a way that the illumination distribution of the outgoing light on a given sphere can be prescribed. We derive a mathematical model and a numerical algorithm based on the iterative solution of a certain minimization problem. Finally, we present a numerical example showing that the algorithm works in practice.

§1. Introduction

We report about a project that has been done for and in cooperation with an Austrian company which designs and manufactures lighting fittings for industrial illumination purposes. These consist of a lamp with known light distribution and a reflector. Our task was to develop a mathematical model and an algorithm (resulting in computer software) for the problem of designing the shape of the reflector in such a way that the illumination distribution on the unit sphere centered at the lamp, obtained by the reflected outgoing light rays which are all thought to emanate from the lamp, behaves in a prescribed way. In the algorithm for this so called *inverse far field problem* (cf. [1]), the reflector should be described in a way that could be directly used for CAM purposes by the company.

This is a very old problem (see [4]). The state of the art seems to be that one can handle 2–dimensional problems in a practically satisfactory way, i.e., cases where the light distribution of the lamp, the desired illumination distribution, and the reflector have either rotational or translational symmetry [7,8]. For a general background about problems in lighting design see e.g. [5,6].

Our aim is to handle the full 3–dimensional problem, where the reflector is not required to have any symmetries. In [1] a model and an algorithm

Curves and Surfaces in Geometric Design
P. J. Laurent, A. Le Méhauté, and L. L. Schumaker (eds.), pp. 335–342.
Copyright © 1994 by A K PETERS, Wellesley, MA
ISBN 1-56881-039-3.

for the computation of such reflectors were derived that were based on the iterative solution of a nonlinear partial differential equation of second order. This algorithm worked quite well on academic test examples, but was not satisfactory for examples from practice. The reason was the following: the model was based on the assumption that parallel outgoing rays will not occur for the starting reflector. Unfortunately, the company was not able to supply us with such a starting reflector. In this paper we describe an algorithm where this assumption is not needed.

The mathematical model and the numerical algorithm are derived in Section 2, and in Section 3 we present a numerical example showing that the algorithm works in practice.

§2. The Mathematical Model and the Numerical Algorithm

For deriving the mathematical model for the inverse far field problem, we need the following Figure 1 (cf. [1]).

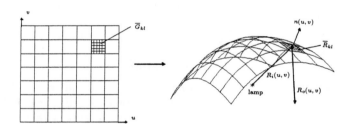

Fig. 1. The parameter domain and the reflector.

In this figure, R_i denotes the direction of the incoming light ray, normed in such a way that $R_i(u, v) = (x(u, v), y(u, v), z(u, v))$ is also the point on the reflector where this ray is reflected. The independent variables (u, v) vary in some parameter domain G. The reflected outgoing light ray R_o may be computed via the reflection condition:

Let $n(u, v)$ denote the normal vector on the reflector and let us assume that R_i is twice continuously differentiable; then the reflection condition means that

$$R_o = R_i - \frac{2\langle R_i, n \rangle}{\|n\|^2} n, \tag{1}$$

with

$$n = \frac{\partial R_i}{\partial u} \times \frac{\partial R_i}{\partial v}, \tag{2}$$

where $\langle \cdot, \cdot \rangle$ and $\| \cdot \|$ denote the Euclidean inner product and norm, respectively.

The basic requirement for the *direct far field problem*, where the light distribution of the lamp and the reflector are known and the illumination distribution on the unit sphere has to be calculated, is the so called *balance condition*. This condition means that the amount of light leaving the lamp in any solid angle increment $d\Omega_i$ should equal the amount of light reflected into the corresponding angle increment $d\Omega_o$ in the illuminated sphere (cf., e.g., [1,7]). Since we want to solve the *inverse far field problem* and since we want to allow parallel outgoing rays, the mathematical meaning of this condition is as follows.

Let ΔB be a sufficiently regular subregion on the unit sphere and let ΔS denote the corresponding subregion of $S := [0, \pi] \times [0, 2\pi]$ in spherical coordinates. Moreover, let ΔR be the largest subregion on the reflector such that for all incoming rays in ΔR the normed reflected outgoing rays are in ΔB. If the corresponding subregion in the parameter domain is denoted by ΔG and if L and I denote the known light distribution of the lamp (up to a constant C) and the desired illumination distribution on the unit sphere, respectively, then the balance condition means

$$\int_{\Delta B} I \, d\Omega_o = C \int_{\Delta R} L \, d\Omega_i,$$

and hence via substitution,

$$\int_{\Delta S} I(\tilde\gamma, \tilde\theta) \sin \tilde\gamma \, d(\tilde\gamma, \tilde\theta) = $$
$$C \int_{\Delta G} L(\gamma(u,v), \theta(u,v)) \frac{|\langle R_i(u,v), n(u,v) \rangle|}{\|R_i(u,v)\|^3} \, d(u,v) \, , \tag{3}$$

where $(\gamma(u,v), \theta(u,v)) \in S$ are the spherical coordinates of $R_i(u,v)$ and $n(u,v)$ is defined as in (2). The expression $\langle R_i(u,v), n(u,v) \rangle$ is not allowed to change sign. This condition means that there are no unattainable points on the reflector (compare (1)). The constant C in (3) may be determined by setting $\Delta S = S$ and $\Delta G = G$. Since the total amount of reflected light only depends on the boundary of the reflector, C can be determined without knowing the reflector, if the boundary of the reflector is fixed as a constraint which is the case in many practical problems.

For deriving a numerical algorithm, we have to describe the reflector by a finite number of variables. Since for CAM purposes, the company interpolates by splines, we have chosen the following bicubic B–spline representation defined on the parameter domain $G := [0, nu] \times [0, nv]$, where nu and nv denote the number of subintervals on the u– and v–axis, respectively:

$$R_i(u,v) = \sum_{k=0}^{nu+2} \sum_{l=0}^{nv+2} cr_{kl} N_k(u) N_l(v). \tag{4}$$

The spline coefficients $cr_{kl} \in \mathbb{R}^3$ are the so called de–Boor control points
and the functions $N_k(u)$ and $N_l(v)$ are the well known cubic basis functions
vanishing in $[0, nu] \setminus [k - 3, k + 1]$ and $[0, nv] \setminus [l - 3, l + 1]$, respectively (cf.
[2]).

The aim is now to determine the coefficients cr_{kl} in (4) so that (3) holds
best possible in the least squares sense. Obviously, one can not check (3) for
all possible subsets ΔS of S. We have chosen a uniform partition of S into
3200 subsets:

$$S_{ij} := [\tilde{\gamma}_{i-1}, \tilde{\gamma}_i] \times [\tilde{\theta}_{j-1}, \tilde{\theta}_j], \qquad i = 1, \ldots, 40, \ j = 1, \ldots, 80, \tag{5}$$

with

$$(\tilde{\gamma}_i, \tilde{\theta}_j) := (i\frac{\pi}{40}, j\frac{\pi}{40}], \qquad i = 0, \ldots, 40, \ j = 0, \ldots, 80. \tag{6}$$

To solve (3) in the least squares sense means to determine the spline coefficients $cr_{kl} \in \mathbb{R}^3$ of the reflector (cf. (4)) as the solution of the minimization
problem

$$\sum_{i=1}^{40} \sum_{j=1}^{80} (ld_{ij} - lc_{ij})^2 \longrightarrow \min, \tag{7}$$

where

$$ld_{ij} = \int_{S_{ij}} I(\tilde{\gamma}, \tilde{\theta}) \sin \tilde{\gamma} \, d(\tilde{\gamma}, \tilde{\theta}) \tag{8}$$

and

$$lc_{ij} = C \int_{G_{ij}} L(\gamma(u, v), \theta(u, v)) \frac{|\langle R_i(u, v), n(u, v) \rangle|}{\|R_i(u, v)\|^3} \, d(u, v). \tag{9}$$

Here G_{ij} is the largest region in G such that for all incoming rays from G_{ij}
the normed reflected outgoing rays are in S_{ij}. Since it is very complicated
to calculate the regions G_{ij}, we determine instead of lc_{ij} an averaged version
$\overline{lc}_{ij}$ of it as follows.

Let $\overline{G}_{kl} := [k - 1, k] \times [l - 1, l]$ ($k = 1, \ldots nu$, $l = 1, \ldots nv$) and let $\overline{R}_{kl}$
denote the corresponding region on the reflector (see Figure 1). Via (3) the
total light $\overline{lr}_{kl}$ being reflected from $\overline{R}_{kl}$ is given by

$$\overline{lr}_{kl} = C \int_{\overline{G}_{kl}} L(\gamma(u, v), \theta(u, v)) \frac{|\langle R_i(u, v), n(u, v) \rangle|}{\|R_i(u, v)\|^3} \, d(u, v).$$

This integral is computed approximately using the Gaussian quadrature formula based on 4×4 nodes. For each of these 16 nodes we calculate the
incoming and outgoing rays via (1), (2) and assign every outgoing ray the
light value $\overline{lr}_{kl}/16$. Since we need in our algorithm that $\overline{lc}_{ij}$ is continuously
differentiable with respect to the spline coefficients cr_{kl}, it is necessary to
distribute the light values $\overline{lr}_{kl}/16$ onto the subregions S_{ij} in a differentiable
way. This is achieved for instance by the following procedure: let $(\tilde{\gamma}, \tilde{\theta})$ be the
spherical coordinates of an outgoing ray with light value $\overline{lr}_{kl}/16$ and define

$$i := [\frac{1}{2} + \frac{\tilde{\gamma}}{h}], \ g := \frac{\tilde{\gamma}}{h} - i \in [-\frac{1}{2}, \frac{1}{2}), \quad j := [\frac{1}{2} + \frac{\tilde{\theta}}{h}], \ t := \frac{\tilde{\theta}}{h} - j \in [-\frac{1}{2}, \frac{1}{2}),$$

where $h := \pi/40$ and $[\cdot]$ denotes the integral part of a real number. Then we distribute $\overline{lr}_{kl}/16$ onto the four subregions S_{ij}, $S_{i+1\,j}$, $S_{i\,j+1}$, $S_{i+1\,j+1}$ (see Figure 2) via

$$\frac{\overline{lr}_{kl}}{16} * \begin{cases} f(g) * f(t) & \text{in } I := S_{i+1\,j+1} \\ f(g) * (1 - f(t)) & \text{in } II := S_{i+1\,j} \\ (1 - f(g)) * f(t) & \text{in } III := S_{i\,j+1} \\ (1 - f(g)) * (1 - f(t)) & \text{in } IV := S_{ij} \end{cases}$$

where

$$f(s) := (1 + 3s - 4s^3)/2, \quad s \in [-\frac{1}{2}, \frac{1}{2}).$$

This definition makes sense, if we define $S_{0j} := S_{1j}$, $S_{41\,j} := S_{40\,j}$, $S_{i0} := S_{i\,80}$ and $S_{i\,81} := S_{i1}$ with S_{ij} as in (5), (6).

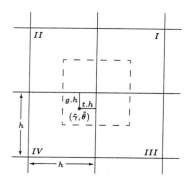

Fig. 2. Distribution of the light value onto four subregions.

The light function $\overline{lc}_{ij}$ for the subregion S_{ij} is then obtained by adding up all light value portions for this subregion. Of course, by the procedure used to determine this light function, it is an averaged version of the *real* light function lc_{ij} defined by (9). In practice this averaging property corresponds to the fact that in reality the lamp is not a point source, but extends over a small region, which has a smoothing effect on the light function anyway.

Instead of solving (7), we now determine the spline coefficients $cr_{kl} \in \mathbb{R}^3$ (cf. (4)) as the solution of the minimization problem

$$\sum_{i=1}^{40} \sum_{j=1}^{80} (ld_{ij} - \overline{lc}_{ij})^2 w_{ij} \longrightarrow \min, \tag{10}$$

where ld_{ij} is as in (8) and $\overline{lc}_{ij}$ is obtained as above. With the weights w_{ij} one can influence the accuracy in the different subregions S_{ij}. Together with problem (10) side conditions of the following type are used: The boundary curve of the reflector can either be fixed or free; in order to control the total size of the reflector, box constraints on the B–spline coefficients are imposed; and finally, the sign condition on $\langle R_i, n \rangle$ mentioned above has to hold.

The resulting constrained minimization problem is solved iteratively with a projected conjugate gradient method due to Powell; the line search is performed via quadratic interpolation (cf. [3]). In each step of this iterative procedure the gradient of the function defined in (10) has to be calculated. This is possible, since one can show that the light function $\overline{l}c_{ij}$ is continuously differentiable with respect to the spline coefficients cr_{kl}, if the known light distribution L of the lamp is continuously differentiable with respect to γ and θ. The exact calculation of this gradient is very complicated and messy, but the time spent for doing this calculation is worth it, since the CPU–time for one iteration can be reduced tremendously compared to the approach, where the gradient is approximated via a forward difference quotient. For instance, for the choice $nu = nv = 10$ or, equivalently, 507 variables the CPU–time was reduced by a factor of 75.

Since we deal with an inverse problem, stability questions arise. Note that we are not looking for some ideal reflector, but just for any reflector that minimizes (10). Therefore, we do not need stability for the coefficients cr_{kl}, but only for the resulting light functions $\overline{l}c_{ij}$. As mentioned above this stability is guaranteed in the C^1 sense. The stability question for the two dimensional inverse far field problem was studied in [1].

Due to confidentiality reasons, more details about the procedure of calculating the reflector cannot be presented.

§3. A Numerical Example

The task was to calculate a *wall–washer*, i.e., a reflector generating a uniform illumination distribution on an infinitely distant plane. In addition the outgoing rays had to be restricted to a certain subdomain of S, which is called interior region in the sequel. This problem can be reformulated as an inverse far field problem.

We solved this problem using the algorithm of Section 2 with $nu = nv = 10$, i.e., with 169 spline coefficients in $I\!R^3$ or, equivalently, with 507 variables. The initial reflector for our algorithm was supplied by the company. The calculations were done on a VAX 3500. The algorithm ended after 360 iterations and 6 hours CPU–time (instead of 18 days, when using the forward difference quotient for approximating the gradient; compare the remark above). The final 360th iterate had the following features: the total light falling into the exterior region was reduced from 8.73 % to 4 % and the average interior deviation from the desired light distribution was reduced by a factor of 3. This result was judged satisfactory by engineers of the company. Note that minimizing the function in (10) is quite difficult, since it has a lot of local minima.

The initial reflector and the final iterate are shown in Figures 3 and 4, respectively. Since the function in (10) is mostly influenced by the second derivative of the reflector, we have to look for changes in the curvature in comparing both figures. A dramatic change of the curvature occurs close to the top of the reflector.

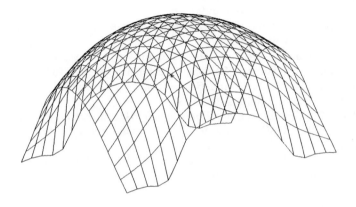

Fig. 3. Initial reflector.

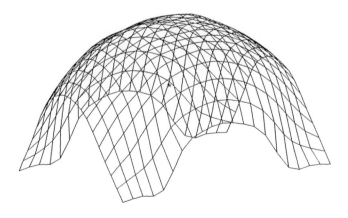

Fig. 4. Final iterate.

References

1. Engl, H. W., and A. Neubauer, Reflector design as an inverse problem, in *Proc. of the Fifth European Conference for Mathematics in Industry*, M. Heiliö (ed.), Kluwer/Teubner, 1991, 13–24.
2. Farin, G., *Curves and Surfaces for Computer Aided Geometric Design*, Academic Press, Boston, 1990.
3. Fletcher, R., *Practical Methods of Optimization*, Vols. 1 & 2, Wiley, Chichester, 1980 & 1981.
4. Halbertsma, N., Die Vorausbestimmung der Lichtverteilungskurve eines spiegelnden Reflektors, Zeitsch. f. techn. Physik **10** (1925), 501–504.
5. Hentschel, H.-J., *Licht und Beleuchtung – Theorie und Praxis der Lichttechnik*, Hüthig, Heidelberg, 1987.
6. Reeb, O., *Grundlagen der Photometrie*, Braun, Karlsruhe, 1962
7. Wolber, W., Optimierung der Lichtlenkung von Leuchten, Dissertation, Universität Karlsruhe, 1973.
8. Wolber, W., Berechung von Reflektoren für beliebige Lichtverteilungen, Lichttechnik **22** (1970), 597–598.

Andreas Neubauer
Institut für Mathematik
Universität Linz
A–4040 Linz, AUSTRIA
neubauer@ indmath.uni-linz.ac.at

Splines with Prescribed Modified Moments

Hans Joachim Oberle and Gerhard Opfer

Abstract. We derive necessary optimality conditions for splines x with prescribed modified moments $\int px$, where p is a given arbitrary polynomial of degree n. The functional to be minimized is $\|x^{(k)}\|$ in the L_2-sense. The result is a unique C^{2k-1}-spline consisting piecewise of polynomials of degree $2k+n$. Besides smoothness and boundary conditions there are also jump conditions for the higher derivatives of x. The area matching case ($n = 0$, $k = 1, 2$) and the first modified moment case ($n = 1$, $k = 1, 2$) are treated as special cases and illustrated by numerical examples.

§1. Introduction

Considerable attention has been paid recently to so-called *histosplines* which match a given area between two consecutive knots. They may be regarded as smooth approximations of a *histogram* which is a graph of a piecewise defined constant function and may also be called *area matching splines*. In particular, J. W. Schmidt and coworkers have contributed to this field. Also constraints of various types have been taken into account. For recent references consult Schmidt & Heß [1993]. However, Schmidt and coworkers usually used splines from a certain prescribed class like cubic splines and tried to give as much smoothness as possible subject to the constraints considered. The unconstrained histosplines were already investigated by Schoenberg [1973].

We approach the problem by searching for those functions which minimize some sort of *mean curvature* or in physical terms *strain energy* and at the same time try to match the given *modified moments* in the intervals which are underlying the splines. The modified moments contain as special cases also the area, and the ordinary moments, so that histosplines and what we would like to call *moment splines* are contained in our presentation.

However, the treatment of other constraints like non-negativity or monotonicity must be postponed to another opportunity. In the first place we are interested in deriving necessary optimality conditions and in the second place in deriving some schemes suitable for computing the corresponding splines.

Curves and Surfaces in Geometric Design
P. J. Laurent, A. Le Méhauté, and L. L. Schumaker (eds.), pp. 343–352.
Copyright © 1994 by A K PETERS, Wellesley, MA
ISBN 1-56881-039-3.

We shall use the following common notations. For a given integer $k \geq 0$ the space $C^k[a, b]$ is the set of all k-times continuously differentiable functions on the interval $[a, b]$, $C^0[a, b]$ will be abbreviated by $C[a, b]$, and the Sobolev space W_2^k is defined by

$$W_2^k[a, b] = \{x \in C^{k-1}[a, b] : x^{(k-1)} \text{ is absolutely continuous, } x^{(k)} \in L_2[a, b]\}.$$

A function f is called *absolutely continuous* if it can be represented by an indefinite integral: $f(t) = \int^t g(\tau)\, d\tau$ of an integrable function g. More details can e. g. be found in Hewitt & Stromberg [1969, §18]. The space $W_2^0[a, b]$ will be identified with $L_2[a, b]$. Norms $\|\ \|$ of functions will be understood always in the L_2-sense.

§2. Statement of the Problem

We start with a given finite *mesh* or *grid* with $N \geq 2$ grid points on the real line:

$$\Delta : \quad a = t_1 < t_2 < \cdots < t_N = b,$$

and two fixed integers $k, n \geq 0$. In addition there are $N - 1$ real numbers α_j, $j = 1, 2, \ldots, N - 1$ given. By Π_n we will always understand the space of all (real) polynomials up to degree n. With these data and with this notation we pose the following

Problem 1. *Given a polynomial $p \in \Pi_n$ with leading coefficient not zero, find $x \in W_2^k[a, b]$ which minimizes*

$$f(x) := \int_a^b \left\{ x^{(k)}(t) \right\}^2 dt =: \|x^{(k)}\|^2, \text{ subject to} \tag{2.1}$$

$$\lambda_j(x) := \int_{t_j}^{t_{j+1}} p(t)x(t)\, dt = \alpha_j, \ j = 1, 2, \ldots, N - 1. \tag{2.2}$$

We call the functional λ_j the *modified n-th moment* in $[t_j, t_{j+1}]$. If $p(t) = t^n$, we call it the *n-th moment*. In the remainder of the paper we omit the appendix $[a, b]$ from the occurring spaces, because all functions used are defined on that interval. The augmented equal-signs ":=" and "=:" indicate a definition, where the expression to be defined is on the side of the colon ":". If we replace W_2^k by C^k, not much is lost, apart from some fine mathematical considerations.

By applying a simple transformation, we can transform the modified moments into a finite expression containing only finitely many values of a function and its derivatives. In Problem 1 we put $x = y^{(n+1)}$ which is possible for all $n \in \mathbb{N}$ and which changes the given formulas (2.1) and (2.2) to

$$f(x) = F(y) := \|y^{(k+n+1)}\|^2, \tag{2.3}$$

$$\lambda_j(x) = \int_{t_j}^{t_{j+1}} p(t)y^{(n+1)}(t)\, dt =: \Lambda_j(y). \tag{2.4}$$

We apply partial integration to the middle part of (2.4) and obtain

$$\int_{t_j}^{t_{j+1}} p(t)y^{(n+1)}(t)\,dt = py^{(n)}\Big|_{t_j}^{t_{j+1}} - \int_{t_j}^{t_{j+1}} p'(t)y^{(n)}(t)\,dt. \qquad (2.5)$$

Because the integral on the right has the same form as the integral on the left side, only n is being reduced by one, we can reduce the whole left integral to a finite sum. We define

$$L(y) := \sum_{\ell=0}^{n} (-1)^\ell p^{(\ell)} y^{(n-\ell)}, \quad L_j(y) := L(y)(t_j),\ j = 1, 2, \ldots, N, \qquad (2.6)$$

and obtain for the integral in (2.5)

$$\Lambda_j(y) = \int_{t_j}^{t_{j+1}} p(t)y^{(n+1)}(t)\,dt = L_{j+1}(y) - L_j(y), j = 1, 2, \ldots, N-1. \qquad (2.7)$$

Because of our requirement (2.2), it follows that $L_{j+1}(y) - L_j(y) = \alpha_j$ and thus, $L_{j+1}(y) = \alpha_j + \alpha_{j-1} + \cdots + \alpha_1 + L_1(y)$. By definition, y is defined only up a polynomial in Π_n. Thus, we are free to choose $L_1(y)$. Our Problem 1 has been changed to

Problem 2. *Given the mesh Δ of Problem 1, two integers κ, n with $\kappa \geq n+1$, N real numbers A_j, and the functionals L_j, $j = 1, 2, \ldots, N$ defined in (2.6) where $p \in \Pi_n$ is a given polynomial whose leading coefficient is not vanishing. Find $y \in W_2^\kappa$ which minimizes*

$$F(y) = \|y^{(\kappa)}\|^2, \ \text{subject to}\ L_j(y) = A_j,\ j = 1, 2, \ldots, N. \qquad (2.8)$$

The connection to Problem 1 is via the following equations: Choose A_1 arbitrary and

$$x = y^{(n+1)}, \kappa = k + n + 1, A_j = A_1 + \alpha_1 + \cdots + \alpha_{j-1}, j = 2, 3, \ldots, N. \quad (2.9)$$

Functionals almost identical with L_j occur in the theory of quasi-interpolation. They are used to construct splines of degree n without solving linear systems of equations. However, in contrast to here, the polynomial p used in quasi-interpolation is chosen in a suitable manner. See de Boor & Fix [1973] or Nürnberger [1989, p. 128].

For $n = 0$, which is equivalent to the area matching problem, the conditions in (2.8) reduce to pure interpolation conditions. In this case Problem 2 is the classical spline interpolation problem with interpolating conditions at the given knots.

Theorem 1. *Let $n = 0$. The solution y of Problem 2 in this case is a natural spline in $C^{2\kappa-2}$ consisting piecewise of polynomials in $\Pi_{2\kappa-1}$ where natural means that the $\kappa - 1$ highest derivatives of y vanish at the two endpoints t_1 and t_N. Unicity is obtained for $\kappa \leq N$. In terms of Problem 1 the solution*

x is a C^{2k-1}-spline consisting piecewise of polynomials in Π_{2k} where the k highest derivatives of x vanish at the two endpoints t_1 and t_N. The spline x is unique if $k \leq N - 1$.

Proof: The first part concerning y is standard, cf. e. g. Holladay [1957] for $\kappa = 2$ and Schoenberg [1964] for the general case. The second part follows by using the relations given in (2.9). ∎

Interestingly, all area matching splines x consist piecewise of polynomials of even degree and an odd degree of smoothness. For the common case $k = 2$ we obtain quartic C^3-splines. The case $k = 1$ yields quadratic C^1-splines. For $k = 2$ and $N = 2$ the above uniqueness condition is violated, and, in fact, the solution is not unique in this case, as we shall see from

Example 1. Let $N = k = 2$ and $x(t) = m(t - (a + b)/2) + \alpha_1/(b - a)$ with arbitrary m. Then $f(x) = 0$, i. e. x minimizes f and $\int_a^b x(t)\,dt = \alpha_1$, hence, x matches the prescribed area for all m. Thus, the minimizer of f is not unique.

§3. Necessary Optimality Conditions

Necessary optimality conditions can be derived by using Problem 1 and Problem 2 as well. Problem 1 has the advantage that the functional to be minimized and the constraints both have integral form. Thus, the application of Lagrange multiplier techniques to Problem 1 seems appropriate. To derive necessary optimality conditions for Problem 2 the application of variational techniques or optimal control is suitable. For these techniques we refer to Opfer & Oberle [1988]. For the derivation of necessary optimality conditions we shall treat Problem 1. For computational purposes it is sometimes easier to use Problem 2.

In order to apply the Lagrange multiplier technique we define for arbitrary real numbers μ_j and for the already defined $\alpha_j, \lambda_j, j = 1, 2, \ldots, N - 1$ the following quantities:

$$\mu := (\mu_1, \mu_2, \ldots, \mu_{N-1})^{\mathrm{T}}, \lambda := (\lambda_1, \lambda_2, \ldots, \lambda_{N-1})^{\mathrm{T}}, \alpha := (\alpha_1, \alpha_2, \ldots, \alpha_{N-1})^{\mathrm{T}},$$
$$L(x, \mu) := f(x) + 2\mu^{\mathrm{T}}(\lambda(x) - \alpha).$$

To obtain necessary optimality conditions, we have for a suitable μ

$$\frac{d}{d\varepsilon} L(x + \varepsilon h, \mu)\Big|_{\varepsilon=0} = 0 \quad \text{for all } h \in W_2^k. \tag{3.1}$$

This result can be found e.g. in Luenberger [1969, p. 185–189]. The quantity L defined above is the Lagrange function. The factor 2 in front of μ^{T} is introduced for later convenience. It has no influence.

Lemma 1. Let $x \in W_2^k$ solve Problem 1. Then there are numbers $\mu_1, \mu_2, \ldots, \mu_{N-1}$ such that

$$\sum_{j=1}^{N-1} \int_{t_j}^{t_{j+1}} (x^{(k)}(t)h^{(k)}(t) + \mu_j p(t)h(t))\,dt = 0 \quad \text{for all} \quad h \in W_2^k. \tag{3.2}$$

Proof: *Equation (3.2) is apart from a factor 2 just the explicit form of equation (3.1).* ∎

We want to apply the following

Lemma 2. *(du Bois-Reymond) Let $l \in L_2[u, v]$ for $u < v$, $\kappa \geq 0$, and define*

$$H = \{\eta \in W_2^\kappa \; : \; \eta^{(\ell)}(u) = \eta^{(\ell)}(v) = 0, \; \ell = 0, 1, \ldots, \kappa - 1\}.$$

Then we have the following implication:

$$\int_u^v l(t)\eta^{(\kappa)}(t) \, dt = 0 \text{ for all } \eta \in H \Longrightarrow l \in \Pi_{\kappa-1}, \; \kappa \geq 1.$$

For $\kappa = 0$ we identify H with L_2 and the implication is $l = 0$.

Proof: For continuous l, see Funk [1970, p. 205], the case treated here may be adapted from Opfer & Oberle [1988, Lemma 2.1]. ∎

In the above Lemma 2, the letter H should be read as capital greek η and not confused with H. If not explicitly mentioned in the sequel, the letter j is to be understood as $j = 1, 2, \ldots, N - 1$. We define

$$H_j = \{h \in W_2^k \; : \; h(t) = 0 \text{ for } t \notin [t_j, t_{j+1}],$$
$$h^{(\ell)}(t_j) = h^{(\ell)}(t_{j+1}) = 0 \text{ for } \ell = 0, 1, \ldots, k - 1\}.$$

If we admit in equation (3.2) only functions $h \in H_j$, we obtain the following set of equations:

$$\int_{t_j}^{t_{j+1}} (x^{(k)}(t)h^{(k)}(t) + \mu_j p(t)h(t)) \, dt = 0, \quad h \in H_j.$$

Repeated partial integration applied to the second part of the integral implies

$$\int_{t_j}^{t_{j+1}} (x^{(k)}(t) + (-1)^k \mu_j q(t))h^{(k)}(t) \, dt = 0, \quad q^{(k)} = p. \tag{3.3}$$

Now the (du Bois-Reymond) Lemma 2 can be applied, and yields

$$x^{(k)} + (-1)^k \mu_j q \in \Pi_{k-1} \text{ in } [t_j, t_{j+1}].$$

This allows differentiation inside the intervals $[t_j, t_{j+1}]$ as often as we want. If we differentiate k times, and multiply for later convenience by $(-1)^k$ we obtain the Euler-Lagrange equations of Problem 1, namely

$$(-1)^k x^{(2k)} + \mu_j p = 0 \text{ in } [t_j, t_{j+1}]. \tag{3.4}$$

Our results up to here say that solutions x of Problem 1 consist piecewise of polynomials in Π_{2k+n}. This implies that we can also apply partial integration to the first part of (3.2) for arbitrary $h \in W_2^k$. With the following abbreviation

$$A(x, h, j) = [x^{(k)}h^{(k-1)} - x^{(k+1)}h^{(k-2)} + \cdots + (-1)^{k-1}x^{(2k-1)}h]\big|_{t_j}^{t_{j+1}}, \quad (3.5)$$

we obtain

$$\int_{t_j}^{t_{j+1}} x^{(k)}(t)h^{(k)}(t)\, dt = A(x, h, j) + (-1)^k \int_{t_j}^{t_{j+1}} x^{(2k)}(t)h(t)\, dt.$$

With these results equation (3.2) becomes for arbitrary $h \in W_2^k$:

$$0 = \sum_{j=1}^{N-1} \int_{t_j}^{t_{j+1}} (x^{(k)}(t)h^{(k)}(t) + \mu_j p(t)h(t))\, dt = \qquad (3.6)$$

$$= \sum_{j=1}^{N-1} \left[A(x, h, j) + \int_{t_j}^{t_{j+1}} ((-1)^k x^{(2k)}(t) + \mu_j p(t))h(t)\, dt \right].$$

and

$$\sum_{j=1}^{N-1} A(x, h, j) = \sum_{\ell=0}^{k-1} (-1)^\ell \left[x^{(k+\ell)}h^{(k-1+\ell)}(t_1) - x^{(k+\ell)}h^{(k-1+\ell)}(t_N) \right. \quad (3.7)$$

$$\left. + \sum_{j=2}^{N-2} (x^{(k+\ell)}(t_j^-) - x^{(k+\ell)}(t_j^+))h^{(k-1+\ell)}(t_j) \right].$$

where $x^{(k+\ell)}(t_j^-)$ is the value of this function at t_j defined on $[t_{j-1}, t_j]$ whereas $x^{(k+\ell)}(t_j^+)$ is the value at t_j of this function on $[t_j, t_{j+1}]$.

Theorem 2. *Let x solve Problem 1. Then x has the following properties.*
(i) *Smooothness conditions:*
   $\operatorname{Res} x\big|_{[t_j, t_{j+1}]} \in \Pi_{2k+n}; \quad x \in C^{2k-1}[a, b],$
(ii) *Boundary conditions:*
   $x^{(k+\ell)}(a) = x^{(k+\ell)}(b) = 0, \; \ell = 0, 1, \ldots, k - 1,$
(iii) *Jump conditions:* *(vacuous for $n = 0$)*
   $x^{(2k+n-i)}(t_j^+) = \dfrac{p^{(n-i)}(t_j)}{p^{(n)}} x^{(2k+n)}(t_j^+), j = 1, 2, \ldots, N-1; \; i = 1, 2 \ldots, n.$

Proof: The first part of (i) has already been shown. The last integral in (3.6) vanishes because of (3.4). Thus, (3.6) implies that the expression given in (3.7) also vanishes for all $h \in W_2^k$. Now we can construct functions h such that for fixed $i = k - 1 + \ell$ we have $h^{(i)}(t_j) \neq 0$ only in a small neighborhood of a knot t_j, $j = 1, 2, \ldots, N$ and $h^{(i')}(t_j) = 0$ for all $i' \neq i$. This can be acchieved by locally defining $h(t) = (t - t_j)^i$. Only one term in (3.7) remains and conditions (i) and (ii) follow. The Euler-Lagrange equations (3.4) imply $(-1)^k x^{(2k+n-i)}(t_j^+) = -\mu_j p^{(n-i)}(t_j), j = 1, 2, \ldots, N-1; i = 0, 1, \ldots, n$, where μ_j are the Lagrange multipliers. Eliminating these multipliers by using the equations with $i = 0$ yields (iii). ∎

Let us count the number of unknowns and the number of conditions contained in the previous theorem. There are $(N-1)(2k+1) + n(N-1)$ coefficients in x. On the other hand there are altogether $(N-1)(2k+1)$ smoothness, boundary and modified moment conditions. The number of jump conditions is $n(N-1)$, so that $(N-1)(2k+n+1)$ is the number of coefficients and of conditions as well.

It should be noted that the above boundary conditions (ii) are not implied if we prescribe the values $x^{(\ell)}(a), x^{(\ell)}(b)$, $\ell = 0, 1, \ldots, k-1$. The following theorem shows that the necessary conditions of Theorem 2 imply also uniqueness.

Theorem 3. *Let $u, v \in W_2^k$ both assume the same prescribed modified moments (2.2). In addition let u have the properties (i)–(iii) of Theorem 2. Then $f(u) \leq f(v)$ where f is defined in (2.1). If $N \geq k+n+1$ equality holds only for $u = v$.*

Proof: We have $0 \leq f(v-u) = f(v) - f(u) - 2\int_a^b (v^{(k)}(t) - u^{(k)}(t))u^{(k)}(t)\,dt$. Integrating by parts, we obtain $\int_a^b (v^{(k)}(t) - u^{(k)}(t))u^{(k)}(t)\,dt = \sum_{j=1}^{N-1}[A(u, v-u, j) + \int_{t_j}^{t_{j+1}} (v(t) - u(t))u^{(2k)}(t)\,dt]$ where $A(u, v-u, j)$ is defined in (3.5). Because of the boundary conditions (ii) of Theorem 2 and $v - u \in C^{k-1}$ the sum involving $A(u, v-u, j)$ vanishes. The Euler-Lagrange equations (3.4) can be used (since they are implied by the jump conditions (iii)) to write $\int_{t_j}^{t_{j+1}} (v(t) - u(t))u^{(2k)}(t)\,dt = (-1)^{k+1}\mu_j \int_{t_j}^{t_{j+1}} (v(t) - u(t))p(t)\,dt = 0$, because u, v have the same prescribed modified moments. Thus, $0 \leq f(v-u) = f(v) - f(u)$ implying $f(u) \leq f(v)$. Let $f(u) = f(v)$, then $f(v-u) = 0$, which means that $u^{(k)} = v^{(k)}$ piecewise on the intervals $[t_j, t_{j+1}]$. It follows that $u = v + q$, $q \in \Pi_{k-1}$ in the subintervals. However, $q = u - v \in C^{k-1}$ and therefore $q \in \Pi_{k-1}$ globally. Our condition is now reduced to $\int_{t_j}^{t_{j+1}} p(t)q(t)\,dt = 0$, $j = 1, 2, \ldots, N-1$. Let $\pi' = pq$, then it follows that $\pi \in \Pi_{n+k}$ and $\pi(t_1) = \pi(t_2) = \cdots = \pi(t_N)$. If $N \geq n+k+1$ we obtain $\pi = \text{const}$ and $\pi' = pq = 0$ implying $q = 0$ and $u = v$. ∎

§4. Area Matching Splines

Let us first construct a quadratic area matching C^1-spline x. We introduce $x_j := x(t_j), x'_j := x'(t_j)$ as unknowns and define $h_j := t_{j+1} - t_j, \Delta x_j := x_{j+1} - x_j, r_j h_j := \alpha_j$ where α_j occurs in (2.2). We introduce further $S_j := \text{Res}\, x\big|_{[t_j, t_{j+1}]}$ and write it in Newton-form with respect to the knot sequence t_j, t_{j+1}, t_{j+1}. We obtain $S_j(t) = x_j + b_j(t - t_j) + c_j(t - t_j)(t - t_{j+1})$ where $b_j = x[t_j, t_{j+1}] = \Delta x_j/h_j$ and $c_j = x[t_j, t_{j+1}, t_{j+1}] = (x'_{j+1} - \Delta x_j/h_j)/h_j$ are the corresponding divided differences. This form already guarantees that $x \in C[a, b]$ regardless of the choice of x_j and x'_j. We have two types of equations, namely smoothness and area matching conditions: $x'_j + x'_{j+1} = 2\Delta x_j/h_j$, $j = 2, 3, \ldots, N-1$, and $(2x_j + 4x_{j+1})/h_j - x'_{j+1} = 6r_j/h_j$, $j = 1, 2, \ldots, N-1$. In addition we have the two boundary conditions (ii) from Theorem 2, namely

$S_1'(t_1) = 0, S_{N-1}'(t_N) = x_N' = 0$. By using the second equations one can eliminate the quantities x_j', x_{j+1}' from the first equations. One obtains $2x_1/h_1 + x_2/h_1 = 3r_1/h_1$, $x_{j-1}/h_{j-1} + 2(1/h_{j-1} + 1/h_j)x_j + x_{j+1}/h_j = 3(r_{j-1}/h_{j-1} + r_j/h_j)$, $j = 2, 3, \ldots, N-1$, $x_{N-1}/h_{N-1} + 2x_N/h_{N-1} = 3r_{N-1}/h_{N-1}$. This is a positive definite $(N \times N)$-system, because it is symmetric and diagonal dominant. This system occurs apart from the boundary conditions and with slightly different meaning in Späth [1990, p. 65, Formula (3.43)]. The boundary conditions (ii) could have been replaced by prescribed values x_1, x_N.

We keep the above introduced notation for $k = 2$, but we use the Taylor form for $S_j(t) = a_j + b_j(t - t_j) + c_j(t - t_j)^2 + d_j(t - t_j)^3 + e_j(t - t_j)^4$, $j = 1, 2, \ldots, N-1$. From the resulting conditions one can eliminate the a_j, d_j, e_j. We define the three $(N-2) \times (N-2)$ matrices

$$\mathbf{A} = \begin{pmatrix} 3h_1 + 2h_2 & h_2 & & & & \\ h_2 & 2(h_2 + h_3) & h_3 & & & \\ & \ddots & \ddots & \ddots & & \\ & & h_{N-3} & 2(h_{N-3} + h_{N-2}) & h_{N-2} & \\ & & & h_{N-2} & 2h_{N-2} + 3h_{N-1} \end{pmatrix},$$

$$\mathbf{B} = \begin{pmatrix} 1/h_2 & -1/h_2 & & & \\ -1/h_2 & 1/h_2 + 1/h_3 & -1/h_3 & & \\ & \ddots & \ddots & \ddots & \\ & & -1/h_{N-3} & 1/h_{N-3} + 1/h_{N-2} & -1/h_{N-2} \\ & & & -1/h_{N-2} & 1/h_{N-2} \end{pmatrix},$$

$$\mathbf{C} = \begin{pmatrix} \frac{9h_1^3 + 4h_2^3}{5} & \frac{7h_2^3}{10} & & & \\ \frac{7h_2^3}{10} & \frac{4(h_2^3 + h_3^3)}{5} & \frac{7h_3^3}{10} & & \\ & \ddots & \ddots & \ddots & \\ & & \frac{7h_{N-3}^3}{10} & \frac{4(h_{N-3}^3 + h_{N-2}^3)}{5} & \frac{7h_{N-2}^3}{10} \\ & & & \frac{7h_{N-2}^3}{10} & \frac{4h_{N-2}^3 + 9h_{N-1}^3}{5} \end{pmatrix}.$$

The system to be solved is then

$$\begin{pmatrix} \mathbf{C} & -\mathbf{A} \\ \mathbf{A} & \mathbf{B} \end{pmatrix} \begin{pmatrix} \mathbf{d} \\ \mathbf{b} \end{pmatrix} = \begin{pmatrix} -\mathbf{r} \\ \mathbf{0} \end{pmatrix}, \tag{3.8}$$

where $\mathbf{d} = (d_2, d_3, \ldots, d_{N-1})^T$, $\mathbf{b} = (b_2, b_3, \ldots, b_{N-1})^T$, $\mathbf{r} = 6(r_2 - r_1, r_3 - r_2, \ldots, r_{N-1} - r_{N-2})^T$, $c_1 = d_1 = d_N = 0$, $b_2 - b_1 = d_2 h_1^2$, $b_N - b_{N-1} = -d_{N-1} h_{N-1}^2$. The remaining unknowns have to be computed according to the following equations: $a_j = r_j - (b_{j+1} + 2b_j)h_j/6 + (7d_{j+1} + 8d_j)h_j^3/60$; $c_j h_j = (b_{j+1} - b_j)/2 - (d_{j+1} + 2d_j)h_j^2/2$; $e_j h_j = (d_{j+1} - d_j)/4$, $j = 1, 2, \ldots, N-1$.

Theorem 4. Let $N \geq 3$. Then the system (3.8) is regular.

Proof: All three matrices are symmetric, $\mathbf{A}$ and $\mathbf{C}$ are diagonal dominant, hence, positive definite, $\mathbf{B}$ is positive semidefinite (and singular because the

columns add up to zero) as can be shown directly. From the first block of
(3.8) we obtain $\mathbf{d} = \mathbf{C}^{-1}(\mathbf{Ab} - \mathbf{r})$. If we insert that into the second block
we obtain $(\mathbf{AC}^{-1}\mathbf{A} + \mathbf{B})\mathbf{b} = \mathbf{AC}^{-1}\mathbf{r}$. Since $\mathbf{C}^{-1}$ is also positive definite,
it follows that $\mathbf{AC}^{-1}\mathbf{A} = \mathbf{A}^T\mathbf{C}^{-1}\mathbf{A}$ and therefore also $\mathbf{AC}^{-1}\mathbf{A} + \mathbf{B}$ are
positive definite. Thus, also the second block equation and the whole system
are uniquely solvable. ■

It should be noticed, that Späth [1990, 189–191] gives the corresponding
solution in terms of our Problem 2. However, due to a sign error, his regularity
proof fails. In Figure 1 we present graphs of a quadratic C^1- and a quartic
C^3-histosplines where the quadratic spline can be identified by the zero slopes
at the endpoints.

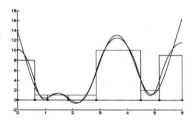

Fig. 1. Area matching splines: Quadratic C^1- and quartic C^3-histosplines.

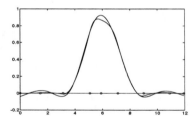

Fig. 2. Modified moment splines: cubic C^1 and quintic C^3 splines.

§5. Splines with Prescribed First Modified Moments

We assume that $p(t) = A + Bt$ is given with $B \neq 0$. The resulting linear
system which consists of the necessary conditions of Theorem 2 and of the
modified moment conditions (2.2) is of order $2(k + 1)(N - 1)$. Examples are
presented in Figure 2. For these examples we chose the knots at random
and prescribed modified moments which coincide with the moments of the
Gaussian bell curve. More precisely, we prescribed the values

$$\alpha_j = r_j h_j = \int_{t_j}^{t_{j+1}} (M - t) \exp(-0.5(t - M)^2)\, dt = \exp(-0.5(t - M)^2)\big|_{t_j}^{t_{j+1}},$$

where M is the midpoint of $[a, b]$. The polynomial p was chosen as $p(t) = M - t$. As one can see, the resulting splines are also bell shaped. Details of the structure of the linear system to be solved are given in Oberle & Opfer [1994].

Acknowledgments. The first draft of this paper was written during a very stimulating Workshop on iterative methods for sparse & structured problems at the Institute for Mathematics and its Applications (IMA) in Minneapolis, 1992. The use of modified moments (rather than moments) was suggested by Gene Golub during that meeting. The figures were drawn with MATLAB 4.0.

References

1. C. de Boor & G. J. Fix, Spline approximation by quasi-interpolants, J. Approx. Theory **8** (1973), 19–45.
2. P. Funk, *Variationsrechnung und ihre Anwendung in Physik und Technik*, 2. Aufl., Springer, Berlin, Heidelberg, New York, 1970.
3. E. Hewitt & K. Stromberg, *Real and Abstract Analysis*, 2nd pr., Springer, Berlin, Heidelberg, New York, 1969.
4. J. C. Holladay, Smoothest curve approximation, Math. Tables Aids Comput. **11** (1957), 233–243.
5. D. G. Luenberger, *Optimization by Vector Space Methods*, Wiley, New York, London, Sydney, Toronto, 1969.
6. G. Nürnberger, *Approximation by Spline Functions*, Springer, Berlin, 1989.
7. Oberle, H.J., and G. Opfer, Splines with prescribed modified moments, Hamburger Beiträge zur Angewandten Mathematik, Reighe A. 78, 1994.
8. G. Opfer & H. J. Oberle, The derivation of cubic splines with obstacles by methods of optimization and optimal control, Numer. Math. **52** (1988), 17–31.
9. J. W. Schmidt & W. Heß, Shape preserving C^2-spline histopolation, J. Approx. Theory **75** (1993), 325–345.
10. I. J. Schoenberg, On interpolation by spline functions and its minimal properties, in P. L. Butzer & J. Korevaar (eds): *Über Approximationstheorie*, International Series of Numerical Mathematics vol. 5, 1964, Birkhäuser, Basel, 109–129.
11. I. J. Schoenberg, Splines and histograms, in A. Meir & A. Sharma (eds.): *Spline Functions and Approximation Theory*, International Series of Numerical Mathematics vol. 21, 1973, Birkhäuser, Basel, 277–327.
12. H. Späth, *Eindimensionale Spline-Interpolations-Algorithmen*, Oldenbourg, München, Wien, 1990.

Hans Joachim Oberle & Gerhard Opfer
Universität Hamburg
Institut für Angewandte Mathematik
Bundesstraße 55, D-20146 Hamburg, GERMANY
opfer@math.uni-hamburg.de

G²-Continuous Cubic Algebraic Splines and Their Efficient Display

M. Paluszny and R. R. Patterson

Abstract. A method is proposed to produce a locally convex G^2-continuous plane cubic algebraic spline which interpolates a sequence of points. This spline is controlled via global parameters, and may be fine tuned using local shape handles. A fast algorithm is offered for the digital generation of this spline. It uses the fact that each of its segments, with respect to some system of barycentric coordinates is given by the same numerical equation.

§1. Introduction

The term *algebraic spline* or A-spline refers to splines which are pieced together from algebraic curves. These were introduced by Sederberg [13,14] and were further studied in [1,7,8,9]. They are cubic if each piece is a segment of an algebraic cubic, which means that it has an equation of the form

$$\sum_{i+j\leq 3} a_{ij}x^i y^j = 0$$

in Euclidean coordinates, or

$$\sum_{i+j+k=3} a_{ijk}s^i t^j u^k = 0$$

in barycentric coordinates, where a_{ij} and a_{ijk} are real constants and $s + t + u = 1$. G^2-continuity means that adjacent segments of the spline have the same tangent lines and curvatures at the joint points. The algebraic splines discussed here are locally convex; that is, free of inflection points.

This paper has two purposes. The first is to describe some defaults that can be assumed when generating the original G^2-spline. This default spline is created as the designer clicks off a sequence of points to be interpolated.

Curves and Surfaces in Geometric Design　　　　　　　　　　　353
P. J. Laurent, A. Le Méhauté, and L. L. Schumaker (eds.), pp. 353–359.
Copyright ⓒ 1994 by A K PETERS, Wellesley, MA
ISBN 1-56881-039-3.

It can be produced rapidly and can be deformed globally in two ways. The second purpose is to describe a tracking algorithm that takes advantage of the barycentric coordinate description of each segment of the curve.

The two global deformation methods are independent and have the following effects. The first pulls the spline into the corners of the control polygon, such as the dashed curve in Figure 1. The second pulls it against the sides of the control polygon, such as the dotted curve in Figure 1.

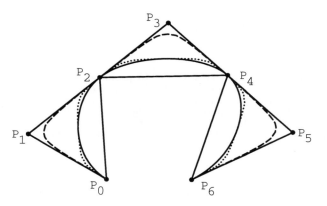

Fig. 1. Global controls of the default spline.

The reason why the default spline can be produced rapidly is that each segment has exactly the same algebraic equation. Under either of the two modes of spline deformation the barycentric coordinates of all the points of the spline may be recomputed by recalculating only those of any one of its segments. This allows for faster tracing of the curve.

Once the spline has been initially defined and perhaps manipulated globally, it can also be controlled locally, each segment individually, preserving G^2-continuity and not perturbing any of the other segments. For this, geometric shape handles have been developed [7,8,9].

§2. Construction

Given three affinely independent points of the plane, $P_0(x_0, y_0), P_1(x_1, y_1)$ and $P_2(x_2, y_2)$, it was observed by Sederberg [13, 14] that an algebraic curve passing through P_0 and P_2 with tangent lines $u = 0$ and $s = 0$ at these points (see Figure 2), has the following form

$$as^2u + bsu^2 - cst^2 - dt^2u + estu - ft^3 = 0, \tag{1}$$

where s, t, u are the barycentric coordinates associated to triangle P_0, P_1, P_2. Here the coefficients have been changed from the a_{ijk}-notation to avoid the proliferation of subscripts. The coefficients can be thought of as associated to the trisection points labelled in Figure 2, and for the particular function

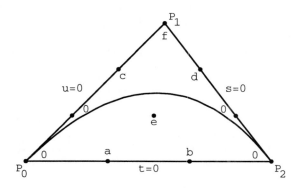

Fig. 2. Graph of $as^2u + bsu^2 - cst^2 - dt^2u + estu - ft^3 = 0$.

graphed, $a = b = 4, c = d = 1, e = f = 0$. The signs are chosen merely for convenience.

Given a control triangle $P_0P_1P_2$, additional conditions have to be imposed on the coefficients a, b, c, d, e and f so that the algebraic cubic given by (1) has a segment joining P_0 and P_2 that is convex and contained inside the triangle. It is proved in [1] and [8] that a sufficient condition is $a, b, c, d > 0$ and $f = 0$. This condition is assumed throughout the rest of this paper, thus the algebraic cubics studied are given in barycentric coordinates by the formula

$$as^2u + bsu^2 - cst^2 - dt^2u + estu = 0 \qquad a, b, c, d > 0. \tag{2}$$

It is proved in [8] that these coefficients can be expressed

$$
\begin{aligned}
a &= \beta t_0^2 u_0 \\
b &= \beta s_0 t_0^2 \\
c &= (1 - \beta)s_0 u_0^2 \\
d &= (1 - \beta)s_0^2 u_0 \\
e &= 2(1 - 2\beta)s_0 t_0 u_0
\end{aligned}
\tag{3}
$$

where $\beta \in (0, 1)$ and (s_0, t_0, u_0) are the barycentric coordinates of an additional interpolated point B_0 in the triangle.

A designer begins constructing a spline by choosing a triangle $P_0P_1P_2$. The default curve is displayed, for which $1/2$ can be used for β and the barycenter $(1/3, 1/3, 1/3)$ can be used for B_0.

The next point selected by the designer is P_4, which must lie on the same side of the P_1P_2 line as P_0. The next segment of the spline will begin at P_2 and end at P_4. It will have a control triangle $P_2P_3P_4$, for some point P_3 which must be computed. The equation of the second segment in terms of the barycentric coordinates of triangle $P_2P_3P_4$ will also be given by equation (2).

It is clear that to guarantee continuity of the tangent line it is enough to constrain P_3 to lie on the ray from P_1 through P_2. In order for the spline to be G^2 the curvatures must agree at P_2.

The curvatures of the curve (2) in triangle $P_0 P_1 P_2$ at the endpoints are given by

$$\kappa_0 = 4\frac{c}{a}\frac{\Delta_0}{g_0^3}, \quad \kappa_1 = 4\frac{d}{b}\frac{\Delta_0}{g_1^3},$$

where g_i is the length of segment $P_i P_{i+1}$ and Δ_i is the area of triangle $P_{2i} P_{2i+1} P_{2i+2}$ [7]. Thus the curvature of the default curve at P_2 will be some known positive number κ_1. To achieve G^2-continuity at P_2 the curvature of the second segment at P_2 must equal that of the first segment. This means

$$4\frac{d}{b}\frac{\Delta_0}{g_1^3} = 4\frac{c}{a}\frac{\Delta_1}{g_2^3}.$$

Because of (3), $\dfrac{d}{b} = \dfrac{c}{a}$ for any choice of β and B_0. It is always possible to choose P_3 along the ray so that the equation is satisfied because $\Delta_1 g_2^{-3}$ is proportional to g_2^{-2}, and g_2 can vary from 0 to infinity.

In this way a G^2-continuous default spline is constructed that interpolates the points $P_0, P_2, P_4, P_6, \ldots$

§3. Global Control Parameters

With the points chosen, the default spline is controlled by the five coefficients a, b, c, d and e. These coefficients depend on the parameters β and $B_0(s_0, t_0, u_0)$. One can use β and t_0 as global shape control parameters, letting $s_0 = u_0 = (1 - t_0)/2$.

It is easy to show that the curve passes through the point $B_i(s_0, t_0, u_0)$ in the i-th triangle with tangent line parallel to the line through P_{2i} and P_{2i+2}. The spline lies in the region determined by these trapezoids (see Figure 3).

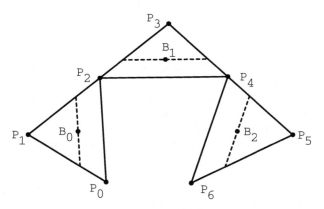

Fig. 3. The interpolated points and the trapezoids.

As t_0 is changed, the corresponding points B_i move collectively, each one in its triangle. In particular, as t_0 tends to 1, each B_i tends to the corner P_{2i+1}, pulling the spline closer to the control polygon — see the dashed curve in Figure 1.

As β tends to 1, the default spline tends to fit more tightly the sides of the sequence of trapezoids — see the dotted curve in Figure 1.

Once global adjustments have been made, [7] shows how local adjustments can be done. Each point B_i can be placed anywhere in its triangle, the slope at B_i can be varied (within limits), and the curve can be pulled toward the trapezoid, all without affecting neighboring curves. In addition the curvature at any of the points P_{2i} can be changed, affecting only the two curves which join there. However once local control methods have been used it becomes necessary to track each segment of the curve separately.

§4. The Tracking Algorithm

For the display of algebraic cubics, several non-rational parametrization methods have been proposed [1,10], but the most commonly utilized methods are tracking algorithms [2,3,4,5,6,11,12]. Here is a new tracking scheme based on the barycentric coordinate description of the curves. The curve can be traced by joining the points where it crosses a triangular grid. Repeated use is made of the fact that because $a, b, c, d > 0$, the function

$$F(s, t, u) = as^2u + bsu^2 - cst^2 - dt^2u + estu$$

is positive at points in the triangle below the curve $F(s, t, u) = 0$ and negative above.

Tracking begins at P_0 by taking a step to the right to a grid point A. One is now in the generic situation encountered while drawing up the left side of the arc: located at the grid point $A(i_0, j_0, k_0)$, on or just below the curve, so that the grid line to the upper left crosses the curve before the next grid point B; see the left side of Figure 4. The curve intersection with the grid line $i = i_0$ can then be found by recursive binary subdivision of AB or by Newton's method. In either case the equation (to be solved for j) is

$$F(i_0, j, 1 - i_0 - j) = 0.$$

If Newton's method is used, the derivative required is merely

$$\frac{dF}{dj} = \frac{\partial F}{\partial t} - \frac{\partial F}{\partial u}.$$

To find the next point, after returning to A, a step is taken to the right and as many steps to the upper left as are necessary to locate the next pair of points A' and B' with $F(A') \geq 0$ and $F(B') < 0$.

If the step to the right crosses the curve it means the top of the arch has been reached. One returns to A and begins to work down the right side of

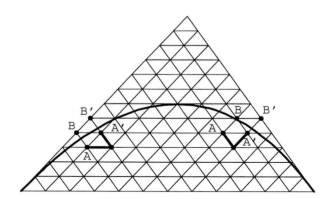

Fig. 4. The tracking algorithm.

the curve. This time from a grid point A one step is taken to the lower right and as many steps as necessary to the upper right to locate A' and B'; see the right side of Figure 4. The intersection point now lies on the line $k = k_0$. The equation to be solved for j is

$$F(1 - j - k_0, j, k_0) = 0.$$

Tracking finishes when the step to the lower right makes j negative. The curve is extended to P_2.

To display the segment the Euclidean coordinates of each of the points must be computed from

$$(x, y) = s(x_0, y_0) + t(x_1, y_1) + u(x_2, y_2).$$

To display the default spline (before local adjustments have been made) it is sufficient to track just one segment and store the barycentric coordinates in a list. Then only the above conversion to Euclidean coordinates needs to be done for each segment.

Acknowledgements. The authors wish to thank P. Marsilio, J. Santana, and O. Villasmil for the development of graphics code to test the algorithms.

References

1. Bajaj, C. and G. Xu, A-Splines: Local interpolation and approximation using C^k-continuous piecewise real algebraic curves, technical report CSD-TR-92-095, Purdue University, 1992.
2. Bajaj, C., C. M. Hoffmann, J. E. Hopcroft and R. E. Lynch, Tracing surface intersections, Computer-Aided Geom. Design **5** (1988), 285–307.

3. Belser, K., Comment on "An improved algorithm for the generation of nonparametric curves", IEEE Trans. on Comput., **C-25** (1976), 103.

4. Chandler, R. E., A tracking algorithm for implicitly defined curves, IEEE Comput. Graph. Appl. **8** (1988), 83–89.

5. Dobkin, D. P., S. V. F. Levy, W. P. Thurston and A. R. Wilks, Contour tracing by piecewise linear approximations, ACM Trans. on Graphics, **9** (1990), 389–423.

6. Jordan, B. W., W. J. Lennon and B. D. Holm, An improved algorithm for the generation of nonparametric curves, IEEE Trans. on Comput., **C-12** (1973), 1052–1060.

7. Paluszny, M. and R. R. Patterson, A family of curvature continuous cubic algebraic splines, preprint.

8. Paluszny, M. and R. R. Patterson, A family of tangent continuous cubic algebraic splines, ACM Trans. on Graphics, **12** (1993), 209–231.

9. Paluszny, M. and R. R. Patterson, Curvas algebraicas de grado tres en modelación geométrica, Quinta Escuela Venezolana de Matemáticas, Univ. de los Andes, Mérida, Venezuela, 1992.

10. Patterson, R. R., Parametrizing and graphing nonsingular cubic curves, Comput. Aided Design **20** (1988), 615–623.

11. Ramot, J., Nonparametric curves, IEEE Trans. on Comput., **C-25** (1976), 103–104.

12. Rubin, F., Generation of nonparametric curves, IEEE Trans. on Comput., **C-25** (1976), 103.

13. Sederberg, T. W., Algorithm for algebraic curve intersection, Comput. Aided Design **21** (1989), 547–554.

14. Sederberg, T. W., Planar piecewise algebraic curves, Comput. Aided Design **1** (1984), 241-255.

M. Paluszny
Escuela de Física y Matemáticas
Universidad Central de Venezuela
Apartado 47809, Los Chaguaramos
Caracas 1041-A, VENEZUELA
paluszny@dino.conicit.ve

R. R. Patterson
Department of Mathematics
Indiana University – Purdue University at Indianapolis
Indianapolis, IN 46202-3216
rpatters@indyvax.iupui.edu

Splines in a Topological Setting

Sinésio Pesco and Geovan Tavares

Abstract. In this paper we describe a topologically based data structure, called *Handle-Edge-Spline*, suited to building interactive geometric modeling environments, where surfaces represented in different forms should interact with each other. Several structural operators are introduced dealing with different types of splines.

§1. Introduction

Since the seminal paper of Baumgart [1], data structures based on topological methods have been advocated as an essential step in flexible computational design environments. The introduction of Boundary Representation (B-Rep) and its associated structural operators, Euler Operators, has established a paradigm to develop algorithms in the area. For a recent survey see [7].

B-Rep and Euler operators have several shortcomings which restrict the range of surfaces they can be applied to:

1) The surface has to be the boundary of a solid, and thus embedded in 3-space.

2) Combined with splines this surface representation is severely limited as pointed out by Chyokura [4].

3) Surface self-intersections are not allowed.

4) Some Euler operators are not clearly topological (e.g. to create and destroy a genus on a surface).

In this paper we place splines in a topological setting by extending Mäntylä's *Half-Edge* data structure [8], to the *Handle-Edge-Spline* data structure; also we introduce new structural operators, called Morse operators [3]. One of the main features of this structure is that it can handle surfaces represented by several modeling techniques like splines, implicit, CSG, etc., in the same environment where interactivity, visualization, design and numerical control can be put together in a unique computational scheme.

A general reference in this area is Chyokura [4], who uses B-Rep and Euler operators to put splines in a topological context.

Curves and Surfaces in Geometric Design
P. J. Laurent, A. Le Méhauté, and L. L. Schumaker (eds.), pp. 361–368.
Copyright © 1994 by A K PETERS, Wellesley, MA
ISBN 1-56881-039-3.

§2. Handlebody Decomposition

The design of the Handle-Edge data structure and its structural operators
introduced in [3] is based on Morse theory [9]. The following theorem is the
main tool we use in applying Morse theory for surfaces.

Theorem. *An orientable compact surface with boundary can be obtained
by attaching handles to a finite sequence of compact surfaces with boundary,
starting with the two-dimensional disc.*

A proof of this theorem can be found in [6] for $C^r, r \geq 1$ surfaces. For the
piecewise-linear case see [11].

The surfaces we work with are defined in the following way. $S \subset \mathbb{R}^m$ is a
surface with boundary if it is provided with a cell decomposition of dimension
2, that satisfy the following conditions:

1) Every edge in S is incident to one or two faces;
2) The link of every vertex in S is homeomorphic to either an interval
   or a circle.

The cells containing a given vertex v in a cell decomposition is called the *star*
of v, denoted st(v). Its boundary $\partial(\text{st}(v))$, is called the *link* of v and denoted
link(v).

§3. Data Structure and Adjacency Relations

The *Handle-Edge* data structure we introduce now is associated with the
intrinsic cell decomposition topology of orientable surfaces with boundary. It
extends the Half-Edge data structure [8] to surfaces with boundary. Also, it
is related to the data structure introduced in [10]. Some of the new features
of the Handle-Edge data structure are:

1) The surface boundary is an essential part of the structure;
2) It does not use the concept of shell, which can now be looked upon
   as a geometric attribute;
3) It does not depends on the dimension of the space where the surface
   is embedded.

The basic nodes of the Handle-Edge data structure reflect the adjacency
relations inherent to a cell decomposition of the surface. They are the following, given in a top-down scheme:

> *Surface* - is an instance of the surface with boundary represented by
> its faces, boundary curves, edges, and vertices.
> *Face* - is represented by a cycle of half-edges which is its boundary.
> For a cycle we mean a finite sequence of half-edges, with the last and
> first half-edges coinciding.
> *Boundary Curve* - is a cycle of boundary edges.
> *Half-Edge* - is an element of a face given by its initial vertex and its
> succeeding and preceding half-edge on the face's cycle.
> *Edge* - two identified half-edges. If one of the half-edges is empty, it
> is called a *Boundary Edge*.

Vertex - represents the vertex of the surface. If it belongs to a *Boundary Edge*, it is called a *Boundary Vertex*.

Figure 3.1 describes the handle-edge data structure, where the arrows indicate the adjacency relations between nodes elements:

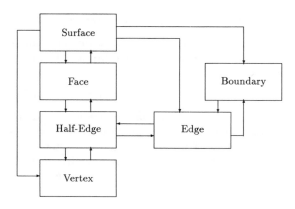

Fig. 3.1. Handle-Edge Data Structure.

§4. Morse Operators

Using the handlebody decomposition for surfaces and the Handle-Edge data structure we define several structural operators. The following notation is used : s for surfaces, f for faces, e for edges, ∂e for boundary edges, ∂c for boundary curves, and v for vertices.

Morse operators have also the purpose of maintaining the topological consistency of the surface to be modeled. They build and unbuild surfaces, faces, boundary curves, edges and vertices. Each operator is identified by a name which syntax has the form: $MxKy$, M for Make and K for Kill and x, y could be: $s, f, \partial c, \partial e, e, v$.

4.1. Instantiation Operator

Making a Disc: $M\{sf\partial ev\partial c\}$.

This operator builds a surface of a disc type, i.e., one face, one boundary curve, one boundary edge, one vertex, and Euler-Poincaré characteristic equal to one.

It is interesting to observe here that, by using the Handle-Edge data structure, the operator $M\{sf\partial ev\partial c\}$ creates all of its nodes, i.e., all topological entities are instantiated.

4.2. Boundary Operators

In order to have cells with an arbitrary number of edges and vertices, we need a second operator which adds to the boundary curve of this cell, one boundary vertex and one boundary edge, splitting one boundary edge. The operator does not change the Euler-Poincaré characteristic.

Adding a vertex and an edge to a boundary curve: $M\{v\partial e\}$

To represent a surface with an arbitrary number of boundary curves and genus, we need operators to make different types of attachment of a handle. The first type is:

Attaching a handle to the same boundary component: $K\{vv\partial e\partial e\}M\{e\partial c\}$

The next operator creates a genus on the surface. Again, in this case, the handle cannot be attached unless each of the boundary curves involved has at least two boundary edges. This operator gives a local control on the genus creation, although the genus of a surface is globally defined. The reason for this is that all of the Morse operators act only on the elements on the boundary.

Attaching a handle to two boundary components on the same surface component: $K\{vv\partial e\partial e\partial c\}M\{eg\}$

In the attachment of a handle to two boundary components on different surface components, each boundary curve involved must have at least two boundary edges.

Attaching a handle to two boundary components on different surface components: $K\{vv\partial e\partial e\partial cs\}Me$

4.3. Closing Operators

With the above presented operators, we can only model surfaces with boundary. In order to obtain a surface without boundary, we have to define an operation to close the surface. First, we glue adjacent boundary edges. Then, we repeat the process again and again until we have only two boundary vertices left, and, consequently, two boundary edges. After that, we glue the remaining boundary edges together.

Gluing two adjacent boundary edges: $K\{v\partial e\partial e\}Me$

Closing a boundary component: $K\{\partial e\partial e\partial c\}Me$

4.4. Inverse Morse Operators

The inverse of the Morse operators can be defined by exchanging M with K in the definitions given above.

§5. The Handle-Edge-Spline Data Structure

In order to to deal with spline generated surfaces, we have to introduce two new nodes in the original Handle-Edge data structure: *Patch* and *Subpatch*. This extended data structure is called *Handle-Edge-Spline*.

The surface is composed by patches which in turn are subdivided into several subpatches. These subpatches are used for example to guarantee C^1 or Visual Continuity [2, 5]. Data structure as this one can deal also with irregular meshes [4].

The data structure should be flexible enough to incorporate one of the main features of the use of splines in computer aided geometric design, the interactivity of surface design via the control polygon. The nodes Patch and Subpatch are structural connection between the control polygon and the spline surface. Figure 5.1 shows the *Handle-Edge-Spline* data structure.

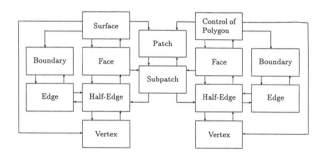

Fig. 5.1. Handle-Edge-Spline Data Structure.

5.1. Insertion Operators

Now we define several insertion operators to assure that the spline surface is C^1 or Visual Continuous. They are tailored to apply to two classes of splines where we actually used our structure: triangular and quadrangular.

Insertion of a vertex in the boundary of a face: apply sucessively the operators $K\{v\partial e\partial e\}Me$, $Mv\partial e$, $Mv\partial e$.

This topological operation is equivalent to the Euler operator Mev [4].

In Figure 5.2 we insert several vertices around the boundary of surface of disc type (we schematically represent it as a triangle). The arrows inside the triangle indicate half-edges.

Next we display schematically two operators to deal with subdivision of a face into subfaces: *inserting a vertex inside a face with no boundary vertices* and *inserting a vertex inside a face with boundary vertices*.

They use the operator Mev defined above and the Euler operator Mef defined as the sequence of operators $Msf\partial ev\partial c$, $Mv\partial e$, $K\{vv\partial e\partial e\partial cs\}Me$.

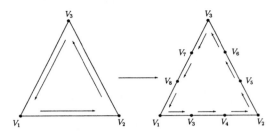

Fig. 5.2. Insertion of vertices along the boundary of a face.

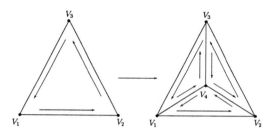

Fig. 5.3. Inserting a vertex inside a face with no boundary vertices.

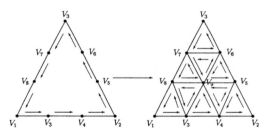

Fig. 5.4. Inserting a vertex inside a face with boundary vertices.

§6. Examples

The first example we give is the torus. It is built with triangular splines using the topological setting devised above and visual continuity [5].

Our last example is the surface with boundary *Oseberg Ship* generated from real data taken at the Viking ship museum in Oslo. The control polygon has been modified with the insertion and deletion of points to enhance the picture quality.

Acknowledgements. We would like to thank Tom Lyche from the University of Oslo for providing us with the *Oseberg Ship* data set, Hélio Lopes for several conversations around the ideas of this paper and Gustavo Nonato for help in handling the ship data. Supported partially by CNPq and IBM/Brazil.

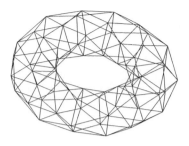

Fig. 6.1. Control Polygon.

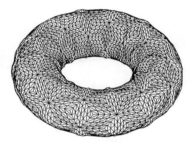

Fig. 6.2. Control Polygon with Visual Continuity Points Inserted.

Fig. 6.3. Triangulated Torus.

References

1. B. Baumgart, A Polyhedron representation for computer vision. AFIPS Proceedings **44** (1975), 589–596.
2. W. Boehm, G. Farin, J. Kahmann, A survey of curve and surface methods in CAGD. Computer-Aided Geom. Design **1** (1984), 1–60.

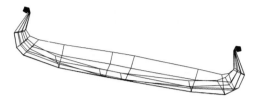

Fig. 6.4. Control Polygon.

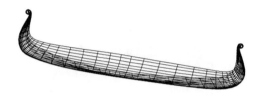

Fig. 6.5. Oseberg Ship.

3. A. Castelo, H. Lopes, G. Tavares, Handlebody representation for surfaces and morse Operators. In J. Warren, Proc. on *Curves and Surfaces in Computer Vision and Graphics III*. Proc. SPIE, (The International Society for Optical Engineering), Boston, (1992), 270–283.

4. H. Chyokura, *Solid Modeling with Designbase*. Adison-Wesley, 1988.

5. G. Farin, Triangular Bernstein-Bézier patches, Computer-Aided Geom. Design **3**, (1986), 83–127.

6. A. Gramain, *Topologie des Surfaces*. Presses Universitaires de France, 1971.

7. C. M. Hoffmann, G. Vaněček, Fundamental techniques for geometric modeling. In C. T. Leondes (ed.) *Advances in Control and Dynamics. Control and Dynamic Systems*, vol. 48. Academic Press, (1991), 101–165.

8. M. Mäntylä, *An Introduction to Solid Modeling*. Computer Science Press, 1988.

9. J. Milnor, *Morse Theory*. Annals of Mathematics Study **51**. Princeton University Press, 1963.

10. J. R. Rossignac, M. A. O'Connor, SGC: A dimension-independent model for pointsets with internal structures and incomplete boundaries. In M. J. Wozny, J. U. Turner, K. Preiss (eds.), *Geometric Modeling for Product Engineering*. North-Holland, 1990, 145–180.

11. C. P. Rourke, B. J. Sanderson, *Introduction to Piecewise-Linear Topology*. Springer-Verlag, 1982.

A Characterization of Connecting Maps
as Nonlinear Roots of the Identity

J. Peters

Abstract. In order to define the smoothness of a piecewise polynomial surface, the domains of adjacent pieces must be related to one another by connecting maps; such maps reparametrize the surface pieces by mapping the domains of adjacent pieces to a joint domain. We characterize the subclass of connecting maps that can be used to surround a point by three or more pieces. The characterization of connecting maps for second order continuity suggests a lower bound on the degree of any curvature continuous surface assembled from polynomial pieces.

§1. Motivation

A popular approach to modeling smooth parametric surfaces is to assemble them from polynomial patches $p_k : \Omega_k \subset \mathbb{R}^2 \mapsto \mathbb{R}^3$. To determine the smoothness of transition from one patch to its neighbor, the domains of adjacent patches must locally be mapped to a joint domain so that directions of differentiation are well defined. Thus, connecting maps $\phi_k : \Omega_{k-1} \mapsto \Omega_k$ play a central role in the construction of smooth parametric surfaces, affecting for example the polynomial degree and the shape of the surfaces. Of particular interest for constructions is the subclass of connecting maps that can be used to smoothly surround a point by three or more patches. The paper characterizes this subclass.

When three or more patches join smoothly at a common point, the pairwise continuity constraints between the patches form a circular system. Correspondingly, the composition of all n connecting maps must map any initial domain to itself and must agree with the identity map, id, at the preimage of the common point up to the given order of continuity (see [2, Theorem 7.1]). This motivates viewing the connecting maps as roots of the identity. In particular, if all connecting maps at the point act identically on their domains, and hence are indistinguishable except for their subscripts, we call them *uniform roots* of the identity.

Curves and Surfaces in Geometric Design
P. J. Laurent, A. Le Méhauté, and L. L. Schumaker (eds.), pp. 369–376.
Copyright © 1994 by A K PETERS, Wellesley, MA
ISBN 1-56881-039-3.

Based on the characterization of uniform and special nonuniform quadratic roots of the identity, we find that a particular directional derivative of a connecting map between two vertices of different degree cannot be linear. Since this derivative features prominently in the continuity constraints between adjacent patches, a simple argument implies that the formal degree of a curvature continuous surface built from polynomial pieces exceeds the degree of the boundary curves by four.

The paper is organized as follows. Section 2 formalizes the notion of connecting maps and gives a closed form expression for the constraint on their composition. Section 3 looks at linear roots of the identity and identifies uniform linear roots as rotations. Sections 4 and 5 characterize second-order uniform and special nonuniform roots. This characterization is used in Section 6 to derive a lower bound on the degree of curvature continuous piecewise polynomial surfaces.

§2. Roots of the Identity

To formalize the constraint on the connecting maps, let $k = 1..n$ and denote by ϕ_k the connecting map between the domain $\Omega_{k-1} \subset \mathbb{R}^2$ of the $(k-1)$st patch p_{k-1} and the domain Ω_k of the adjacent patch p_k. Circularity implies that $\Omega_0 = \Omega_n$ and that $\phi_n \circ \phi_{n-1} \circ \ldots \circ \phi_1$ maps from Ω_n to Ω_n. Let $D_i := \frac{\partial}{\partial x_i}$ be the derivative in the direction of the ith unit vector e_i. To avoid repetition, we follow the convention that *all functions are evaluated at the origin* $\mathbf{0} := \begin{bmatrix} 0 \\ 0 \end{bmatrix}$ which is the parameter value corresponding to the common point of all patches. Thus

$$J_r \phi := (D_1^m D_2^n \phi)_{m+n \leq r}$$

is an ordered collection of Taylor coefficients of a connecting-map ϕ expanded at $\mathbf{0}$ up to the rth Taylor term. The composition constraint on admissible ϕ_l is

$$J_r \text{ id} = J_r(\circ_{l=1}^n \phi_l) := J_r(\phi_n \circ \phi_{n-1} \circ \ldots \circ \phi_1), \qquad \text{(C)}$$

where $\circ$ is the symbol for composition. Note that by the above convention, both sides of constraint (C) are evaluated at $\mathbf{0}$. We normalize ϕ_l such that $\phi_l(0,0) = \mathbf{0}$ for $l = 1..n$. Thus, for $r = 0$, (C) is

$$\mathbf{0} = \circ_{l=1}^n \phi_l. \qquad (2.0)$$

Denote the components of any connecting map ϕ as $\phi^{[1]}$ and $\phi^{[2]}$ and define $D\phi := \begin{pmatrix} D_1\phi^{[1]} & D_2\phi^{[1]} \\ D_1\phi^{[2]} & D_2\phi^{[2]} \end{pmatrix}$. For $r = 1$ and $i \in \{1, 2\}$, since the derivative of the identity map at $\mathbf{0}$ is the identity matrix and the ith column of the identity matrix is the ith unit vector e_i, the additional constraints are by the chain rule

$$e_i = D_i(\circ_{l=1}^n \phi_l) = (\prod_{l=1}^n D\phi_l)e_i. \qquad (2.1)$$

For $r = 2$, and $i, j \in \{1, 2\}$, since id has no quadratic terms, the chain rule and the product rule yield

$$0 = D_j D_i(\circ_{l=1}^n \phi_l) = \sum_{k=1}^n (\prod_{l<k} D\phi_l) D^2\phi_k((\prod_{l>k} D\phi_l)e_i, (\prod_{l>k} D\phi_l)e_j). \quad (2.2)$$

Here $\langle\rangle$ indicates that each of the two components of the Hessian $D^2\phi_k\langle,\rangle$ is a bilinear form with two vector-valued arguments.

Since Ω_k must share an edge with $\phi_k(\Omega_{k-1})$, it is reasonable to stipulate that Ω_k and $\phi_k(\Omega_{k-1})$ share a coordinate direction corresponding to the common edge. This implies that the edge is traced with a common orientation and parameter v:

$$\phi(0, v) = \begin{bmatrix} v \\ 0 \end{bmatrix} \quad (A_1)$$

and that the transversal derivative (with respect to u) of $\phi^{[2]}$ is constant for varying v:

$$D_1 D_2 \phi^{[2]} = 0. \quad (A_2)$$

§3. Uniform Linear Roots of the Identity

We first consider the case $r = 1$, the characterization of the linear components of the connecting-maps. For now, we assume that the connecting maps are *uniform*, that is $\phi_l = \phi$. Since the neighborhood of the origin is to be covered exactly once, the linear part of ϕ is a rotation by $\theta := 2\pi/n$.

Proposition 3. *If $\phi_l = \phi$, $l = 1..n$, and (A$_1$) holds, then (2.0) and (2.1) hold if and only if*

$$\phi = \begin{bmatrix} \phi^{[1]} \\ \phi^{[2]} \end{bmatrix} := \begin{bmatrix} 2\cos\theta & 1 \\ -1 & 0 \end{bmatrix} \begin{bmatrix} u \\ v \end{bmatrix} + \text{higher order terms}.$$

Proof: The assumption $\phi(0, v) = (v, 0)$ implies $D\phi = \begin{bmatrix} u_1 & 1 \\ u_2 & 0 \end{bmatrix}$. Since $D\phi$ has to have the eigenvalues $e^{\pm\iota\theta}$, $\iota := \sqrt{-1}$, of a rotation matrix,

$$e^{\pm 2\iota\theta} - u_1 e^{\pm\iota\theta} - u_2 = 0$$

must hold. From this constraint, $u_1 = 2\cos\theta$, $u_2 = -1$ follows. ∎

The linear part of ϕ can be diagonalized:

$$D\phi := \begin{bmatrix} 2\cos\theta & 1 \\ -1 & 0 \end{bmatrix} = \begin{bmatrix} -e^{-\iota\theta} & -e^{\iota\theta} \\ 1 & 1 \end{bmatrix} \begin{bmatrix} e^{-\iota\theta} & 0 \\ 0 & e^{\iota\theta} \end{bmatrix} \begin{bmatrix} -e^{-\iota\theta} & -e^{\iota\theta} \\ 1 & 1 \end{bmatrix}^{-1}$$

implying that

$$D(\circ_{k=1}^{l}\phi) = \prod_{k=1}^{l} D\phi = \begin{bmatrix} 2\cos\theta & 1 \\ -1 & 0 \end{bmatrix}^{l}$$

$$= \begin{bmatrix} -e^{-\iota\theta} & -e^{\iota\theta} \\ 1 & 1 \end{bmatrix} \begin{bmatrix} e^{-\iota\theta} & 0 \\ 0 & e^{\iota\theta} \end{bmatrix}^{l} \begin{bmatrix} -e^{-\iota\theta} & -e^{\iota\theta} \\ 1 & 1 \end{bmatrix}^{-1}$$

$$= \frac{1}{F(1)} \begin{bmatrix} F(l+1) & F(l) \\ -F(l) & -F(l-1) \end{bmatrix},$$

where

$$F(l) := e^{\iota\theta l} - e^{-\iota\theta l} = 2\iota\sin(\theta l).$$

Example. If $l = n$, then $\theta l = 2\pi$, and hence

$$D(\circ_{k=1}^{n}\phi) = \begin{bmatrix} 1 & 0 \\ 0 & 1 \end{bmatrix}$$

as required for a first-order connecting-map by condition (2.1). ∎

§4. Uniform Quadratic Roots of the Identity

We now apply the calculus of the previous section to the case $r = 2$. In particular, we want to characterize the uniform nonlinear connecting maps that satisfy (2.0)–(2.2).

Theorem 4. *If $\phi_l = \phi$, $l = 1..n$ and (A_1) and (A_2) hold, then (2.0), (2.1) and (2.2) hold if and only if*

$$\phi := \begin{bmatrix} 2\cos\theta & 1 \\ -1 & 0 \end{bmatrix} \begin{bmatrix} u \\ v \end{bmatrix}$$
$$+ \frac{1}{2} [\, u \quad v \,] \begin{bmatrix} x_1 & a \\ a & 0 \end{bmatrix} \begin{bmatrix} u \\ v \end{bmatrix} e_1$$
$$+ \frac{1}{2} [\, u \quad v \,] \begin{bmatrix} x_2 & 0 \\ 0 & 0 \end{bmatrix} \begin{bmatrix} u \\ v \end{bmatrix} e_2$$
$$+ h.o.t.$$

for certain constants x_1, x_2, a. If $n > 3$, then x_1, x_2, a can be chosen independently and arbitrarily. If $n = 3$, then $x_1 = x_2 = -2a$ must hold.

Proof: The proof is structured as follows. First we express (2.2) in terms of $F(l) := 2\iota\sin(\theta l)$ and use the diagonalization derived in the previous section. Then we show that the sums of $F(l)$ that multiply the constants x_1, x_2, and a vanish except if $n = 3$. The case $n = 3$ is analyzed separately.

Since $D_2^2\phi = \mathbf{0}$ and $D_1 D_2 \phi^{[2]} = 0$ by (A_1) and (A_2), the kth summand of the right hand side (2.2) can be expressed using the constants $A_{ij}(k)$ and $B_{ij}(k)$:

$$(D\phi)^{k-1} D^2 \phi \langle (D\phi)^{n-k} e_i, (D\phi)^{n-k} e_j \rangle \tag{4.1}$$

$$= F^{-3}(1) \begin{bmatrix} F(k) & F(k-1) \\ -F(k-1) & -F(k-2) \end{bmatrix} (D_1^2 \phi A_{i,j}(k) + D_1 D_2 \phi B_{i,j}(k)).$$

The constant $A_{i,j}(k)$ is the entry in the ith row and jth column of the matrix $A(k)$ which tabulates all possible combinations of the first entry of $(\prod_{l=k+1}^n D\phi)e_i$ and $(\prod_{l=k+1}^n D\phi)e_j$ since these multiply $D_1^2\phi$. Since $F(0) = F(n) = F(n/2) = 0, F(n+l) = F(l)$ and $F(-l) = -F(l)$,

$$A(k) := \begin{bmatrix} F(n-k+1) \\ F(n-k) \end{bmatrix} [F(n-k+1) \quad F(n-k)]$$

$$= \begin{bmatrix} F^2(n-k+1) & F(n-k)F(n-k+1) \\ F(n-k)F(n-k+1) & F^2(n-k) \end{bmatrix}$$

$$= \begin{bmatrix} F^2(k-1) & F(k)F(k-1) \\ F(k)F(k-1) & F^2(k) \end{bmatrix}.$$

Similarly $B_{i,j}(k)$ is the i, j entry of

$$B(k) := \begin{bmatrix} F(n-k+1) \\ F(n-k) \end{bmatrix} [-F(n-k) \quad -F(n-k-1)]$$

$$+ \begin{bmatrix} -F(n-k) \\ -F(n-k-1) \end{bmatrix} [F(n-k+1) \quad F(n-k)].$$

$$= - \begin{bmatrix} 2F(k)F(k-1) & F(k+1)F(k-1) + F^2(k) \\ F(k+1)F(k-1) + F^2(k) & 2F(k)F(k+1) \end{bmatrix}.$$

Combining all the summands, we have $D_j D_i(\circ_{l=1}^n \phi) =$

$$(F(1))^{-3} \sum_{k=1}^n \begin{bmatrix} F(k)x_1 + F(k-1)x_2 \\ -F(k-1)x_1 - F(k-2)x_2 \end{bmatrix} A_{i,j}(k) + a \begin{bmatrix} F(k) \\ -F(k-1) \end{bmatrix} B_{i,j}(k).$$

Now we note that the multiplyers of a, x_1 and x_2 are of the form

$$\sum_{k=1}^n \sin(k\theta) \sin((k+l)\theta) \sin((k+m)\theta)$$

$$= \alpha_1 \sum_{k=1}^n \sin(k\theta) + \alpha_2 \sum_{k=1}^n \sin^3(k\theta) + \alpha_3 \sum_{k=1}^n \cos(k\theta) + \alpha_4 \sum_{k=1}^n \cos^3(k\theta)$$

for constants α_1, α_2 α_3 and α_4.

Claim: If $n > 3$, then

$$\sum_{k=1}^{n} \sin(k\theta) = \sum_{k=1}^{n} \cos(k\theta) = \sum_{k=1}^{n} \sin^3(k\theta) = \sum_{k=1}^{n} \cos^3(k\theta) = 0.$$

proof of claim: Let $\Im a$ denote the imaginary part of a and $\Re a$ the real part of a. Then

$$\sum_{k=1}^{n} \sin(k\theta) = \sum_{k=0}^{n-1} \Im \, e^{\iota k\theta} = \Im \left(\frac{e^{\iota n\theta} - 1}{e^{\iota\theta} - 1} \right) = 0,$$

since $n\theta = 2\pi$ and similarly $\sum_{k=1}^{n} \cos(k\theta) = 0$. Since

$$4\cos^3 \alpha = \cos 3\alpha + 3\cos \alpha, \quad 4\sin^3 \alpha = -\sin 3\alpha + 3\sin \alpha,$$

$$\sum_{k=1}^{n} \cos^3(k\theta) = \frac{1}{4} \sum_{k=0}^{n-1} \Re \, e^{\iota k 3\theta} = \Re \frac{e^{\iota n 3\theta} - 1}{e^{\iota 3\theta} - 1} = 0$$

and similarly $\sum_{k=1}^{n} \sin^3(k\theta) = 0$. *End of proof of claim.*

The claim proves the theorem for $n > 3$. If $n = 3$, then $\sum_{k=1}^{3} \cos^3(k\theta_3) = \Re 3 \neq 0$ and we need to analyze (2.2), $\mathbf{0} = D_j D_i(\phi \circ \phi \circ \phi)$, in detail. We list the three cases $i = j = 1$, $i \neq j$, and $i = j = 2$ of (2.2), one per column below, and use the fact that $F(0) = 0$ and $F(2) = -F(1)$.

$$\begin{bmatrix} 0 & 0 & 0 \\ 0 & 0 & 0 \end{bmatrix} = -\begin{bmatrix} 2 & 0 & -2 \\ 2 & -2 & 0 \end{bmatrix} a + \begin{bmatrix} -1 & 1 & 0 \\ 0 & 1 & -1 \end{bmatrix} x_1 + \begin{bmatrix} 0 & -1 & 1 \\ -1 & 0 & 1 \end{bmatrix} x_2.$$

We see that (2.2) holds if and only if $x_1 = x_2 = -2a$. ∎

§5. Nonuniform Quadratic Roots of the Identity

Next, we characterize first order uniform, but second order nonuniform connecting maps that satisfy (2.2) for the special case of fourth roots.

Proposition 5. *If $n = 4$ and $J_1\phi_l = J_1\phi$ for $l = 1..4$, then (2.0–2.2) hold if and only if*

$$\phi_k := \begin{bmatrix} 2\cos\theta & 1 \\ -1 & 0 \end{bmatrix} \begin{bmatrix} u \\ v \end{bmatrix}$$

$$+ \frac{1}{2} [u \quad v] \begin{bmatrix} x_{1,k} & a_k \\ a_k & 0 \end{bmatrix} \begin{bmatrix} u \\ v \end{bmatrix} e_1$$

$$+ \frac{1}{2} [u \quad v] \begin{bmatrix} x_{2,k} & 0 \\ 0 & 0 \end{bmatrix} \begin{bmatrix} u \\ v \end{bmatrix} e_2$$

$$+ h.o.t.$$

and

$$a_1 = a_3, \quad a_2 = a_4, \quad x_{i,j} = -x_{i,j+2}, \quad \text{for } i,j \in \{1,2\}.$$

Proof: We follow the structure of the proof of Theorem 4. Since $F(1) = 1 = -F(3)$, $F(0) = F(2) = F(4) = 0$,

$$A(k) := \begin{bmatrix} F^2(k-1) & 0 \\ 0 & F^2(k) \end{bmatrix}, \quad B(k) := (-1)^{k+1} \begin{bmatrix} 0 & 1 \\ 1 & 0 \end{bmatrix}.$$

Setting

$$a := \begin{bmatrix} a_1 \\ a_2 \\ a_3 \\ a_4 \end{bmatrix}, \quad x_1 := \begin{bmatrix} x_{1,1} \\ x_{1,2} \\ x_{1,3} \\ x_{1,4} \end{bmatrix}, \quad x_2 := \begin{bmatrix} x_{2,1} \\ x_{2,2} \\ x_{2,3} \\ x_{2,4} \end{bmatrix}$$

the equations (2.2) and (4.1) simplify to

$$D_1 D_2(\circ_{l=1}^4 \phi) = \begin{bmatrix} 0 & 1 & 0 & -1 \\ -1 & 0 & 1 & 0 \end{bmatrix} a$$

$$D_1 D_1(\circ_{l=1}^4 \phi) = \begin{bmatrix} 0 & 0 & 0 & 0 \\ -1 & 0 & -1 & 0 \end{bmatrix} x_1 + \begin{bmatrix} 1 & 0 & 1 & 0 \\ 0 & 0 & 0 & 0 \end{bmatrix} x_2$$

$$D_1 D_2(\circ_{l=1}^4 \phi) = \begin{bmatrix} 0 & 1 & 0 & 1 \\ 0 & 0 & 0 & 0 \end{bmatrix} x_1 + \begin{bmatrix} 0 & 0 & 0 & 0 \\ 0 & -1 & 0 & -1 \end{bmatrix} x_2.$$

Setting the expressions to zero proves the claim. ∎

The next section uses the following simple extension of this proposition.

Corollary 5.2. *If $n = 8$ and every odd connecting map is the identity, then Proposition 5 applies to the even numbered connecting maps.*

§6. Degree Bounds for Curvature Continuous Surfaces

We now apply the theorems developed in Sections 4 and 5 to estimate the minimal degree of polynomial pieces necessary for building free-form surfaces that follow the outline of a mesh, where neither the degree of its vertices nor the number of vertices to a mesh cell is restricted. To improve our chances of fitting a low degree surface, we may decrease the combinatorial complexity of the input mesh by inserting a midpoint on every edge and connecting the midpoints of a cell to its centroid (see [3]). After this refinement every original vertex is surrounded by vertices of degree four and all cells are quadrilateral. To mimic the quartic C^2 box spline with directions e_1, e_1, e_2, e_2, e_1+e_2, e_1-e_2 and keep the *total degree* of the surface pieces low, we split each quadrilateral into four triangles and connect the patch domains by the identity across the splitting edges.

Since we are interested in a worst case analysis, we can cook up data. In particular, let P be one of the midpoints generated above surrounded by 8 patches and with original mesh point neighbors P_i, $i = 1..4$. We may assume that P_1 and P_3 have different degree, but that the data at both points are *locally symmetric*. That is, the data relevant to the determination of each connecting map are indistinguishable under rotation. Thus any rotation invariant construction must use uniform roots of the identity at P_1 and P_3. Since the degrees of freedom at P are maximal when the linear part of the connecting map is uniform, we stipulate that the connecting maps at P are uniform up to first order.

Let ϕ_i be the connecting map associated with the edge PP_i and $\lambda_i(v) := (D_2\phi_i^{[1]})(0, v)$. Then by Corollary 5.2 either λ_1 or λ_3 has to be at least quadratic if (2.0)–(2.2) are to hold. For if both λ_i were linear, then their derivative at P is determined differently depending on the number of patches meeting at P_i and thus $a_1 \neq a_3$.

Now consider the two patches $p(u, v)$ and $q(u, v)$ with a common boundary curve $\gamma(u)$ such that $\gamma(0) = P_1$ and $\gamma(1) = P$. Let $\phi := \phi_1$ be the connecting map between p and q, $\lambda := \lambda_1$ be quadratic and choose the data such that symmetry implies $D_2\phi_i^{[2]} \equiv -1$. If d is the degree of γ, then the left hand side of the G^1 constraints (cf. [1])

$$\lambda D_1\gamma = D_2p + D_2q \tag{G_1}$$

is of degree $d - 1 + 2$ and hence D_2p and D_2q are formally of degree $d + 1$. We say formally, since D_2p and D_2q could be degree-raised polynomials. The terms $\lambda D_1 D_2p$ and $\lambda D_1 D_2q$ in the G^2 constraints

$$D_2^2p - D_2^2q - \lambda D_1 D_2p + \lambda D_1 D_2q = \tfrac{1}{2}(D_2q - D_2p)D_2^2\phi^{[2]}. \tag{G_2}$$

are therefore of degree $d+1-1+2$. Unless we have cancellation, D_2^2p and D_2^2q must therefore be of degree $d+2$ and , if none of the intermediate polynomials are degree-raised, p and q must be of degree $d + 4$.

Matching this bound, a curvature continuous surface spline that generalizes the C^2 box spline with directions $e_1, e_1, e_2, e_2, e_1+e_2, e_1-e_2$ and boundary curves of degree $d = 4$ has been developed and implemented by the author.

Acknowledgements. This work was supported by NSF grant CCR-9211322

References

1. Gregory, J.A., Geometric continuity, in *Mathematical Methods in Computer Aided Geometric Design*, T. Lyche and L. L. Schumaker (eds.), Academic Press, New York, 1989, 353–371.
2. Hahn, J, Geometric continuous patch complexes, Computer-Aided Geom. Design **6** (1989), 55–67.
3. Peters, J., C^1 free-form surface splines, SIAM J. Numer. Anal. , to appear.

Jörg Peters
Department of Computer Science
Purdue University
W-Lafayette, IN 47907-1398
jorg@cs.purdue.edu

Applications of the Dual Bézier Representation of Rational Curves and Surfaces

Helmut Pottmann

Abstract. The dual Bézier representation of rational curves and surfaces facilitates the development of efficient geometric algorithms for the design of certain special functional forms. This is demonstrated for developable surfaces, rational curves and surfaces with rational offsets and rational canal surfaces.

§1. The Dual Bézier Representation

The dual form of a planar curve or a surface in 3–space describes this object as the set of its tangents or tangent planes, respectively. For a rational curve or surface, the dual representation is also rational, and therefore may be expressed in Bézier form. The resulting dual Bézier representation has been studied by Hoschek [8], particularly in connection with shape interrogation algorithms [9].

In the following, we will work in the projective extension P^d of real Euclidean d-space E^d ($d = 2, 3$). We use homogeneous cartesian coordinates $(x_0, \ldots, x_d)$, collected in the vector $\mathbf{X}$. The one-dimensional subspace of $\mathbb{R}^{d+1}$ spanned by $\mathbf{X}$ is a point in P^d. This point will also be denoted by $\mathbf{X}$ if no ambiguity can result. For points not at infinity, i.e. $x_0 \neq 0$, the corresponding inhomogeneous cartesian coordinates are $x = x_1/x_0$, $y = x_2/x_0$, $z = x_3/x_0$. The points $\mathbf{X}$ of a line in P^2 or a plane in P^3 satisfy a linear homogeneous equation $\langle \mathbf{U}, \mathbf{X} \rangle = 0$, where the coefficient vector $\mathbf{U}$ contains the homogeneous line or plane coordinates, respectively.

The *dual Bézier representation* of a planar rational curve segment $\mathbf{C}$ constructs a tangent line $\mathbf{U}(t)$ to each parameter value $t \in [0, 1]$:

$$\mathbf{U}(t) = \sum_{i=0}^{m} B_i^m(t) \mathbf{B}_i^*. \tag{1}$$

Curves and Surfaces in Geometric Design
P. J. Laurent, A. Le Méhauté, and L. L. Schumaker (eds.), pp. 377–384.

The vectors $\mathbf{B}_i^*$ are the line coordinate vectors of the *Bézier lines*. Since $\mathbf{B}_i^*$ and any multiple $\lambda\mathbf{B}_i^*$, $\lambda \neq 0$, represent the same line, the curve is not determined by the lines $\mathbf{B}_i^*$ alone. Analogously to the familiar point representation, a complete description can be based on the use of weights w_i. However, this makes the scheme dependent on the origin. Therefore, we will use the duals to the 'auxiliary points' of G. Farin [4]. These are the *Farin lines* $\mathbf{F}_i^*$, which are concurrent with $\mathbf{B}_i^*$ and $\mathbf{B}_{i+1}^*$, and represented by the vectors

$$\mathbf{F}_i^* = \mathbf{B}_i^* + \mathbf{B}_{i+1}^*. \tag{2}$$

Given the lines $\mathbf{B}_i^*$ $(i = 0, \ldots, m)$ and the lines $\mathbf{F}_i^*$ $(i = 0, \ldots, m-1)$ we can normalize the homogeneous coordinate vectors $\mathbf{B}_i^*$ such that (2) holds. Then (1) defines a unique dual curve $\mathbf{U}$. Equation (2) also shows the projective invariance of the Farin lines.

Applying the principle of duality of projective geometry, we can deduce properties of the dual form [8,13]. Here, we just mention a sufficient *convexity condition*: if the Bézier lines contain the edges of a convex domain D which possesses the intersections $\mathbf{B}_i^*\mathbf{B}_{i+1}^*$ as vertices and the Farin lines $\mathbf{F}_i^*$ as supporting lines, then the curve $\mathbf{C}$ is convex and lies outside D (Fig. 2).

For the *computation* of $\mathbf{C}$, we apply the algorithm of de Casteljau to the vectors $\mathbf{B}_i^*$, $(i = 0, \ldots, m)$ until there are just 2 vectors $\mathbf{U}_0^{m-1}(t), \mathbf{U}_1^{m-1}(t)$ left. These vectors represent lines through the desired curve point $\mathbf{C}(t)$ which is therefore computable as the exterior product $\mathbf{C}(t) = \mathbf{U}_0^{m-1}(t) \wedge \mathbf{U}_1^{m-1}(t)$. We observe that the computation has the same efficiency as that of an ordinary Bézier curve of degree m.

The above procedure immediately leads to the following *formula for conversion* from the dual form $\mathbf{U}$ of a Bézier curve to its standard representation as point set $\mathbf{C}$:

$$\mathbf{C}(t) = \sum_{k=0}^{2m-2} B_k^{2m-2}(t)\mathbf{B}_k, \tag{3_1}$$

with

$$\mathbf{B}_k = \frac{1}{\binom{2m-2}{k}} \sum_{i+j=k} \binom{m-1}{i}\binom{m-1}{j}\mathbf{B}_i^* \wedge \mathbf{B}_{j+1}^*. \tag{3_2}$$

We see that $\mathbf{C}(t)$ possesses the degree $2m - 2$. The curve $\mathbf{C}$ is therefore in general a rational algebraic curve of order $n = 2m - 2$. *Degree reductions* occur for inflections and certain higher order singularities of the tangent set ($\mathbf{U}_0^{m-1}(t_0)$ and $\mathbf{U}_1^{m-1}(t_0)$ linearly dependent), since they introduce a base point $\mathbf{C}(t_0) = (0, 0, 0)$. This makes a degree reduction possible [14].

Exchanging the meaning of line and point coordinates, formula (3) also describes the change from standard to dual form. Note that for a rational curve $\mathbf{C}$ whose algebraic class m is lower than its algebraic order n, the dual form (1) is preferable from the view point of computational efficiency.

These ideas can be extended to *rational surfaces* [8]. However, we have to distinguish between surfaces with a two parameter set of tangent planes (which are dual to surfaces in P^3) and those with just a one parameter set of tangent planes. The latter are the *developable surfaces*. They are important in certain applications [11], and are discussed in the following section.

§2. Developable Rational Bézier Surfaces

Supposing sufficient differentiability, *developable surfaces* are ruled surfaces and may be characterized as envelopes of a one parameter set $\mathbf{U}$ of planes. Hence, the dual Bézier representation of a rational developable surface is already given by (1). The appearing vectors are now in $\mathbb{R}^4$ and plane coordinate vectors. The dual control structure is defined by the Bézier planes $\mathbf{B}_i^*$ and the Farin planes $\mathbf{F}_i^*$. For the latter ones, the program may supply defaults (e.g. bisecting planes) such that the user input consists of the Bézier planes only. For interactive design, it is useful to intersect with two planes α, β, particularly if the patch to be designed is bounded by curves in these planes (Fig. 1).

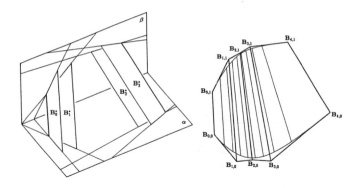

Fig. 1. Dual control structure and Bézier net of a developable patch.

Planar intersections are simple to deal with [13]: *The planar intersection* $\mathbf{C}$ *of a developable rational surface* $\mathbf{S}$ *is a rational curve. The Bézier and Farin lines of the dual representation of* $\mathbf{C}$ *are the planar intersections of the corresponding Bézier and Farin planes in the dual control figure of* $\mathbf{S}$.

Developable patches with boundary curves in planes α, β are easily convertible to *tensor product form*. We just use formulae (3) to convert the intersection curves with α and β into standard form. Let $\mathbf{B}_{i,0}, \mathbf{B}_{i,1}$ ($i = 0, \ldots, 2m - 2$) be the resulting Bézier points. For each parameter value t, we get points on both curves which belong to the same generator. This shows that the *Bézier net* of such a developable rational surface patch is formed by the array $\mathbf{B}_{i,j}$ ($i = 0, \ldots, 2m - 2$; $j = 0, 1$) (Fig. 1).

For purposes of fast evaluation, it might be better to use directly the dual form. In the next to last step of the de Casteljau algorithm we get two planes passing through the generator of $\mathbf{S}$ to the chosen parameter value t. Note that equations (3) (with $\wedge$ as Grassmann product) provide the formulae for the Pluecker coordinate vectors $\mathbf{C}(t) \in \mathbb{R}^6$ of the rulings.

If the vectors $\mathbf{B}_i^*$ are contained in a 3-dimensional subspace of $\mathbb{R}^4$, all control planes pass through a point $\mathbf{V}$ and the resulting surface is a *cone* with vertex $\mathbf{V}$, and is a cylinder if $\mathbf{V}$ is at infinity.

A non–conical developable surface $\mathbf{S}$ possesses a *line of regression*. This curve $\mathbf{R}$, whose tangents form $\mathbf{S}$, may directly be computed with the algorithm of de Casteljau. One just has to run it until there are only three vectors left. These vectors $\mathbf{U}_i^{m-2}(t)$ $(i = 0, 1, 2)$ describe planes through the point of regression $\mathbf{R}(t) = \mathbf{U}_0^{m-2}(t) \wedge \mathbf{U}_1^{m-2}(t) \wedge \mathbf{U}_2^{m-2}(t)$. The mathematical formulation of this procedure leads to the *formula for the line of regression*:

$$\mathbf{R}(t) = \sum_{l=0}^{3m-6} B_l^{3m-6}(t) \mathbf{B}_l, \tag{4_1}$$

with

$$\mathbf{B}_l = \frac{1}{\binom{3m-6}{l}} \sum_{i+j+k=l} \binom{m-2}{i}\binom{m-2}{j}\binom{m-2}{k} \mathbf{B}_i^* \wedge \mathbf{B}_{j+1}^* \wedge \mathbf{B}_{k+2}^*. \tag{4_2}$$

Hence, $\mathbf{R}(t)$ is in general a rational curve of order $3m - 6$. Degree reductions occur in the presence of points $\mathbf{R}(t_0)$ with vanishing torsion and for other points with linearly dependent vectors $\mathbf{U}_i^{m-2}(t_0)$ $(i = 0, 1, 2)$.

For more details and further algorithms, the reader is referred to [13]. We should point out that some aspects of the present dual approach have been mentioned by Bodduluri and Ravani [2]. The dual form is superior to previous treatments of developable Bézier surfaces [1,10] which rely on the solution of a system of nonlinear equations.

§3. Rational Curves and Surfaces with Rational Offsets

Let us consider a rational curve with the inhomogeneous rational representation $\mathbf{c}(t)$. With $\mathbf{n}(t)$ as unit normal vector of $\mathbf{c}$, the offset at distance d is

$$\mathbf{c}_d(t) = \mathbf{c}(t) + d\mathbf{n}(t).$$

Since $\mathbf{n}(t)$ is in general not rational, this is in general not a rational curve. The importance of offset curves, e.g. in the description of milling processes, motivated Farouki and Sakkalis to introduce *Pythagorean–hodograph curves*. These are polynomial curves with rational $\mathbf{n}(t)$. Although a series of papers (cf. [5] and the references therein) made these curves accessible for practical use, some drawbacks such as the high degree of the offsets are inherent in the approach. Therefore, we propose to use the entire class of *rational curves with rational offsets.* After a brief review of the major results from [12], we illustrate their practicality by new rational quartic G^1 splines with quartic offsets. The main results can be extended to *surfaces* where again the dual form is preferable [12].

If $\mathbf{c}_d(t)$ shall be a rational parameterization, $\mathbf{n}(t) = (n_1(t), n_2(t))$ must be a rational representation of the unit circle $x^2 + y^2 = 1$ and therefore have the form

$$n_1(t) = \frac{2a(t)b(t)}{a^2(t) + b^2(t)}, \ n_2(t) = \frac{a^2(t) - b^2(t)}{a^2(t) + b^2(t)}, \tag{5}$$

with polynomials $a(t), b(t)$ [5]. The tangent $g(t)$ at $\mathbf{c}(t)$ has the line coordinate vector

$$\mathbf{U}(t) = (-h(t), n_1(t), n_2(t)).$$

The function $h(t)$ must be rational since $\mathbf{c}(t)$ lies on the line. Because $\|\mathbf{n}\| = 1$, $h(t)$ is the signed distance of the oriented line $g(t)$ from the origin. Hence, replacing h by $h + d$ gives the dual representation of the offset $\mathbf{c}_d$. Without loss of generality, we may set $h = -e/[g(a^2 + b^2)]$ and obtain the following result (for an explicit standard representation, see [7,12]).

Theorem 1. *Any rational curve* $\mathbf{c}(t)$ *with rational offsets* $\mathbf{c}_d(t)$ *can be expressed in the dual form*

$$\mathbf{U}(t) = (e, 2abg, (a^2 - b^2)g), \tag{6}$$

where a, b, e, g *are arbitrary polynomials in* t. *The dual representation of the offset at distance* d *is*

$$\mathbf{U}(t) = (e - d(a^2 + b^2)g, 2abg, (a^2 - b^2)g). \tag{7}$$

The dual representation (6) reveals a remarkably simple construction of the dual Bézier control structure. For this, we first observe that $e \equiv -g(a^2 + b^2)$ in (6) gives a dual representation of the unit circle. For any t, we get exactly the tangent at the point $\mathbf{n}(t)$. The circle representation does not differ from (6) in the second and third coordinate. After the choice of an interval and conversion to Bernstein form of the same degree, we see that the Bézier and Farin lines of the circle segment $\bar{\mathbf{c}} \subset \mathbf{n}$ are parallel to the corresponding Bézier and Farin lines of $\mathbf{c}$. Because of the arbitrary choice of e in (6) this characterizes rational curves with rational offsets.

Theorem 2. *The dual control structure of a rational curve with rational offsets is characterized by the property that its Bézier and Farin lines are parallel to the corresponding lines in the control structure of a dual representation of a circular arc* $\bar{\mathbf{c}}$.

The simplest nontrivial example is obtained by the use of a dual cubic representation of a circular arc $\bar{\mathbf{c}}$ (cf. Fig. 2). The obtained curve segments of order 4 may contain a cusp, but never possess an inflection point.

We will now show how to use these curves for the solution of the G^1 *Hermite interpolation problem*: given 2 points $\mathbf{p}_0, \mathbf{p}_1$ and non-parallel unit normal vectors $\mathbf{n}_0, \mathbf{n}_1$, construct a non–inflecting interpolating arc.

The vectors $\mathbf{n}_0, \mathbf{n}_1$ determine two segments of the unit circle, one of which has to be chosen as normal image $\bar{\mathbf{c}}$ of the desired solution. Any dual quadratic parameterization of $\bar{\mathbf{c}}$ possesses the tangent at $\mathbf{n}_0$, the line $\mathbf{n}_0\mathbf{n}_1$ and the tangent at $\mathbf{n}_1$ as Bézier lines. For a symmetric parameterization we have to choose the bisectors $\mathbf{F}_i^{q*}$ of consecutive Bézier lines as Farin lines (Fig. 2). Let $\mathbf{Q}_i^*$ be the coordinate vectors of the Bézier lines, normalized according to (2).

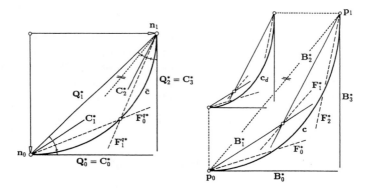

Fig. 2. Geometry of G^1 Hermite interpolation.

Now degree elevation, i.e. multiplication with a linear polynomial, say $(1 - \alpha)t + \alpha(1 - t)$, yields the Bézier lines $\mathbf{C}_i^*$ of the dual cubic representations of $\bar{\mathbf{c}}$:

$$\mathbf{C}_0^* = \alpha\mathbf{Q}_0^*, \ \mathbf{C}_1^* = [(1 - \alpha)\mathbf{Q}_0^* + 2\alpha\mathbf{Q}_1^*]/3,$$
$$\mathbf{C}_2^* = [2(1 - \alpha)\mathbf{Q}_1^* + \alpha\mathbf{Q}_2^*]/3, \ \mathbf{C}_3^* = (1 - \alpha)\mathbf{Q}_2^*. \tag{8}$$

To avoid cusps in the interpolant, we compute with (8) the interval $I \subset (0, 1)$ for α such that the Bézier lines of the interpolant fulfil the convexity condition.This amounts to a condition for either $\mathbf{C}_1^*$ or $\mathbf{C}_2^*$ ($\mathbf{C}_2^*$ in Fig. 2) depending on the direction of the line $\mathbf{p}_0\mathbf{p}_1$. We then just have to modify the first coordinates of $\mathbf{C}_i^*$ such that the resulting lines $\mathbf{B}_i^*$ pass through $\mathbf{p}_0$ or $\mathbf{p}_1$. In this way we get a one parameter family of Hermite interpolants (see Fig. 3, where the bold curve corresponds to the midpoint α_m of the interval I). If a solution with a circular arc is possible, all interpolants collapse into this arc. The closer the data are to being circular, the thinner the region covered by the interpolants. Offset generation requires only appropriate translations of the control lines (Fig. 2).

These Hermite elements yield *rational quartic G^1 splines with quartic offsets*. Using the dual form, we even get the computational efficiency of cubics. Note that C^1 splines may be obtained by appropriate reparameterizations of the segments.

Using the above methods, interpolation and approximation of a given set of tangents is greatly simplified by the fact that the corresponding parameter values are apriori computable from the directions of the given lines. In a separate investigation it will be shown how to construct G^2 splines with rational offsets. It turns out that the dual representation of the osculating circles provides simple algorithms.

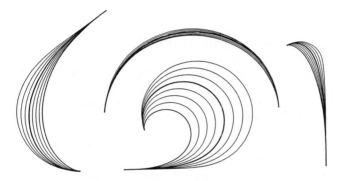

Fig. 3. Some sets of Hermite interpolants.

§4. Rational Pythagorean–Hodograph Space Curves

The dual form also provides the key for the extension of polynomial Pythagorean-hodograph (PH) space curves [6] to the rationals. By definition, the unit tangent vectors $e(t)$ of a rational PH curve $c(t)$ are rational in t. They describe a rational curve e on the unit sphere Ω. From the center of Ω, the curve is projected by a rational cone Γ whose tangent planes are parallel to the osculating planes of c. Hence, we may start with a rational curve e on the sphere [3] and proceed as follows.

Theorem 3. *To construct a rational PH space curve, perform the following steps:*

(i) Construct a rational Bézier curve e on the sphere Ω.

(ii) Project e from the center of Ω by a cone Γ and get its Bézier representation.

(iii) Use (3) to convert Γ to its dual form.

(iv) Take a dual control structure of a developable surface S whose Bézier and Farin planes are parallel to the corresponding control planes of Γ.

(v) With (4) compute the Bézier representation of the line of regression of S to get a rational PH space curve c.

In their treatment of (special) polynomial PH space curves, Farouki and Sakkalis [6] proved that these curves **c** are spine curves of *rational canal surfaces* with constant radius. The canal surfaces are defined as envelopes of congruent spheres centered at the points of **c** and may be considered as analogues to the offsets of planar curves. In practical applications, they appear for example as rolling ball blends. With arguments similar to those in [6] we can extend the remarkable result of Farouki and Sakkalis.

Theorem 4. *The set of congruent spheres which are centered at the points of a rational PH curve possesses a rational surface as envelope.*

References

1. Aumann, G., Interpolation with developable Bézier patches, Comp. Aided Geom. Design **8** (1991), 409–420.
2. Bodduluri, R. M. C., and B. Ravani, Geometric design and fabrication of developable surfaces, ASME Adv. Design Automation **2** (1992), 243–250.
3. Dietz, R., J. Hoschek, and B. Juettler, An algebraic approach to curves and surfaces on the sphere and on other quadrics, Comp. Aided Geom. Design **10** (1993), 211–229.
4. Farin, G., *Curves and Surfaces for Computer Aided Geometric Design*, Academic Press, 1993.
5. Farouki, R., Pythagorean–hodograph curves in practical use, in *Geometry Processing for Design and Manufacturing*, R. E. Barnhill (ed.), SIAM, Philadelphia, 1990, 3–33.
6. Farouki, R. and T. Sakkalis, Pythagorean–hodograph space curves, Advances in Comp. Math., to appear.
7. Fiorot, J. C., and T. Gensane, Characterizations of the set of rational curves with rational offsets, in *Curves and Surfaces in Geometric Design*, P.-J. Laurent, A. Le Méhauté, and L. L. Schumaker (eds.), A K Peters, Wellesley, 1994, 151–158.
8. Hoschek, J., Dual Bézier curves and surfaces, in *Surfaces in Computer Aided Geometric Design*, R. E. Barnhill and W. Boehm (eds.), North Holland, 1983, 147–156.
9. Hoschek, J., Detecting regions with undesirable curvature, Comp. Aided Geom. Design **1** (1984), 183–192.
10. Lang, J., and O. Roeschel, Developable $(1, n)$–Bézier surfaces, Comp. Aided Geom. Design **9** (1992), 291–298.
11. Mancewicz, M. J. and W. H. Frey, Developable surfaces: properties, representations and methods of design, General Motors, 1992, GMR–7637.
12. Pottmann, H., Rational curves and surfaces with rational offsets, Comp. Aided Geom. Design , to appear.
13. Pottmann, H. and G. Farin, Developable rational Bézier and B–spline surfaces, preprint.
14. Warren, J., Creating multisided rational Bézier surfaces using base points, ACM Trans. Graphics **11** (1992), 127–139.

Bifurcation Phenomenon
in a Tool Path Computation

J. F. Rameau

Abstract. This paper examines a problem arising in numerical control. The case studied is specific: a toroidal cutter, with vertical axis, moves in contact with a horizontal cylinder. The axis of the torus is guided by a given curve. The cutter climbs the cylinder until it reaches the ridge, at which point it must translate horizontally before descending. Without this translation, gouging of the cylindrical surface will occur. The geometrical analysis shows that this situation involves a bifurcation phenomenon. The bifurcation theory provides both the nature and the shape of the singularity. Thanks to this information, the complete and continuous trajectory can be computed.

§1. Introduction

The art of numerical control is to compute the trajectory of a tool so that it machines a mechanical part with a given apriori geometry. The part is modelled by its boundary: a set of adjacent faces. The guided cutting principle is to compute each tool path by guiding the tool axis on a given curve. The tool is free to move along its axis and has to stay in contact with the surface of the part.

We examine here the mathematical model of the guided cutting when the cutting part of the tool is a torus, and when the part is bounded by a parametrised surface. In particular, we study the model involving the implicit equation of the torus. This study is a extension of [9] which was restricted to particular surfaces.

The implicit torus model explains the sliding of the tool on a crest-shaped surface. This phenomenon, quite troublesome in the applications, is actually a simple bifurcation of the solutions of a nonlinear system. In the same situation, another model based on the explicit geometry of the torus features a discontinous trajectory [9].

After setting up the equations of the model, we recall the fundamental result. The proof can be found in [1,2,3]. The fundamental theorem is then applied to the nonlinear system, and the physical interpretation is given.

Curves and Surfaces in Geometric Design
P. J. Laurent, A. Le Méhauté, and L. L. Schumaker (eds.), pp. 385–392.

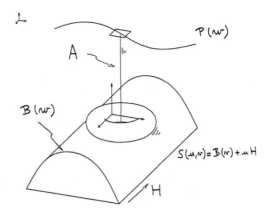

Fig. 1. Guided cutting geometry.

§2. Model of Guiding Curve and Contact Surface

The crest-shaped configuration introduced earlier is modelled by a cylindrical surface: $S(u, v) = B(v) + uH$. The guiding curve $P = P(w)$ is plane and perpendicular to the axis A of the torus (Figure 1).

In order to discuss the sliding phenomenon, we suppose that the axis is perpendicular to the extrusion direction of the contact surface: $\langle A, H \rangle = 0$. For geometrical consistency, we suppose that the basic contour of the cylindrical surface is not tangent to the extrusion direction: $\langle B'(v), H \rangle \neq 0$.

§3. Geometrical Model

3.1. Equations

The torus shaped tool is represented by an implicit equation: $g(x, y, z) = 0$. Noting $\Omega(w, t) = P(w) + tA$, the position of the torus center, the contact with the surface S is written

$$g(S(u, v) - \Omega(w, t)) = 0.$$

Furthermore, the tangency condition at $S(u, v)$ implies that the normal of the torus $g'(S(u, v) - \Omega(w, t))$ is perpendicular to the tangent plane of surface S (generated by the partial derivatives S_u and S_v). This gives the equations

$$\langle g'(S(u, v) - \Omega(w, t)), S_u(u, v) \rangle = 0$$

and

$$\langle g'(S(u, v) - \Omega(w, t)), S_v(u, v) \rangle = 0.$$

A more suitable formulation involves the scalar function

$$\alpha(u, v, w, t) = g(S(u, v) - \Omega(w, t))$$

We then have

$$\alpha_u(u, v, w, t) = \langle g'(S(u, v) - \Omega(w, t)), S_u(u, v) \rangle$$

and

$$\alpha_v(u, v, w, t) = \langle g'(S(u, v) - \Omega(w, t)), S_v(u, v) \rangle.$$

The standard form of the problem is now: find the zeros of the mapping F from $(u, v, w, t) \in [0, 1]^4$ to $\mathbb{R}^3$:

$$F(u, v, w, t) = \begin{bmatrix} \alpha(u, v, w, t) \\ \alpha_u(u, v, w, t) \\ \alpha_v(u, v, w, t) \end{bmatrix}.$$

The problem has one more unknown than equations. The solution is generally a parametrised curve in a four dimensional space:

$$\sigma \longrightarrow (u(\sigma), v(\sigma), w(\sigma), t(\sigma))$$

and $(u(\sigma), v(\sigma))$ is the contact edge on the surface, $\Omega(w(\sigma), t(\sigma))$ is the trajectory of the torus center.

3.2. Notes

1) The implicit equation $g(\cdot) = 0$ of the tool is involved, so the model can support any tool shape, provided it is defined by an implicit equation.

2) The contact points are not always physical. Indeed, a contact point on the inside portion of the torus is possible. But this degree of freedom allows the understanding of the sliding phenomenon.

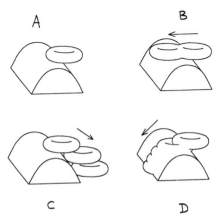

Fig. 2. Sliding torus.

3.3. Sliding torus: intuitive approach

At the top of the crest (Figure 2A), the torus "has a choice" between:

1) Going down on the side of the tool, following a physical trajectory (Figure 2C). The tool is going back.
2) Going down on the other side and plunging into the material (Figure 2D), this trajectory is not physical but is a solution of the geometric model.
3. Sliding horizontally on the top of the crest up to a point where the symmetric choice is possible (Figure 2B), this trajectory is physical.

The sliding phenomenon is well known from the applications where it features a jumping phenomenon. The appropriate mathematical modelling, however, is new: it is a powerful application of the bifurcation theory.

§4. Background of Bifurcation Theory

This mathematical theory concerns the study of the solutions of a non linear system with n equations and $n+1$ unkowns. We consider a mapping F defined on $\mathbb{R}^{n+1}$ into $\mathbb{R}^n$. The arguments of F are $(\lambda, U) \in \mathbb{R} \times \mathbb{R}^n$. The aim of the theory is to characterise the set of the solutions (λ, U) of $F(\lambda, U) = 0$ in the neighbourhood of a known point (λ_0, U_0).

4.1. Regular point

A point (λ_0, U_0) such that $F(\lambda_0, U_0) = 0$ is regular if the linear mapping $F_U(\lambda_0, U_0)$ from $\mathbb{R}^n$ to $\mathbb{R}^n$ is invertible. Under these conditions, there exists locally a unique mapping $U = U(\lambda)$ such that $U(\lambda_0) = U_0$ and $F(\lambda, U(\lambda)) = 0$.

4.2. Simple bifurcation point

A point (λ_0, U_0) is a simple bifurcation point if:

1) It is a solution of the problem: $F(\lambda_0, U_0) = 0$.
2) The linear mapping $F_U(\lambda_0, U_0)$, denoted by $(F_U)^\circ$ has a one dimensional null space. This null space and the image are supplementary. We write ϕ for a generator of the null space, and ϕ^* for a normal vector to the image. Because of the supplementarity condition, we can choose: $\langle \phi^*, \phi \rangle = 1$.
3) The vector $F_\lambda(\lambda_0, U_0)$ written as $(F_\lambda)^\circ$ belongs to the image, that is, $\langle \phi^*, (F_\lambda)^\circ \rangle = 0$.
4) The coefficients a, b, c defined by

$$a = \langle \phi^*, (F_{\lambda\lambda})^\circ + 2(F_{\lambda U})^\circ \psi + (F_{UU})^\circ (\psi, \psi) \rangle$$
$$b = \langle \phi^*, (F_{\lambda U})^\circ \phi + (F_{UU})^\circ (\phi, \psi) \rangle$$
$$c = \langle \phi^*, (F_{UU})^\circ (\phi, \phi) \rangle$$

satisfy $b^2 - ac > 0$. The vector ψ is the unique vector in the image such that

$$(F_\lambda)^\circ + (F_U)^\circ \psi = 0$$

The main result ([1,2,3,4]) is the following *bifurcation theorem*:

Theorem. *If (λ_0, U_0) is a simple bifurcation point, then the zeros of F are locally two curves, crossing at (λ_0, U_0). These curves are parametrised as follows. If $c \neq 0$, then*

$$U(\lambda) = U_0 + \left(\left(-\frac{b \pm \sqrt{b^2 - ac}}{c} \right) \phi + \psi \right) (\lambda - \lambda_0) + (\lambda - \lambda_0)\varepsilon(\lambda - \lambda_0).$$

If $c = 0$, then

$$U(\lambda) = U_0 + (\psi - \frac{a}{2b}\phi)(\lambda - \lambda_0) + (\lambda - \lambda_0)\varepsilon(\lambda - \lambda_0),$$

and

$$\lambda(t) = \lambda_0 + t\varepsilon(t) \quad \text{and} \quad U(t) = U_0 + t\phi + t\varepsilon(t),$$

where $\varepsilon(t)$ vanishes with t.

§5. Application to the Sliding Torus

5.1. Definition of a critical contact

A contact point at S on the torus centered at Ω is *parabolic* if $\|S - \Omega\|^2 = R^2 + r^2$, *elliptic* if $\|S - \Omega\|^2 > R^2 + r^2$, *hyperbolic* if $\|S - \Omega\|^2 < R^2 + r^2$. If (u, v, w, t) is an elliptic or hyperbolic contact point then $\langle S(u, v) - \Omega(w, t), S_u(u, v) \rangle = 0$. Indeed, we develop the term $\alpha_u(u, v, w, t)$ and use $\langle A, S_u \rangle = 0$. We can now state the definition of a *critical contact*. A contact point (u, v, w, t) is critical if: $\|S(u, v) - \Omega(w, t)\|^2 = R^2 + r^2$ and $\langle S(u, v) - \Omega(w, t), S_u(u, v) \rangle = 0$.

These two conditions characterise the critical contact point of Figure 1.

5.2. Application of bifurcation theorem

The application of the theorem requires three steps: check that the critical point is a genuine simple bifurcation point (Lemmas 1 and 2), compute the tangent vectors to the crossing curves (Lemma 3), and interpret these tangents (given in $\mathbb{R}^4$) by considering their projections in more realistic subspaces. The proofs require a lot of algebra and cannot be reproduced in this short paper.

Lemma 1. *Given (u_0, v_0, w_0, t_0) a critical point, by choosing $\lambda = u$ and $U = (v, w, t) \in \mathbb{R}^3$, the linear mapping (of $\mathbb{R}^3$) $(F_U)^\circ$ has a one dimensional null space spanned by $\phi = (-\alpha_{vw}/\alpha_{vv}, 1, 0)$ and an image (of co-dimension one) normal to $\phi^* = (0, 1, 0)$. Furthermore $(F_u)^\circ$ belongs to this image, i.e., $\langle \phi^*, (F_u)^\circ \rangle = 0$.*

Lemma 2. *The critical point (u_0, v_0, w_0, t_0) is a simple bifurcation point, and*

$$a = 0$$

$$b = \alpha_{uuw}$$

$$c = \left(\frac{\alpha_{vw}}{\alpha_{vv}} \right)^2 - 2\frac{\alpha_{vw}}{\alpha_{vv}} + \alpha_{uvw}.$$

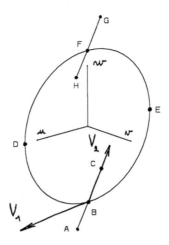

Fig. 3. Bifurcation diagram.

The term b is nonzero if the guiding curve is not tangent to S_u, i.e., if $\langle S - \Omega, P' \rangle \neq 0$. If $\langle S_u, S_v \rangle = 0$, the term c is simplified as $c = \alpha_{uww}$. We have $c = 0$ if in addition $\langle P', S_u \rangle = 0$, i.e. the guiding curve is perpendicular to the extrusion direction of surface S.

Lemma 3. *At the simple bifurcation point* (u_0, v_0, w_0, t_0), *the two crossing curves have the tangent vectors, if* $c \neq 0$,

$$T_1 = \left(1, \left(\frac{b - |b|}{c}\right) \frac{\alpha_{vw}}{\alpha_{vv}}, \frac{|b| - b}{c}, 0\right)$$

and

$$T_2 = \left(1, \left(\frac{b + |b|}{c}\right) \frac{\alpha_{vw}}{\alpha_{vv}}, \frac{-b - |b|}{c}, 0\right),$$

or if $c = 0$,

$$T_1 = (1, 0, 0, 0) \quad \text{and} \quad T_2 = \left(0, -\frac{\alpha_{vw}}{\alpha_{vv}}, 1, 0\right).$$

5.3. Physical Interpretation

The tangent vectors T_1 and T_2 are given in the four dimensional space of the unknowns (u, v, w, t). For a better understanding, we must consider a three dimensional subspace of the space parameters: (u, v, w) (Figure 3). This interpretation is given in the case $c = 0$. The same arguments hold when $c \neq 0$, but the geometry is less intuitive.

In the (u, v, w) space, at the critical point (point B in Figure 3) the (projections of the) tangent vectors are $V_1 = (1, 0, 0)$ and $V_2 = (0, -\alpha_{vw}/\alpha_{vv}, 1)$. The vector V_1 corresponds to the sliding loop in the $v = 0$ plane, and the vector V_2 is the transition from the elliptic contact to the hyperbolic contact.

Figure 4 illustrates the following features of passing bifurcation:

A. The contact is regular, the torus goes up on the $y > 0$ side.

B. First bifurcation at point B. The first choice is a constant z sliding on the top of the crest, which corresponds to the branches BD and BE in the $v = 0$ plane. The second choice is the hyperbolic contact of the branch BC.

D and E. The half torus has slid on the crest. The points D and E represent the two parabolic contacts of the torus on the crest.

F. End of the sliding and second bifurcation at F: There is a choice between the branch FH (going down on the $y > 0$ side, with a hyperbolic contact), and the branch FG (going down on the $y < 0$ side with an elliptic contact).

G. Following of the regular motion at G on the $y < 0$ side.

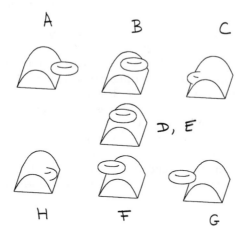

Fig. 4. Passing the bifurcation.

§6. Conclusions

Bifurcation theory is applied to the guided cutting model with the implicit torus. This gives a rigorous treatment and a complete understanding of the sliding torus. In particular, low order derivatives are required (order two for the surface and three for the tool): the degeneracy is not as serious.

However, it must be noted that this theory is efficient only if one knows a priori the singular point. Numerical computation of such a point requires an iterative process ([5,7]), which is still an open question. The added value of this analysis is to show what price must be paid to solve the problem. Future work will deal with fly detection and the computation of critical points in numerical procedures for tool path approximation.

References

1. Brezzi, F., J. Rappaz, and P. A. Raviart, Finite Dimensionnal Approximation of Nonlinear Problems. Part I: Branches of Nonsingular Solutions, Numer. Math. **36** (1980), 1–25.
2. Brezzi, F., J. Rappaz, and P. A. Raviart, Finite Dimensionnal Approximation of Nonlinear Problems. Part II: Limit points, Numer. Math. **37** (1981), 1–28.
3. Brezzi, F., J. Rappaz, and P. A. Raviart, Finite Dimensionnal Approximation of Nonlinear Problems. Part III: Simple bifurcation points, Numer. Math. **38** (1981), 1–30.
4. Crouzeix, M. and J. Rappaz, *On Numerical Approximation in Bifurcation Theory*, Masson Springer-Verlag, 1990.
5. Janovski, V., A Note on Computing Simple Bifurcation Points, *Computing* **43** (1989), 27–36.
6. Jepson, A. D., and A. Spence, On a reduction process for nonlinear equations, *SIAM J. Math. Anal.* **20** (1989).
7. Keller, H. B., Numerical Solution of Bifurcation and Nonlinear Eigenvalue Problems, in *Applications of Bifurcation Theory*, P. Rabinowitz (ed), Academic Press, New York, 1977, 359–384.
8. Rabier, P., Etude locale des problèmes non linéaires perturbés, J. Math. pures et appl. **4** (1982), 311–343.
9. Rameau, J. F., Tore à plat et bifurcation, Mécanique Matériaux Electricité **447** (1993).
10. Rameau, J. F., and C. Schmidt-Lainé, Perturbed bifurcation phenomenon in a diffusion flame problem, Applied Numerical Mathematics **5** (1989), 237–255.

Jean F. Rameau
Dassault Systemes
24-28, Av du Gal de Gaulle
BP - 310
92156 Suresnes cedex FRANCE
Tel. (33.1) 40.99.40.99

Interpolation with an Arc Length Constraint

John A. Roulier and Bruce Piper

Abstract. Theory and algorithms will be presented for the generation of a parametric curve $\mathbf{f}$ into $\mathbb{R}^2$ or $\mathbb{R}^3$ which interpolates a finite set of data points $P_0, P_1, \ldots, P_n$, preserves the shape of the data if possible, and attains a specified arc length L. We allow for specification of tangent directions at all of the data points. This work extends and uses recent work by Roulier and by Roulier and Piper on length specification for the case of two data points and unit tangent vectors. In this work, the arc length is reduced to a function $L(\alpha)$ of a single variable α.

§1. Introduction

We present theory and algorithms for the generation of a parametric curve $\mathbf{f}$ into $\mathbb{R}^2$ or $\mathbb{R}^3$ which interpolates a finite set of data points $P_0, P_1, \ldots, P_n$, preserves the shape of the data if possible, and attains a specified arc length L. We allow for specification of tangent directions at all of the data points. Of course, we must have $L \geq \sum_{k=0}^{n-1} |P_k P_{k+1}|$ in order for such an $\mathbf{f}$ to exist since the shortest curve interpolating the data is the piecewise linear curve connecting the consecutive data points. Furthermore, if L is too large, the shape indicated by the data may not be preserved since inflections or loops may be forced on $\mathbf{f}$.

This work extends and uses recent work by Roulier [1] and by Roulier and Piper [2,3] on length specification for the case of two data points and unit tangent vectors. In this work, the arc length is reduced to a function $L(\alpha)$ of a single variable α. Conditions are given in which $L(\alpha)$ is strictly concave upward on the real line and tends to $+\infty$ as $\alpha \to \pm\infty$. The goal then, is to find a value of α so that $L(\alpha) = L$. Once such an α is found, $\mathbf{f}$ is determined by that α.

The form of the function $L(\alpha)$ allows the use of the secant method to rapidly produce an acceptable value of α. An algorithm will be presented which implements these ideas for our case. No user input of starting values for the secant method is required. Several examples will be given.

Curves and Surfaces in Geometric Design
P. J. Laurent, A. Le Méhauté, and L. L. Schumaker (eds.), pp. 393–400.

§2. The Problem

In this section we will describe the problem to be studied and introduce some of the notation to be used in the rest of the paper.

Definition 1. *The arc length of a parametric function* $\mathbf{f}$ *on parameter interval* $[a, b]$ *will be denoted by:*

$$L(\mathbf{f}, a, b) = \int_a^b |\mathbf{f}'(t)| \, dt. \tag{1}$$

We now describe the problem to be solved.

Problem: *Given points* $P_0, P_1, \ldots, P_n \in \mathbb{R}^3$, *corresponding unit vectors* $\mathbf{u}_0, \mathbf{u}_1, \ldots, \mathbf{u}_n$, *and length* $L > \sum_{j=0}^{n-1} |P_j P_{j+1}|$, *produce a parametric curve* $\mathbf{f}$ *on some parameter interval* $[a, b]$ *so that*

$$L(\mathbf{f}, a, b) = L, \tag{2}$$

and for some parameter values $a = t_0 < t_1 < \cdots < t_n = b$

$$\mathbf{f}(t_j) = P_j \tag{3}$$

and

$$\mathbf{f}'(t_j) = a_j \mathbf{u}_j \text{ where } a_j > 0 \tag{4}$$

for $j = 0, 1, \ldots, n$.

Notes.

- $\mathbf{f}$ should preserve the shape indicated by the data and unit vectors if possible for the given value of L.

- We do not insist on a continuous tangent vector at each data point. We only insist on a continuous *unit* tangent vector at each data point. If a continuous tangent vector is desired, then this may be easily incorporated into the algorithm.

§3. General Approach to the Problem: Divide and Conquer

In this section we will describe a method which will allow us to break the problem into smaller subproblems which are more easily solved. The general approach is as follows:

1) Distribute the length L into smaller lengths L_j proportionally among the segments $P_j P_{j+1}$ so that $L = \sum_{j=0}^{n-1} L_j$.

2) On each segment $P_j P_{j+1}$ for $j = 0, 1, \ldots, n - 1$, produce a parametric curve $\mathbf{r}_j$ with arc length L_j so that $\mathbf{r}_j(0) = P_j$, $\mathbf{r}_j(1) = P_{j+1}$, $\mathbf{r}_j'(0) = a_j \mathbf{u}_j$, and $\mathbf{r}_j'(1) = a_{j+1} \mathbf{u}_{j+1}$ where $a_j > 0$ and $a_{j+1} > 0$, and so that $\mathbf{r}_j$ has the shape indicated by the end conditions if possible. This uses results of Roulier [1], and of Roulier and Piper [2,3].

3.1. A quadrilateral frame for each segment.

Assume that points P_1 and P_2 and unit vectors $\mathbf{u}_1$ and $\mathbf{u}_2$ are given. Let *basic lengths* $d_1 > 0$ and $d_2 > 0$ be given. These will be used to describe allowable ranges for the interior control points for a Bézier polynomial or rational curve in order to preserve the shape.

An algorithm to produce these basic lengths is given in [4]. We give a brief description of how to calculate the basic lengths for most cases. Assume that neither $\mathbf{u}_1$ nor $\mathbf{u}_2$ is perpendicular to the line through P_1 and P_2. Let ℓ_1 be the line through P_1 parallel to $\mathbf{u}_1$ and let ℓ_2 be the line through P_2 parallel to $\mathbf{u}_2$. Let Π be a plane normal to the segment P_1P_2, and let R be the point at which Π intersects the segment P_1P_2. Let Q_1 and Q_2 be the points where ℓ_1 and ℓ_2 respectively intersect the plane Π. Let $h_1 = \mid Q_1R \mid$ and $h_2 = \mid Q_2R \mid$. We now find the point $\overline{R}$ in the segment P_1P_2 for which the plane $\overline{\Pi}$ through $\overline{R}$ perpendicular to P_1P_2 produces Q_1 and Q_2 with $h_1 = h_2$. For this $\overline{R}$ let $d_1 = min(\mid P_1Q_1 \mid, \zeta \mid P_1P_2 \mid)$ and $d_2 = min(\mid P_2Q_2 \mid, \zeta \mid P_1P_2 \mid)$. $\zeta > 0$ is a user definable constant which can be chosen to assure that the resulting values for d_1 and d_2 are not too large compared to $\mid P_1P_2 \mid$. For example, $\zeta = 2$ is a reasonable choice. If $\mathbf{u}_1$ or $\mathbf{u}_2$ is perpendicular to P_1P_2, then a special case is used to find the basic lengths. We will not present this here. We now define the following variables:

$$G_1 = P_1 + d_1\mathbf{u}_1 \tag{5}$$

$$G_2 = P_2 - d_2\mathbf{u}_2 \tag{6}$$

$$m = \mid P_1P_2 \mid \tag{7}$$

$$M = d_1 + \mid G_1G_2 \mid + d_2 \tag{8}$$

m is the lower limit on the length of a curve from P_1 to P_2, and M will be considered to be the upper limit on the length of a reasonable curve from P_1 to P_2 which is consistent with the shape indicated by the unit vectors $\mathbf{u}_1$ and $\mathbf{u}_2$. The general idea here is that the points P_1, G_1, G_2, and P_2 form the extremal points of a frame which will restrict the control polygon which will be generated so that the sum of the consecutive chord lengths will be less than M.

3.2. Distributing the length.

For the given data points $P_0, \ldots, P_n$, unit vectors $\mathbf{u}_0, \ldots, \mathbf{u}_n$, and desired length L, proceed as follows:

- For each $j = 0, 1, \ldots, n - 1$ compute the basic lengths $d_{j,1}$, $d_{j,2}$, points $G_{j,1}$, $G_{j,2}$, and values m_j and M_j as above for points P_j and P_{j+1} and unit vectors $\mathbf{u}_j$ and $\mathbf{u}_{j+1}$.
- Let $\mu = \sum_{j=0}^{n-1} m_j$ and $\mathcal{M} = \sum_{j=0}^{n-1} M_j$, and calculate $\gamma > 0$ so that $L = (1 - \gamma)\mu + \gamma\mathcal{M}$. ($\gamma = (L - \mu)/(\mathcal{M} - \mu)$.)
- For each $j = 0, 1, \ldots, n - 1$ compute the length L_j for segment P_jP_{j+1} by $L_j = (1 - \gamma)m_j + \gamma M_j$.

- Note that $L = \sum_{j=0}^{n-1} L_j$.

This assures that the L_j are placed with respect to their respective boundaries $m_j < M_j$ as L is placed relative to $\mu < \mathcal{M}$.

We have now reduced the problem to producing a curve $\mathbf{f}_j$ on each segment $P_j P_{j+1}$ with the appropriate shape so that $L(\mathbf{f}_j, 0, 1) = L_j$. A suitable reparametrization will produce the composite curve $\mathbf{f}$ which satisfies the original problem.

§4. Solving the Problem on Each Segment

In this section, we assume that points P_1 and P_2 and unit vectors $\mathbf{u}_2$ and $\mathbf{u}_2$ are given. The following theorem appears in Roulier and Piper [2].

Theorem 1. *Suppose that* $\mathbf{F}$ *and* $\mathbf{G}$ *are continuous parametric curves with piecewise continuous, bounded derivatives on* $[a, b]$ *and*

$$\text{there is at least one } u \in [a, b] \text{ such that } \mathbf{F}' \text{ and } \mathbf{G}' \tag{9}$$
$$\text{are defined and continuous at } u \text{ and } \mathbf{F}'(u) \neq \mathbf{G}'(u).$$

Let

$$\mathbf{H}(t, \alpha) = (1 - \alpha)\mathbf{F}(t) + \alpha\mathbf{G}(t). \tag{10}$$

Then $L(\mathbf{H}(t, \alpha), a, b)$ *is a convex function of* α *and*

$$\lim_{\alpha \to \pm\infty} L(\mathbf{H}(t, \alpha), a, b) = \infty.$$

If we replace the assumption (9) *by the stronger assumption*

$$\text{there is at least one } u \in [a, b] \text{ such that } \mathbf{F}' \text{ and } \mathbf{G}' \tag{11}$$
$$\text{are defined and continuous at } u \text{ and } \mathbf{F}'(u) \not\parallel \mathbf{G}'(u),$$

then we get $L(\mathbf{H}(t, \alpha), a, b)$ *is a strictly convex function of* α.

4.1. Rational Bézier curves in $\mathbb{R}^3$.

Here, we obtain results for degree 3 rational Bézier curves. We use the quadrilateral $P_1 G_1 G_2 P_2$ in $\mathbb{R}^3$ described above.

Notation. Define the control points as a function of parameter α:

$$\begin{aligned}
D_0(\alpha) &= P_1 \\
D_1(\alpha) &= (1 - \alpha)P_1 + \alpha G_1 \\
D_2(\alpha) &= (1 - \alpha)P_2 + \alpha G_2 \\
D_3(\alpha) &= P_2
\end{aligned} \tag{12}$$

Recall that G_1, G_2, m, and M are given in (5)–(8).

Definition 2. *Define the rational Bézier cubic curve*

$$\mathbf{X}_3(t, \alpha) = \frac{\sum_{k=0}^{3} w_k \binom{3}{k} t^k (1-t)^{3-k} D_k(\alpha)}{\sum_{k=0}^{3} w_k \binom{3}{k} t^k (1-t)^{3-k}}. \tag{13}$$

Define the terms

$$\mathbf{f}(t) = (1-t)G_1 + tG_2 \tag{14}$$

$$D(t) = \mid P_1\mathbf{f}(t) \mid + \mid \mathbf{f}(t)\mathbf{f}(1-t) \mid + \mid \mathbf{f}(1-t)P_2 \mid \tag{15}$$

Lemma 1. *Let $\eta > 0$ be given and assume that $m < L < L + \eta < M$ and that $\mid G_1G_2 \mid > 0$. If*

$$0 < \varepsilon < min(\frac{1}{2}, \frac{M - L - \eta}{4 \mid G_1G_2 \mid}) \tag{16}$$

then

$$L + \eta < D(\varepsilon) < M. \tag{17}$$

Proof: The right side of (17) is easily seen to be true for $0 < \varepsilon < 1/2$. The left side of (17) follows from the facts that $\mathbf{f}(\varepsilon) - \mathbf{f}(1-\varepsilon) = (1 - 2\varepsilon)(G_1 - G_2)$, $\mid P_1\mathbf{f}(\varepsilon) \mid \geq \mid P_1G_1 \mid - \mid G_1\mathbf{f}(\varepsilon) \mid$, and $\mid P_2\mathbf{f}(1-\varepsilon) \mid \geq \mid P_2G_2 \mid - \mid G_2\mathbf{f}(1-\varepsilon) \mid$. The details are left to the reader. ∎

Let the weights w_j satisfy:

$$w_0 = w_3 = 1, \quad w_1 = w_2 = \omega \tag{18}$$

Then the expression for $\mathbf{X}_3(t, 1)$ which has control points P_1, G_1, G_2, and P_2 becomes:

$$\mathbf{X}_3(t, 1) = \frac{(1-t)^3 P_1 + 3\omega(1-t)t\mathbf{f}(t) + t^3 P_2}{(1-t)^3 + 3\omega(1-t)t + t^3} \tag{19}$$

This gives the following:

Theorem 2. *Let $G_1 \neq G_2$ and let ε satisfy (16) where $\eta > 0$ is as in the previous lemma. If ω satisfies*

$$\omega > \frac{((1-\varepsilon)^3 + \varepsilon^3)(\mid P_1P_2 \mid + L)}{3\varepsilon(1-\varepsilon)\eta} \tag{20}$$

then

$$L(\mathbf{X}_3(t, 1), 0, 1) > L. \tag{21}$$

We actually get

$$\mid P_1\mathbf{X}_3(\varepsilon, 1) \mid + \mid \mathbf{X}_3(\varepsilon, 1)\mathbf{X}_3(1 - \varepsilon, 1) \mid + \mid \mathbf{X}_3(1 - \varepsilon, 1)P_2 \mid > L. \tag{22}$$

If $C = G_1 = G_2$ then we may replace (20) by

$$\omega > \frac{L}{3(\mid P_1 C \mid + \mid P_2 C \mid - L)} \tag{23}$$

Proof: The proof of (22) follows by first using Lemma 1 to show that the left side of (22) is greater than $A(\omega) = \frac{3\omega(1-\varepsilon)\varepsilon(L+\eta)-((1-\varepsilon)^3+\varepsilon^3)\mid P_1 P_2\mid}{(1-\varepsilon)^3+\varepsilon^3+3\omega(1-\varepsilon)\varepsilon}$ and then solving $A(\omega) > L$ for ω. We easily get (21) from (22). If $G_1 = G_2$, then $\varepsilon = 1/2$ and (23) is shown to imply (22) in [3]. We omit the details. ∎

Theorem 2 shows us how to use (18) to select weights so that for some $\alpha \in (0, 1)$, $L(\mathbf{X}_3(t, \alpha), 0, 1) = L$.

Theorem 3. *For control points given by (12), let*

$$L(\alpha) = L(\mathbf{X}_3(t, \alpha), 0, 1). \tag{24}$$

Then $L(\alpha)$ is a strictly convex increasing function of α for $\alpha \in [0, +\infty)$. Moreover, if the weights are chosen as in the previous theorem, and if $\eta > 0$ satisfies $m < L < L + \eta < M$, then

$$m = L(0) < L < L(1) < M. \tag{25}$$

In this case, there is a unique $\overline{\alpha} \in (0, 1)$ for which $L(\overline{\alpha}) = L$.

Proof: The first part follows from Theorem 1 by observing that $\mathbf{X}_3(t, \alpha) = (1-\alpha)\mathbf{X}_3(t, 0) + \alpha\mathbf{X}_3(t, 1)$. It is easy to see that (11) is satisfied since $\mathbf{X}_3'(t, 0)$ is parallel to segment $P_1 P_2$ for all $t \in (0, 1)$ and $\mathbf{X}_3'(t, 1)$ is not. The inequalities $L(0) < L < L(1)$ follow immediately from Lemma 1 and Theorem 2. The proof of $L(1) < M$ uses deCasteljau subdivision. We omit the details. ∎

4.2. How do we find an $\overline{\alpha}$ such that $L(\overline{\alpha}) = L$?

Roulier and Piper [2] show that if a function $g(x)$ is convex and increasing on interval $[a, +\infty)$ and if $a < b$ with $g(a) < 0 < g(b)$, then the secant method will produce a sequence of numbers $\{x_k\}$ for which $\lim_{k\to\infty} x_k = \overline{x}$ (the unique solution of $g(x) = 0$ in $[a, b]$) if the starting values are chosen to be $x_0 = b + \delta$ and $x_1 = b$ for some $\delta > 0$.

Moreover, the secant method iterates converge monotonically to the solution as fast as the Newton Raphson iterates starting with x_0 in the sense that the secant iterate after every two steps is always closer to the solution than the Newton Raphson iterate after every step. These require the same number of function evaluations, and the secant method does not require a derivative.

Here, we let $g(\alpha) = L(\alpha) - L$. Note that each evaluation of $g(\alpha)$ requires the evaluation of an integral. For this we use a numerical integration formula. For speed and accuracy, we have employed a 15 point composite Gaussian quadrature formula. Generally, one 15 point segment is used. For greater speed, lower order Gaussian quadrature can be used with some additional loss of accuracy.

4.3. An Algorithm – Putting it all together.

Let the points P_j, unit vectors $\mathbf{u}_j$, and length L as above be given. Assume that $L > \sum_{j=0}^{n-1} | P_j P_{j+1} |$.

1) Compute the length L_j and the quantities $G_{j,1}$, $G_{j,2}$ and M_j for each $j = 0, 1, \ldots, n-1$ so that $L = \sum_{j=0}^{n-1} L_j$ as in section 3. Let $\mathcal{M} = \sum_{j=0}^{n-1} M_j$. If $L \geq \mathcal{M}$, then use Algorithm 2 from Roulier [1] to produce a Bézier cubic polynomial curve for each $j = 0, 1, \ldots, n-1$.

2) For each $j = 0, 1, \ldots, n-1$, produce weights $w_{j,k}$ for $k = 0, 1, 2, 3$ using Theorem 2 and (18) so that the rational Bézier cubic $\mathbf{X}_{j,3}(t, 1)$ with these weights and control points $P_j, G_{j,1}, G_{j,2}, P_{j+1}$ corresponding to the segment $P_j P_{j+1}$ satisfies $L(\mathbf{X}_{j,3}(t, 1), 0, 1) > L_j$.

3) For each $j = 0, 1, \ldots, n-1$, set up the rational Bézier cubic $\mathbf{X}_{j,3}(t, \alpha)$ with the weights found in the previous step and control points $D_{j,k}(\alpha)$ for $k = 0, 1, 2, 3$. Set $g_j(\alpha) = L(\mathbf{X}_{j,3}(t, \alpha), 0, 1) - L_j$.

4) For each $j = 0, 1, \ldots, n-1$, use the secant method with $x_0 = 1 + \delta$ and $x_1 = 1$ to find the unique zero $\overline{\alpha}_j$ of $g_j(\alpha)$ in the interval $[0, 1]$. Use this value to generate the control points $D_{j,k}(\overline{\alpha}_j)$ for $k = 0, 1, 2, 3$ for the rational Bézier cubic $\mathbf{X}_{j,3}(t, \overline{\alpha}_j)$ with weights $w_{j,k}$ for $k = 0, 1, 2, 3$ produced above.

§5. Examples and Conclusions

We present examples of the application of the algorithm to one set of data points to be interpolated. The unit tangent vectors for this data were calculated using the method in [4]. The data points to be interpolated are:

$$P_0 = (0,0,0), P_1 = (1,0,0), P_2 = (1,1,0), P_3 = (1,1,1)$$

This data gives $\mu = \sum_{j=0}^{2} m_j = 3$ and $\mathcal{M} = \sum_{j=0}^{2} M_j = 4.59$. Therefore, we will use values of L between these two values if we want to preserve the shape of the data.

The first example produces a piecewise rational Bézier cubic curve of length $L = 3.2$. The graph is shown in Figure 1.

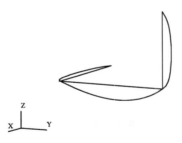

Fig. 1. Piecewise rational Bézier cubic curve of length 3.2.

The second example produces a piecewise rational Bézier cubic curve of length $L = 4.0$. The graph is shown in Figure 2. Note the convexity of the end segments which are still within the allowable range.

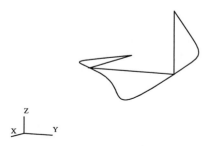

Fig. 2. Piecewise rational Bézier cubic curve of length 4.0.

These examples illustrate the use of the algorithm in this paper. If we had chosen a length $L \geq \mathcal{M}$, then Algorithm 2 from Roulier [1] would have been used to produce a piecewise Bézier cubic curve, and the shape of the data might not have been preserved. Note that the strict convexity of the arc length function given in Theorem 3 was not necessary for the algorithm to work.

References

1. Roulier, J., Specifying the arc length of Bézier curves, Computer-Aided Geom. Design **10** (1993), 25–56.
2. Roulier, J., and Piper, B., Prescribing the length of parametric curves, preprint.
3. Roulier, J., and Piper, B., Prescribing the length of rational Bézier curves, preprint.
4. Shen, T., Roulier, J., and Foley, T., A general algorithm for automatic generation of control points for Bézier curves, Univ. of Conn. Computer Science and Engineering Tech. Rept. CSE-TR-90-22.

John A. Roulier
Department of Computer Science and Engineering U-155
University of Connecticut
Storrs, CT 06269-3155
jrou@ brc.uconn.edu

Bruce Piper
Department of Mathematics
Rensselaer Polytechnic Institute
Troy, NY 12180-3590
piperb@ rpi.edu

Curve Reconstruction

Jean-Christophe Roux

Abstract.   An approach for curve reconstruction is presented here, based on local curve arc approximation by circles. Circles are computed by a constrained least square method, which is presented here. Using point projection on these circles, it is then possible to set up an order on the points in relation with the natural one induced by a regular parametrization, in a neighbourhood of a point. Then, we can reconstruct arcs of the curve, before linking them to obtain the curve.

§1. Introduction

The problem we examine in this paper can be stated as follows: *given n points which lie on a curve, how to reconstruct the curve ?*, *i.e.*, how to find an order on the points in accordance with the one induced by a natural parametrization of the curve, like arc length parametrization by example. Theoretical results related to this problem are always qualitative, and so not really useful in practice. On the other hand, the method we present allows to compute noisy data sets, self-intersections or cusps, and a great number of points.

The idea we apply is to calculate a simple algebraic curve, given in its implicit form, and approximating the curve ; we use circles to approximate the curve arcs.

Why this choice ? Firstly, circles give a geometrical meaning to the approximation as we will see below. Secondly, we will obtain theoretical results in relation to curvature for point projection and order-preserving.

Several authors have already developed different methods for curve and surface reconstruction. Here we cite Dedieu & Favardin [3], who used algorithms based on proximity and tangent variation, and Boissonnat [1] and T. DeRose et al. [2] for surfaces.

First, we present the minimization problem we use to solve the problem of approximating the circle, and its geometrical meaning. Next we discuss point projection on circles and order-preserving. Finally, we describe the method we use for curve reconstruction.

Curves and Surfaces in Geometric Design

P. J. Laurent, A. Le Méhauté, and L. L. Schumaker (eds.), pp. 401–408.

Copyright © 1994 by A K PETERS, Wellesley, MA

ISBN 1-56881-039-3.

§2. Circle Approximation

Our goal is to find a circle which best fits the set of points and which preserves the natural order induced by the parametrization, when the points are projected on it. Let $\mathcal{C}$ be a circle given by

$$F(x,y) = a_1(x^2 + y^2) + a_2 x + a_3 y + a_4 = 0 \ ,$$
$$\text{or } F(u) = a_1{}^t uu + {}^t u\omega + a_4 = 0 \text{ with } {}^t u = (x,y) \text{ and } {}^t \omega = (a_2, a_3) \ .$$

Then the problem is, for a given set of points $u_1, \ldots, u_n$,

$$\underset{a_1, \omega, a_4}{\text{Min}} \sum_{i=1}^{n} (a_1{}^t u_i u_i + {}^t u_i \omega + a_4)^2 \quad \text{with constraint } a_2^2 + a_3^2 - 4a_1 a_4 = 1 \ \ (P)$$

We will see later why we take this constraint, and will discuss its geometric meaning. We have

$$\underset{a_1, \omega, a_4}{\text{Min}} \sum_{i=1}^{n} (a_1{}^t u_i u_i + {}^t u_i \omega + a_4)^2$$

$$= \underset{p}{\text{Min}} \sum_{i=1}^{n} {}^t p \, M_i \, p \ ,$$

where ${}^t p = (a_1, a_2, a_3, a_4)$, and

$$M_i = \begin{pmatrix} ({}^t u_i u_i)^2 & x_i({}^t u_i u_i) & y_i({}^t u_i u_i) & {}^t u_i u_i \\ x_i({}^t u_i u_i) & x_i^2 & x_i y_i & x_i \\ y_i({}^t u_i u_i) & x_i y_i & y_i^2 & y_i \\ {}^t u_i u_i & x_i & y_i & 1 \end{pmatrix} .$$

Let $M = \sum_{i=1}^{n} M_i$. Then (P) becomes

$$\underset{p}{\text{Min}} \ {}^t pMp \quad \text{with constraint } \varphi(p) = a_2^2 + a_3^2 - 4a_1 a_4 = 1 \qquad (P)$$

Using Lagrange multipliers to solve (P), if $J(p) = {}^t p \, M \, p$, a necessary condition for the existence of a minimum is $\nabla J = \lambda \nabla \varphi$, i.e.,

$$2Mp = \lambda \nabla \varphi(p) \ .$$

In other words, ${}^t(\nabla \varphi(p)) = (-4a_4, 2a_2, 2a_3, -4a_1) = {}^t(Ap)$, and thus

$$2Mp = \lambda Ap \Longleftrightarrow 2A^{-1}Mp = \lambda p \Longleftrightarrow Bp = \lambda p$$

Therefore, the solution vector $\bar{p}_{min}$ is an eigenvector of the matrix B, and it is easy to show that it corresponds to the smallest eigenvalue λ_{min}.

Geometrical Interpretation

Consider the circle $(\mathcal{C})$ defined by $F(M) = 0$, with normalized equation $a_2^2 + a_3^2 - 4a_1 a_4$, which means:

$$\|\nabla F(M)\|^2 = 1, \forall M \in (\mathcal{C}).$$

For any $M \in \mathbb{R}^2$, there exists $M_0 \in (\mathcal{C})$ and $t \in \mathbb{R}$ such that $M = M_0 + t\nabla F(M_0)$ (unique in the general case). Then,

$$F(M) = F(M_0 + t\nabla F(M_0)) \ ,$$

and in a neighbourhood of the point M_0 (*i.e.*, near the circle), $|t|$ is small. Hence,

$$F(M) \simeq F(M_0) + t\nabla F(M_0) \cdot \nabla F(M),$$
$$= F(M_0) + t\|\nabla F(M_0)\|^2 \ .$$

Consequently,

$$F(M) - F(M_0) \simeq t\|\nabla F(M_0)\|^2 = t \ .$$

Hence, in a neighbourhood of $\mathcal{C}$, we have

$$|F(M)| \simeq \mathrm{d}\,(M, M_0) = \underset{\overline{M} \in (\mathcal{C})}{\mathrm{Min}}\ \mathrm{d}\,(M, \overline{M}) = \mathrm{d}\,(M, \mathcal{C}) \ .$$

We can say then that the computed circle is at minimal distance from the set of points $u_i = (x_i, y_i)$. We also remark that this constraint is a geometrical invariant. In practice this gives us an efficient and robust method to calculate a circle which best fit a set of points, even if the data are perturbed or lie practically on a line.

§3. Point Projection and Order-preserving

Our goal is here to find conditions on the curve to guarantee order-keeping when we project points onto the approximating circle. Let's define first the order relation:

Definition 1. *Let $g : I \to \mathbb{R}$ be a regular C^2 parametrization of a curve arc γ and let $M_1 = g(\theta_1)$, $M_2 = g(\theta_2)$ two points of γ. We say that M_1 is the predecessor of M_2 on γ if and only if $\theta_1 < \theta_2$. We denote this by $M_1 \prec M_2$.*

The following condition is natural:

Condition 1. *The variation of the angle of the tangent must not exceed 2π.*

Now let C_0 be the osculating circle at point M_0 to a curve arc γ, and let Ω_0 be the centre of C_0. Then we state another condition:

Condition 2. $\forall M \in \gamma$, Ω_0 *does not lie on the tangential line to γ at point M.*

Obviously, we have a parametrization of C_0 so that if P_1 and P_2 are the projected points of M_1 and M_2 onto C_0, then

$$M_1 \prec M_2 \Longleftrightarrow P_1 \prec P_2 \; ,$$

and we can finaly deduce (keeping the same notations):

Theorem 3. *The order on the points of γ induced by the relation $\prec$ is the same as the one on the projected points onto C_0, under the conditions 1 and 2 stated above.*

Proof: If the point $M \in \gamma$ is defined in complex coordinates in a suitable way, we show that angle θ is a change of parametrization for γ ; hence we can define γ by the parametric representation: $f : [0, 2\pi[\rightarrow \mathbb{R}^2, \; \theta \mapsto r(\theta)\vec{u}(\theta)$, where $\vec{u}(\theta)$ is a suitable unit vector. ∎

Remark. In practice, γ is given by a finite number of points. The osculating circle is replaced by the approximating circle of §2, and we use the projection onto that circle. This intuitive idea proves itself to be very robust in practice, although a mathematical proof is available only in the limit case, when all the points converge to a given point M_0 on γ (see [4] for a complete proof).

§4. Reconstruction Algorithms

Our method proceeds in four general steps which we now described.

- Step I: Cutting up: *point subsets creation.*

    In this step we use a quadtree to cut up the set of points, in accordance to a cutting up threshold in relation to approximation quality by circles in each leaf of the quadtree. Also we look at projected arc length, to garantee, as we have seen previously, the order-preserving by projection. We stop cutting the points sets if projected arc length becomes less than π. Quadtree is then recursively computed under these conditions.

- Step II: Approximation: *circle computation for each subset.*

    An approximating circle is computed in each non-empty cell of the quad-tree which contain at least 3 points. Then cells are labeled *ordinary* or *complex*, in accordance with approximation quality: good or bad, respectively. As we will see further, we will use this label into the linking step, to compute self-intersections or cusps if needed.

- Step III: Arc calculation: *local reconstruction of the curve.*

    In each ordinary labeled cell, we project the subset of points onto the approximating circle, and sort the projected points by ascending order in relation to angles.

Sorted points are next put in their initial positions, and a polygonal line corresponding to the curve arc is got by cutting contour at the longest edge (that can be done, remembering that the projected arc's length is $< \pi$).

- Step IV: Links: *reconstruction of the whole curve.*

At this step, we have to separate the case of simple curves and the one of curves with possible self-intersections or cusps, which needs a different computation. This case will appear if cells are labeled complex.

Case 1: Only ordinary cells: links between neighbour arcs:

We use the quadtree's structure and an associated "neighbour" relation to link arc extremities ; a threshold in relation to distance between extremities is also applied to avoid links between too far points. The algorithm is then the following:

```
Remove empty cells
Take the first cell in the quadtree as the main cell.
Iterate until quadtree is empty
        Select the neighbour current cell, which contain
                the nearest extremity holding on the threshold.
        Link the two lists of points into the main cell,
                and update the main cell.
        Remove the current cell.
End iterate.
```

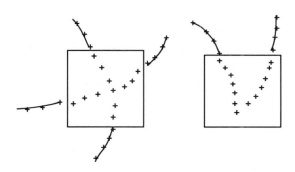

Fig. 1. Self-intersection and cusp computation.

Case 2: Complex cells: computing cusps and self-intersections:

In this case, we first compute the ordinary cells, and so obtain a set of closely related components, not all linked because of complex cells. We next use the information given by these components in the neighbourhood of the complex cell to calculate one or several approximating circles based on these neighbour extremities. In fact, we enumerate all the cases, taking extremities alone or by pairs, and calculate the corresponding circles.

This enumeration is the key of the cusps or self-intersections detection, and it allows us to cope with such cases, as it is illustrated in Figure 1: we can compute a self-intersection by taking two pairs of extremities, or a cusp by taking two extremities alone.

We first compute one or several main cells (we call it composite main cell) using algorithm 1, and then the following:

```
Iterate until no more complex cell.
    Select a complex cell.
    Search for the neighbour extremities among ordinary cells.
    Enumerate cases.
    For each case:
        Compute circles based on each combination of extremeties.
        Project points onto the nearest circle.
        Compute error.
    Choose the best case, and compute the subsets of points
        corresponding to each circle.
    For each subset:
        Compute corresponding arc.
        Link arc with neighbour extremities of ordinary cells,
            and update the composite main cell.
    Remove the complex cell.
End iterate
```

Property 1. *Given n points in the plane, the complexity of the algorithm we describe here is $\mathcal{O}(n \log n)$.*

Proof: : ([4]) Time $\mathcal{O}(n \log n)$ is due to quadtree computation and point sort. All other steps need time $\mathcal{O}(n)$ or $\mathcal{O}(m \log n)$, where m is the number of cells. ∎

Remark: A large number of points (over one million) can be processed quickly (in about 10 minutes).

§5. Examples

We now give some examples, with self-intersection, noisy data, or cusps. The progress of the curve reconstruction is illustrated in Figure 1. In Figure 2, it can be seen that algorithm doesn't fail with noisy data, because of a smoothing effect due to the approximating circle. Finally, Figures 3 and 4 illustrate the cases with complex cells, where curve arcs are not yet computed at step 3. We remark that as shown in Figure 4, in practice, we can reconstruct the curve even if it is not C^2.

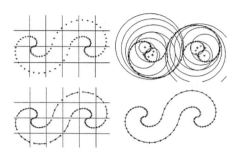

Fig. 2. The four steps of reconstruction.

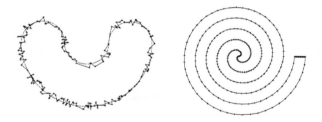

Fig. 3. Noisy data (on left) and spiral.

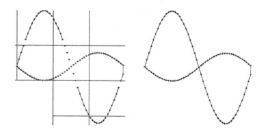

Fig. 4. Self-intersection: steps 3 and 4.

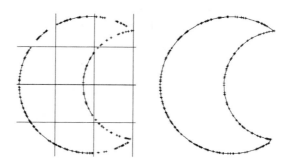

Fig. 5. Cusps: steps 3 and 4.

§6. Conclusion

We present here a composite approach for curve reconstruction based on proximity and subset computation. We need only a few cells to reconstruct the curve from cutting up, and since circle approximation is robust, we get time effiency which enables us to process a great number of points. We also get good performance in the case of noisy data.

Acknowledgements. I wish to thank Bernard LACOLLE who supervise me in that work.

References

1. J.-D. Boissonat,Geometric Structures for Three-Dimensional Shape Representation, ACM Trans. Graphics, 3 (1984), 266–286.
2. T. DeRose et al., Surface reconstruction from unorganized points, Computer Graphics **26** (1992), 71–78.
3. Dedieu, J. P., and C. Favardin, How to draw a curve using geometrical data, in *Curves and Surfaces*, P.-J. Laurent, A. Le Méhauté, and L. L. Schumaker (eds.), Academic Press, New York, 1991, 135–138.
4. Roux, J.-C., Thesis, Université Joseph Fourier, Grenoble, 1994.

Jean-Christophe Roux
Laboratoire LMC-IMAG
Université Joseph Fourier
BP 53X, 38041 Grenoble cedex FRANCE
roux@imag.fr

The Ubiquitous Ellipse

G. Sapiro and A. M. Bruckstein

Abstract. We discuss three different affine invariant evolution processes for smoothing planar curves. The first one is derived from a *geometric heat-type flow*, both the initial and the smoothed curves being continuous. The second smoothing process is obtained from a discretization of this affine heat equation. In this case the curves are represented by planar *polygons*. The third process is based on *B-spline* approximations. For this process, the initial curve is a planar polygon, and the smoothed curves are continuous. We show that, in the limit, all three affine invariant smoothing processes collapse any initial curve into an *elliptic point*.

§1. Introduction

Multiscale descriptions of signals have been the subject of extensive research. A possible formalism for this topic comes from the idea of multiscale filtering that was introduced by Witkin [27], and developed in a variety of frameworks over the past decade [2, 16, 28]. The basic idea of scale-space representations is to filter the signal $\Phi_0(\vec{X}) : \mathbb{R}^n \to \mathbb{R}^m$ with a kernel $\mathcal{K}(\vec{X}, t) : \mathbb{R}^n \to \mathbb{R}^m$, where $t \in \mathbb{R}^+$ represents the scale.

A classical example of a scale-space kernel is the Gaussian one. In this case, the scale-space is linear, and the filter is defined via convolution. The Gaussian kernel is one of the most studied in the theory of scale-spaces [2, 16, 28]. It has some very interesting properties, one of them being that the signal $\Phi(\vec{X}, t)$ obtained from it is the solution of the classical heat equation.

One of the lessons from the Gaussian example, is that the scale-space can be obtained as the solution of a partial differential *evolution equation*. The idea of connecting multiscale representations to evolution equations was developed in [1, 15, 23] with a view to various applications.

We describe three different affine invariant multiscale representations of planar curves (boundaries of planar shapes), representations that give increasingly smooth curves. The first one is derived from a *geometric heat-type flow* [21, 22, 23, 24], where both the initial and the smoothed curves are continuous. The second one is obtained from a discretization of this affine heat

Curves and Surfaces in Geometric Design
P. J. Laurent, A. Le Méhauté, and L. L. Schumaker (eds.), pp. 409–418.
Copyright © 1994 by A K PETERS, Wellesley, MA
ISBN 1-56881-039-3.

equation [6]. In this case the curves are represented by *polygons*. The third
process is based on *B-spline* approximations [19]. For this process, the initial
curve is given by a polygon, and the smoothed curves are continuous. As a
nice consequence of affine invariance, we show that all these processes deform
an arbitrary initial curve into an *elliptic point*. See the mentioned references
and [20] for details.

§2. The Affine Geometric Heat Flow

We assume in this section that the curves are sufficiently smooth, so that the
derivatives are well defined. Consider a family of parameterized planar curves
$C(u, t) : [a, b] \times [0, \tau) \to \mathbb{R}^2$, defined via the evolution equation

$$\frac{\partial C}{\partial t} = \frac{\partial^2 C}{\partial p^2}, \tag{1}$$

with the initial condition $C(u, 0) = C_0(u)$.

If $p \equiv u$, then (1) becomes the classical heat equation discussed in the
introduction. If however $p \equiv v$, where v is the *Euclidean arc-length*, we obtain
the *Euclidean shortening flow*, or *Euclidean geometric heat flow*, see [12, 14].
Gage and Hamilton [12] proved that any simple convex curve converges into a
circular point when evolving according to the Euclidean geometric flow. Then
Grayson [14] proved that any simple non-convex curve converges into a convex
one. Therefore, any simple curve evolves into a circular point when evolving
according to the Euclidean geometric heat flow. This flow defines a geometric
Euclidean invariant multiscale representation [1, 15].

A natural question is whether one can obtain a multiscale representation,
similar to that obtained via the Euclidean heat flow, invariant under the group
of affine transformations. In [21, 22, 24], it was shown that if $p \equiv s$ in (1),
where s is the *affine arc-length*, i.e., the basic affine invariant parameterization
[4, 5, 21], then the *affine shortening flow*, or *affine geometric heat flow*, is
obtained. The main result of [21] is:

Theorem 1. *Any convex, smooth, and embedded initial curve, remains convex and smooth, and converges to an elliptical point when evolving according
to (1), with $p \equiv s$. The convergence is in the sense that the normalized dilated
curves converge in the Hausdorff metric to an ellipse.*

These results were also extended for non-convex curves. Since C_{ss} is not
defined at inflection points [4], this involves the study of the following flow
[22, 24]:

$$\frac{\partial C}{\partial t} = \begin{cases} C_{ss} & \text{non inflection points} \\ 0 & \text{inflection points.} \end{cases} \tag{2}$$

Equation (2) is the natural extension of (1) (with $p \equiv s$), and for this
flow, the following result holds [22, 24]:

Theorem 2. *Let $\mathcal{C}(\cdot, 0) : S^1 \to \mathbb{R}^2$ be a smooth embedded curve in the plane. Then there exists a family $\mathcal{C} : S^1 \times [0, T) \to \mathbb{R}^2$ satisfying (2), such that $\mathcal{C}(\cdot, t)$ is smooth and embedded for all $t < T$, and moreover there is a $t_0 < T$ such that for all $t > t_0$, $\mathcal{C}(\cdot, t)$ is smooth and convex.*

Therefore, by the results in [21, 22, 24], any simple smooth curve converges to an elliptic point (becoming convex first) when evolving according to the affine geometric heat flow, being (2) the affine invariant analogue of the Euclidean heat flow.

The affine flow (2) can be implemented using an efficient numerical algorithm for curve evolution proposed by Sethian and Osher in [18], and based on this, a geometric affine invariant multiscale smoothing for planar curves is available [23]. Fig. 1 presents an example of outlines of a *hand*, related by affine transformations, evolving according to (2).

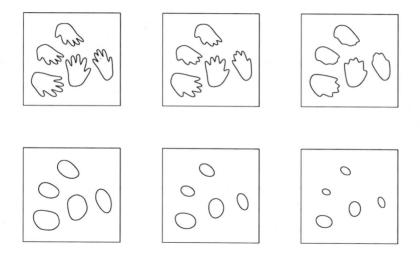

Fig. 1. *Hands*, related by affine transformations evolving via the *affine heat flow*.

§3. Polygonal Affine Invariant Evolution

Consider a planar polygon P with N vertices. P may be non-convex and even self intersecting. Each one of the vertices P_i of P, $i = 0, 1, ..., N - 1$, can be represented by a point in the complex plane, i.e., the polygon P is an N-dimensional vector over the complex plane, $P = [P_0, P_1, ..., P_{N-1}]^T$.

A general linear evolution of the polygon is described by

$$P(n) = MP(n - 1), \tag{3}$$

with the initial condition $P(0) = P$, where M is a constant $N \times N$ complex matrix, and $n \in \mathbb{N}^+$ is the discrete time. The linear polygonal evolution given by equation (3), is affine invariant if M is real, i.e., if the points of $P(n)$ and $\tilde{P}(n)$ are related by an affine mapping, and $P(n)$ evolves according to (3), $\tilde{P}(n)$ also evolves according to (3), with the same evolution matrix M.

Next we shall show that a polygonal version of the *affine heat flow* (2) takes the form of (3). (For details, and general results on linear polygonal evolutions, see [6].) We set the parameterization p of the polygon P to be consecutive integers (modulo N) at the vertices, so that for $i \in \{0, 1, ..., N-1\}$, $P(i, n) \equiv P_i(n)$ (the i-th vertex of the polygon $P(n)$). Note that since polygon vertices (i.e., curve breakpoints) are affine invariant, a straightforward discrete affine arc-length may be chosen so that at the i-th vertex this arc-length is i. With this naturally affine invariant parameterization, a straightforward discretization of (2) leads to a linear evolution of type (3), where M is a real circulant matrix with first row m given by

$$m = \left[1 - c, \frac{c}{2}, 0, ..., 0, \frac{c}{2}\right], \tag{4}$$

where $c \in (0, 1]$. We see that the evolution of each polygon vertex is a step towards the local weighted center of mass.

A well-known property of circulant matrices is that they can be represented as $M = U\Lambda U^{-1}$, where U is the orthogonal Fourier matrix with columns W_i, and Λ is the diagonal matrix of eigenvalues λ_i of M. It is easy to check that $U^{-1} = \frac{1}{N}U^*$ (where U^* stands for the conjugate transpose of U). Now, we obtain

$$P(n) = M^n P = \frac{1}{N}U\Lambda^n U^* P = \frac{1}{N}\sum_{i=0}^{N-1}(\lambda_i)^n DFT_i(P)W_i.$$

Assume that the matrix M in (3) is normalized such that $\max_i |\lambda_i| = 1$, and define

$$P^\infty(n) := \frac{1}{N}\sum_{\{i:|\lambda_i|=1\}} \exp\{jn \arg(\lambda_i)\} DFT_i(P)W_i, \tag{5}$$

where $\arg(x)$ stands for the complex argument of x. We clearly have that $P(n)$ converges to $P^\infty(n)$ in the sense that $\lim_{n\to\infty} |P_i(n) - P_i^\infty(n)| = 0$.

In the case of m as in (4) (i.e., the discretized affine heat flow), it is easy to show that λ_i is real for all i. Also, the biggest eigenvalue is $\lambda_0 = 1$. Therefore, P^∞ is simply the centroid of the initial polygon. Since the limiting polygon P^∞ is a point, we can ask about the shape $P(n)$ takes while approaching P^∞. In order to investigate this, consider the polygon $P(n) - P^\infty$ normalized as follows:

$$B(n) := \frac{1}{N(\max_{i\neq 0}\{|\lambda_i|\})^n}\sum_{i\neq 0}(\lambda_i)^n DFT_i(P)W_i.$$

Define

$$B^\infty(n) := \frac{1}{N} \sum_{\{i:|\lambda_i|=\max_{i \neq 0}\{|\lambda_i|\}\}} \exp\{jn\arg(\lambda_i)\}DFT_i(P)W_i, \qquad (6)$$

and $B^\infty(n)$ provides the geometric behavior of the polygon when $n \to \infty$ (i.e., when $P(n)$ converges to P^∞). From equation (6) it is clear that the shape of the polygon, when approaching P^∞, is governed by the second greatest eigenvalue. For the evolution defined by (4), we have [6]

Theorem 3. *Let $P(n)$ be a polygon evolving according to the evolution equation (4), with $P(0) = P$. Assume that $N \neq 4$, $0 < c \leq \frac{2}{3}$, and $DFT_1(P) \neq 0$ or $DFT_{N-1}(P) \neq 0$. Then $P(n)$ converges to the centroid of the initial polygon, and the normalized polygon $B^\infty(n)$ converges to a fixed polygonal approximation of an ellipse, which is an affine transformation of a regular polygon.*

The result in Theorem 3 is not unexpected in the light of the observation that this evolution is a discretized polygonal version of the affine curve evolution studied in [21, 22, 23, 24]. It is also interesting to note that while the discrete analog is simple to analyze, the study of the continuous affine geometric heat flow requires advanced methods from the theory of partial differential equations and affine differential geometry. Fig. 2 shows examples of this polygonal evolution.

We end this section by pointing out that after we studied this subject of the discrete version of the affine geometric heat flow, we learned that the topic of linear polygonal evolutions has an extensive literature, starting with a beautiful paper [8] by Darboux written in 1878. Subsequently, long after the results by Darboux were forgotten, other researchers re-discovered some of these results. Among them we mention I. J. Schoenberg [25] in 1950, J. H. Cadwell [7] in 1953, E. R. Berlekamp *at al.* [3] in 1965, L. Fejes Tóth [11] in 1969. Many other researchers worked on this and related fascinating problems connecting Fourier analysis with basic geometry. See [6, 20] for an extended reference list of old and new related works.

§4. The B-Spline based Representation

We shall next discuss yet another affine smoothing process. Note that in the first example, the curves are continuous, and in the second one, they are represented by planar polygons. In the smoothing process presented below, the original curve is a polygon, while the evolved smoothed curves are continuous.

We briefly review the theory of B-spline approximations. For details see [9, 10, 26]. Let $\mathcal{C}(u) : [a, b] \to \mathbb{R}^2$ be a planar curve with Cartesian coordinates $[x(u), y(u)]$. Polynomials are computationally efficient to work with, but it is not always possible to describe well enough a curve $\mathcal{C}$ using single polynomials for x and y. Therefore, in applications, the curve is described

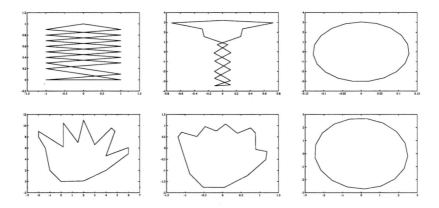

Fig. 2. Two examples of the polygonal evolution.

as a sequence of segments, each one defined by a given polynomial. The segments are joined together to form a *piecewise polynomial* curve. The joints between the polynomial segments occur at special curve points called *knots*. The sequence u_1, u_2, ... of knots is required to be nondecreasing. The "distance" between two consecutive knots can be constant or not. Two successive polynomial segments are joined together at a given knot u_j in such a way that the resulting piecewise polynomial has d continuous derivatives. Of course, the order of the polynomials depends on d.

Formally, the curve $\mathcal{C}$ is a *B-spline* approximation of the series of points $V_i = [x_i, y_i]$, $1 \leq i \leq N$, called *control vertices*, if it can be written as

$$\mathcal{C}_k(u) = \sum_{i=1}^{N} V_i B_{i,k}(u), \tag{7}$$

where $B_{i,k}(\cdot) = B(\cdot\,;\, u_i,\, u_{i+1},\, ...,\, u_{i+k})$ is the i-th *B-spline basis* of order k for the knot sequence $[u_1,\, ...,\, u_{N+k}]$. In particular, $B_{i,k}$ is a piecewise polynomial function of degree $< k$, with breakpoints $u_j,\, ...,\, u_{j+k}$. The multiplicity of the knots governs the smoothness. If a given number τ occurs r times in the knot sequence $[u_i,\, ...,\, u_{i+k}]$, then the first $k-r-1$ derivatives of $B_{j,k}$ are continuous at the breakpoint τ. For properties of the basis $B_{i,k}$ see for example [9, 10, 26].

Observe that from (7), the affine invariant property of the B-spline representation is immediate. If $\{\tilde{V}_i\}_1^N$ is obtained from $\{V_i\}_1^N$ by an affine transformation (A, T) (A being a real 2×2 matrix and $T \in \mathbb{R}^2$ a translation vector), then the corresponding B-splines are related by the same affine transformation. Based on this, we can define a B-spline based, affine invariant, multiscale shape representation (*BAIM*) of the polygon described by the points $\{V_i\}_1^N$, as the family of curves $\mathcal{C}_k$ obtained from (7) for $k = 2, 3, ...$ [19]. Note that in contrast with the multiscale representations described in Section 2, the *BAIM* is discrete in the scale parameter ($k = 2, 3, ...$).

The examples presented here were implemented using the Matlab Spline Toolbox [10]. Fig. 3 presents the first *BAIM* example. The polygon contains 12 points. In the left figure, the initial polygon is given, together with the corresponding *BAIM* for $k = 2^i$, $i = 1, 2, 4, 6, 7$. In the right figure, the initial polygon is obtained via an affine transformation of the polygon in the top. Due to the affine invariant property, the corresponding *BAIM* is related to the one in the top by the same affine transformation.

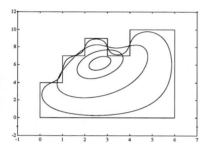 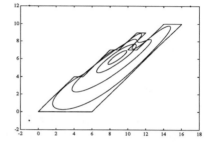

Fig. 3. Affine invariant property of the *BAIM*.

The following theorem shows the behavior of the B-spline approximations as the order k increases [20] (see Fig. 3).

Theorem 4. *As k increases ($k \to \infty$), the B-spline representation converges to the centroid of the control points $\{V_i\}_1^N$, its shape becoming elliptical.*

Fig. 4 gives examples showing the smoothing property of the *BAIM*. Actually, it can be proven formally that the *BAIM* is a smoothing process, (see for example [9, 13, 17, 19, 26]).

§5. Concluding Remarks

In this paper, we presented three different affine invariant smoothing processes for planar curves. The first one is derived from a *geometric heat-type flow*, were both the initial and the smoothed curves are continuous. The second one is obtained from a discretization of this affine heat equation. In this case the curves are represented by *polygons*. The third process is derived from *B-spline* approximations. For this process, the initial curve is given by a polygon, and the smoothed curves are continuous.

Note also that in the first model, the time scale (smoothing scale), is continuous ($t \in [0, \tau]$). In the second one, this scale is discrete, but c can be taken as small as required, and as c decreases, we approach a continuous time evolution process. In the last model, the smoothing scale is related to the B-spline order, therefore, it is strictly discrete ($k = 2, 3, ...$). In the first

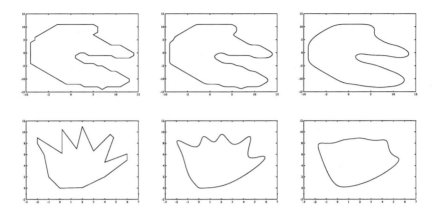

Fig. 4. Two examples of the *BAIM*.

case we could also propose a "discrete time" evolution process, where $\mathcal{C}(n, p)$ is obtained by averaging over a given constant affine arc length neighborhood of $\mathcal{C}(n-1, p)$. This is, for small averaging neighborhoods, an affine invariant numerical approximation of the affine geometric heat flow. To complete the picture, it would be interesting to also find a "continuous time," B-spline based, smoothing process. This can probably be achieved by postulating continuous changes in the influence of $B_{i,k}(u)$ in the weighted average given by (7). Together with Prof. A. Cohen from CEREMADE-Paris, we are currently investigating the use of a continuous-scale B-spline basis, i.e, a basis in $\mathbf{C}^{k+r}$, $r \in [0, 1]$, based on sub-division schemes. The basis reduces to the classical B-spline for $r = 0$, and has final support for any r as well.

We showed that the three processes discussed deform any initial curve into an *elliptic point*. This nice result is hardly unexpected, since the processes are affine invariant smoothing operations, and the ellipse is the smoothest affine invariant shape. What is interesting to note is that each type of smoothing process requires a different approach to prove the result.

References

1. Alvarez, L., F. Guichard, P. L. Lions, and J. M. Morel, Axioms and fundamental equations of image processing, Arch. for Rational Mechanics, to appear.
2. Babaud, J., A. P. Witkin, M. Baudin, and R. O. Duda, Uniqueness of the Gaussian kernel for scale-space filtering, IEEE Trans. Pattern Anal. Machine Intell. **8** (1986), 26–33.
3. Berlekamp, E. R., E. N. Gilbert, and F. W. Sinden, A polygon problem, The American Mathematical Monthly **72** (1965), 233–241.
4. Blaschke, W. *Vorlesungen über Differentialgeometrie II,* Springer, Berlin, 1923.

5. Bruckstein, A. M., and A. N. Netravali, On differential invariants of planar curves and recognizing partially occluded planar shapes, in *Proc. of Visual Form Workshop*, Capri, Plenum Press, 1991.

6. Bruckstein, A. M., G. Sapiro, and D. Shaked, Evolutions of planar polygons, CIS Report **9202**, Department of Computer Science, Technion, I. I. T., Haifa 32000, Israel, 1992, submitted.

7. Cadwell, J. H., A property of linear cyclic transformations, Math. Gaz. **37** (1953), 85–89.

8. Darboux, M. G., Sur un problème de géométrie élémentaire, Bull. Sci. Math **2** (1878), 298–304.

9. de Boor, C., *A Practical Guide to Splines*, Applied Mathematical Sciences **27**, Spinger-Verlag, New York, 1978.

10. de Boor, C., *Spline Toolbox for use with MATLAB*^{TM}, The MathWorks, Inc., Natick, 1990.

11. Fejes Tóth, L., Iteration methods for convex polygons (in Hungarian), Mat. Lapok **20** (1969), 15–23.

12. Gage, M., and R. S. Hamilton, The heat equation shrinking convex plane curves, J. Differential Geometry **23** (1986), 69–96.

13. Goodman, T. N. T., Inflections on curves in two and three dimensions, Computer Aided Geometric Design **8** (1991), 37–50.

14. Grayson, M., The heat equation shrinks embedded plane curves to round points, J. Differential Geometry **26** (1987), 285–314.

15. Kimia, B. B., A. Tannenbaum, and S. W. Zucker, Toward a computational theory of shape: An overview, in *Lecture Notes in Computer Science* **427**, Springer-Verlag, New York, 1990.

16. Koenderink, J. J., The structure of images, Biological Cybernetics (1984) **50**, 363–370.

17. Lane, J. M., and R. F. Riesenfeld, A geometric proof for the variation diminishing property of B-spline approximation, J. Approx. Theory **37** (1983), 1–4.

18. Osher, S. J., and J. A. Sethian, Fronts propagation with curvature dependent speed: Algorithms based on Hamilton-Jacobi formulations, Journal of Computational Physics **79** (1988), 12–49.

19. Sapiro, G., and A. M. Bruckstein, A B-spline based affine invariant multiscale shape representation, CIS Report **9303**, Department of Computer Science, Technion, I. I. T., Haifa 32000, Israel, January 1993, submitted.

20. G. Sapiro and A. M. Bruckstein, The ubiquitous ellipse, CIS Report **9304**, Department of Computer Science, Technion, I. I. T., Haifa 32000, Israel, January 1993, submitted.

21. Sapiro, G., and A. Tannenbaum, On affine plane curve evolution, J. Functional Analysis, Jan. 1994.

22. Sapiro, G., and A. Tannenbaum, Affine shortening of non-convex plane curves, EE Publication **845**, Department of Electrical Engineering, Technion, I. I. T., Haifa 32000, Israel, July 1992, submitted.

23. Sapiro, G., and A. Tannenbaum, Affine invariant scale-space, Int. J. Computer Vision **11**, 1 (1993), 25–44.

24. Sapiro, G., and A. Tannenbaum, On invariant curve evolution and image analysis, Indiana J. Math. **42**, 3 (1993), 985–1009.
25. Schoenberg, I. J., The finite Fourier series and elementary geometry, Amer. Math. Monthly **57** (1950), 390–404.
26. Schoenberg, I. J., *Cardinal Spline Interpolation*, SIAM Press, Philadelphia, 1973.
27. Witkin, A. P., Scale-space filtering, in *Int. Joint. Conf. Artificial Intelligence* (1983), 1019–1021.
28. Yuille and T. A. Poggio, A. L., Scaling theorems for zero crossings, IEEE Trans. Pattern Anal. Machine Intell. **8** (1986), 15–25.

Guillermo Sapiro and Alfred M. Bruckstein
Technion-Israel Institute of Technology
Haifa, ISRAEL 32000
guille@ techunix.technion.ac.il
freddy@ cs.technion.ac.il

Axial Convexity:
A Well-shaped Shape Property

Thomas Sauer

Abstract. The paper is concerned with axial convexity, a weaker generalization of univariate convexity, which is, in contrast to classical convexity, preserved by multivariate Bernstein polynomials. In the first part, similarities and differences between axial and classical convexity are established as well as the close tie between axial convexity and parallel subsimplices, while the second part considers the applicability of axial convexity for design purposes.

§1. Introduction

It was pointed out by Schmid [8] as early as 1975 and later rediscovered by Chang and Davis [1] that Bernstein polynomials in two variables do not preserve convexity any more. Posing the question whether there are generalizations of univariate convexity which are preserved by bivariate Bernstein polynomials, Schmid arrived at a property weaker than convexity, called axial convexity. Moreover, he showed that axial convexity is not only preserved by Bernstein polynomials, it even causes the sequence of Bernstein polynomials to converge montonically towards the approximated function.

In this paper we point out some more properties of axial convexity, in particular some similarities and differences between axial convexity and convexity, and establish the close tie between axial convexity and parallel subsimplices.

In order to formulate our results, we first have to introduce some notation for Bernstein polynomials on multidimensional simplices. Let

$$\Delta_m := \{u = (u_0, \ldots, u_m) : u_k \geq 0, u_0 + \cdots + u_m = 1\}$$

denote the m–dimensional unit simplex in barycentric coordinates and let e^j, $j = 0, \ldots, m$, be the unit vectors in Δ_m. Then for $u \in \Delta_m$ and any multiindex $\alpha \in \mathbb{N}_0^{m+1}$ the *Bernstein–Bézier basis polynomial* B_α is given as

$$B_\alpha(u) := \frac{|\alpha|!}{\alpha!} u^\alpha = \frac{(\alpha_0 + \cdots + \alpha_m)!}{\alpha_0! \cdots \alpha_m!} u_0^{\alpha_0} \cdots u_m^{\alpha_m}.$$

Curves and Surfaces in Geometric Design

P. J. Laurent, A. Le Méhauté, and L. L. Schumaker (eds.), pp. 419–425.

Copyright © 1994 by A K PETERS, Wellesley, MA

ISBN 1-56881-039-3.

Using coefficients $f_\alpha \in \mathbb{R}$, $|\alpha| = n$, (throughout the whole paper we will restrict to the functional case) we define the *Bernstein–Bézier polynomial* as

$$B_n f(u) := \sum_{|\alpha|=n} f_\alpha B_\alpha(u), \tag{1}$$

and if $f \in C(\Delta_m)$, we always take $f_\alpha = f(\alpha/|\alpha|)$. With this convention (1) denotes the classical Bernstein polynomial, cf. [5].

A useful tool from CAGD is the so–called degree raising formula which reads as

$$B_n f = \sum_{|\alpha|=n} f_\alpha B_\alpha = \sum_{|\alpha|=n+1} \hat{f}_\alpha B_\alpha, \quad \hat{f}_\alpha := \sum_{k=0}^{m} \frac{\alpha_k}{n+1} f_{\alpha-e^k}. \tag{2}$$

In order to describe derivatives of Bernstein–Bézier polynomials, the shift operators $E_j f_\alpha = f_{\alpha+e^j}$, $j = 0, \ldots, m$, have proved to be very convenient, mainly because of the relation

$$D_{e^j-e^k} B_n f = n \sum_{|\alpha|=n-1} (E_j - E_k) f_\alpha B_\alpha.$$

Next we formally define the property under consideration:

Definition 1. *A function $f \in C(\Delta_m)$ is called axially convex, if*

$$f(\lambda u + (1-\lambda)v) \leq \lambda f(u) + (1-\lambda)f(v), \quad \lambda \in [0,1],$$

whenever $u, v \in \Delta_m$ lie on a line parallel to one of the edges (or "axes") of Δ_m; i.e., if $u - v = \rho(e^j - e^k)$ for appropriate $\rho \neq 0$ and $0 \leq j < k \leq m$.

Obviously, any convex function is axially convex, too.

A (possibly lower dimensional) subsimplex $\Delta \subset \Delta_m$, written as the convex hull $[u^0, \ldots, u^\mu]$ of its affinely independent vertices $u^0, \ldots, u^\mu$, $1 \leq \mu \leq m$, will be called *parallel* if all of its edges are parallel to edges of Δ_m; i.e., for any $0 \leq i < j \leq \mu$ there exist $0 \leq k \leq l \leq m$ and $\rho \neq 0$ such that

$$u^i - u^j = \rho(e^k - e^l). \tag{3}$$

From this definition one can prove that a subsimplex $\Delta = [u^0, \ldots, u^m]$ (i.e., Δ is of "full" dimension m) is parallel if and only if there exists some $\sigma \neq 0$ such that $u^j - u^k = \sigma(e^j - e^k)$, $0 \leq j, k \leq m$, at least after numbering the vertices properly. It should be noticed that this means that the number ρ in (3) has to be independent of j and k. This in turn yields that, considering $|\sigma|$ to be a "size" parameter, there are essentially two types of parallel m–dimensional subsimplices, according to the sign of σ. This situation is depicted in Fig. 1 for $m = 3$.

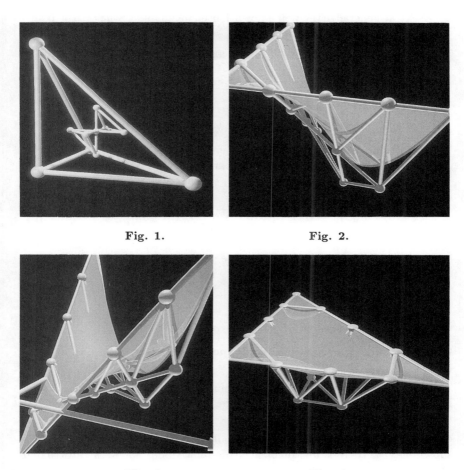

Fig. 1.   Fig. 2.

Fig. 3.   Fig. 4.

§2. Axially Convex Functions

First we give some characterizations of axially convex functions which can be found in [6]:

Theorem 2. *For $f \in C(\Delta_m)$ the following statements are equivalent:*

(i) *f is axially convex.*

(ii) *For $|\alpha| = n - 2$ and $0 \le j, k \le m$*

$$(E_j - E_k)^2 f_\alpha \ge 0.$$

(iii) *If, in addition, $f \in C^2(\Delta_m)$,*

$$D^2_{e^j - e^k} f(u) \ge 0,$$

for $u \in \Delta_m$ and $0 \le j, k \le m$.

(iv) *$B_n f$ is axially convex for all $n \in \mathbb{N}$.*

Next we generalize the well–known Jensen inequality to the case of axially convex functions:

Theorem 3. *A function $f \in C(\Delta_m)$ is axially convex if and only if*

$$f(u) \leq \sum_{k=0}^{\mu} v_k f(u^k), \quad u = \sum_{k=0}^{\mu} v_k u^k, \quad v \in \Delta_\mu, \tag{4}$$

holds for any parallel subsimplex $\Delta = [u^0, \ldots, u^\mu]$.

Recall that we obtain the classical Jensen inequality; i.e., a description of convexity, if we just drop the word "parallel" in the above theorem.

Using the degree raising formula (2) to compute

$$B_{n-1}f - B_n f = \sum_{|\alpha|=n} \left(\sum_{k=0}^{m} \frac{\alpha_k}{n} f_{\alpha - e^k} - f_\alpha \right) B_\alpha$$

and noticing that the points $u^k = (\alpha - e^k)/(|\alpha| - 1)$ form a parallel simplex and satisfy the condition

$$\sum_{k=0}^{m} \frac{\alpha_k}{|\alpha|} u^k = \frac{\alpha}{|\alpha|},$$

we obtain (as an immediate consequence of Theorem 3):

Corollary 4. *If $f \in C(\Delta_m)$ is axially convex, then $B_n f \geq B_{n+1} f$.*

With some more work, we can also deduce a strong maximum principle from Theorem 3. It seems worthwhile mentioning here that once again the difference with the corresponding result for convex functions lies just in the restriction to parallel subsimplices.

Corollary 5. *Let $f \in C(\Delta_m)$ be axially convex and let Δ be a parallel subsimplex. If there is one point u in the relative interior of Δ such that in (4) equality holds, then f is linear on Δ.*

Thus one sees that axially convex functions are essentially determined by their behavior on parallel simplices. Conversely, it is also possible to characterize parallel simplices using axially convex functions, namely

Theorem 6. *The restriction of any axially convex function to a subsimplex Δ is again axially convex (with respect to the edges of Δ), if and only if Δ is parallel.*

Proofs of the results presented in this section can be found in [7].

Looking at Theorem 3 and Corollary 5, one might have the idea that axial convexity could not be very far away from convexity. On the other hand, there are bivariate, axially convex functions being formed like the front part of a ship (Fig. 2), having saddle points (Fig. 3) or being nonconvex at any of the three vertices (Fig. 4). Hence, axial convexity is really quite a lot weaker than convexity.

§3. Axially Convex Bézier Nets

For construction purposes it is convenient to join the data points f_α, $|\alpha| = n$, by a piecewise linear function, the so–called *Bézier net*. This is obtained by first joining any of the points $\alpha/n \in \Delta_m$ with its neighbors $(\alpha + e^j - e^k)/n$, $0 \leq j, k \leq m$. For $m = 2$ this induces the canonical triangulation of Δ_2 (in fact, that is exactly the way the data points are connected in Figs. 2–4), while for $m > 2$ one has to introduce further edges (for details cf. [2,6]). Nevertheless, in either case we obtain a triangulation of Δ_m with vertices α/n and define the *Bézier net* according to f_α as the unique function that is linear on any simplex of the above triangulation and interpolates f_α at α/n.

We call the Bézier net of f_α, $|\alpha| = n$, *axially convex* if

$$(E_j - E_k)^2 f_\alpha \geq 0, \quad |\alpha| = n - 2, \quad 0 \leq j, k \leq m. \tag{5}$$

It is an important fact that the above condition does not mean the Bézier net, taken as a piecewise linear continuous function, is axially convex in the sense of Definition 1, for this can be shown to require that the Bézier net itself is already convex, which is much too strong. Note, moreover, that (5) is independent of the chosen triangulation if $m > 2$. On the other hand, the above definition is reasonable, particularly in view of

Theorem 7. *Axial convexity of the Bézier net is preserved by degree raising.*

Proof: It is easily seen that (5) holds if and only if the symmetric second order difference

$$\nabla_{j,k} f_\alpha := f_{\alpha+(e^j-e^k)} - 2f_\alpha + f_{\alpha-(e^j-e^k)}$$

is nonnegative for any $|\alpha| = n$, $\alpha_j, \alpha_k \geq 1$ and $0 \leq j, k \leq m$.

Incorporating the degree raising process (2) we compute for $|\beta| = n + 1$

$$\hat{f}_{\beta+(e^j-e^k)} = \frac{1}{n+1} \sum_{i=0}^{m} (\beta_i + \delta_{ij} - \delta_{jk}) f_{\beta-e^i+(e^j-e^k)}$$

$$= \frac{1}{n+1} \left(\sum_{i=0}^{m} \beta_i f_{\beta-e^i+(e^j-e^k)} + f_{\beta-e^j+(e^j-e^k)} - f_{\beta-e^k+(e^j-e^k)} \right).$$

Hence, if $\beta_j, \beta_k \geq 1$,

$$\nabla_{j,k} \hat{f}_\beta = \frac{1}{n+1} \Big(\sum_{i=0}^{m} \beta_i \nabla_{j,k} f_{\beta-e^i} + f_{\beta-e^k} - f_{\beta-e^k+(e^j-e^k)}$$

$$+ f_{\beta-e^j} - f_{\beta-e^j+(e^k-e^j)} \Big).$$

Since

$$f_{\beta-e^k+(e^j-e^k)} - f_{\beta-e^k} = f_{\beta-e^k+(e^j-e^k)} - 2f_{\beta-e^k} + f_{\beta-e^k-(e^j-e^k)}$$

$$+ f_{\beta-e^k} - f_{\beta-e^k-(e^j-e^k)}$$

$$= \nabla_{j,k} f_{\beta-e^k} + f_{\beta-e^k} - f_{\beta-e^j},$$

and, symmetrically,

$$f_{\beta - e^j + (e^k - e^j)} - f_{\beta - e^j} = \nabla_{j,k} f_{\beta - e^j} + f_{\beta - e^j} - f_{\beta - e^k},$$

we obtain that

$$\nabla_{j,k} \hat{f}_\beta = \frac{1}{n+1} \left(\sum_{i=0}^m \beta_i \nabla_{j,k} f_{\beta - e^i} - \nabla_{j,k} f_{\beta - e^k} - \nabla_{j,k} f_{\beta - e^j} \right)$$

$$= \frac{1}{n+1} \left(\sum_{i=0}^m (\beta_i - \delta_{ij} - \delta_{ik}) \nabla_{j,k} f_{\beta - e^i} \right) \geq 0,$$

since we assumed that $\beta_j, \beta_k \geq 1$. Thus the Bezier net of $\hat{f}_\beta$, $|\beta| = n + 1$, is also axially convex. ∎

If we apply degree raising to $\hat{f}_\beta$, $|\beta| = n + 1$, and so on, we obtain a sequence of axially convex Bézier nets converging uniformly to $B_n f$. Thus,

Proposition 8. *If the Bézier net f_α, $|\alpha| = n$ is axially convex, then so is $B_n f$.*

Proof: By virtue of Theorem 2, this is is almost obvious since

$$D^2_{e^j - e^k} B_n f = n(n-1) \sum_{|\alpha| = n-2} (E_j - E_k)^2 f_\alpha B_\alpha \geq 0. \quad \blacksquare$$

On the other hand, axial convexity of bivariate Bézier nets can be very easily checked by tracing the piecewise linear curves joining

$$f_\alpha, \; f_{\alpha + (e^j - e^k)}, f_{\alpha + 2(e^j - e^k)}, \ldots, f_{\alpha + \alpha_k(e^j - e^k)}$$

for $|\alpha| = n$ and $0 \leq j, k \leq m$. If all of these curves are convex, then so is the Bézier net and hence the surface, too. Thus, there is a "visible" sufficient condition for the axial convexity of a functional Bézier surface.

Another important point in CAGD is subdivision of Δ and restriction of the surface to the respective subdomains. Since, by virtue of Theorem 6, we need parallel subsimplices to preserve axial convexity one should use a subdivision method which produces parallel subsimplices. If $m = 2$ this is essentially the so–called "midpoint subdivision" introduced by Goodman [3], which is also the only type of subdivision preserving convexity of the Bézier net (see [4]), while for $m > 2$ there is no subdivision method which is able to produce parallel subsimplices.

To explain this somewhat unexpected result, we first note that the vertices of parallel subsimplices can always be numbered such that they satisfy $u^j = u^k + \sigma(e^j - e^k)$, $j = 0, \ldots, m$, for any fixed $0 \leq k \leq m$. Using that bad lack of necessary flexibility, we proceed as follows: first we notice that the only parallel simplex fitting to the vertex, say e^1, has to be of the form $[u^0, \ldots, u^m]$ where $u^0 = \sigma_0 e^0 + (1 - \sigma_0)e^1$ and $u^k = u^0 + \sigma_0(e^k - e^0)$, $k = 1, \ldots, m$, for some $\sigma_0 > 0$. Let us call this simplex Δ^0. Now any parallel subsimplex $\Delta^1 = [v^0, \ldots, v^m]$ that shares with Δ^0 at least a part of the face $[u^0, u^2, \ldots, u^m]$ must satisfy (after permuting the vertices) $v^0 = u^0$, $v^k = v^0 + \sigma_1(e^k - e^0)$, $k = 2, \ldots, m$, with $0 < \sigma_1 \leq \sigma_0$; note that this significantly depends from the assumption that $m > 2$, since otherwise one can simply set $v^0 = u^2$ and $v^2 = u^0$, obtaining some $\sigma_1 < 0$. But now parallelity of Δ implies $v^1 = v^0 + \sigma_1(e^1 - e^0)$ and hence $\Delta^1 \subset \Delta^0$ which is definitely not what one expects of a triangulation.

Conclusion. From the mathematical point of view axial convexity combines a high degree of flexibility (see again Figs. 2–4) with some controlled convexity behavior. Moreover, axial convexity of the Bézier net is a simple, geometric and "visible" property which is (at least for $m = 2$) compatible with a "nice" subdivision process as well as with degree raising. Nevertheless, the question of whether there is any practical use for it must be left to the practitioners.

References

1. Chang, G. and P. J. Davis, The convexity of Bernstein polynomials over triangles, J. Approx. Theory **40** (1984), 11–28.
2. Dahmen, W. and C. A. Micchelli, Convexity of multivariate Bernstein polynomials and box spline surfaces, Studia Sci. Math. Hungar. **23** (1988), 265–287.
3. Goodman, T. N. T., Convexity of Bézier nets on triangulations, Comp. Aided Geom. Design **8** (1991), 175–180.
4. Gregory, J. A. and Zhou, J., Convexity of Bézier nets on sub–traingles, Comp. Aided Geom. Design **8** (1991), 207–211.
5. Lorentz, G. G., *Bernstein polynomials*, University of Toronto Press, 1953.
6. Sauer, T., Multivariate Bernstein polynomials and convexity, Comp. Aided Geom. Design **8** (1991), 465–478.
7. Sauer, T., Parallel subsimplices and convexity, Computer-Aided Geom. Design , to appear.
8. Schmid, H. J., Bernsteinpolynome, manuscript, 1975.

Thomas Sauer
Mathematical Institute
University of Erlangen–Nuremberg
Bismarckstr. $1\frac{1}{2}$
91054 Erlangen, GERMANY
sauer@ mi.uni-erlangen.de

Variation Diminution and Blossoming
for Curves and Surfaces

Gerd Schmeltz

Abstract. This article presents an idea to extend the notion of variation diminution to surfaces. While it is known that direct translations of the geometric definition result in properties not shared by Bézier surfaces (which seems inappropriate), another approach is used here: First, generalized curve blossoms are reviewed which exist iff the underlying curve representation scheme is variation diminishing. Then analogous blossom structures for surfaces are formulated and investigated. They turn out to be far more restrictive than in the case of curves. It is shown that only rational Bézier triangles with positive weights have this property.

§1. Introduction

The variation diminishing property of some curve representation schemes has proven useful in many ways. A generalization to surfaces is therefore desirable but has not yet been found. Since direct geometric approaches have failed [6], this article extends a blossom structure, which is shared by all sufficiently smooth variation diminishing curve schemes, to surfaces. It turns out that while variation diminishing curve schemes exist abundantly [4,5] the only surfaces that have the investigated blossom structure are rational triangular Bézier surfaces. Details that had to be omitted here can be found in [8]; some of the results on curves are also described in [5] from a different point of view. The remainder of this section introduces some notation and includes the definition of the variation diminishing property. Section 2 deals with curves and Section 3 with surfaces.

All curves and surfaces in this article are of the form

$$X(t) = \sum_{j=I} \phi_j(t)P_j, \qquad \text{with } \sum_{j \in I} \phi_j(t) \equiv 1 . \tag{1}$$

Here, I is an index set, ϕ_j are linear independent scalar functions on a domain D and P_j are points in the image space $\mathbb{R}^d$. The ϕ_j are called the *basis*

Curves and Surfaces in Geometric Design
P. J. Laurent, A. Le Méhauté, and L. L. Schumaker (eds.), pp. 427–434.

functions or simply the *basis*; the P_j are referred to as *control points*. For curves, $I = \{0, \ldots, n\}$ for some $n \in \mathbb{N}$ and $D = [a, b]$ with $a, b \in \mathbb{R}$, while for surfaces, $I = \{(i, j, k) \in \mathbb{N}_0^3 : i + j + k = n\}$ and $D = \triangle UVW$ is a triangle with vertices $U, V, W \in \mathbb{R}^2$. In any case n will be called the *order* of the curve or surface, respectively.

Definition 1. *A system of basis functions is called* variation diminishing *(VD) if a curve given by (1) never has more intersection points with a hyperplane than its control polygon, i.e., with the polygonal arc $[P_0, \ldots, P_n]$. For simplicity, hyperplanes that are tangent to the curve or contain a control point are excluded here.*

§2. Blossoming for Curves

In this section, the domain of X is $D = [a, b] \subset \mathbb{R}$ so that X is a parametrized curve. The main objective of the section is to illustrate the connection between the VD property (of curves) and a certain blossoming structure.

Definition 2. *A blossom of order 1 is a function $\widetilde{X} : D \to \mathbb{R}^d$ that satisfies*

$$t \in \text{cvx}\{t^-, t^+\} \Rightarrow \widetilde{X}(t) \in \text{cvx}\{\widetilde{X}(t^-), \widetilde{X}(t^+)\} \quad \text{for } t^-, t^+ \in D . \qquad (2)$$

The notation has been chosen so that the similarity to the surface case becomes apparent. "cvx" stands for "convex hull of". For emphasis, blossoms will usually be referred to as either curve or surface blossoms.

Definition 3. *A blossom of order n is a symmetric function $\widetilde{X} : D^n \to \mathbb{R}^d$ such that for every choice of $t_2, \ldots, t_n$, the function $\widetilde{X}(\cdot, t_2, \ldots, t_n)$ is an order 1 blossom.*

Definition 4. *When fixed values are chosen for some of the arguments of a blossom, a blossom of lower order is produced. This lower order blossom is called a* subblossom *of the higher order blossom.*

The values of a curve blossom can be found by the following *de Casteljau-like procedure*. Define *control points* as

$$P_i := \widetilde{X}(\underbrace{a, \ldots, a}_{n-i}, \underbrace{b, \ldots, b}_{i}) . \qquad (3)$$

Now assume $t \notin \{a, b\}$ and consider the relation (cf. (2))

$$\widetilde{X}(t, t_2, \ldots, t_n) \in \text{cvx}\{\widetilde{X}(a, t_2, \ldots, t_n), \widetilde{X}(b, t_2, \ldots, t_n)\} .$$

Then each of the "$\widetilde{X}$" on the right side of the "$\in$" has more arguments out of the set $\{a, b\}$ than the "$\widetilde{X}$" on the left side. By iteratively using this relation, the n arguments can be inserted one by one.

With every curve blossom $\widetilde{X}$, there is associated a curve X through *diagonalization*:

$$X(t) = \widetilde{X}(t, \ldots t)$$

The computation of a curve point with the generalized de Casteljau Algorithm can be viewed as a *corner-cutting algorithm* [2] as is illustrated in Figure 1 for $n = 3$. The polygonal arc $[\widetilde{X}(0,0,0), \widetilde{X}(0,0,t), \widetilde{X}(0,t,t), \widetilde{X}(t,t,t), \widetilde{X}(t,t,1),$ $\widetilde{X}(t,1,1), \widetilde{X}(1,1,1)]$ can be obtained from the control polygon by repeatedly chopping off its corners.

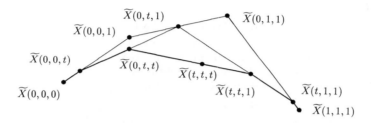

Fig. 1. Corner cutting with a curve blossom – a generalization de Casteljau's Algorithm.

Theorem 5. *A curve that is obtained from a curve blossom has the VD property with respect to control points defined by (3).*

Proof: Let us consider the polygon

$$[\widetilde{X}(s_0, \ldots, s_0, s_0), \widetilde{X}(s_0, \ldots, s_0, s_1), \ldots, \widetilde{X}(s_0, s_1, \ldots, s_1), \widetilde{X}(s_1, \ldots, s_1),$$
$$\widetilde{X}(s_1, \ldots, s_1, s_2), \ldots, \widetilde{X}(s_2, \ldots, s_2), \ldots, \widetilde{X}(s_3, \ldots, s_3), \ldots, \widetilde{X}(s_N, \ldots, s_N)] \tag{4}$$

with $a = s_0 \leq \cdots \leq s_N = b$. If $N = 1$, this is the control polygon; if $N = 2$, this is the result of one application of de Casteljau's Algorithm using $s = s_1$. In the general case, use this algorithm several times: first with $s = s_1$ on the control polygon, then with $s = s_2$ on the part of the resulting polygon between $\widetilde{X}(s_1, \ldots, s_1)$ and $\widetilde{X}(s_N, \ldots, s_N)$ and so on until we end up with Polygon (4). Because de Casteljau's Algorithm is a corner-cutting procedure, our polygon cannot intersect any hyperplane more often than the control polygon. Since it has an arbitrarily chosen number of curve points $\widetilde{X}(s_i)$ among its vertices, the same is true of the curve itself. ■

The idea to use corner cutting as a method of proof in this way has been introduced by Goodman [3]. With every curve X of the form (1) that satisfies certain smoothness and non-degeneracy conditions (see below) there is associated a curve blossom $\widetilde{X}$ through

$$\widetilde{X}(\underbrace{t_1, \ldots, t_1}_{m_1}, \ldots, \underbrace{t_k, \ldots, t_k}_{m_k}) := \bigcap_{i=1}^{k} E^{n-m_i}(t), \tag{5}$$

where the t_i are from D and distinct, and $E^{n-m_i}(t)$ is the $n - m_i$'th osculating space of X at parameter t. The convention $E^0(t) = \{X(t)\}$ is used.
Furthermore, $\widetilde{X}$ is stipulated to be symmetric. The set on the right hand side
is identified with its one element. The following theorem was proved in [8].

Theorem 6. *Let X be a curve of the form (1) with a C^{2n} variation diminishing basis and such that*

$$\forall t \in D : \dim E^n(t) = n .$$

*Then the function $\widetilde{X}$ associated with the curve X through Equation (5) is
well-defined and a blossom.*

Under the hypotheses of Theorem 6, the basis is an ETP system in the
sense of [4] and [9]. On the basis of a theorem of Scherk [7], Pottmann comes
to the conclusion that the annoying extra differentiability is not needed when
a geometric definition of the term "osculating space" is used rather than the
more common one using derivatives of the curve [5].

Remark 7. *Obviously, the curve associated with the curve blossom associated with the curve X is again X. In [8], it is proved that the curve blossom
associated with the curve associated with the curve blossom $\widetilde{X}$ is again $\widetilde{X}$
provided $X \in C^{2n}$ and the control points are affinely independent.*

In the preceding section a close relation between curve blossoms and VD
curve schemes has been established.

§3. Blossoming for Surfaces

Now we are dealing with surfaces, so $D = \triangle UVW \subset \mathbb{R}^2$. Let us first have
a look at how to define surface blossoms of the lowest possible order: $n =
1$. Since there is only one argument, in this case $\widetilde{X}(t) = X(t)$. Recalling
that order 1 curves are line segments (with endpoints P_0 and P_1), we want
order 1 surfaces to be planar. It is possible to *reuse* Definitions 2, 3 and 4
verbatim for the surface case! To exclude certain degeneration phenomena,
only the additional condition is imposed that the image set of an order 1
surface blossom not be contained in a straight line.

Remark 8.
  − D has been chosen to be a triangle only because there is some connection
    to triangular Bézier surfaces as will be shown. Actually it would suffice
    to make D a convex set.
  − The non-degeneracy condition that all subblossoms of order 1 have image
    sets not contained in a line could be weakened. But then we would have
    to prove that the one given here follows from it. This is done in [8].

Theorem 9. *Every first order blossom $\widetilde{X}$ maps line segments into line segments. If $\widetilde{X}$ is continuous, line segments are mapped onto line segments,
preserving endpoints.*

Proof: Condition (2) states that a point t on the line segment $\mathrm{cvx}\{t_-, t_+\}$
is mapped into the line segment $\mathrm{cvx}\{\widetilde{X}(t_-), \widetilde{X}(t_+)\}$. Continuity implies that

the image is connected; thus the image is all of the line segment. Preservation of the endpoints is trivial. ■

Now the surface blossom will be characterized. Again, I start with order 1.

Theorem 10. *A continuous blossom* $\widetilde{X} : D \to \mathbb{R}^2$ *of order 1 is a projective map defined on the projective closure* $\mathbb{P}^2$ *of the affine plane* $\mathbb{R}^2$ *restricted to the triangle* D.

Proof: According to Theorem 9, $\widetilde{X}$ maps line segments into line segments. Now, there is a theorem that states that every continuous map from an open subset of $\mathbb{P}^2$ to $\mathbb{P}^2$ that sends line segments to line segments and whose range has non-empty interior is a restriction of a projective map. ■

For a proof of the stated theorem, see [8]; see also [1] where it is proved with the additional assumption that the map is injective.

Theorem 11. *A continuous blossom* $\widetilde{X} : D^n \to \mathbb{R}^2$ *of order* n *has a representation as a multilinear map with restricted domain* D^n *using homogeneous coordinates in both* D *and* $\widetilde{X}(D)$.

Proof: Since according to Theorem 10 every order 1 subblossom is the restriction of a projective map, it is linear with respect to homogeneous coordinates. But this is just another way of saying that $\widetilde{X}$ is linear in each of its arguments. ■

The theorem implies that $\widetilde{X}$ (and thus X) is a rational function, since a polynomial representation using homogeneous coordinates can always be transformed into a rational representation using Cartesian coordinates (see Eq. (10)). Thus, the surface $X(t) = \widetilde{X}(t, \ldots, t)$ can be represented as a rational triangular Bézier surface over D. The next step will be to relate the control points and weights to the blossom. The general form of a multilinear representation is given by

$$\widetilde{X}(t_1, \ldots, t_n) \,\hat{=}\, \sum_{\tau_1=1}^{3} \cdots \sum_{\tau_n=1}^{3} a_{\tau_1 \cdots \tau_n} u_{\tau_1 1} \cdots u_{\tau_n n}, \tag{6}$$

where

$$t_i = u_{1i}U + u_{2i}V + u_{3i}W \,, \qquad u_{1i} + u_{2i} + u_{3i} = 1$$

and every $a_{\tau_1 \cdots \tau_n}$ is a $d+1$-tuple. I write "$\hat{=}$" instead of "$=$" because the expression on the left hand side is in Cartesian and the one on the right in homogeneous coordinates. I also have normalized the homogeneous coordinates in D to barycentric coordinates.

Substituting vertices of the domain for the arguments in Equation (6) yields

$$\widetilde{X}(\underbrace{U, \ldots, U}_{i}, \underbrace{V, \ldots, V}_{j}, \underbrace{W, \ldots, W}_{k}) \,\hat{=}\, a_{\underbrace{1 \cdots 1}_{i} \underbrace{2 \cdots 2}_{j} \underbrace{3 \cdots 3}_{k}} =: \widetilde{P}_{ijk} \,, \quad i+j+k = n \,,$$

$$\tag{7}$$

which is analogous to the curve case $\bigl($cf. Eq. (3)$\bigr)$ since the P_{ijk} as defined here will play the role of the control points. The symmetry of $\widetilde{X}$ can be seen in the relation:

$$a_{\tau_1 \cdots \tau_n} = a_{\tau_{\sigma(1)} \cdots \tau_{\sigma(n)}} \tag{8}$$

for every permutation σ of the set $\{1, \ldots, n\}$. Taking all $t_i = t = uU + vV + wW$ and collecting identical terms, turns (6) into the well-known representation of a triangular Bézier-surface

$$X(t) \cong \sum_{i+j+k=n} \widetilde{P}_{ijk} \frac{n!}{i!j!k!} u^i v^j w^k =: \sum_{i+j+k=n} \widetilde{P}_{ijk} B^n_{ijk}(u, v, w) . \tag{9}$$

Now, let us be more specific about coordinates. In the domain, I have already used barycentric coordinates with respect to $\triangle UVW$ i.e., $U \cong (1, 0, 0)$, $V = (0, 1, 0)$, and $W = (0, 0, 1)$ with the line at infinity described by $u+v+w = 0$ $\bigl($see Eq. (6)$\bigr)$. For the range, I use the convention that the points at infinity have their first coordinate zero. We obtain

$$(\rho, \tilde{x}, \tilde{y}, \tilde{z}) \cong \tfrac{1}{\rho} (\tilde{x}, \tilde{y}, \tilde{z}) \tag{10}$$

as general relation between homogeneous coordinates and the corresponding Cartesian coordinates and for the coordinates of the control points

$$\widetilde{P}_{ijk} = (\rho_{ijk}, \rho_{ijk} P_{ijk}) . \tag{11}$$

The factor ρ_{ijk} is called the *weight* of the control point P_{ijk}. Even though we are dealing with homogeneous coordinates here, it cannot be chosen freely but holds information about the surface. This is possible because the weighted averaging in Formula (6) is not invariant with respect to these factors. Note that zero weights are ruled out because $(0, \ldots, 0)$ are not valid coordinates. A vanishing first coordinate is also forbidden for *any* value of the blossom since points at infinity are not in its range. Using (10), Equation (9) can be put into the Cartesian form

$$X(t) = \sum_{i+j+k=n} \phi_{ijk}(u, v, w) P_{ijk},$$

with rational basis functions $\phi_{ijk}(u, v, w)$, which also depend on the weights but not on the control points P_{ijk}.

Theorem 12. *Continuous surface blossoms of order n describe rational triangular Bézier surfaces of degree n where all weights have the same definite sign.*

Proof: A representation of the diagonalization of such a surface blossom as a Bézier surface has already been given in the text. Since zero weights have also been ruled out, I only have to show that there cannot be weights of opposite signs. Let us assume there are weights of all signs. Then there exist two weights whose indices differ only by one. Let us denote them w.l.o.g. by $P_{i_0 j_0 k_0}$ and $P_{i_0-1,j_0+1 k_0}$. Those are blossom values where the corresponding argument tuples differ only in one place, the former having a U where the latter has V. This means that there is a first order subblossom $\widetilde{Y}$ of $\widetilde{X}$ such that

$$\widetilde{Y}(U) = P_{i_0 j_0 k_0} \qquad \text{and} \qquad \widetilde{Y}(V) = P_{i_0-1,j_0+1 k_0} .$$

By the linearity of $\widetilde{Y}$ there would be a point between U and V with its first coordinate zero which is impossible. ∎

Theorem 13. *Every rational triangular Bézier surface with positive weights can be produced by a continuous surface blossom.*

Proof: From given control points P_{ijk} and weights ρ_{ijk} homogeneous control points are formed according to (11). Using (8) we can set up a representation for a candidate blossom $\widetilde{X}$ as in (6). I now have to show Property (2). Fixing $n-1$ arguments reduces $\widetilde{X}$ to one of its subblossoms $\widetilde{Y}$ which clearly has the form

$$\widetilde{Y}(t) = b_1 u + b_2 v + b_3 w \qquad \text{where} \qquad t = uU + vV = wW .$$

If $t \in \text{cvx}\{t_-, t_+\}$ then $t = (1-\lambda)t_- + \lambda t_+$ with $0 \le \lambda \le 1$ yielding

$$\widetilde{Y}(t) = (1-\lambda)\widetilde{Y}(t_-) + \lambda \widetilde{Y}(t_+) .$$

Converting to Cartesian coordinates shows that $\widetilde{Y}$ is in the line segment $\text{cvx}\{\widetilde{Y}(t_-), \widetilde{Y}(t_+)\}$ iff the first homogeneous coordinates of $\widetilde{Y}(t_-)$ and $\widetilde{Y}(t_+)$ have the same sign. This is true since by Equation (9), the first coordinates of all blossom values are nonnegative combinations of the weights taking in account that the coordinates of the t's are all nonnegative. ∎

Not many geometric properties of surface blossoms have been derived. The most important ones are a version of de Casteljau's Algorithm and the convex hull property. Both will be now be discussed in the framework of surface blossoms. Note that the condition on the weights in Theorems 12 and 13 is exactly the one that insures the convex hull property.

Proposition 14. *A surface blossom of order 1 satisfies*

$$t \in \text{cvx}\{t_U, t_V, t_W\} \Rightarrow \widetilde{X}(t \in \text{cvx}\{\widetilde{X}(t_U), \widetilde{X}(t_V), \widetilde{X}(t_W)\} . \tag{12}$$

Proof: There is a point $s \in \text{cvx}\{t_U, t_V\}$ such that $t \in \text{cvx}\{s, t_W\}$. Use Relation (2) twice and the proof is complete. ∎

With the help of Proposition 14, a de Casteljau-type algorithm can be devised which finds the values of surface blossoms. The following is a direct consequence of (12):

$$\widetilde{X}(t, t_2, \ldots, t_n)$$
$$\in \text{cvx}\{\widetilde{X}(U, t_2, \ldots, t_n), \widetilde{X}(V, t_2, \ldots, t_n), \widetilde{X}(W, t_2, \ldots, t_n)\}$$

Now, let $t \notin \{U, V, W\}$. Then each of the "$\widetilde{X}$" on the right side of the "$\in$" has more arguments out of the set $\{U, V, W\}$ than the $\widetilde{X}$ on the left side. By iteratively using this relation, the n arguments can be inserted one by one. From this follows directly the convex hull property.

Theorem 15. *For any surface blossom* $\widetilde{X} : D^n \to \mathbb{R}^d$ *with control points* P_{ijk} *defined by (7) and (10),*

$$X(D) \subset \widetilde{X}(D^n) \subset \text{cvx}\{P_{ijk} \mid i + j + k = n \text{ and } i, j, k \geq 0\}$$

holds.

Proof: Apply de Casteljau's Algorithm as outlined above. ∎

References

1. Brauner, H., Die erzeugendentreuen geodätischen Abbildungen aus Regelflächen, Mh. Math. **99** (1985), 85–103.
2. deBoor, C., Cutting corners always works, Computer-Aided Geom. Design **4** (1987), 125–131.
3. Goodman, T.N.T., Shape preserving representations, in *Mathematical Methods in Computer Aided Design*, T. Lyche and L. L. Schumaker (eds.), Academic Press, New York, 1989, 333–351.
4. Karlin, S., *Total Positivity I*, Stanford University Press, Stanford California, 1968.
5. Pottmann, H., The geometry of Tchebycheffian splines, Computer-Aided Geom. Design **10** (1993), 181–210.
6. Prautzsch, H., and T. Gallagher, Is there a geometric variation diminishing property for B-spline or Bézier surfaces?, Computer-Aided Geom. Design **9** (1992), 119–124.
7. Scherk, P., Über differenzierbare Kurven und Bögen II, Časopis pro pěstovani matematiky a fysiky **66** (1937), 172–191.
8. Schmeltz, G., Variationsreduzierende Kurvendarstellungen und Krümmungskriterien für Bézierflächen, dissertation, Darmstadt University of Technology, 1992.
9. Schumaker, L. L., *Spline Functions: Basic Theory*, Wiley, New York, 1981.

Gerd Schmeltz
Institute of Computer Science
Hebrew University
Givat Ram, Ross Building
91904 Jerusalem, ISRAEL
schmeltz @ cs.huji.ac.il

Approximation with Helix Splines

G. Seemann

Abstract. In this paper algorithms for an approximation of a set of points in $\mathbb{R}^3$ by a helix spline curve minimizing a least squares objective function are developed. Additionally, an approximative representation by a rational Bézier curve is introduced.

§1. Introduction

Helix splines are spline curves with piecewise constant curvature and constant torsion. Therefore helix splines might be of interest for milling machines. Although interpolation has been addressed in [3], an approach for approximation is unknown. A helix can be interpreted as the trajectory of a screw motion of a point $\mathbf{P}$: $\mathbf{P}$ rotates around an axis $\vec{\mathbf{a}}$ while it additionally is translated proportional to the rotation angle into the direction of the axis. The helix lies on a circular cylinder with axis $\vec{\mathbf{a}}$. The projection of a helix segment into a plane which is perpendicular to the axis yields a circular arc. If it contains the first point of the helix segment, we refer to this circular arc as a *basis circle*. In this paper these circular arcs are restricted to have an opening angle $2\varphi < 2\pi$. We call 2φ the *opening angle* of the helix segment.

In the next section, representations of a helix segment are discussed. Since a helix is a transcendental curve, it cannot be described by polynomial or rational curves. To deal with the problem in systems based on rational polynomials, an approximating rational Bézier curve is introduced. In Section 3 a set of parameters is introduced to uniquely define a helix spline. This will be done by interpolating points. For the geometric construction the *principal normal* $\vec{\mathbf{h}}$ is used. The principal normal $\vec{\mathbf{h}}$ is the derivative with respect to arc length parameter of the tangent $\vec{\mathbf{t}}$ and is perpendicular to $\vec{\mathbf{t}}$. The vectors $\vec{\mathbf{t}}$ and $\vec{\mathbf{h}}$ span the so-called *osculating plane* ([1]). In Section 4 the approximation process is described and two examples are presented.

Curves and Surfaces in Geometric Design
P. J. Laurent, A. Le Méhauté, and L. L. Schumaker (eds.), pp. 435–442.

§2. Representation of a Helix Segment

With the above observations we get a canonical representation

$$\mathbf{X}_c(\psi) = \mathbf{e}_0 + \vec{\mathbf{e}}_1 \cos\psi + \vec{\mathbf{e}}_2 \sin\psi + \vec{\mathbf{e}}_3\psi, \qquad \psi \in [0, 2\varphi], \tag{1}$$

with $\mathbf{e}_0$ being the centre of the basis circle, $\vec{\mathbf{e}}_1$, $\vec{\mathbf{e}}_2$, $\vec{\mathbf{e}}_3$ mutually perpendicular and $|\vec{\mathbf{e}}_1| = |\vec{\mathbf{e}}_2| = r$, the radius of the helix and $|\vec{\mathbf{e}}_3| = p$, the screw parameter ($2\pi p$ is the pitch)[1]. Figure 1a shows a helix segment together with the cylinder it is lying on. The parameter ψ in (1) is proportional to the arc length. In [7] this representation is used to construct affine Tschebycheffian splines.

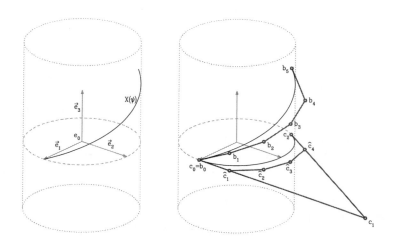

Fig. 1. a) Helix segment, representation (1)
b) Basis circle and helix segment, control polygon of the approximant (4).

Due to the fast and stable algorithms of the Bézier technique, many CAGD systems use Bézier curves. Therefore it would be desirable to have a Bézier-like representation of a helix segment. For the special case $p = 0$, i.e. a circular arc, a representation with a rational quadratic Bézier curve is well known [2,3,4,5]:

$$\mathbf{C}(t) = \frac{\mathbf{c}_0 B_0^2(t) + w\mathbf{c}_1 B_1^2(t) + \mathbf{c}_2 B_2^2(t)}{B_0^2(t) + wB_1^2(t) + B_2^2(t)}, \qquad t \in [0, 1], \tag{2}$$

where $\mathbf{c}_0$, $\mathbf{c}_2$ are the first, resp. last point of the circular arc, $\mathbf{c}_1$ is the intersection point of the tangents at $\mathbf{c}_0$ and $\mathbf{c}_2$, and $w = \cos\varphi$ where 2φ is the opening angle of the circular arc. If $p \neq 0$, the helix is transcendental, so no rational representation exists. The parameter t in (2) is not proportional to the arc length, so we can get the angle ψ between $\mathbf{c}_0$ and $\mathbf{C}(t)$ with help of the arctangent function only. Thus we have

$$\mathbf{X}_b(t) = \mathbf{C}(t) + \arctan\left(\frac{\langle \vec{e}_2, \mathbf{C}(t) - \mathbf{e}_0 \rangle}{\langle \vec{e}_1, \mathbf{C}(t) - \mathbf{e}_0 \rangle}\right) \cdot \vec{e}_3, \qquad t \in [0,1], \qquad (3)$$

with $\langle \cdot, \cdot \rangle$ denoting the scalar product and $\mathbf{e}_0$, $\vec{e}_1$, $\vec{e}_2$ and $\vec{e}_3$ as in (1). They can be calculated with the help of $\mathbf{c}_0$, $\mathbf{c}_1$ and $\mathbf{c}_2$.

The above disadvantage of mixing the Bézier representation with another class of functions can be overcome if we use an approximate representation by a rational Bézier curve. In [6], a special degree-4-approximant is found with help of rational axial motions. We now point out a method to get a general degree-5-approximant with small deviation from the helix, as well as small deviations from curvature and torsion. This method is discussed in detail in [8]. Since we can represent a circle exactly, we use curves which lie on the same cylinder as the helix. The approximants are found by degree elevating the basis circle by expanding it with a polynomial factor. The resulting degrees of freedom are used to achieve the desired properties. The degree elevation is performed by expanding $\mathbf{C}(t)$ in (2) with a polynomial in Bézier representation. Using the product formulas for Bernstein polynomials we get control points $\hat{\mathbf{c}}_i$ and weights w_i of the degree elevated rational curve.

In Figure 1b the control polygon $\mathbf{c}_i$ of the basis circle is shown together with the control polygon $\hat{\mathbf{c}}_i$ of a degree-5-representation. The control points $\hat{\mathbf{c}}_i$ are then moved by translations T_i in direction of the cylinder axis, resulting in the control points $\mathbf{b}_i$ of the approximant (4). For a degree-5-curve, the factor $\rho(t) = B_0^3(t) + \lambda B_1^3(t) + \lambda B_2^3(t) + B_3^3(t)$ is used, leading to

$$\mathbf{X}_a(t) = \frac{\displaystyle\sum_{i=0}^{5} w_i \mathbf{b}_i B_i^5(t)}{\displaystyle\sum_{i=0}^{5} w_i B_i^5(t)}, \qquad t \in [0,1]. \qquad (4)$$

The translations can be determined using the following conditions: $\mathbf{b}_0$, $\mathbf{b}_5$ are the first (resp. last) point of the segment, and $\mathbf{b}_1$, $\mathbf{b}_4$ lie on the tangent in $\mathbf{b}_0$, $\mathbf{b}_5$ resp.. For symmetry reasons , the translations T_2, T_3 for $\mathbf{b}_2$, $\mathbf{b}_3$ resp. use the same parameter τ: $T_2 = \tau \vec{e}_3$, $T_3 = (2\varphi - \tau)\vec{e}_3$. The parameters τ and λ are fixed so that the curvature of the approximant in $t = 0$ (and $t = 1$) is identical with the curvature of the helix and the tangent direction in $\mathbf{X}(\frac{1}{2})$ is identical to the tangent direction of the helix at this point. The fact that the approximant has three non-coplanar tangents in common with the helix segment makes it easy to regain the axis direction from the curve, since all tangents of a helix lie on one circular cone. If only two tangent directions are given, the axis direction can only be found by solving a nonlinear equation as we show in the next section.

§3. The Construction of the Helix Spline

We construct the helix spline by interpolating given points $\mathbf{P}_i$, $(i = 0, \ldots, n)$ with helix segments. Due to the tangent continuity the tangent direction $\vec{\mathbf{t}}_i$ at $\mathbf{P}_i$, $i = 1, \ldots, n-1$, is given from the preceding segment except for the first segment where $\vec{\mathbf{t}}_0$ can be freely chosen. So we have to find a helix segment that connects $\mathbf{P}_i$ with $\mathbf{P}_{i+1}$ and has the tangent direction $\vec{\mathbf{t}}_i$ at $\mathbf{P}_i$. There is only one degree of freedom available for this segment, while an independent choice of the axis direction would require two degrees of freedom. Thus, the axis direction depends upon the given data. To get a (with respect to our assumption unique) helix segment we use this degree of freedom as parameter for the osculating plane in $\mathbf{P}_i$, introducing the parameter β_i as angle between the principal normal of the circular arc connecting $\mathbf{P}_i$ and $\mathbf{P}_{i+1}$ and the principal normal $\vec{\mathbf{h}}_i$ of the helix. For $\beta_i = 0$ the helix segment is a circular arc that can be directly constructed with simple geometric operations. For other values of β_i, we only know that $\vec{\mathbf{a}}_i$ is perpendicular to $\vec{\mathbf{h}}_i$.

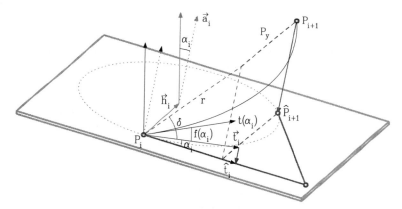

Fig. 2. Construction of a helix segment, error function $f(\alpha_i)$.

If we know the axis direction $\vec{\mathbf{a}}_i$, we can construct the helix by projecting $\mathbf{P}_{i+1}$ and $\vec{\mathbf{t}}_i$ into the plane which is perpendicular to $\vec{\mathbf{a}}_i$ and contains $\mathbf{P}_i$ (yielding $\hat{\mathbf{P}}_{i+1}$, $\hat{\mathbf{t}}_i$ resp., see Figure 2). The projection of the helix into this plane is a circular arc, so $\hat{\mathbf{P}}_{i+1}$ is the end point of this circular arc, $\mathbf{P}_i$ the first point and $\hat{\mathbf{t}}_i$ the tangent at this point. With this data the circular arc is well defined (cf. [3,4]). Additionally, the screw parameter is given by $p = \frac{|\hat{\mathbf{P}}_{i+1} - \mathbf{P}_i|}{2\varphi}$. The resulting helix interpolates $\mathbf{P}_i$, $\mathbf{P}_{i+1}$, $\vec{\mathbf{h}}_i$, but might not interpolate the given tangent direction $\vec{\mathbf{t}}_i$.

   To find the appropriate $\vec{\mathbf{a}}_i$, we perform the above process symbolically while $\vec{\mathbf{a}}_i$ is given by an angle α_i (see Figure 2). The resulting helix has the tangent $\mathbf{t}(\alpha_i)$ at $\mathbf{P}_i$. As error $f(\alpha_i)$ between the two vectors $\mathbf{t}(\alpha_i)$ and $\vec{\mathbf{t}}_i$ we use the angle they define. With the quantities as defined in Figure 2, we get the error function

$$f(\alpha_i) = \frac{r|P_y|\cos\alpha_i \sin(\delta + \alpha_i)}{\sqrt{r^2 \cos^2(\delta + \alpha_i) + P_y^2} \cdot \arccos\left(\frac{r\cos(\delta + \alpha_i)}{\sqrt{r^2 \cos^2(\delta + \alpha_i) + P_y^2}}\right)}$$
$$- \sin\alpha_i \sqrt{r^2 \cos^2(\delta + \alpha_i) + P_y^2}. \qquad (4)$$

This error function has exactly one root in the allowed interval $[0, \frac{\pi}{2}]$ which can be calculated easily with the Newton method. In the plane $(\mathbf{P}_i, \vec{t}_i, \mathbf{P}_{i+1})$, the point $\mathbf{P}_{i+1}$ has planar coordinates (x, y) with respect to $\vec{t}_i$ and $\vec{t}_i^\perp$. Using the parameter β_i, we get

$$r = \sqrt{x^2 + y^2 \sin^2\beta_{i-1}}, \quad \delta = \arctan\left(\frac{-y}{x}\sin\beta_i\right) \quad P_y = y\cos\beta_i. \qquad (4a)$$

For the special case $\beta_i = 0$ the solution is seen easily to be $\alpha_i = 0$. As the main result of this section we can formulate the following theorem.

Theorem 1. *Given $n + 1$ points $\mathbf{P}_i$, $i = 0, \ldots, n$, the tangent direction $\vec{t}_0$ in $\mathbf{P}_0$, and n angles between the principal normal of the circular arc connecting $\mathbf{P}_i$ and $\mathbf{P}_{i+1}$ and the principal normal $\mathbf{h}_i$ of the helix β_i, $i = 0, \ldots, n-1$, there exists a unique helix spline interpolating these data.*

We should remark that this result is independent from any chosen representation out of Section 2.

§4. Approximation

Let $\mathbf{F}_k$, $k = 0, \ldots, N_F$, be points in the Euclidian space $\mathbb{R}^3$. These points are to be approximated by a helix spline $\mathbf{X}(t)$ consisting of n segments ($n < N_F$). We assume that the first and the last point are interpolated. With parameter values t_k for the points $\mathbf{F}_k$ we have to minimize the distance function d.

$$d = \sum_{k=1}^{N_F - 1} (\mathbf{F}_k - \mathbf{X}(t_k))^2 \to \min \qquad (5)$$

With Theorem 1 we know that the objective function then has $(4n - 1)$ parameters. These unknowns must be determined by an optimization algorithm. By courtesy of the author, we used an implementation of a BFGS-method described in [9].

Every nonlinear optimization program needs a good initial guess for the unknowns. In our approach we first approximate the points with an integral cubic B-spline $\mathbf{S}(t)$ (any other curve which solves (5) via linear equations would be possible). The parameter values are found by chordal parametrization and are then modified by parameter correction (cf. [5]). The initial guess

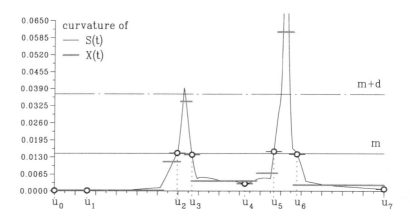

Fig. 3. Curvature for construction of initial guess together
with curvature of approximating helix spline.

for the tangent direction $\vec{t}_0$ is chosen as the direction of the first edge of the
control polygon of $S(t)$. Then we investigate the curvature of $S(t)$, bearing
in mind that the curvature of a helix spline is piecewise constant. Our aim
is to find parameter values u_i such that $S(u_i)$ is an initial guess for P_i. For
this purpose we calculate the mean curvature m and the standard deviation
d, using about 500 discrete points of the curvature. Among these points we
are looking for parameter values u_i where

1) the curvature has a local minimum in regions less than m. (See u_1, u_4 in
   Figure 3).
2) the curvature changes from values greater than m to values less than m.
   (See u_3, u_6 in Figure 3).
3) the curvature changes from values less than m to values greater than m
   and values greater than $m + d$ follow before decreasing again. (See u_2,
   u_5 in Figure 3).

The parameter values u_i, $i = 0, \ldots, n$, (including $u_0 = t_0$ and $u_n = t_{N_F}$)
shall form a knot vector U for the helix spline. The above criteria are due
to automation, an interactive user could sometimes make a more intelligent
choice. The number of segments needed for an approximating helix spline
is one less than the number of parameter values u_i. The parameters β_i are
chosen to $\beta_i = 0$, $i = 1, \ldots, (n - 1)$, i.e. the intial guess is a circular spline.

In a first step we only approximate with circular splines (i.e. the β_i are not
changed). Then a parameter correction is performed and the approximation
is repeated.

In the second step all $(4n - 1)$ unknowns are used for the approximation.
Again the approximation is repeated after parameter correction. Omitting
the first step of approximating with circular splines in general leads to the
same result but computation time increases.

The length of the parameter interval Δu_i for the i-th spline segment is

fixed by the knot vector U. It might not be optimal due to the construction
of U. For this purpose we introduce additional parameters Δu_i, $i = 1, \ldots, n$,
for the first step of the approximation. We allow any positive value for Δu_i,
the sum $\sum_{i=1}^{n} \Delta u_i$ is then mapped onto the interval $[t_0, t_{N_F}]$. In the second
step of the approximation we keep the former Δu_i fixed. Were they to be
changed one might have to face an increase of computation time.

Fig. 4. a) Initial guess.

Fig. 4. b) Approximating helix spline.

Figure 4 shows an example with 57 points. The initial guess (Figure 4a)
was found as described above with help of the curvature of $\mathbf{S}(t)$ shown in
Figure 3. The approximated points are drawn as grey full circles, the end
points $\mathbf{P}_i$ of the 7 segments are drawn as circles. The largest distance vector
is marked with an arrow. The approximation result is shown in Figure 4b. In
this picture we included the control polygon of the basis circle (see Section
2) for the first and last segment of the helix spline. Additionally, there is
an arrow pointing from the last point of the basis circle to the last point of
the helix spline segment. The total process for this example took 58 CPU-
sec on a HP 9000/735, and at the resulting curve the distance function d
evaluated to 17.08. At the initial guess its value is 1256.95 (For comparison:
the distance between $\mathbf{F}_0$ and $\mathbf{F}_{N_F}$ is 217.72). The piecewise constant curvature
of the approximating helix spline is included as horizontal grey lines in Figure
3. A similar effect can be observed for the torsion.

Another example is shown in Figure 5. The 63 points which are approximated were taken from one helix. In the picture we included a part of the axis
for every segment. The approximation with 9 segments took 64.33 CPU-sec,
the distance function d evaluated to $6.8 * 10^{-8}$. The segments of the spline all
have the same radius, screw parameter and axis as the helix the points were
taken from. In this sense we can say that the algorithm reproduces helices.

We tested all our examples with all of the three representations given in
Section 2. No major differences occurred.

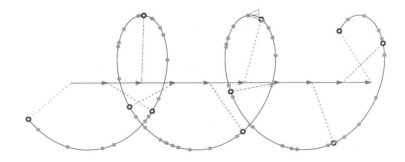

Fig. 5. Helix spline of 9 segments approximating points on a helix.

§6. Conclusion

In this paper we showed that we can achieve good approximation results with helix splines. The usage of nonlinear optimization programs leads to long computation time, but computers will become faster, and future research might give a better initial guess and an optimization program that "knows" more about helices, so the method could be accelerated.

References

1. Do Carmo, M. P., *Differential Geometry of Curves and Surfaces*, Prentice-Hall, New Jersey, 1976.
2. Farin, G., *Curves and Surfaces for Computer Aided Geometric Design. A Practical Guide.*, second edition, Academic Press, 1990.
3. Fuhs W., and H. Stachel, Circular pipe-connections, Computer & Graphics **12**, (1988), 53–57.
4. Hoschek, J., Circular splines, Comput. Aided Design **24**, (1992), 611–618.
5. Hoschek, J., and D. Lasser, *Fundamentals of Computer Aided Geometric Design*, AK Peters, Wellesley, 1993.
6. Mick, S., and O. Röschel, Interpolation of helical patches by kinematic rational Bézier patches, Computer & Graphics **14**, (1990), 275–280.
7. Pottmann, H., and M. G. Wagner, Helix splines as an example of affine Tschebycheffian splines, to appear in Advances for Computational Mathematics, 1993.
8. Seemann, G., Approximating a helix segment with a rational Bézier curve, to appear as a preprint of TH Darmstadt, 1994.
9. Spellucci, P., *Numerische Verfahren der nichtlinearen Optimierung*, Birkhäuser, Basel, 1993.

Gerald Seemann
Fachbereich Mathematik, Technische Hochschule Darmstadt
Schloßgartenstraße 7
D–64289 Darmstadt, GERMANY
seemann@ mathematik.th-darmstadt.de

Simplex Splines Support Surprisingly Strong Symmetric Structures and Subdivision

H.-P. Seidel and A.H. Vermeulen

Abstract. The simplex spline recurrence is symmetric. This leads to a simple redevelopment of simplex splines from scratch and facilitates the construction of the DMS simplex spline space. The subdivision property for DMS splines is established and explicit formulas for computing the coefficients are given.

§1. Introduction

A new simplex spline space has recently been developed by Dahmen, Micchelli, and Seidel in [5]. The development of this space, which we refer to as DMS simplex spline space, has been based on the combination of simplex splines with polar forms, and thus the DMS space exhibits several symmetry properties that facilitate computation and have been helpful for the development of new results [13]. Up to now, these symmetry properties have generally been perceived as properties of the DMS space, not as properties of the individual basis functions.

In this paper we correct this common misunderstanding and show that the standard simplex spline recurrence [3,8] is symmetric. This leads to a simple redevelopment of simplex splines (Section 1) and of the DMS spline space (Section 2) from scratch. We then use our machinery to establish the subdivision property for DMS splines and give explicit formulas for computing the coefficients (Section 3).

§2. Simplex Splines and Polar Simplex Splines

In this section we show that the recurrence relation for simplex splines is well-defined and symmetric. From this, most standard properties of simplex splines can be deduced very easily.

Curves and Surfaces in Geometric Design
P. J. Laurent, A. Le Méhauté, and L. L. Schumaker (eds.), pp. 443–455.
Copyright © 1994 by A K PETERS, Wellesley, MA
ISBN 1-56881-039-3.

Definition 1. *Let* $v, u_1, \ldots, u_n, t_0, \ldots, t_{n+2} \in \mathbb{R}^2$ *be given. The polar simplex spline* $m_v(u_1, \ldots, u_n | t_0, \ldots, t_{n+2})$ *is defined recursively as*

$$m_v(. | t_0 t_1 t_2) = \frac{\chi_{[t_0 t_1 t_2)}}{|d(t_0, t_1, t_2)|} \qquad (1)$$

and

$$m_v(u_1, \ldots, u_n | t_0, \ldots, t_{n+2}) = \sum_{i=0}^{n+2} \lambda_i m_v(u_1, \ldots, u_{n-1} | t_0, \ldots, \hat{t}_i, \ldots, t_{n+2}) \quad (2)$$

where $u_n = \sum_{i=0}^{n+2} \lambda_i t_i$ *with* $\sum_{i=0}^{n+2} \lambda_i = 1$. *The value* v *is the local parameter,* $u_1, \ldots, u_n$ *are the polar arguments of* m, *and* $t_0, \ldots, t_{n+2}$ *are the knots of* m.

The following theorem shows that the above recursion is well-defined and symmetric.

Theorem 2. *The polar simplex spline* $m_v(u_1, \ldots, u_n | t_0, \ldots, t_{n+2})$ *is well-defined, i.e., independent of the specific choice of the* λ_i, *and* n-*affine and symmetric in the polar arguments* $u_1, \ldots, u_n$.

Proof: We proceed by induction on n. The statement is trivial for $n = 1$. Let us now assume that the statement is true for n.

(i) We first show that $m_v(u_1, \ldots, u_{n+1} | t_0, \ldots, t_{n+3})$ is well-defined. In order to do so we first observe that

$$m_v(u_1, \ldots, u_{n-1} t_j | t_0, \ldots, t_{n+2}) = m_v(u_1, \ldots, u_{n-1} | t_0, \ldots, \hat{t}_j, \ldots, t_{n+2}).$$

This follows from Equation (2) by setting $\lambda_i = 1$ for $i = j$ and $\lambda_i = 0$ otherwise (note that $m_v(u_1, \ldots, u_n | t_0, \ldots, t_{n+2})$ is well-defined by the induction hypothesis). A double application of this equation then yields

$$m_v(u_1, ..., u_{n-1} t_j | t_0, ..., \hat{t}_i, ..., t_{n+3}) = m_v(u_1, ..., u_{n-1} t_i | t_0, ..., \hat{t}_j, ..., t_{n+3}).$$

Suppose now that $u_{n+1} = \sum_{i=0}^{n+3} \lambda_i t_i$ with $\sum_{i=0}^{n+3} \lambda_i = 1$. We wish to show that the expression $\sum_{i=0}^{n+3} \lambda_i m_v(u_1, \ldots, u_n | t_0, \ldots, \hat{t}_i, \ldots, t_{n+3})$ is independent of the specific choice of the λ_i. Let $u_n = \sum_{j=0}^{n+3} \mu_j t_j$ with $\sum_{j=0}^{n+3} \mu_j = 1$. Since $m_v(u_1, \ldots, u_n | t_0, \ldots, \hat{t}_i, \ldots, t_{n+3})$ is affine in u_n (by induction hypothesis) we obtain

$$\sum_{i=0}^{n+3} \lambda_i m_v(u_1, \ldots, u_{n-1}, u_n | t_0, \ldots, \hat{t}_i, \ldots, t_{n+3})$$

$$= \sum_{j=0}^{n+3} \mu_j \sum_{i=0}^{n+3} \lambda_i m_v(u_1, \ldots, u_{n-1}, t_j | t_0, \ldots, \hat{t}_i, \ldots, t_{n+3})$$

$$= \sum_{j=0}^{n+3} \mu_j \sum_{i=0}^{n+3} \lambda_i m_v(u_1, \ldots, u_{n-1}, t_i | t_0, \ldots, \hat{t}_j, \ldots, t_{n+3})$$

$$= \sum_{j=0}^{n+3} \mu_j m_v(u_1, \ldots, u_{n-1}, u_{n+1} | t_0, \ldots, \hat{t}_j, \ldots, t_{n+3})$$

which is in fact independent of the λ_i.

(ii) Next we show that $m_v(u_1, ..., u_{n+1}|t_0, ..., t_{n+3})$ is affine in $u_1, ..., u_{n+1}$. Equation (2) together with the induction hypotheses implies that $m_v(u_1, ..., u_{n+1}|t_0, ..., t_{n+3})$ is affine in in $u_1, ..., u_n$. In order to see that it is affine in u_{n+1} as well we consider $x, y \in \mathbb{R}^2$, $x = \sum_{i=0}^{n+3} \xi_i t_i$, $y = \sum_{i=0}^{n+3} \eta_i t_i$, $\sum_{i=0}^{n+3} \xi_i = \sum_{i=0}^{n+3} \eta_i = 1$, and let $\lambda_i = \alpha \xi_i + (1 - \alpha)\eta_i$. Then $\sum_{i=0}^{n+3} \lambda_i = 1$ and $\alpha x + (1 - \alpha)y = \sum_{i=0}^{n+3} \lambda_i t_i$, and we get

$$m_v(u_1, \ldots, u_n, \alpha x + (1 - \alpha)y|t_0, \ldots, t_{n+3})$$

$$= \sum_{i=0}^{n+3} \lambda_i m_v(u_1, \ldots, u_n|t_0, \ldots, \hat{t}_i, \ldots, t_{n+3})$$

$$= \alpha \sum_{i=0}^{n+3} \xi_i m_v(u_1, \ldots, u_n|t_0, \ldots, \hat{t}_i, \ldots, t_{n+3})$$

$$+ (1 - \alpha) \sum_{i=0}^{n+3} \eta_i m_v(u_1, \ldots, u_n|t_0, \ldots, \hat{t}_i, \ldots, t_{n+3})$$

$$= \alpha \, m_v(u_1, \ldots, u_n, x|t_0, \ldots, t_{n+3})$$

$$+ (1 - \alpha) \, m_v(u_1, \ldots, u_n, y|t_0, \ldots, t_{n+3})$$

which proves our claim.

(iii) We are left with symmetry. Since $m_v(u_1, \ldots, u_{n-1}, t_i, t_j|t_0, \ldots, t_{n+3})$ $= m_v(u_1, \ldots, u_{n-1}, t_j, t_i|t_0, \ldots, t_{n+3})$, the multiaffine property implies that $m_v(u_1, \ldots, u_{n+1}|t_0, \ldots, t_{n+3})$ is symmetric in u_n and u_{n+1}. The induction hypothesis completes the proof. ∎

Some additional properties of polar simplex splines are summarized in the following theorem:

Theorem 3. *Polar simplex splines have the following properties:*

1) *Polar simplex splines are symmetric in the knots and satisfy*

$$m_v(u_1, ..., u_n|t_0, ..., t_{n+2}) = \sum_{i=0}^{n+2} \lambda_i m_v(u_1, ..., u_n|t_0, ..., \hat{t}_i, ..., t_{n+2}, t), \quad (3)$$

where $t = \sum_{i=0}^{n+2} \lambda_i t_i$ with $\sum_{i=0}^{n+2} \lambda_i = 1$.

2) *If the same parameter t appears both as a polar argument and as a knot, then it can be eliminated, i.e.,*

$$m_v(u_1, \ldots, u_n, t|t_0, \ldots, t_{n+2}, t) = m_v(u_1, \ldots, u_n|t_0, \ldots, t_{n+2}). \quad (4)$$

3) *If the local parameters v_1, v_2 are not separated by any knot line, then*

$$m_{v_1}(u_1, \ldots, u_n|t_0, \ldots, t_{n+2}) = m_{v_2}(u_1, \ldots, u_n|t_0, \ldots, t_{n+2}). \quad (5)$$

Proof: Equations (4),(5) are immediate from the recurrence. Then Equation (3) follows from

$$m_v(u_1, \ldots, u_n | t_0, \ldots, t_{n+2}) = m_v(u_1, \ldots, u_n, t | t_0, \ldots, t_{n+2}, t)$$

$$= \sum_{i=0}^{n+2} \lambda_i m_v(u_1, \ldots, u_n | t_0, \ldots, \hat{t}_i, \ldots, t_{n+2}, t)$$

which completes the proof. ∎

We can now define simplex splines as the diagonal of a polar simplex spline.

Definition 4. *Let $u, t_0, \ldots, t_{n+2} \in \mathbb{R}^2$ be given. The simplex spline $M(u|t_0, \ldots, t_{n+2})$ is defined as*

$$M(u|t_0, \ldots, t_{n+2}) = m_u(u, \ldots, u | t_0, \ldots, t_{n+2}). \tag{6}$$

An immediate consequence of this definition is the standard recurrence for simplex splines [3,8]:

Corollary 5. *Simplex splines satisfy the recurrence*

$$M(u|t_0, t_1, t_2) = \frac{\chi_{[t_0 t_1 t_2]}}{|d(t_0, t_1, t_2)|}, \tag{7}$$

and

$$M(u|t_0, \ldots, t_{n+2}) = \sum_{i=0}^{n+2} \lambda_i M(u|t_0, \ldots, \hat{t}_i, \ldots, t_{n+2}), \tag{8}$$

where $u = \sum_{i=0}^{n+2} \lambda_i t_i$ with $\sum_{i=0}^{n+2} \lambda_i = 1$.

Proof: This follows immediately from Definition 1 and Definition 4. ∎

Another consequence is knot exchange:

Corollary 6. *Simplex splines are symmetric in the knots and satisfy*

$$M(u|t_0, \ldots, t_{n+2}) = \sum_{i=0}^{n+2} \lambda_i M(u|t_0, \ldots, \hat{t}_i, \ldots, t_{n+2}, t) \tag{9}$$

where $t = \sum_{i=0}^{n+2} \lambda_i t_i$ with $\sum_{i=0}^{n+2} \lambda_i = 1$.

Proof: This follows immediately from Definition 4 and Theorem 3. ∎

Corollary 7. *The directional derivative of a simplex spline w.r.t. a vector $\xi \in \mathbb{R}^2$ is*

$$D_\xi M(u|t_0, \ldots, t_{n+2}) \quad = n\, m_u(u, \ldots, u, \xi | t_0, \ldots, t_{n+2})$$

$$= n \sum_{i=0}^{n+2} \mu_i M(u|t_0, \ldots, \hat{t}_i, \ldots, t_{n+2}), \tag{10}$$

where $\xi = \sum_{i=0}^{n+2} \mu_i t_i$ with $\sum_{i=0}^{n+2} \mu_i = 0$.

Proof: The first equation follows from $D_\xi F(u) = n f(u, \dots, \xi)$ for every polynomial F with polar form f [9]. The second equation follows from the fact that $f(u, \dots, u, \xi)$ is linear in the vector ξ, and from Theorem 3. ∎

We conclude this section with a characterization of simplex splines by their minimal support property:

Theorem 8. *Simplex splines satisfy the following properties:*

1) *Piecewise Polynomial:* $M(u|t_0, \dots, t_{n+2})$ *is a piecewise polynomial of degree* n *over the partition* $\mathcal{P}$ *induced by the knots* $\{t_0, \dots, t_{n+2}\}$.

2) *Local Support:* $M(u|t_0, \dots, t_{n+2}) = 0$ *for* $u \notin [t_0, \dots, t_{n+2}]$.

3) *Positivity:* $M(u|t_0, \dots, t_{n+2}) \geq 0$.

4) *Smoothness: If at most* μ *knots are collinear,* $2 \leq \mu \leq n + 2$, *then* $M(u|t_0, \dots, t_{n+2})$ *is* $C^{n+1-\mu}$-*continuous along this line. Hence, if the knots are in general position, then* $M(u|t_0, \dots, t_{n+2})$ *is* C^{n-1}-*continuous everywhere.*

Moreover, $M(u|t_0, \dots, t_{n+2})$ *is uniquely characterized by the above four properties up to multiplication with a positive constant.*

Proof: In order to keep our presentation simple, we assume that the knots are in general position (although this is not essential for the argument that follows). Properties (1)–(3) are immediate from the recurrence. In order to prove (4), we consider a parameter u on a knot line, say $u \in [t_i, t_j]$. Let $v_1, v_2 \in \mathbb{R}^2$ be only separated by $[t_i, t_j]$, but by no other knot line. Hence v_1 and v_2 describe the same regions within the knot sets $\hat{T}_i = \{t_0, \dots, \hat{t}_i, \dots, t_{n+2}\}$ and $\hat{T}_j = \{t_0, \dots, \hat{t}_j, \dots, t_{n+2}\}$. Writing $u = \alpha t_i + (1 - \alpha)t_j$, we obtain

$$
\begin{aligned}
&m_{v_1}(u_1, \dots, u_{n-1}, u | t_0, \dots, t_{n+2}) \, \text{\textbullet} \\
&= \alpha \, m_{v_1}(u_1, \dots, u_{n-1} | t_0, \dots, \hat{t}_i, \dots, t_{n+2}) \\
&\quad + (1 - \alpha) \, m_{v_1}(u_1, \dots, u_{n-1} | t_0, \dots, \hat{t}_j, \dots, t_{n+2}) \\
&= \alpha \, m_{v_2}(u_1, \dots, u_{n-1} | t_0, \dots, \hat{t}_i, \dots, t_{n+2}) \\
&\quad + (1 - \alpha) \, m_{v_2}(u_1, \dots, u_{n-1} | t_0, \dots, \hat{t}_j, \dots, t_{n+2}) \\
&= m_{v_2}(u_1, \dots, u_{n-1}, u | t_0, \dots, t_{n+2}),
\end{aligned}
$$

which is equivalent to the C^{n-1}-continuity of $M(u|t_0, \dots, t_{n+2})$ at u [9].

It remains to show that $M(u|t_0, \dots, t_{n+2})$ is actually characterized by Properties (1)–(4). We proceed by induction on n. The statement is trivial for $n = 0$. Assume that the statement is true for $n - 1$, and let F be a piecewise polynomial of degree n that satisfies (1)-(4). We consider the sets $\hat{T}_i = \{t_0, \dots, \hat{t}_i, \dots, t_{n+2}\}$ and the functions $F_{\hat{T}_i}$ defined by $F_{\hat{T}_i}(u) = f_u(u, \dots, u, t_i)$, $i = 0, \dots, n + 2$. It is easy to check that $F_{\hat{T}_i}$ is a piecewise polynomial of degree $n - 1$ over the Partition $\mathcal{P}_{\hat{T}_i}$ induced by the set of knots

$\hat{T}_i$ that is C^{n-2}-continuous everywhere and is supported on $\hat{T}_i$. Hence, by induction hypothesis, there exist constants $\gamma_{\hat{T}_i} \in \mathbb{R}$ s.t. $F_{\hat{T}_i}(u) = \gamma_{\hat{T}_i} M(u|\hat{T}_i)$. Assume that $\gamma_{\hat{T}_i} \neq \gamma_{\hat{T}_j}$. We then consider the line $L_{ij} = \langle t_i, t_j \rangle$. For $u \in L_{ij}$, $u = \lambda_i(u)t_i + \lambda_j(u)t_j$ with $\lambda_i + \lambda_j = 1$, we then obtain

$$
\begin{aligned}
F(u) =& f_u(u, \ldots, u) \\
=& \lambda_i(u)f_u(u, \ldots, u, t_i) + \lambda_j(u)f_u(u, \ldots, u, t_j) \\
=& \gamma_{\hat{T}_i}\lambda_i(u)M(u|\hat{T}_i) + \gamma_{\hat{T}_j}\lambda_j(u)M(u|\hat{T}_j) \\
=& \gamma_{\hat{T}_i}(\lambda_i(u)M(u|\hat{T}_i) + \lambda_j(u)M(u|\hat{T}_j)) \\
& + (\gamma_{\hat{T}_j} - \gamma_{\hat{T}_i})\lambda_j(u)M(u|\hat{T}_j) \\
=& \gamma_{\hat{T}_i}M(u|t_0, \ldots, t_{n+2}) + (\gamma_{\hat{T}_j} - \gamma_{\hat{T}_i})\lambda_j(u)M(u|\hat{T}_j).
\end{aligned}
$$

Since F and $M(.|t_0, \ldots, t_{n+2})$ are both C^{n-1}-continuous, the assumption $\gamma_{\hat{T}_i} \neq \gamma_{\hat{T}_j}$ implies that the simplex spline $M(u|\hat{T}_j)$ of degree $n-1$ is C^{n-1}-continuous along the line L_{ij}, a contradiction. Therefore $\gamma_{\hat{T}_i} = \gamma_{\hat{T}_j} = \gamma$, and for $u = \sum_{i=0}^{n+2} \lambda_i t_i$ with $\sum_{i=0}^{n+2} \lambda_i = 1$ we obtain

$$
\begin{aligned}
F(u) &= \sum_{i=0}^{n+2} \lambda_i F_{\hat{T}_i}(u) = \gamma \sum_{i=0}^{n+2} \lambda_i M(u|t_0, \ldots, \hat{t}_i, \ldots, t_{n+2}) \\
&= \gamma\, M(u|t_0, \ldots, t_{n+2}),
\end{aligned}
$$

which completes the proof. ∎

§3. The DMS Simplex Spline Space

We now apply the results of the previous section to the study of a certain simplex spline space. A first simplex spline space has been constructed independently by Höllig [7] and by Dahmen and Micchelli [4]. A different spline space has recently been constructed by Dahmen, Micchelli, and Seidel [5]. In this paper we are interested in the second space which we refer to as DMS spline space. The goal of this section is to give a simple redevelopment of DMS splines from scratch using the results of the preceding section.

We start with the basic setup of DMS splines [5]. Let T be a triangulation of $\mathbb{R}^2$ where every triangle $[r_0, s_0, t_0] \in T$ is oriented counter-clockwise. With every vertex t_0 of T we associate a cloud of knots $t_0, \ldots, t_n$ such that for every triangle $[r_0, s_0, t_0] \in T$ all the triangles $[r_i, s_j, t_k]$ are non-degenerate. For every triangle $[r_0, s_0, t_0] \in T$ we then select the knot sets

$$
V_{ijk}^{[r_0, s_0, t_0]} = \{r_0, \ldots, r_i, s_0, \ldots, s_j, t_0, \ldots, t_k\}, \tag{11}
$$

and consider the simplex splines

$$
M_{ijk}^{[r_0, s_0, t_0]}(u) = M(u|r_0, \ldots, r_i, s_0, \ldots, s_j, t_0, \ldots, t_k) \tag{12}
$$

as well as the normalized splines

$$N_{ijk}^{[r_0,s_0,t_0]}(u) = d(r_i, s_j, t_k) M_{ijk}^{[r_0,s_0,t_0]}(u). \tag{13}$$

A DMS spline is then defined as a linear combination

$$F(u) = \sum_{[r_0,s_0,t_0]\in T} \sum_{|\vec{\imath}|=n} c_{ijk}^{[r_0,s_0,t_0]} N_{ijk}^{[r_0,s_0,t_0]}(u). \tag{14}$$

The corresponding polar DMS spline is

$$f_v(u_1,\ldots,u_n) = \sum_{[r_0,s_0,t_0]\in T} \sum_{|\vec{\imath}|=n} c_{ijk}^{[r_0,s_0,t_0]} n_{v,ijk}^{[r_0,s_0,t_0]}(u_1,\ldots,u_n). \tag{15}$$

The coefficients $c_{ijk}^{[r_0,s_0,t_0]}$ are called control points. The following properties of DMS splines are immediate from the definition:

Theorem 9. *A DMS spline is a piecewise polynomial of degree n over the partition $\mathcal{P}$ induced by the knot sets $V_{ijk}^{[r_0,s_0,t_0]}$, $[r_0,s_0,t_0]\in T$, $|\vec{\imath}|=n$, that is C^{n-1}-continuous if the knots are in general position. If at most μ knots within any $V_{ijk}^{[r_0,s_0,t_0]}$ are collinear, $2 \le \mu \le n+2$, then a DMS spline is $C^{n+1-\mu}$-continuous along this line. Furthermore, every control point $c_{ijk}^{[r_0,s_0,t_0]}$ only affects the surface locally.*

Proof: This follows immediately from the definition. ∎

Equation (15) expresses DMS splines as a sum over the triangles of the triangulation T. It is sometimes convenient to express DMS splines as a sum over the edges of the triangulation T instead. In order to do so, we adopt the following notation: Let E be the set of (oriented) edges of the triangulation T. Given an (oriented) edge $[r_0 s_0]$ we let t_0^+ and t_0^- be those two vertices such that the triangles $[r_0 s_0 t_0^+], [r_0 t_0^- s_0] \in T$ share the edge $[r_0 s_0]$ and are both oriented counter-clockwise.

Proposition 10. *A DMS spline $F = \sum_{[r_0,s_0,t_0]\in T} \sum_{|\vec{\imath}|=n} c_{ijk}^{[r_0,s_0,t_0]} N_{ijk}^{[r_0,s_0,t_0]}$ can be written as*

$$\begin{aligned}
F(u) = \sum_{[r_0 s_0]\in E} \sum_{|\vec{\imath}|=n} &(c_{ijk}^{[r_0 s_0 t_0^+]} d(r_i, s_j, u) M_{i,j,k-1}^{[r_0,s_0,t_0^+]}(u) \\
&- c_{ikj}^{[r_0 t_0^- s_0]} d(r_i, s_j, u) M_{i,k-1,j}^{[r_0,t_0^-,s_0]}(u).
\end{aligned} \tag{16}$$

We call this the edge representation of F.

Proof: We start by expressing u in barycentric coordinates w.r.t. $[r_i s_j t_k]$,

$$u = \frac{d(u, s_j, t_k)}{d(r_i, s_j, t_k)} r_i + \frac{d(r_i, u, t_k)}{d(r_i, s_j, t_k)} s_j + \frac{d(r_i, s_j, u)}{d(r_i, s_j, t_k)} t_k.$$

The simplex spline recurrence (8) then yields

$$F(u) = \sum_{[r_0,s_0,t_0]\in T} \sum_{|\vec{\imath}|=n} c_{ijk}^{[r_0,s_0,t_0]} d(r_i,s_j,t_k) M_{ijk}^{[r_0,s_0,t_0]}(u)$$

$$= \sum_{[r_0,s_0,t_0]\in T} \sum_{|\vec{\imath}|=n} c_{ijk}^{[r_0,s_0,t_0]} \{ d(u,s_j,t_k) M_{i-1,j,k}^{[r_0,s_0,t_0]}(u)$$

$$+ d(r_i,u,t_k) M_{i,j-1,k}^{[r_0,s_0,t_0]}(u) + d(r_i,s_j,u) M_{i,j,k-1}^{[r_0,s_0,t_0]}(u) \}.$$

Since every edge $[r_0 s_0]$ belongs to exactly two triangles, every term such as $d(r_i,s_j,u) M(u|r_0,\ldots,r_i,s_0,\ldots,s_j,t_0,\ldots,t_{k-1})$ apears twice in the above sum, once with positive, and once with negative sign, since $d(r_i,s_j,u) = -d(r_i,u,s_j)$. Collecting terms we thus obtain

$$F(u) = \sum_{[r_0 s_0]\in E} \sum_{|\vec{\imath}|=n} (c_{ijk}^{[r_0 s_0 t_0^+]} d(r_i,s_j,u) M_{i,j,k-1}^{[r_0,s_0,t_0^+]}(u)$$

$$- c_{ikj}^{[r_0 t_0^- s_0]} d(r_i,s_j,u) M_{i,k-1,j}^{[r_0,t_0^+,s_0]}(u)$$

which completes the proof. ∎

Corollary 11. *The normalized DMS splines form a partition of unity, i.e.,*

$$\sum_{[r_0,s_0,t_0]\in T} \sum_{|\vec{\imath}|=n} N_{ijk}^{[r_0,s_0,t_0]}(u) \equiv 1. \tag{17}$$

In particular, the relationship between a DMS spline and its control points is affinely invariant.

Proof: We proceed by induction on n. The assertion is trivial for $n = 0$. Assume it is true for $n - 1$. Using the edge representation of Proposition 10 we get

$$\sum_{[r_0,s_0,t_0]\in T} \sum_{|\vec{\imath}|=n} N_{ijk}^{[r_0,s_0,t_0]}(u)$$

$$= \sum_{[r_0 s_0]\in E} \sum_{|\vec{\imath}|=n} \{ d(r_i,s_j,u) M_{i,j,k-1}^{[r_0,s_0,t_0^+]}(u) - d(r_i,s_j,u) M_{i,k-1,j}^{[r_0,t_0^-,s_0]}(u) \}.$$

Since terms with $k = 0$ cancel, this can be rewritten as

$$= \sum_{[r_0 s_0]\in E} \sum_{|\vec{\imath}|=n-1} \{ d(r_i,s_j,u) M_{i,j,k}^{[r_0,s_0,t_0^+]}(u) - d(r_i,s_j,u) M_{i,k,j}^{[r_0,t_0^-,s_0]}(u) \}$$

$$= \sum_{[r_0,s_0,t_0]\in T} \sum_{|\vec{\imath}|=n-1} N_{ijk}^{[r_0,s_0,t_0]}(u) = 1$$

by the induction hypothesis. ∎

Next we turn to the representation of polynomials and piecewise polynomials [13].

Theorem 12. *Let F be a C^q-continuous piecewise polynomial of degree n over T. Assume that every vertex t_0 of T has multiplicity $\mu = n - q$, i.e., $t_0 = \cdots = t_{n-q-1}$. Then F can be represented as a DMS spline and we get $F(u) = \sum_{[r_0,s_0,t_0] \in T} \sum_{|\vec{\imath}|=n} c_{ijk}^{[r_0,s_0,t_0]} N_{ijk}^{[r_0,s_0,t_0]}(u)$ with*

$$c_{ijk}^{[r_0,s_0,t_0]} = f_{[r_0,s_0,t_0]}(r_0,\ldots,r_{i-1},s_0,\ldots,s_{j-1},t_0,\ldots,t_{k-1}), \qquad (18)$$

where $f_{[r_0,s_0,t_0]}$ is the polar form of the restriction $F_{[r_0,s_0,t_0]}$ of F to the triangle $[r_0,s_0,t_0]$.

Proof: We show that

$$F(u) = \sum_{[r_0,s_0,t_0] \in T} \sum_{|\vec{\imath}|=l} c_{ijk}^{[r_0,s_0,t_0]}(u) N_{ijk}^{[r_0,s_0,t_0]}(u)$$

with

$$c_{ijk}^{[r_0,s_0,t_0]}(u) = f_{[r_0,s_0,t_0]}(r_0,\ldots,r_{i-1},s_0,\ldots,s_{j-1},t_0,\ldots,t_{k-1},u^{n-l})$$

for $l = 0,\ldots,n$. The statement is trivial for $l = 0,\ldots,n-q-1$ (since every vertex has multiplicity $\mu = n - q$, the simplex splines $M_{ijk}^{[r_0,s_0,t_0]}$ of degree $\leq n-q-1$ reduce to the Bernstein polynomials, and the above is nothing but the de Casteljau Algorithm). Assume the statement is true for $l-1 \leq n-q-1$. We show that it is true for l. First, Proposition 10 yields

$$\sum_{[r_0,s_0,t_0] \in T} \sum_{|\vec{\imath}|=l} c_{ijk}^{[r_0,s_0,t_0]}(u) N_{ijk}^{[r_0,s_0,t_0]}(u)$$

$$= \sum_{[r_0 s_0] \in E} \sum_{|\vec{\imath}|=l} \{ c_{ijk}^{[t_0,s_0,t_0^+]}(u) d(r_i,s_j,u) M_{i,j,k-1}^{[r_0,s_0,t_0^+]}$$

$$- c_{ikj}^{[r_0,t_0^-,s_0]}(u) d(r_i,s_j,u) M_{i,k-1,j}^{[r_0,t_0^-,s_0]}(u) \}.$$

Since $l \geq n-q$, the C^q-continuity of F along $[r_0,s_0]$ then implies $c_{i,j,0}^{[r_0,s_0,t_0^+]}(u) = c_{i,0,j}^{[r_0,t_0^-,s_0]}(u)$ i.e., terms with $k = 0$ cancel in the above, and the above can be rewritten as a sum over $|\vec{\imath}| = l - 1$,

$$\sum_{[r_0 s_0] \in E} \sum_{|\vec{\imath}|=l-1} \{ c_{ijk}^{[r_0,s_0,t_0^+]}(u) d(r_i,s_j,u) M_{ijk}^{[r_0,s_0,t_0^+]}(u)$$

$$- c_{ikj}^{[r_0,t_0^-,s_0]}(u) d(r_i,s_j,u) M_{i,k,j}^{[r_0,t_0^-,s_0]}(u) \}$$

$$= \sum_{[r_0,s_0,t_0] \in T} \sum_{|\vec{\imath}|=l-1} N_{ijk}^{[r_0,s_0,t_0]}(u),$$

which equals $F(u)$ by induction hypothesis. ∎

Note that we have not imposed any restrictions on the knot clouds beyond that every triangle $[r_i, s_j, t_k]$ is non-degenerate. In particular, we have not demanded that the sets Ω_n considered in [5] are non-empty (which would ensure that all triangles $[r_i, s_j, t_k]$ are oriented consistently). Thus the determinant $d(r_i, s_j, t_k)$ and hence the normalized splines $N_{ijk}^{[r_0, s_0, t_0]}$ may become negative if the knots are placed arbitrarily. However, even in this general setting, the normalized splines still form a partition of unity, and it is possible to represent any piecewise polynomial over T as a DMS spline. For an explicit discussion of DMS splines with arbitrary knots in the spirit of [5], see [10].

§4. Subdivision

We now wish to establish the subdivision property for DMS splines: Let T be a triangulation of $\mathbb{R}^2$ and let $\tilde{T}$ be a refinement of T. We show that every DMS spline over T can also be represented as a DMS spline over $\tilde{T}$. Since $\tilde{T}$ is obtained from T by inserting a finite sequence of vertices, and since vertex insertion is a local process, this follows from the following theorem:

Theorem 13. *Consider a triangle $[r_0, s_0, t_0]$ and a new vertex $v_0 \in [r_0, s_0, t_0]$, and let $F(u) = \sum_{|\vec{n}|=n} c_{ijk}^{[r_0, s_0, t_0]} N_{ijk}^{[r_0, s_0, t_0]}(u)$ be a DMS spline over $[r_0, s_0, t_0]$ (all other coefficients zero). Let $C(u)$ be the unique polynomial (B-patch) with*

$$c(r_0, \ldots, r_{i-1}, s_0, \ldots, s_{j-1}, t_0, \ldots, t_{k-1}) = c_{ijk}^{[r_0, s_0, t_0]}, \tag{19}$$

where c is the polar form of C, and define

$$c_{ijk}^{[r_0, s_0, t_0]}(v_0, \ldots, v_{l-1}) = c(r_0, ..., r_{i-1}, s_0, ..., s_{j-1}, t_0, ..., t_{k-1}, v_0, ..., v_{l-1}) \tag{20}$$

for $l = 0, \ldots, n$. Then

$$
\begin{aligned}
F(u) = &\sum_{l+j+k=n} c_{0,j,k}^{[r_0, s_0, t_0]}(v_0, \ldots, v_{l-1}) N_{l,j,k}^{[v_0, s_0, t_0]}(u) \\
&+ \sum_{i+l+k=n} c_{i,0,k}^{[r_0, s_0, t_0]}(v_0, \ldots, v_{l-1}) N_{i,l,k}^{[r_0, v_0, t_0]}(u) \\
&+ \sum_{i+j+l=n} c_{i,j,0}^{[r_0, s_0, t_0]}(v_0, \ldots, v_{l-1}) N_{i,j,l}^{[r_0, s_0, v_0]}(u),
\end{aligned}
\tag{21}
$$

i.e., F can be represented as a DMS spline over the subtriangles $[v_0, s_0, t_0]$, $[r_0, v_0, t_0]$, and $[r_0, s_0, v_0]$.

Proof: We use induction over l to show that

$$
\begin{aligned}
F(u) = \sum_{h=0}^{l} \{ &\sum_{h+j+k=n} c_{0,j,k}^{[r_0,s_0,t_0]}(v_0,...,v_{h-1}) N_{h,j,k}^{[v_0,s_0,t_0]}(u) \\
&+ \sum_{i+h+k=n} c_{i,0,k}^{[r_0,s_0,t_0]}(v_0,...,v_{h-1}) N_{i,h,k}^{[r_0,v_0,t_0]}(u) \\
&+ \sum_{i+j+k=n} c_{i,j,0}^{[r_0,s_0,t_0]}(v_0,...,v_{h-1}) N_{i,j,h}^{[r_0,s_0,v_0]}(u) \} \\
+ \sum_{i+j+k+l+1=n} \{ &c_{i,j,k}^{[r_0,s_0,t_0]}(v_0,...,v_l) d(r_i,s_j,t_k) \\
&M(u|r_0,...,r_i,s_0,...,s_j,t_0,...,t_k,v_0,...,v_l) \}
\end{aligned}
$$

for $l = -1,\ldots,n$. The statement is trivial for $l = -1$ (since the first sum vanishes, and the second term reduces to the definition of $F(u)$). Assume the statement is true for $l-1$. We start by expressing v_l in barycentric coordinates w.r.t. $[r_i,s_j,t_k]$,

$$
v_l = \frac{d(v_l,s_j,t_k)}{d(r_i,s_j,t_k)} r_i + \frac{d(r_i,v_l,t_k)}{d(r_i,s_j,t_k)} s_j + \frac{d(r_i,s_j,v_l)}{d(r_i,s_j,t_k)} t_k.
$$

Corollary 6 then yields

$$
\begin{aligned}
&d(r_i,s_j,t_k) M(u|r_0,\ldots,r_i,s_0,\ldots,s_j,t_0,\ldots,t_k,v_0,\ldots,v_{l-1}) \\
&= d(v_l,s_j,t_k) M(u|r_0,\ldots,r_{i-1},s_0,\ldots,s_j,t_0,\ldots,t_k,v_0,\ldots,v_l) \\
&+ d(r_i,v_l,t_k) M(u|r_0,\ldots,r_i,s_0,\ldots,s_{j-1},t_0,\ldots,t_k,v_0,\ldots,v_l) \\
&+ d(r_i,s_j,v_l) M(u|r_0,\ldots,r_i,s_0,\ldots,s_j,t_0,\ldots,t_{k-1},v_0,\ldots,v_l),
\end{aligned}
$$

and we obtain

$$
\begin{aligned}
F(u) = \sum_{h=0}^{l-1} \{ &\sum_{h+j+k=n} c_{0,j,k}^{[r_0,s_0,t_0]}(v_0,\ldots,v_{h-1}) N_{h,j,k}^{[v_0,s_0,t_0]}(u) \\
&+ \sum_{i+h+k=n} c_{i,0,k}^{[r_0,s_0,t_0]}(v_0,\ldots,v_{h-1}) N_{i,h,k}^{[r_0,v_0,t_0]}(u) \\
&+ \sum_{i+j+k=n} c_{i,j,0}^{[r_0,s_0,t_0]}(v_0,\ldots,v_{h-1}) N_{i,j,h}^{[r_0,s_0,v_0]}(u) \} \\
+ \sum_{i+j+k+l=n} \{ &c_{0,j,k}^{[r_0,s_0,t_0]}(v_0,\ldots,v_{l-1}) \, d(v_l,s_j,t_k) M_{l,j,k}^{[v_0,s_0,t_0]}(u) \\
&+ c_{i,0,k}^{[r_0,s_0,t_0]}(v_0,\ldots,v_{l-1}) \, d(r_i,v_l,t_k) M_{i,l,k}^{[r_0,v_0,t_0]}(u) \\
&+ c_{i,j,0}^{[r_0,s_0,t_0]}(v_0,\ldots,v_{l-1}) \, d(r_i,s_j,v_l) M_{i,j,l}^{[r_0,s_0,v_0]}(u) \}
\end{aligned}
$$

$$+ \sum_{i+jk+l+1=n} \{d(v_l, s_j, t_k) \; c_{i+1,j,k}^{[r_0,s_0,t_0]}(v_0, \ldots, v_{l-1})$$

$$+ d(r_i, v_l, t_k) \; c_{i,j+1,k}^{[r_0,s_0,t_0]}(v_0, \ldots, v_{l-1})$$

$$+ d(r_i, s_j, v_l) \; c_{i,j,k+1}^{[r_0,s_0,t_0]}(v_0, \ldots, v_{l-1})\}$$

$$M(u|r_0, \ldots, r_i, s_0, \ldots, s_j, t_0, \ldots, t_k, v_0, \ldots, v_l).$$

Since

$$d(r_i, s_j, t_k)c_{ijk}^{[r_0,s_0,t_0]}(v_0, \ldots, v_l)$$

$$= d(v_l, s_j, t_k)c_{i+1,j,k}^{[r_0,s_0,t_0]}(v_0, \ldots, v_{l-1})$$

$$+ d(r_i, v_l, t_k)c_{i,j+1,k}^{[r_0,s_0,t_0]}(v_0, \ldots, v_{l-1})$$

$$+ d(r_i, s_j, v_l)c_{i,j,k+1}^{[r_0,s_0,t_0]}(v_0, \ldots, v_{l-1})$$

This proves our claim for l. The theorem then follows from $l = n$. ∎

Note that Theorem 13 poses no restrictions on the position of the new vertex v_0 inside the triangle $[r_0, s_0, t_0]$. In particular, v_0 is not restricted to lie inside the regions Ω_n considered in [5]. Moreover, Equation (21) even stays valid if v_0 is placed on the boundary of $[r_0, s_0, t_0]$, say $v_0 \in [r_0 s_0]$. However, although the subtriangle $[r_0, s_0, v_0]$ is degenerate in this situation, the simplex spline $N_{i,j,l}^{[r_0,s_0,v_0]}$ will in general be non-zero and has to be taken into account.

References

1. C. de Boor. Multivariate piecewise polynomials. Acta Numerica (1993), 65–109.
2. P. de Casteljau. *Formes à Pôles.* Hermès, Paris, 1985.
3. W. Dahmen. Multivariate B-splines - recurrence relations and linear combinations of truncated powers. in *Multivariate Approximation Theory,* W. Schempp and K. Zeller (eds.), Birkhäuser, 1979, 64–82.
4. W. Dahmen and C. A. Micchelli. On the linear independence of multivariate B-splines I. Triangulations of simploids. SIAM J. Numer. Anal. **19** (1982), 993–1012.
5. W. Dahmen, C. A. Micchelli, and H.-P. Seidel. Blossoming begets B-splines built better by B-patches. Math. Comp. **59** (1992), 97–115.
6. R. Gormaz. Floraisons Polynomiales: Applications à l'étude des B-splines à plusieurs variables. PhD Thesis, Universite Joseph Fourier, Grenoble, 1993.
7. K. Höllig. Multivariate splines. SIAM J. Numer. Anal. **19** (1982), 1013–1031.
8. C. A. Micchelli. A constructive approach to Kergin interpolation in $\mathbb{R}^k$, multivariate B-splines and Lagrange interpolation. Rocky Mt. J. Math. **10** (1980), 485–497.
9. L. Ramshaw. Blossoms are polar forms. Computer-Aided Geom. Design **6** (1989), 323–358.

10. T. Sauer. Multivariate B-splines with (almost) arbitrary knots. Technical report, Universität Erlangen, 1993.
11. H.-P. Seidel. A new multiaffine approach to B-splines. Computer-Aided Geom. Design **6** (1989), 23–32.
12. H.-P. Seidel. Symmetric recursive algorithms for surfaces: B-patches and the de Boor algorithm for polynomials over triangles. Constr. Approx. **7** (1991), 257–279.
13. H.-P. Seidel. Representing piecewise polynomials as linear combinations of multivariate B-splines, in *Mathematical Methods in Computer Aided Geometric Design II* T. Lyche and L. L. Schumaker (eds.), Academic Press, Boston, 1992, 559–566.

Hans-Peter Seidel
Universität Erlangen
Graphische Datenverarbeitung
Am Weichselgarten 9
D-91058 Erlangen, GERMANY
seidel@informatik.uni-erlangen.de

Al Vermeulen
Rogue Wave Software
P.O. Box 2328
Corvallis, OR 97339
alv@roguewave.com

Object-Oriented Framework
for Curves and Surfaces
with Applications

Philipp Slusallek, Reinhard Klein,
Andreas Kolb, and Günther Greiner

Abstract. In computer graphics and geometric modeling one generally faces the problem to integrate a variety of curve and surface types into a single program. Object-oriented design offers the opportunity to use the inherent hierarchical structure of curves and surfaces to solve this problem. This paper presents an object-oriented framework together with its C++ implementation that starts from an abstract class of general differentiable curves and surfaces and in turn refines this design to various parametric representations of curves and surfaces. This design includes all of the standard curve and surface types and provides a powerful and uniform interface for applications. Examples from differential geometry and blending illustrate the approach.

§1. Introduction

In this paper we present an object-oriented framework together with its C++ implementation that starts from an abstract class of general differentiable curves and surfaces and in turn refines this design to various parametric representations of curves and surfaces [19, 26]. This design includes all of the standard curve and surface types and provides a powerful and uniform interface for applications.

In Section 2 after a short introduction into object-oriented design and classes we present our approach to order the types of curves and surfaces into a hierarchical structure and review implementation features and selected curve and surface classes.

The main issue will be to extract the operations that identify a certain class of objects and set them apart from objects of other classes. The hierarchical structure will serve as a reference for the derivation of the set of C++

Curves and Surfaces in Geometric Design   457
P. J. Laurent, A. Le Méhauté, and L. L. Schumaker (eds.), pp. 457–466.
Copyright ⓒ 1994 by A K PETERS, Wellesley, MA
ISBN 1-56881-039-3.

classes, which implement this hierarchy. Here, we assume that the reader is familiar with object-oriented design and C++ in general.

The presented object-oriented framework is supposed to support a wide range of application programs. Some examples are given in Section 3 which illustrates the power of this approach. These examples include visualization of differential geometry properties and the design of blending surfaces.

Section 4 summarizes the achievements of this object-oriented approach and discusses extensions and further research areas.

§2. Design

We start with a general overview of curves and surfaces and how they can be grouped into a hierarchical scheme. We will use this scheme to derive an implementation as a set of abstract C++ classes. These abstract classes are used to derive the classes of actual curve and surface types like B-spline curve or a specific tensor-product-patch.

In this paper we restrict ourselves to curves and surfaces in $\mathbb{R}^3$, although most of the presented design also applies to higher dimensions.

Object-Oriented Design and Classes

In object-oriented design the main issue is to identify the operations that can be applied to a certain class of objects. These operations, often called methods, describe objects of a class completely, when seen from its environment – the internal structure is hidden from the user of an object (encapsulation). A class can be derived from another class, by which it inherits all the methods from its superclass (inheritance, code sharing). An object of a derived class can be substituted for one of its super classes and respond to the same set of methods. In this case derived classes can use a different algorithm to implement the same method (polymorphism, virtual methods in C++).

The following section describes our approach to apply these design methods to curves and surfaces. We start with a coarse description of the main curve classes. The description for curves then easily carries over to surfaces.

Overview

An overview of our class hierarchy is shown in Figure 1. The important abstract classes together with some classes of special curve or surface types are given. Some less important classes have been removed from this figure to clarify the approach. Solid arrows mark derivations from superclass to subclass. Dotted arrows mark classes that take a references to other classes, which either implement certain parts of this object (e.g. `ParameterRegion` for a surface) or which specify the class of objects that this class can operate on (e.g. `CompositeCurve` has `ParamCurves` as sub curves).

The whole framework is implemented in C++ [5, 25], which enables an efficient and simple translation of the theoretical results to program code. At this point C++ is used almost exclusively for all projects within our group.

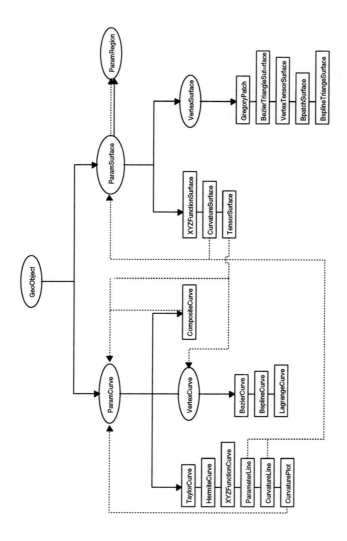

Fig. 1. Schematic view of the class hierarchy.

Parameterized Curves

A parameterized curve is a mapping of an interval I to $\mathbb{R}^3$. This type of curve is so common that a class is certainly required, which we called `ParamCurve`. The most fundamental methods for this type of curves are to obtain the parameter range I and to evaluate the curve at a given parameter $t \in I$ to derive a point $C(t)$ on the curve.

Applications in Computer Aided Geometric Design (CAGD) and other areas often require methods to obtain the derivatives $C^r(t)$, curvature, torsion, arc length or the Frenét Frame at a given parameter value t. These methods can be implemented numerically using the point evaluation method of the curve.

To offer all this functionality for any curve that at least knows how to evaluate a point $C(t)$, we have chosen to implement this functionality for the abstract class `ParamCurve`. At this level the methods use numerical approximation techniques, to obtain results for a general differentiable curve. If a subclass of a curve provides better or faster algorithms to obtain these results, then these methods can always be overwritten.

This implementation on the abstract level frees the implementor of a new curve class from the burden to implement all these probably difficult algorithms and instead rely on numerical approximation. Other algorithms could then be substituted at a later stage of the design process.

Since we also want to visualize the curve, we need a way to output the curve to a graphics display. We have therefore implemented a method to generate a piecewise linear approximation. The accuracy that should be met by the approximation is specified by the user or the application program. This accuracy description is a separate class, with methods to query for criteria like 'flatness', number of segments, etc., whichever is more appropriate for the given curve or surface type.

Vertex Curves

In CAGD many of the standard curve schemes are based on geometric control vertices. The shape of the curve is then derived from these control points by approximation or interpolation techniques. This common property of many curve types motivates another abstract class derived from `ParamCurve`, called `VertexCurve`.

This class handles methods like management and user interaction of the control points already on the abstract level. Thus instantiations like Bézier- or Lagrange curves do not need to handle those operations explicitly, but could still do so by overwriting certain methods, if they have special needs that are not covered by the abstract methods. Only the specific algorithm to calculate a point on the curve using the control points needs to be implemented for these derived classes.

Meta Classes

The application programmer that will use this framework often has many additional methods that he would like to implement for all surfaces. Changing the framework might not be the best way to do this, due to a probable inflation of methods. Instead we have chosen to implement this functionality using *meta classes*, which operate on objects of other classes.

There is a large set of meta classes, which do not define any actual curves or surfaces by themselves, but reference other curves or surfaces and visualize their properties. For instance the class `CurvaturePlot` is a planar curve that plots the curvature of the associated curve over its arc length. Since `CurvaturePlot` is itself a `ParamCurve` any method of this class also works on it. Thus a `CurvaturePlot` applied to a `CurvaturePlot` is simple.

We found that this concept of meta classes provides so much flexibility and functionality – while being simple and quick to implement – that we have used it throughout the framework.

Parameterized Surfaces

Parameterized surfaces are a bit more difficult and interesting. They are a mapping of a subset $D \subset \mathbb{R}^2$ to $\mathbb{R}^3$ and are implemented in a class called `ParamSurface`. All methods that evaluate the surface at a single point, or near a single point for a numerical approximation, are nearly identical to the curve methods, except that we now have to calculate partial derivatives, etc. Problems arise when a non-local method needs to be applied to a surface, since the method needs knowledge about the whole parameter region over which the surface is defined. This is more difficult than in the one dimensional case.

Again the most fundamental method is to obtain a point on the surface given its parameter $u \in D$. As in the curve class, all the other local methods are implemented at the abstract level in the class hierarchy as numerical approximations using point evaluation. These methods include calculation of partial derivatives and cross derivatives, normal vectors, Gaussian-, minimal and maximal curvatures, and the Fundamental Forms [4, 8] of the surface. This functionality is offered for any derived class, but can again be overwritten, if better algorithms are available for a specific derived class.

Vertex Surface

Again a large set of surface types use geometric control points to specify the geometry of a surface, which is adequately reflected as a separate abstract class called `VertexSurface`. But as with the parameter region, this is more difficult for surfaces than for curves, due to the non-linear topology of these control vertices. Common topologies are rectangular, triangular, or regular triangulations.

Our implementation of this abstract class offers only linear access to the set of control vertices. This normally suffices to implement user interaction and other general operations, and we can access any set of control points regardless of their topology. The classes for the other topologies are derived from this class and offer other access methods and storage implementations to efficiently implement special arrangements.

Special instantiations of these surface types are tensor-product surfaces, triangular Bézier patches [6], and multivariate B-splines [10].

Parameter Region

Non-local operations require knowledge about the domain of the surface. The problem is that the domain region is not a one-dimensional interval, but a region in two dimensions that could even be non-connected as is often the case for a trimmed surface. To make matters worse, the region is often quite complicated and bounded by free-form curves.

In order to solve this problem, we have developed an abstract description of the parameter region of a surface. This class can be queried for such information as point in region, the border as a set of bounding curves, a 2D-bounding box, etc.

The region can also return a tessellation of itself. The tessellation can either be a simpler representation (bounded by piecewise linear curves) or the region can be tessellated into a set of triangles which offers support for many standard algorithms. Note that this tessellation does not operate on the surface but only on the parameter region. Similarly, meshing methods are implemented for surfaces, which then have access to the surface properties to obtain an adaptive tessellation e.g. based on the surface curvature.

§3. Applications

In this section we illustrate the benefits of the above object-oriented approach by applying our framework to the visualization of differential geometry and to the construction of blend-surfaces.

Application 1: Visualizing Differential Geometry

The above framework and the available methods to obtain derivatives can be used to visualize differential geometry properties of curves and surfaces.

The curvature κ of a curve at a given point can be visualized by displaying the curvature circle, also called the osculating circle. This circle lies in the osculating plane spanned by the tangent t and the main normal n and has the radius $\frac{1}{\kappa}$. In an analog way we visualize the torsion τ of a curve by a cylinder through the point on the surface and with axis parallel to the binormal and having radius $\frac{1}{\tau}$. Animating the curvature circle, the torsion cylinder or the Frenét frame along the curve results in a method for displaying their variation.

A simple and convenient method for displaying the variation of the scalar curvature and torsion values is by means of a color-coded map. Another method for displaying the curvature is through curvature plots. The curvature plot is a two dimensional graph which plots the curvature over the arc length.

For surfaces, sectional curvature is visualized through curvature circles. In order to visualize the variation of the curvature as a function of direction we can also animate these circles.

The variation of the scalar-valued Gaussian- and mean curvature, as well as minimal and maximal curvature over the surface can again be visualized by means of a color-coded map. There are several other elaborated techniques which produce good results [2, 8].

An informative method for analyzing the variation of the principal directions across the surface is to incorporate a family of lines of curvature [2] into the display. A line of curvature is a curve on the surface whose tangent direction at each point coincides with one of the principal directions.

Application 2: Constructing Blending Surfaces

Another application that uses this framework is the construction and visualization of blending surfaces. Given two *primary surfaces* the problem is to construct a smooth transitional surface. Such a surface is called a *blend surface*. Our method [15, 16, 17] is based on a variational principle or an optimization problem. These methods have become quite popular in recent years in different areas in computer graphics [22, 23, 27]. The main idea to construct the blend surface is as follows:

1) it gives a smooth transition to the primary surface at the boundaries,

2) a *fairing functional*, which somehow measures total mean curvature, is minimized.

The boundary curves and the derivatives along those curves are given by special curve objects, that live on each of the blended surfaces. They describe the geometry of the problem completely and together with the functional ensure a unique solution to the blending problem.

So far this method has been implemented for tensor-product B–spline surfaces (TPS). For two primary TPS and specified boundaries (which are B-spline curves), a TPS is constructed such that it meets the primary surfaces at the boundaries. In addition, the cross-boundary derivatives of blend surface and a primary surface coincide at the boundary where they meet. The fairing functional J which will be minimized is of the form

$$J(F) = \int_\Omega \sum_{i\,\alpha\,\beta} w_{i\alpha\beta} \frac{\partial^\alpha S_i}{(\partial u)^\alpha} \frac{\partial^\beta S_i}{(\partial v)^\beta},$$

where α and β are multi-indices of order ≤ 2 , and S_i $(i = 1, 2, 3)$, denote x- y- and z component of the surface S. The weight functions $w_{i\alpha\beta}$ depend on the geometry of the region to be blended and are chosen via a parameter transformation between the parameter space of the TPS (rectangle) and a more natural parameter space. This has the effect that the fairing functional J is a good approximation for the total mean curvature (in mean square sense). The details are given in [17].

Since the fairing functional is quadratic, the problem reduces to a least square problem for the (inner) control points of the blend surface. Thus the control points of the blend surface are obtained as the solution of a linear system. An example of a blending surface and its curvature distribution is given in Figure 2.

The implementation of the blending operation relies on a set of classes that describe the boundary curves and the derivatives along those curves.

The boundary curves all lie on the blended surfaces. So they are implemented using curves that map a parameter interval to the two dimensional parameter region of the surface. This is the same technique that is used for trimming curves of parametric surfaces. This class **SurfaceCurve** of curves on surfaces has additional methods to calculate derivatives of the surface along the curve. A `SurfaceCurve` object can be queried for a derivative and will return a new object of a class derived from SurfaceCurve called `SurfaceDeriv`. Evaluating a point on this curve results is the requested derivative.

Encapsulating the derivatives in another object allows us to trade accuracy for speed without changing any other algorithm in the framework. The class `SurfaceDeriv` can query the surface for derivatives at a few points and can then use interpolation to obtain intermediate results, which can result in large speedups. But all this is invisible to the blending algorithm using this object.

Fig. 2. A blending surface and its curvature plot.

§4. Conclusion and Further Work

We have presented an object-oriented framework for applications that work on parametric curves and surfaces. Only the most fundamental method, point evaluation, must be supplied in order to integrate a general new curve or surface into the scheme. All the other methods are already implemented in abstract base classes. Thus the programmer is free to experiment without worrying about details such as implementing derivatives or similar operations, but still has the ability to use better methods as they become available.

A complete set of methods for curve and surface analysis with support for blending of arbitrary surfaces, differential geometry, scattered data interpolation, tessellation, display, and user interaction is provided.

The support for surface manipulation based on differential geometry or other local operations that do not directly work on control vertices but on the surface itself, have not been studied. This is certainly a very interesting research area, but it is yet unclear if and how these operators can be applied to arbitrary abstract surface classes.

References

1. Bartels, R. H., J. C. Beatty, and B. A. Barsky, *An Introduction to Splines for Use in Computer Graphics and Geometric Modelling*, Morgan Kaufman Publisher, 1987.

2. Beck, J., R. Farouki, and J. Hinds, Surface analysis methods, Computer Graphics & Applications **12** (1986), 18–38.

3. Bloomenthal, J., Polygonization for implicit surfaces, Computer Aided Geometric Design **5** (1988), 341–355.

4. do Carmo, M. P., *Differential Geometry of Curves and Surfaces*, Prentice Hall, Englewood Cliffs, N.J., 1976.

5. Ellis M. A., and B. Stroustrup, *The Annotated C++ Reference Manual*, Addison Wesley, 1990.

6. Farin, G. E., Triangular Bernstein–Bézier patches, Computer Aided Geometric Design **3** (1986), 83–127.

7. Farin, G. E., *Curves and Surfaces for Computer Aided Geometric Design*, Academic Press, New York, 2. edition, 1990.

8. Farouki, R. T., Graphical methods for surface differential geometry, in *The Mathematics of Surfaces II*, Martin, R. R. (ed.), Oxford Science Publications, Oxford, 1987, 363–385.

9. A. Forrest. Interactive interpolation and approximation by Bézier polynomials, Computer Journal **15** (1972), 71–79.

10. Fong, Ph., and H.-P. Seidel, An implementation of triangular B-spline surfaces over arbitrary triangulations Computer Aided Geometric Design **10** (1993), 267–275.

11. Fortune, S., Voronoi Diagrams and Delaunay triangulations, in *Computing in Euclidean Geometry*, D. Z. Du, F.Hwang (eds.), World Scientific Publ., 1992, 193–223.

12. Franke, R., and G. Nielson, Scattered data interpolation: A tutorial and survey, in *Geometric Modelling: Methods and Applications*, H. Hagen, D. Roller (eds.), Springer Verlag, New York, 1991, 131–160.

13. Georgiades, P. N., and D. P. Greenberg, Locally manipulating the geometry of curved surfaces. Computer Graphics & Applications **1** (1992), 54–64.

14. Gregory, J., Smooth interpolation without twist constraints, *Computer Aided Geometric Design*, R. Barnhill and W. Riesenfeld (eds.), Academic Press, New York, 1974.

15. Greiner, G., Blending Techniques based on variational principles, in *Curves and Surfaces in Computer Vision and Graphics III*, J. Warren (ed.), Proc. SPIE 1830, 1992, 174–184.

16. Greiner, G, and H.-P. Seidel, Curvature continuous blend surfaces, in *Modeling in Computer Graphics*, B. Falcidieno, T. L. Kunii (eds.), Springer Verlag, 1993, 309–317.

17. Greiner, G., Surface constructions based on variational principles, in *Wavelets, Images, and Surface Fitting*, P.-J. Laurent, A. Le Méhauté, and L. L. Schumaker (eds.), A K Peters, Wellesley, 1994, 277–286.

18. Kallay, M., Constrained optimization in surface design, in *Modeling in Computer Graphics*, B. Falcidieno, T. L. Kunii (eds.), Springer Verlag, 1993, 85–94.

19. Klein, R., Ph. Slusallek, An Object-Oriented Framework for Curves and Surfaces, in *Curves and Surfaces in Computer Vision and Graphics III*, J. Warren (ed.), Proc. SPIE 1830, 1992, 284–295.

20. Klass, R., Correction of local surface irregularities using reflection lines. Computer Aided Design **2** (1980), 73–76.

21. Kolb, A., Interpolating scattered data with C^2 surfaces, preprint, Universität Erlangen, 1993.

22. Lounsbery, M., S. Mann, S. and T. deRose, Parametric Surface Interpolation, Computer Graphics & Applications **9**, 1992, 97–115.

23. Moreton, H. P., and C. H. Séquin, Functional optimization for fair surface design, in *Computer Graphics*, E. E. Catmull (ed.), ACM Siggraph, ACM Press, 1992, 167–176.

24. Schumaker, L. L., Triangulation in CAGD, Computer Graphics & Applications **13**, 1993, 47–52.

25. Stroustrup, B., *The C++ Programming Language*, Addison Wesley, 2. edition, 1991.

26. Vermeulen A. H., and R. H. Bartels, C++ spline classes for prototyping, in *Curves and Surfaces in Computer Graphics II*, J. D. Warren (ed.), Proc. SPIE 1830, 1991, 121–131.

27. Welch, W., and A. Witkin, Variational surface modeling, in *Computer Graphics*, E. E Catmull (ed.), ACM Siggraph, ACM Press, 1992, 157–166.

Philipp Slusallek
Andreas Kolb
Günther Greiner
Universität Erlangen
IMMD IX- Graphische Datenverarbeitung
Am Weichselgarten 9
D-91058 Erlangen, GERMANY
slusallek,kolb,greiner@informatik.uni-erlangen.de

Reinhard Klein
Universität Tübingen, WSI/GRIS
Auf der Morgenstelle 10, C9
D-72076 Tübingen, GERMANY
reinhard@gris.informatik.uni-tuebingen.de

Designing a Progressive Lens

Mohammed Tazeroualti

Abstract. We introduce a new method for designing a *Progressive Lens*. This lens is represented by a sufficiently regular surface. On this surface we impose conditions on its principal curvatures in some regions (far vision region and near vision region) and other conditions on its principal directions of curvature in other regions (nasal region and temporal region). The surface is described with tensor product B-splines of degree four. For its computation, we have to minimize a non-quadratic operator. This minimization is then processed by an iterative method for which quick convergence is numerically tested.

§1. Introduction

The amplitude of accommodation of the eye is its faculty of both near and far sightedness. This accommodation is obtained by changing the curvature of the cristalline. The amplitude of accomodation decreases with age; in general, after age forty, the eye cannot clearly perceive near objects: this phenomenon is called *presbyopia*.

For correcting presbyopia we use corrective glasses with a far vision region above and a near vision region below. The progressive lens is a lens for correcting presbyopia; its properties allow the eye to move progressively from the far vision to the near vision region. One of the sides of the progressive lens is a surface with a smoothly varying curvature.

§2. Modelling

Let $\Omega = [-40, 40] \times [-40, 40]$ be a rectangular domain which is subdivided into eight regions H, HP, I, B, BP, T, N and L as shown in Figure 1.

The regions H and HP correspond to far vision. The regions B and BP correspond to near vision. I is the intermediate region where the eye moves progressively from far to near vision. The rest of Ω is subdivided into regions T, *temporal*, N, *nasal* and L, *lateral*. Our problem is to find a function $\sigma : \Omega \longrightarrow \mathbb{R}$ such that :

Curves and Surfaces in Geometric Design
P. J. Laurent, A. Le Méhauté, and L. L. Schumaker (eds.), pp. 467–474.
Copyright © 1994 by A K Peters, Wellesley, MA
ISBN 1-56881-039-3.

- In $HP \cup H$: σ is spheric with curvature c_1.
- In $BP \cup B$: σ is spheric with curvature c_2.
- In I : Both principal curvatures of σ are rather near and pass smoothly from c_1 to c_2.
- In T and N: One of the principal directions of σ is close to a coordinate axis direction.

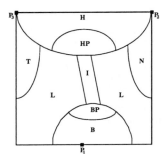

Fig. 1. Subdivision of Ω into regions.

In searching for the function σ, we solve a smoothing problem using tensor products of B-splines of degree four. This problem is reduced to minimizing an operator Q defined later.

Definition 1. *We define a weight function on Ω as a null or positive function, constant in each region of Ω.*

Next, let us assume that we have N_d points $(x_i, y_i)_{i=1}^{N_d}$ scattered over all regions of Ω. For the construction of our operators, we use the Gauss notations E, F, G, L, M, and N (see [1]). These quantities are functions of the point, so we write E_i instead of $E(x_i, y_i)$ and the same for F_i, G_i, L_i, M_i and N_i.

§3. Construction of the Operators

Let f be a $\mathcal{C}^3$ function over Ω. For the construction of the quantity to be minimized, we define several operators. Each operator will serve to impose a condition on the solution, on its principal directions, or on its curvatures.

3.1. Curvature Approximation Operator

The horizontal curvature of f in a point (x, y) is defined as

$$c_h(x, y) = \frac{L(x, y)}{E(x, y)},$$

and so we define an operator of horizontal curvature approximation $\mathcal{A}_c^h$ by

$$\mathcal{A}_c^h(f) = \sum_{i=1}^{N_d} \omega_h(x_i, y_i)(c_h(x_i, y_i) - c(x_i, y_i))^2,$$

where ω_h is the weight function relative to the horizontal curvature approximation and c the desired curvature.

Similarly we define an operator of vertical curvature approximation $\mathcal{A}_c^v$ by

$$\mathcal{A}_c^v(f) = \sum_{i=1}^{N_d} \omega_v(x_i, y_i)(c_v(x_i, y_i) - c(x_i, y_i))^2,$$

where ω_v is the weight function relative to the vertical curvature approximation.

We define the curvature approximation operator $\mathcal{A}_c$ by

$$\mathcal{A}_c(f) = \mathcal{A}_c^h(f) + \mathcal{A}_c^v(f).$$

3.2. Sphericity Operator

We have local sphericity in (x, y) if and only if

$$M(x, y)E(x, y) = L(x, y)F(x, y)$$
$$N(x, y)E(x, y) = L(x, y)G(x, y)$$
$$M(x, y)G(x, y) = N(x, y)F(x, y).$$

Thus we define the sphericity operator $\mathcal{S}$ by

$$\mathcal{S}(f) = \sum_{i=1}^{N_d} \omega_s(x_i, y_i) \left[(M_i E_i - L_i F_i)^2 + (N_i E_i - L_i G_i)^2 + (M_i G_i - N_i F_i)^2 \right],$$

where ω_s is the weight function of sphericity.

3.3. Principal Directions Operator

One of the principal directions of f in (x, y) is the direction of the y axis if and only if

$$G(x, y)M(x, y) = F(x, y)N(x, y).$$

Thus, we define the principal directions operator $\mathcal{D}$ by

$$\mathcal{D}(f) = \sum_{i=1}^{n_d} \omega_d(x_i, y_i)(G_i M_i - F_i N_i)^2,$$

where ω_d is the weight function of the principal directions.

3.4. Tension Operator

One of the implicitly imposed conditions on the solution is that its curvatures do not oscillate. The ideal solution for eliminating the undesirable oscillations of the curvatures is to introduce a tension over it by minimizing the norm of its derivatives. In our case it is quite difficult to do this because the curvatures are a complicated combination of the first and the second derivatives of f (see [1]).

Our purpose is to control the oscillations of the curvatures by minimizing the norm of the third derivatives of f. Thus, we define the tension operator $\mathcal{T}$ by

$$\mathcal{T}(f) = \sum_{i=1}^{n_d} \left(\tau_{30} \left(f_{x^3}^{(3)}(x_i, y_i) \right)^2 + \tau_{21} \left(f_{x^2 y}^{(3)}(x_i, y_i) \right)^2 \right.$$
$$\left. + \tau_{12} \left(f_{xy^2}^{(3)}(x_i, y_i) \right)^2 + \tau_{03} \left(f_{y^3}^{(3)}(x_i, y_i) \right)^2 \right).$$

Here, τ_{30}, τ_{21}, τ_{12}, and τ_{03} are the tension parameters. Table 1 shows how to control the oscillations of the horizontal and vertical curvatures in the horizontal or vertical axis direction by these parameters.

Curvature	Direction	Derivative	Parameter
Horizontal	Horizontal	$\dfrac{\partial^3 f}{\partial^3 x}$	τ_{30}
Horizontal	Vertical	$\dfrac{\partial^3 f}{\partial^2 x \partial y}$	τ_{21}
Vertical	Horizontal	$\dfrac{\partial^3 f}{\partial x \partial^2 y}$	τ_{12}
Vertical	Vertical	$\dfrac{\partial^3 f}{\partial^3 y}$	τ_{03}

Table 1. Control of the curvatures oscillations.

3.5. Border Operator

All the operators we have defined impose conditions only on the principal directions and the curvatures of the solution, which will be determined to within a polynomial of degree 1. For obtaining uniqueness, we propose to fix it in three non aligned points of the border of Ω, for example

$$p_1 = (0, 40, 0), p_2 = (40, -40, 0) \text{ and } p_3 = (-40, -40, 0)$$

Thus we define the border operator

$$\mathcal{B}(f) = \mu \left((f(0, 40))^2 + (f(40, -40))^2 + (f(-40, -40))^2 \right),$$

where μ is the border parameter.

3.6. Smoothness Operator

This operator will serve to regularize our problem and to obtain a smooth solution. We define it as :

$$J(f) = \rho \sum_{i=1}^{n_d} \left[\frac{\partial^8 f}{\partial x^4 \partial y^4}(x_i, y_i) \right]^2,$$

where ρ is the smoothness parameter.

3.7. Final Operator

With the operators defined above, we construct our final operator Q by

$$Q(f) = J(f) + B(f) + T(f) + A_c(f) + S(f) + D(f)$$

The function c, the theoretical curvature, in $A_c(f)$ is taken equal to c_1 in $HP \cup H$, to c_2 in $BP \cup B$, and to 0 elsewhere. The weight functions will be chosen according to the importance of the corresponding conditions in each region (and equal to zero outside of $HP \cup H \cup BP \cup B$).

§4. Algorithm

We search for the solution σ in the form of a tensor product of B-splines of degree 4:

$$\sigma(x, y) = \sum_{j=-4}^{n_x-1} \sum_{k=-4}^{n_y-1} \alpha_{jk} B_j(x) B_k(y). \tag{3.1}$$

For the computation of σ we minimize the operator Q over S_4, the space of functions in the form (3.1)

$$Q(\sigma) = \min_{f \in S_4} Q(f). \tag{$\mathcal{P}$}$$

By writing the elements of S_4 in the form (3.1), we reduce the problem ($\mathcal{P}$) to a minimization problem (P) in $\mathbb{R}^{n_x+4 \times n_y+4}$ of a functional Q.

The functional Q is not quadratic, because we have terms such as

$$\left(\sqrt{1 + f_x'^2 + f_y'^2} \right)$$

in the denominator of $A_c(f)$ (see the definition of Gauss notations in [1]). To circumvent this problem, we iterate over the first derivatives. If we suppose $f_x' = p$ and $f_y' = q$, where p and q are known functions, the functional Q becomes quadratic in $(\alpha_{ij})_{ij}$, see [3].

4.1. Iteration over the First Derivatives

We supposed the functions p and q to be known. This does not hold in practice. So we iterate over these functions by the following algorithm:

. Iteration 0 $p = q = 0$ Solve (P) hence σ_0

. Iteration 1 $p = (\sigma_0)'_x$ and $q = (\sigma_0)'_y$ Solve (P) hence σ_1

$$\vdots \qquad\qquad\qquad\qquad \vdots \qquad\qquad\qquad \vdots$$

. Iteration $k + 1$ $p = (\sigma_k)'_x$ and $q = (\sigma_k)'_y$ Solve (P) hence σ_{k+1}

Stop if:

$$\sum_{i=1}^{n_d} \left(c_h^k(x_i, y_i) - c_h^{k+1}(x_i, y_i) \right)^2 + \left(c_v^k(x_i, y_i) - c_v^{k+1}(x_i, y_i) \right)^2 \leq \varepsilon$$

We have not proved the theoretical convergence of this process. But in all the numerical experiments done, we have noticed its quick convergence. These experiments involved different examples and different parameter choices.

§5. Numerical Experimentation

Definition 2. *The astigmatism of a surface in a point is a quantity proportional to the difference between the maximal curvature and minimal curvature of the surface in this point.*

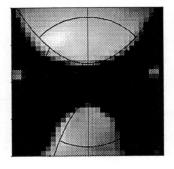 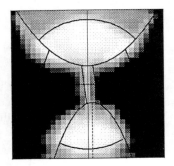

Fig. 2. Sphericity Operator: small ω_s left, large ω_s right.

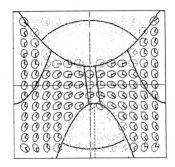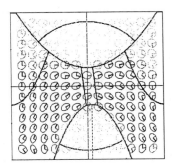

Fig. 3. Principal Directions Operator: small ω_d left, large ω_d right.

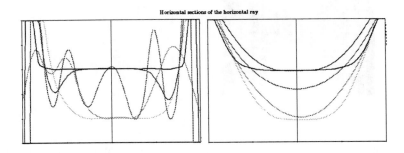

Fig. 4. Tension Operator: small τ_{11} left, large τ_{11} right.

§6. Good Lens Designed by this Method

We have used this method for designing a progressive lens with curvatures equal to 80 in $HP \cup H$ and to 60 in $BP \cup B$. After adjusting our parameters we have obtained the following results.

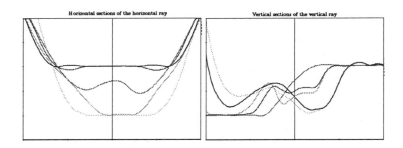

Fig. 5. Good Lens Design: Horizontal radius left, vertical radius right.

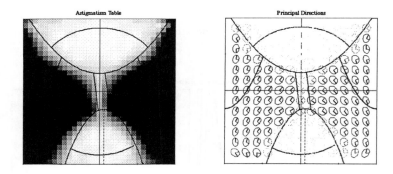

Fig. 6. Good Lens Design: Astigmatism left, principal directions right.

References

1. Lelong Ferand, Arnaudès, Géométrie et cinématique, Dunod, Paris, 1975.
2. P. J. Laurent, Courbes ouvertes et fermées par B-Splines régularisées, R.R. 652. IMAG, Université Joseph Fourier, Grenoble, 1987.
3. M. Tazeroualti, Modélisation de surfaces à l'aide de fonctions splines. Conception d'un verre progressif, Thèse université Joseph Fourier, Grenoble, 1993.

Mohammed Tazeroualti
LMC-IMAG, Université Joseph Fourier
BP 53X, 38041 Grenoble, FRANCE
tazer@busard.imag.fr

Multiplication as a General Operation for Splines

Kenji Ueda

Abstract. Multiplication, which is a basic B-spline function operation, can make different rational B-splines compatible to have a common weight function. This paper presents a blossoming method to obtain spline function products and shows its equivalence to Mørken's algorithm based on discrete B-splines. Degree raising and knot insertion, which are useful to make different B-spline curves and surfaces compatible, can be recognized as special cases of the multiplication.

§1. Introduction

NURBs curves and surfaces [11, 12] play important parts in computer aided geometric design. To increase the flexibility of the defining curve (or surface) basis, degree raising [6] and knot insertion [2, 5] are the most important tools to make different NURBs curves and surfaces compatible; i.e., they are of the same degree and are defined over the same knot sequence(s). Multiplication of splines is another important tool to combine two different NURBs curves or surfaces to one; i.e., they have common weight functions, which are the denominators of the rational B-splines.

Blossoming [13,14] (also known as polar forms [4]) solves the change of representation problems of B-splines, such as degree raising and knot insertion. In this paper blossoming is applied to multiplication of two B-spline functions. In the process of the multiplication, the compatibility techniques are required implicitly. In other words, multiplication of splines is a general operation, which includes knot insertion and degree raising, to make NURBs curves and surface compatible.

Discrete B-splines [9] can also be used to solve the problem of transforming B-splines from one knot sequence to another [5,6]. It has been shown by Mørken [10] that the product of B-splines can be expressed as a linear combination of B-splines in the form of discrete B-splines.

Curves and Surfaces in Geometric Design
P. J. Laurent, A. Le Méhauté, and L. L. Schumaker (eds.), pp. 475–482.
Copyright © 1994 by A K PETERS, Wellesley, MA
ISBN 1-56881-039-3.

§2. Blossoms or Polar Forms

The *blossom* or *polar form* of a polynomial $P(u)$ of degree d is a multivariable function $p(u_1, u_2, \ldots, u_d)$, which exists uniquely for the polynomial $P(u)$. The blossom $p(u_1, u_2, \ldots, u_d)$ has the following properties:

1) *symmetry*:

$$p(u_1, \ldots, v, \ldots, w, \ldots, u_d) = p(u_1, \ldots, w, \ldots, v, \ldots, u_d). \tag{1}$$

2) *multi-affine*:

$$\begin{aligned} p(u_1, \ldots, \alpha v + \beta w, \ldots, u_d) = \\ \alpha p(u_1, \ldots, v, \ldots, u_d) + \beta p(u_1, \ldots, w, \ldots, u_d), \quad (\alpha + \beta = 1). \end{aligned} \tag{2}$$

3) *diagonal*:

$$P(u) = p(u, u, \ldots, u). \tag{3}$$

Blossoming solves the representation problem of Bézier and B-spline curves and surfaces by applying the above properties [8,15]. The articles [1,7,16] concisely explain the blossoms or polar forms.

§3. B-spline Functions and B-spline Coefficients

A B-spline function $F(u)$ of B-spline coefficients $\{f_i\}$ is a piecewise polynomial expressed as

$$F(u) = \sum_{i=0}^{n} N_{i,d}^{K}(u) f_i, \tag{4}$$

where $N_{i,d}^{K}(u)$ is the normalized basis function of degree d over a knot sequence K. The knot sequence is a non-decreasing sequence of knots k_i:

$$\begin{aligned} K &= [k_0, k_1, \ldots, k_{n+d}, k_{n+d+1}] & (k_i \leq k_{i+1}) \\ &= [\underbrace{k^0, \ldots, k^0}_{\kappa_0 = d+1}, \underbrace{k^1, \ldots, k^1}_{\kappa_1 \text{ times}}, \cdots, \underbrace{k^\nu, \ldots, k^\nu}_{\kappa_\nu = d+1}] & (k^i < k^{i+1}), \end{aligned} \tag{5}$$

where κ_i is the multiplicity of k^i in K and $\kappa_i < d+1$ $(i = 1, \ldots, \nu - 1)$.

The value of ith B-spline basis function is computed by means of the following recurrence relation for B-splines [3]:

$$N_{i,d}^{K}(u) = \frac{u - k_i}{k_{i+d} - k_i} N_{i,d-1}^{K}(u) + \frac{k_{i+d+1} - u}{k_{i+d+1} - k_{i+1}} N_{i+1,d-1}^{K}(u), \tag{6}$$

and

$$N_{i,0}^{K}(u) = \begin{cases} 1 & \text{if } k_i \leq u < k_{i+1}, \\ 0 & \text{otherwise.} \end{cases} \tag{7}$$

The blossoming of polynomials can be extended to piecewise polynomials. Each polynomial segment of a piecewise polynomial has a corresponding blossom segment. The blossom of a B-spline function not only has properties of the blossom of polynomial, but also a fourth property

4) *dual functional:*

$$f_j = f(k_{j+1}, k_{j+2}, \ldots, k_{j+d}). \tag{8}$$

This means that the coefficients of a B-spline can be obtained by evaluating the corresponding blossom segment at the appropriate d consecutive knots. This property applies not only to the original knot sequence, but also to any refinement of that knot sequence. If K' is a refinement of K, then the coefficients f'_j expressing the function $F(u)$ over the refined knot sequence are obtained by evaluating the blossoms:

$$f'_j = f(k'_{j+1}, k'_{j+2}, \ldots, k'_{j+d}). \tag{9}$$

§4. Symbolic Evaluation of Blossoms

The discussion in the previous two sections indicates that the goal is to evaluate a blossom to an expression that is composed of the known coefficients of B-spline functions. With this in mind, the value of a blossom $f(u_1^*, u_2^*, \ldots, u_d^*)$ is computed symbolically in the following manner.

The parameters of the blossom are sorted beforehand in ascending order $u_1^* \leq u_2^* \leq \cdots \leq u_d^*$, to match the parameters of the blossom with the original knot sequence. Note that this is an application of the symmetry property. If the parameters $u_1^*, u_2^*, \ldots, u_j^*$ match with a subsequence $k_{i+1}, k_{i+2}, \ldots, k_{i+j}$ of the original knot sequence K, that is,

$$f(u_1^*, u_2^*, \ldots, u_d^*) = f(k_{i+1}, \ldots, k_{i+j}, u_{j+1}^*, u_{j+2}^*, \ldots, u_d^*), \tag{10}$$

and $k_{i+j} \leq u_{j+1}^* < k_{i+j+1}$, then u_{j+1}^* is expressed by the knots k_i and k_{i+j+1}, which are adjacent elements to the matched subsequence in K:

$$u_{j+1}^* = \frac{k_{i+j+1} - u_{j+1}^*}{k_{i+j+1} - k_i} k_i + \frac{u_{j+1}^* - k_i}{k_{i+j+1} - k_i} k_{i+j+1}. \tag{11}$$

Hence the blossom is transformed to the sum of two products by applying the multi-affine property:

$$f(k_{i+1}, \ldots, k_{i+j}, u_{j+1}^*, u_{j+2}^*, \ldots, u_d^*)$$

$$= \frac{k_{i+j+1} - u_{j+1}^*}{k_{i+j+1} - k_i} f(k_i, k_{i+1}, \ldots, k_{i+j}, u_{j+2}^*, \ldots, u_d^*) \tag{12}$$

$$+ \frac{u_{j+1}^* - k_i}{k_{i+j+1} - k_i} f(k_{i+1}, \ldots, k_{i+j}, k_{i+j+1}, u_{j+2}^*, \ldots, u_d^*).$$

This procedure must be applied recursively, until all blossoms appear in the transformed expression become the coefficients of the original B-spline function.

§5. Computing the B-spline Coefficients of Products of Splines

Consider two piecewise polynomials, $F(u)$ and $G(u)$, of the B-spline form and corresponding knot sequences, R and S:

$$F(u) = \sum_{i=0}^{l} N_{i,p}^{R}(u)f_i, \quad R = [\underbrace{r^0,\ldots,r^0}_{\rho_0=p+1}, \underbrace{r^1,\ldots,r^1}_{\rho_1}, \cdots, \underbrace{r^\lambda,\ldots,r^\lambda}_{\rho_\lambda=p+1}],$$

$$G(u) = \sum_{i=0}^{m} N_{i,q}^{S}(u)g_i, \quad S = [\underbrace{s^0,\ldots,s^0}_{\sigma_0=q+1}, \underbrace{s^1,\ldots,s^1}_{\sigma_1}, \cdots, \underbrace{s^\mu,\ldots,s^\mu}_{\sigma_\mu=q+1}],$$

(13)

where $s^0 = r^0$ and $s^\sigma = r^\lambda$. The piecewise blossom of $F(u)$ is $f(u_1, u_2, \ldots, u_p)$ and that of $G(u)$ is $g(u_1, u_2, \ldots, u_q)$.

When these two piecewise polynomials are multiplied, a third piecewise polynomial $H(u)$ of the following form is obtained as a result.

$$H(u) = \sum_{i=0}^{n} N_{i,p+q}^{T}(u)h_i = F(u)G(u).$$

(14)

With given B-spline coefficients of $f_0, f_1, \ldots, f_l$ and $g_0, g_1, \ldots, g_m$, it is required to compute $h_0, h_1, \ldots, h_n$ in the following manner.

First, the knot sequence R is refined to R' and, S, to S', to satisfy the condition of the dual functional property as in degree raising of B-splines [10].

$$R' = [\underbrace{r^0,\ldots,r^0}_{\rho_0+q}, \underbrace{r^1,\ldots,r^1}_{\rho_1+q}, \cdots, \underbrace{r^\lambda,\ldots,r^\lambda}_{\rho_\lambda+q}].$$

$$S' = [\underbrace{s^0,\ldots,s^0}_{\sigma_0+p}, \underbrace{s^1,\ldots,s^1}_{\sigma_1+p}, \cdots, \underbrace{s^\mu,\ldots,s^\mu}_{\sigma_\mu+p}].$$

(15)

Then the refined knot sequences R' and S' are merged into the new knot sequence T, which is a knot sequence of the B-spline function $H(u)$.

$$T = R' \cup S' = [t_0, t_1, \ldots, t_{n+p+q+1}]$$
$$= [\underbrace{t^0,\ldots,t^0}_{\tau_0=p+q+1}, \underbrace{t^1,\ldots,t^1}_{\tau_1}, \cdots, \underbrace{t^\nu,\ldots,t^\nu}_{\tau_\nu=p+q+1}].$$

(16)

The multiplicity τ_i of knot t^i is $\max(\rho_j, \sigma_k)$, when $t^i = r^j = s^k$.

The piecewise blossom $h(u_1, u_2, \ldots, u_p, u_{p+1}, \ldots, u_{p+q})$ of the function $H(u)$ is constructed by symmetrizing the product of the blossoms of $F(u)$ and $G(u)$, so that the blossom is symmetric and multi-affine:

$$h(u_1, u_2, \ldots, u_p, u_{p+1}, \ldots, u_{p+q})$$

$$= \frac{\displaystyle\sum_{(i_1, i_2, \ldots, i_p)} f(u_{i_1}, u_{i_2}, \ldots, u_{i_p}) g(\overline{u_{i_{p+1}}}, \overline{u_{i_{p+2}}}, \ldots, \overline{u_{i_{p+q}}})}{\dbinom{p+q}{p}},$$

(17)

where the summation is taken for all indices $(i_1, i_2, \ldots, i_p)$. Each index is chosen from the set $\{1, \ldots, p+q\}$, no indices are equal, and the variables $\{\overline{u_{i_{p+1}}}, \overline{u_{i_{p+2}}}, \ldots, \overline{u_{i_{p+q}}}\}$ are the difference

$$\{u_1, u_2, \ldots, u_p, u_{p+1}, \ldots, u_{p+q}\} \setminus \{u_{i_1}, u_{i_2}, \ldots, u_{i_p}\}. \tag{18}$$

Now the B-spline coefficients h_i of B-spline function $H(u)$ are computed as the values of the following blossoms:

$$
\begin{aligned}
h_0 &= h(t_1, t_2, \ldots, t_{p+q}), \\
h_1 &= h(t_2, t_3, \ldots, t_{p+q+1}), \\
&\quad \cdots \\
h_n &= h(t_{n+1}, t_{n+2}, \ldots, t_{n+p+q}).
\end{aligned}
\tag{19}
$$

This procedure is computationally expensive, and thus the naive approach is not acceptable. Therefore we use an associative structure in our implementation of this procedure to avoid evaluating the same blossom twice.

§6. B-spline Coefficients and Discrete B-splines

The discrete B-spline [9] is another tool to solve representation problems of B-spline functions. Each B-spline basis function over a knot sequence K is, in a unique way, expressed in terms of functions over a refinement K' of K:

$$N_{i,d}^K(u) = \sum_{j=0}^{m} \alpha_{i,d}^{K,K'}(j) N_{j,d}^{K'}(u), \tag{20}$$

where $\alpha_{i,d}^{K,K'}(j)$ is a discrete B-spline of degree d on K' with knots K. According to (4), (9) and (20), the new B-spline coefficients f_j' have the following relation:

$$f_j' = f(k_{j+1}', k_{j+2}', \ldots, k_{j+d}') = \sum_{i=0}^{n} \alpha_{i,d}^{K,K'}(j) f_i. \tag{21}$$

To put this more precisely, f_j' is obtained as a sum of at most d discrete B-splines [9].

As a discrete B-spline form, equivalent to a blossom, exists, Equation (17) may be rewritten as the following discrete B-spline form:

$$\frac{\displaystyle\sum_{(i_1, i_2, \ldots, i_p)} \sum_{l_1 = i_{j_1} - p}^{i_{j_1}} \alpha_{l_1, p}^{R,T}(j_1) f_{l_1} \sum_{l_2 = i_{j_2} - q}^{i_{j_2}} \alpha_{l_2, q}^{S,T}(j_2) g_{l_2}}{\dbinom{p+q}{p}}. \tag{22}$$

The knot sequence T and this rewritten form are equivalent to the formulas (3.1) and (3.2) respectively in [10].

§7. Practical Considerations

As described in [10], general degree raising of a B-spline function $F(u)$ is achieved by computing the product of $F(u)$ and constant function $G(u) = 1$. Suppose the constant function $G(u)$ and its knot sequence S are written as follows:

$$G(u) = \sum_{i=0}^{q} N_{i,q}^{S}(u), \quad S = [\underbrace{s^0, \ldots, s^0}_{\sigma_0 = q+1}, \underbrace{s^1, \ldots, s^1}_{\sigma_1 = q+1}]. \tag{23}$$

Because $s^0 = r^0$ and $s^1 = r^\lambda$, $N_{i,q}^{S}(u)$ is the Bernstein polynomial of degree q over the interval $[r^0, r^\lambda]$.

The knot sequence T for the product of $F(u)$ and $G(u)$ is obtained as

$$T = R' = [\underbrace{r^0, \ldots, r^0}_{\rho_0 + q}, \underbrace{r^1, \ldots, r^1}_{\rho_1 + q}, \cdots, \underbrace{r^\lambda, \ldots, r^\lambda}_{\rho_\lambda + q}]. \tag{24}$$

Because $g(\overline{u_{i_{p+1}}}, \overline{u_{i_{p+2}}}, \ldots, \overline{u_{i_{p+q}}}) = 1$, the blossom of the product $H(u)$ is expressed as

$$h(u_1, u_2, \ldots, u_p, u_{p+1}, \ldots, u_{p+q}) = \frac{\displaystyle\sum_{(i_1, i_2, \ldots, i_p)} f(u_{i_1}, u_{i_2}, \ldots, u_{i_p})}{\dbinom{p+q}{p}}. \tag{25}$$

For $q = 1$, this form is well known formula as degree raising by blossoming [15].

The B-spline of degree 0 is not only a constant function, but also recognized as a knot sequence. Multiplication with a B-spline function of degree 0 is a knot insertion procedure. Consider the function $G()$ of degree 0 and its knot sequence S are written as follows:

$$G(u) = \sum_{i=0}^{m} N_{i,0}^{S}(u), \quad S = [\underbrace{s^0, \ldots, s^0}_{\sigma_0}, \underbrace{s^1, \ldots, s^1}_{\sigma_1}, \cdots, \underbrace{s^\mu, \ldots, s^\mu}_{\sigma_\mu}]. \tag{26}$$

Since the continuity is not changed by multiplication with a function of degree 0, it is not required to increase the multiplicities of the knot sequence R and $T = R \cup S$.

The blossom of the product $H(u)$ is expressed as

$$h(u_1, u_2, \ldots, u_p) = \frac{\displaystyle\sum_{(i_1, i_2, \ldots, i_p)} f(u_{i_1}, u_{i_2}, \ldots, u_{i_p}) g()}{\dbinom{p}{p}} = f(u_1, u_2, \ldots, u_p). \tag{27}$$

This is merely a new B-spline coefficient for a refined knot sequence [8]. Furthermore, multiplication of B-spline functions of degree 0 and value 1, only merge their knot sequences.

§8. Multiplication of Bivariate B-spline Functions

To combine B-spline surfaces, bivariate B-spline functions must be multiplied. A bivariate B-spline function $F(u, v)$ of degree $p \times q$ is a tensor-product of univariate B-spline functions that takes the form

$$F(u, v) = \sum_{i=0}^{m^{F_1}} \sum_{j=0}^{n^{F_2}} N_{i,p}^{F_1}(u) N_{j,q}^{F_2}(v) f_{i,j}. \tag{28}$$

The blossom of this function takes the form of $f(u_1, u_2, \ldots, u_p; v_1, v_2, \ldots, v_q)$. In this blossom u_i and v_j are not interchangeable.

The blossom of the bivariate function $H(u, v)$, which is the product of bivariate function $F(u, v)$ of degree $p \times q$ and $G(u, v)$ of degree $r \times s$, is constructed analogously to the univariate B-splines:

$$h(u_1, \ldots, u_p, u_{p+1}, \ldots, u_{p+r}; v_1, \ldots, v_q, v_{q+1}, \ldots, v_{q+s})$$

$$= \frac{\displaystyle\sum_{\substack{(i_1, i_2, \ldots, i_p; \\ j_1, j_2, \ldots, j_q)}} f(u_{i_1}, \ldots, u_{i_p}; v_{j_1}, \ldots, v_{j_q}) g(\overline{u_{i_{p+1}}}, \ldots, \overline{u_{i_{p+r}}}; \overline{v_{j_{q+1}}}, \ldots, \overline{v_{j_{q+s}}})}{\dbinom{p+r}{p} \dbinom{q+s}{q}},$$

$$\tag{29}$$

where $g(u_1, u_2, \ldots, u_r; v_1, v_2, \ldots, v_s)$ is the blossom of $G(u, v)$.

A bivariate B-spline function of degree $0 \times n$ or $m \times 0$ is considered as a univariate B-spline function, since a univariate B-spline function of degree 0 is a constant.

NURBs surfaces that are the combination of NURBs surfaces, e.g., ruled surfaces created by interpolating NURBs curves with different weights linearly, and the Boolean sum of NURBs surfaces, can be constructed using multiplications.

§9. Conclusion

In this paper, a blossoming method for multiplication of two B-spline functions and its relation to the algorithm based on discrete B-splines are presented. It is also shown that degree raising and knot insertion, which make different B-spline curves and surfaces compatible, can be recognized as special cases of the multiplication.

Since the multiplication of B-splines is computationally expensive, efficiency must be considered when implementing this operation. Our implementation adopts an associative structure to avoid computing the same blossom twice.

The multiplication of splines can make different rational B-splines compatible. Therefore, the flexibility of NURBs curves and surfaces in computer aided geometric design is increased by this operation.

References

1. Barry, P. J., An introduction to blossoming, in *Knot Insertion and Deletion Algorithms for B-spline Curves and Surfaces*, R. N. Goldman and T. Lyche (eds.), SIAM, 1993, 1–10.
2. Boehm, W., Inserting new knots into a B-spline curve, Comp. Aided Design **12** (1980), 50–62.
3. de Boor, C., *A Practical Guide to Splines*, Springer-Verlag, 1978.
4. de Casteljau, P., *Shape Mathematics and CAD*, Kogan Page, 1986.
5. Cohen, E., T. Lyche and R. F. Riesinfeld, Discrete B-splines and subdivision techniques in computer aided geometric design and computer graphics, Computer Graphics and Image Processing **14** (1980), 87–111.
6. Cohen, E., T. Lyche and L. L. Schumaker, Algorithms for degree raising of splines, ACM The Transactions of Graphics **4**, 3 (1985), 171–181.
7. DeRose, T., and R. Goldman, A tutorial introduction to blossoming, in *Geometric Modeling: Methods and Applications*, H. Hagen and D. Roller (eds.), Springer-Verlag, 1991, 267–286.
8. Goldman, R. N., Blossoming and knot insertion algorithms for B-splines, Comp. Aided Geom. Design **7** (1990), 69–81.
9. Lyche, T., Discrete B-splines and conversion problem, in *Computations of Curves and Surfaces*, W. Dahmen, M. Gasca and C. A. Miccheli (eds.), Kluwer Academic Publishers, 1990, 117-134.
10. Mørken, K., Some identities for products and degree raising of splines, Constructive Approximation **7** (1991), 195-208.
11. Piegl, L. and W. Tiller, Curve and surface constructions using rational B-splines, Comp. Aided Design **19**, 9 (1987), 485-498.
12. Piegl, L. A., Rational B-spline curves and surfaces for CAD and graphics, in *State of the Art in Computer Graphics: Visualization and Modeling*, D. F. Rogers and R. A. Earnshaw (eds.), Springer-Verlag, 1991, 225–269.
13. Ramshaw, L., *Blossoming: A Connect-the-Dots Approach to Splines*, System Research Center, DEC, 1987.
14. Ramshaw, L., Béziers and B-splines as multiaffine maps, in *Theoretical Foundations of Computer Graphics and CAD*, R. A. Earnshaw (ed.), Springer-Verlag, 1988, 757–776.
15. Seidel, H.-P., Computing B-spline control points, in *Theory and Practice of Geometric Modeling*, W. Straßer and H.-P., Seidel (eds.), Springer-Verlag, 1989, 17–32.
16. Seidel, H.-P., An introduction to polar forms, IEEE Computer Graphics & Applications **13**, 1 (1993), 38–46.

Kenji Ueda
Ricoh Company, Ltd.
1-1-17, Koishikawa, Bunkyo-ku
Tokyo, 112, JAPAN
ueda@ src.ricoh.co.jp

Symmetric Tchebycheffian
B–Spline Schemes

Michael G. Wagner and Helmut Pottmann

Abstract. Tchebycheffian B–spline schemes are a natural generalization of the polynomial B–spline scheme and possess essentially all properties which are considered important for CAGD. In this paper we determine all affine Tchebycheffian spline segments with the following symmetry property: over each interval, the Tchebycheffian Bézier curve to the control points $\mathbf{b}_0 \ldots \mathbf{b}_m$ is the same as the curve to $\mathbf{b}_m \ldots \mathbf{b}_0$. The question is solved via the corresponding normal curve which turns out to be an affine W–curve. The underlying function space is the null space of a self–adjoint linear differential operator. In case of order three, one gets cubic spline curves, exponential spline segments in tension, and helix spline segments associated with the function space $\{1, t, \sin(t), \cos(t)\}$ as the only possible solutions.

§1. Introduction

The theory of Tchebycheffian B–splines (cf. [10]), is based on the theory of Tchebycheffian systems (cf. [5]). Recently, the importance of these free form schemes was shown by Lempel and Seroussi (cf. [7]), who gave an explicit derivation of general spline bases over function spaces closed under differentiation. In this paper we use a more geometric approach. Based on the normal curve of the associated function space, which was first introduced for B–splines by Seidel (cf. [11]), one can define a blossoming method and a de Casteljau algorithm for the construction of a Tchebycheffian B–spline segment (cf. [8]). First we give a brief introduction to these methods. The main part of the paper discusses symmetric Tchebycheffian splines. We present an explicit derivation of the possible spline bases and give some geometric interpretations of the achieved results.

Curves and Surfaces in Geometric Design 483
P. J. Laurent, A. Le Méhauté, and L. L. Schumaker (eds.), pp. 483–490.
Copyright © 1994 by A K PETERS, Wellesley, MA
ISBN 1-56881-039-3.

§2. Affine Tchebycheffian Bézier curves

In this paper we deal with certain affine free-form curves called *Tchebycheffian Bézier* (TB–) *curves* in an affine space A^d where the functions T_i are elements of an *extended Tchebycheff* (ET–) *space* $\mathcal{U}^{m+1}$ over a real interval I (cf. [10]):

$$\mathbf{c}(t) = \sum_{i=0}^{m} T_i(t)\mathbf{b}_i, \quad \mathbf{b}_i \in A^d, \qquad t \in I' = [c,d] \subset I. \tag{1}$$

Of course, the T_i have to sum up to 1, which means that $\mathcal{U}^{m+1}$ has to contain 1. Let $U = (1, u_1(t), \ldots, u_m(t))$ therefore be a basis in $\mathcal{U}^{m+1}$. Then $\mathbf{u}(t) = (u_1(t), \ldots, u_m(t))^T$ can be regarded as a point in affine space A^m. The point set $\mathbf{u} := \{\mathbf{u}(t) | t \in I\}$ is a parameterized curve in A^m, the so-called *normal curve* of $\mathcal{U}^{m+1}$. As shown in [8], a curve $\mathbf{u}$ is a normal curve of an ET–system if and only if it is of *geometric order* m, hence it is intersected by all hyperplanes in at most m points. Here, intersections have to be counted with multiplicities up to order m. Now $\mathbf{c}$ is an affine image of a segment of the normal curve $\mathbf{u}$ to an interval $I' = [c,d] \subset I$ using an affine transformation $\alpha : A^m \to A^d$.

The *blossom* of an affine TB–curve can be defined in terms of its normal curve $\mathbf{u}$ (cf. [11]):

$$\mathbf{b}(t_1, t_2, \ldots, t_m) := \bigcap_{i=1}^{d} \Omega^{m-l_i}(\tau_i), \tag{2}$$

where the arguments $(t_1, t_2, \ldots, t_m)$ are sorted to $(t_{i_1}, \ldots, t_{i_m})$ with

$$(t_{i_1}, t_{i_2}, \ldots, t_{i_m}) = (\underbrace{\tau_1, \ldots, \tau_1}_{l_1}, \ldots, \underbrace{\tau_d, \ldots, \tau_d}_{l_d}), \quad \tau_j < \tau_{j+1},$$

and $\Omega^{m-l_i}(\tau_i)$ denotes the osculating $(m - l_i)$–space of $\mathbf{u}$ at τ_i. Here and in the sequel we only deal with normal curves whose blossom is a map from I^m to A^m, which is the case if and only if the tangent surface of $\mathbf{u}$ intersects the hyperplane at infinity in a curve of geometric order $m - 1$. The set of osculating hyperplanes of $\mathbf{u}$ forms a curve in dual projective space, the so-called *dual curve* of $\mathbf{u}$. We also have to assume that the dual curve of $\mathbf{u}$ is a C^m curve. As shown in [8], the blossom $\mathbf{b}$ of $\mathbf{u}$ is symmetric, injective in each argument, and continuous with $\mathbf{b}(\underbrace{t, \ldots, t}_{m}) = \mathbf{u}(t)$.

We can also define the blossom, again denoted by $\mathbf{b}$, for a curve $\mathbf{c} \subset A^d$ which is the image of a segment $\bar{\mathbf{c}} \subset \mathbf{u}$ under an affine map $\alpha : A^m \to A^d$. We just define $\mathbf{b}$ to be the composition of the blossom of $\bar{\mathbf{c}}$ with α. From a Bézier–like representation of $\mathbf{c}$ we expect that the k–th derivative at $t = c$ and $t = d$ only depends on $\mathbf{b}_0, \ldots, \mathbf{b}_k$ and $\mathbf{b}_m, \ldots, \mathbf{b}_{m-k}$, respectively. This leads to

Lemma 1. *The Bézier points* $\mathbf{b}_i$ *of the segment* $\bar{\mathbf{c}}$ *of the normal curve* $\mathbf{u}$ *to the parameter interval* $I' = [c, d]$ *are the vertices of the osculating simplex of* $\mathbf{u}$ *to the end points* $\mathbf{u}(c)$ *and* $\mathbf{u}(d)$. *For any TB–curve* $\mathbf{c}$ *defined over* $[c, d]$ *the Bézier points are:*

$$\mathbf{b}_i = \mathbf{b}(\underbrace{c, \ldots, c}_{m-i}, \underbrace{d, \ldots, d}_{i}), \quad i = 0, \ldots, m. \tag{3}$$

A curve $\mathbf{c} \subset A^d$, represented in the form (1) with control points (3) is called an affine TB–curve. Then (3) shows:

Lemma 2. *(de Casteljau algorithm for affine TB–curves). Consider the following triangular array of points to a parameter value* $t \in I'$:

$$\mathbf{b}_i^k(t) := \mathbf{b}(\underbrace{c, \ldots, c}_{m-i-k}, \underbrace{t, \ldots, t}_{k}, \underbrace{d, \ldots, d}_{i}), \quad k = 0, \ldots, m, \quad i = 0, \ldots, m - k. \tag{4}$$

Then the point $\mathbf{b}_0^m(t)$ is the point $\mathbf{c}(t)$. Moreover, the segment $\mathbf{c}^-$ of $\mathbf{c}$ corresponding to the left interval $I^- = [c, t]$ has the Bézier points $\mathbf{b}_0^k(t)$, $k = 0, \ldots, m$. Analogously, the points $\mathbf{b}_i^{m-i}(t)$ are the Bézier points of the right segment $\mathbf{c}^+$ corresponding to the interval $I^+ = [t, d]$.

Proofs can be found in [8]. As the de Casteljau algorithm turns out to be a corner cutting procedure, affine TB–curves possess the *variation diminishing property* (hence also the convex hull property) with respect to their Bézier polygon $\mathbf{b}_0 \ldots \mathbf{b}_m$.

§3. Symmetric TB–curves

Definition 1. *Consider an affine Tchebycheffian Bézier curve* $\mathbf{c}$ *over* I *with the following symmetry property: over each interval* $I' \subseteq I$, *the Tchebycheffian Bézier curve corresponding to the control points* $\mathbf{b}_0 \ldots \mathbf{b}_m$ *is the same as the curve corresponding to* $\mathbf{b}_m \ldots \mathbf{b}_0$. *Then* $\mathbf{c}$ *is called a symmetric TB–curve.*

Symmetric Tchebycheffian B–spline curves are those whose segments are formed of symmetric TB–curves (cf. [8,9]). Therefore we restrict our investigations to the derivation of the possible symmetric TB–curves.

Theorem 1. *An affine TB–curve scheme of order* m *is symmetric if and only if its normal curve* $\mathbf{u}$ *can be parameterized as a normal curve of an ET–space which contains 1, and is the null space of a self–adjoint linear differential operator* L *of order* $m + 1$ *with constant coefficients.*

Proof: Consider a normal curve $\mathbf{u}(t), t \in [c, d]$, of an affine TB–scheme of order m. As any TB–curve of this scheme is the affine image of $\mathbf{u}$, the associated free–form scheme is symmetric only if $\mathbf{u}$ is symmetric and therefore preserved under a certain affine transformation ρ with

$$\mathbf{b}_i = \rho(\mathbf{b}_{m-i}), \quad i = 0, \ldots, m. \tag{5}$$

The midpoints of the straight line segments $\mathbf{b}_i\mathbf{b}_{m-i}$ span the fixed space of ρ which is of dimension $[m/2]$ where $[\ ]$ denotes the Gaussian brackets. In the sequel we call such an affine transformation an *affine reflection* of the segment from $\mathbf{u}(c)$ to $\mathbf{u}(d)$. It is possible to choose a special parameterization of $\mathbf{u}$ which satisfies

$$\mathbf{u}(-a+t) = \rho\big(\mathbf{u}(a-t)\big), \quad t \in I = [-a, a], \quad a = (d-c)/2. \tag{6}$$

Here the parameter is again denoted by t. After one subdivision at the parameter value $\hat{t}$, the two segments of $\mathbf{u}$ must have the same properties as $\mathbf{u}$. Hence, there exists an affine reflection $\hat{\rho}$ of the segment from $\mathbf{u}(-a)$ to $\mathbf{u}(\hat{t})$. We now define

$$\mathbf{u}(t^*) := \rho\big(\hat{\rho}(\mathbf{u}(t))\big). \tag{7}$$

Note that if we choose two parameter values t_0 and t_1 with $t_0 < t_1 (t_0, t_1 \in I)$ we get $t_0^* < t_1^*$ as ρ and $\hat{\rho}$ are affine transformations and $\mathbf{u}$ is parameterized according to (6). This forces

$$t^* = t + \phi(t), \quad \frac{dt^*}{dt} > 0, \quad \text{i.e.} \quad 1 + \phi'(t) > 0. \tag{8}$$

Using the parameter transformation $t = t(u)$ with

$$t'(u) = \frac{1}{1 + \phi'(t(u))}, \tag{9}$$

equation (8) can be written as

$$t^* = u_0 + u, \tag{10}$$

which is a translation on the u–axis. Note that the differential equation (9) has a unique solution due to (8), and that u_0 converges to 0 if $\hat{t}$ converges to a. Now we are able to state the equations of the affine transformation $(\rho \circ \hat{\rho})$ which depends on u_0:

$$\mathbf{u}(t^*) = \mathbf{u}(u_0 + u) = \mathbf{m}(u_0) + M(u_0)\mathbf{u}(u), \tag{11}$$

where $M(u_0)$ is an (m, m)–matrix. This yields

$$\lim_{\hat{t} \to a} \frac{\mathbf{u}(u_0 + u) - \mathbf{u}(u)}{u_0} = \lim_{u_0 \to 0} \frac{\mathbf{u}(u_0 + u) - \mathbf{u}(u)}{u_0} =$$

$$= \lim_{u_0 \to 0} \frac{\mathbf{m}(u_0) - \mathbf{m}(0)}{u_0} + \lim_{u_0 \to 0} \left(\frac{M(u_0) - M(0)}{u_0}\right)\mathbf{u}(u)$$

$$\dot{\mathbf{u}}(u) = \dot{\mathbf{m}}(0) + \dot{M}(0)\mathbf{u}(u).$$

We see that the coordinate functions of $\mathbf{u}$ lie in the null space $\mathcal{L}$ of a linear differential operator L with constant coefficients

$$L = D^{m+1} + \sum_{i=1}^{m} a_i D^i, \quad a_i = \text{const} \in \mathbb{R}, \tag{12}$$

using the standard notation $D := d/dt$.

u has to be preserved under the affine reflection ρ, i.e., a function $f(u)$ is a solution of $Lf = 0$ if and only if $f(-u)$ is a solution. Therefore, $f(u)$ has to lie in the null space $\mathcal{L}^*$ of the operator

$$L^* = (-1)^{m+1} D^{m+1} + \sum_{i=1}^{m} (-1)^i a_i D^i, \quad a_i = \text{const} \in \mathbb{R}, \qquad (13)$$

which is the so–called *adjoint operator* of L (cf. [10]). Hence, $L = L^*$ has to be a self–adjoint differential operator. Note that $\mathcal{L}^*$ contains the coordinate functions of the dual curve of **u**. This condition is sufficient as **u** is preserved under a continuous one parameter group of affine transformations due to (11) (**u** is a so–called affine W–curve). Any affine reflection of any segment of **u** can therefore be reduced to an affine transformation of this group and the reflection ρ.

Finally, we see that the possible solution spaces are null spaces of a differential operator of the form

$$L = \begin{cases} D^{m+1} + a_{m-1}D^{m-1} + a_{m-3}D^{m-3} + \ldots + a_1 D & \text{if } m \text{ is even,} \\ D^{m+1} + a_{m-1}D^{m-1} + a_{m-3}D^{m-3} + \ldots + a_2 D^2 & \text{if } m \text{ is odd,} \end{cases} \qquad (14)$$

$a_i = \text{const} \in \mathbb{R}$, and vice versa. ∎

Remark 1. The associated spline bases contain the following groups of functions:

- $\{t^k\}, \quad k = 0,\ldots,l, \quad l > 0$
- $\{t^k \sin(\beta t), t^k \cos(\beta t)\}, \quad \beta \neq 0, \quad k = 0,\ldots,l, \quad l \geq 0$
- $\{t^k \sinh(\beta t), t^k \cosh(\beta t)\}, \quad \beta \neq 0, \quad k = 0,\ldots,l, \quad l \geq 0$
- $\{t^k e^{\rho t} \sin(\beta t), t^k e^{\rho t} \cos(\beta t), t^k e^{-\rho t} \sin(\beta t), t^k e^{-\rho t} \cos(\beta t)\}, \quad \rho, \beta \neq 0,$
  $k = 0,\ldots,l, \quad l \geq 0$

Remark 2. The general determination of the largest interval I over which the obtained function spaces are ET–spaces remains open. If the resulting spline basis contains no sine or cosine functions, we get $I = \mathbb{R}$. Besides, I is only known for special cases (cf. [5]).

To get a more geometric interpretation of Theorem 1, we look at the projective closure P^m of the affine space A^m which is generated by the points of A^m and the points at infinity. The hyperplanes of P^m form the so–called *dual projective space* $(P^m)^*$ of P^m. A bijective linear mapping from P^m onto $(P^m)^*$ is called a *projective correlation*. The following property of the resulting normal curves is already well–known for rational curves (cf. [1]).

Corollary 1. *Consider the normal curve* $\mathbf{u} \subset A^m$ *of an affine TB–scheme of order* m *to the interval* I. *Further let* P^m *be the projective closure of* A^m, *and* $(P^m)^*$ *the dual projective space of* P^m. *Then* **u** *is a normal curve of a* *symmetric TB–scheme if and only if* **u** *is an affine W–curve, and there exists a projective correlation* $\kappa : P^m \to (P^m)^*$ *that maps the point* **u**(t) *onto the osculating hyperplane of* **u** *at* **u**(t) *for all* $t \in I$.

Proof: Let **u**$^*(t)$ be the dual curve of **u**(t). The coordinate functions of **u**$^*(t)$ then form a basis in $\mathcal{L}^*$. As $L = L^*$ is self–adjoint, there exists a bijective linear mapping with **u**$(t) \mapsto$ **u**$^*(t)$ for all $t \in I$ and vice versa. ■

As any affine reflection of any segment of **u** preserves the osculating simplex $\mathcal{B}$ to its endpoints, it clearly preserves the absolute value of the volume of $\mathcal{B}$, and **u** has to be point path of a one parameter group of equiaffine transformations. Let

$$s = \int_{t_0}^{t_1} \det \left(\dot{\mathbf{u}}(t), \ddot{\mathbf{u}}(t), \ldots, \mathbf{u}^{(m)}(t) \right)^{\frac{2}{m(m+1)}} dt$$

denote the affine arc length, and $\mathbf{u}' = d\mathbf{u}/ds$. An affine arc length parameterization of **u** implies

$$\det(\mathbf{u}', \mathbf{u}'', \ldots, \mathbf{u}^{(m)}) = 1,$$
$$\det(\mathbf{u}', \mathbf{u}'', \ldots, \mathbf{u}^{(m-1)}, \mathbf{u}^{(m+1)}) = 0,$$

and furthermore

$$\mathbf{u}^{(m+1)} + \kappa_1 \mathbf{u}^{(m-1)} + \kappa_2 \mathbf{u}^{(m-2)} + \ldots + \kappa_{m-1} \mathbf{u}' = \mathbf{o}. \tag{15}$$

Here the coefficients $\kappa_1, \ldots \kappa_{m-1}$ are the so–called *affine curvatures* of **u** (cf. [2]).

Theorem 2. *An affine TB–curve scheme of order m is symmetric if and only if its normal curve* **u** *has constant affine curvatures* $\kappa_1, \ldots, \kappa_{m-1}$ *with* $\kappa_i = 0$ *if i is even.*

Proof: Let **u**(t) be the normal curve of a symmetric TB–curve scheme. As **u**$(t) =$ **m**$(t) + M(t)$**u**(0) is point path of a one parameter group of equiaffine transformations,

$$\det \left(\dot{\mathbf{u}}(t), \ddot{\mathbf{u}}(t), \ldots, \mathbf{u}^{(m)}(t) \right) = c = \text{const}, \quad \forall t \in \mathbb{R}. \tag{16}$$

Hence, the parameter t is proportional to the affine arc length s. Using L from (14) with $ds/dt = c^{2/(m(m+1))} =: k$, **u**$(t)$ satisfies

$$L\mathbf{u}(s) = \mathbf{o},$$

which yields the affine curvatures κ_i of **u**:

$$\kappa_i = a_{m-i} k^{-i-1}, \quad i = 1, \ldots, m - 1. \tag{17}$$

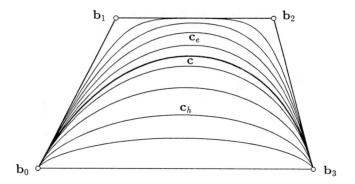

Fig. 1. Symmetric TB–curves of order three.

The κ_i are constant with $\kappa_i = 0$ if i is even.

On the other hand, let $\mathbf{u}(s)$ be normal curve of an affine TB–scheme, parameterized with respect to the affine arc length s, which satisfies

$$\mathbf{u}^{(m+1)} + \sum_{i=1}^{m-1} \kappa_i \mathbf{u}^{(m-i)} = \mathbf{o}, \quad \kappa_i = \text{const} \in \mathbb{R}, \kappa_i = 0 \text{ for } i \text{ even.} \qquad (18)$$

The coordinate functions of $\mathbf{u}$ lie in the null space $\mathcal{K}$ of a self–adjoint linear differential operator with constant coefficients

$$K = \begin{cases} D^{m+1} + \kappa_1 D^{m-1} + \ldots + \kappa_{m-3} D^3 + \kappa_{m-1} D & \text{if } m \text{ is even,} \\ D^{m+1} + \kappa_1 D^{m-1} + \ldots + \kappa_{m-4} D^4 + \kappa_{m-2} D^2 & \text{if } m \text{ is odd.} \end{cases}$$

As $\mathcal{K}$ must be an ET–space over I which contains 1, the TB–scheme is symmetric by Theorem 1. ∎

Example 1. There exist three types of symmetric TB–spline schemes of *second order*, all of them describing conics:

1) $\{1, t, t^2\}$
2) $\{1, \sinh(t), \cosh(t)\}$
3) $\{1, \sin(t), \cos(t)\}$

Example 2. In case of *order three* one gets the following types:

1) $\{1, t, t^2, t^3\}$
2) $\{1, t, \sinh(t), \cosh(t)\}$ (exponential splines in tension, cf. [6]) The shape only depends on the length of the interval I. As for all curve schemes

under consideration, the limit shape for $|I| \to 0$ is the Bézier curve **c** to the given control polygon (cf. [8]). For $|I| \to \infty$ the exponential curves $\mathbf{c}_e$ converge to the control polygon (Fig. 1).

3) $\{1, t, \sin(t), \cos(t)\}$ (helix–splines) Here, the length of the interval I has to be less than 2π. The scheme provides exact representations of straight lines, circles and euclidean helix curves in an arc length parameterization. For $|I| \to 2\pi$ the helix spline segments $\mathbf{c}_h$ converge to the straight line segment $\mathbf{b}_0 \mathbf{b}_3$ (Fig. 1). The corresponding tensor product surfaces contain helicoidal surfaces, surfaces of revolution and patches on all types of quadrics. A detailed discussion of this free–form scheme can be found in [9].

The normal curves are exactly the paths of one parameter groups of motions in 3–dimensional affine Cayley–Klein spaces (cf. [3]).

References

1. Clifford, W. K., On the classification of loci, Phil. Transactions of the Royal Society (1878), 663–681.
2. Blaschke, W., *Vorlesungen über Differentialgeometrie*, Chelsea Publishing Company, New York, 1967.
3. Giering, O., *Vorlesungen über höhere Geometrie*, Vieweg, Braunschweig/ Wiesbaden, 1982.
4. Goldman, R. and Heath, D. C., Linear subdivision is strictly a polynomial phenomenon, Computer-Aided Geom. Design **1** (1984), 269-278.
5. Karlin, S. and Studden, W. J., *Tchebycheff Systems. With Applications in Analysis and Statistics*, Wiley–Interscience, New York, 1966.
6. Koch, P. E. and Lyche, T., Exponential B–splines in tension, in *Approximation Theory VI*, C. K. Chui, L. L. Schumaker and J. D. Ward (eds.), Academic Press, New York, 1989, 361–364.
7. Lempel, A. and Seroussi, G., Systematic derivation of spline bases, Comp. Aided Geom. Design **9** (1992), 349-363.
8. Pottmann, H., The geometry of Tchebycheffian splines, Computer-Aided Geom. Design **10** (1993), 181–210.
9. Pottmann, H. and Wagner, M. G., Helix splines as an example of affine Tchebycheffian splines, Adv. in Comp. Math., to appear.
10. Schumaker, L. L., *Spline Functions: Basic Theory*, Wiley, New York, 1981.
11. Seidel, H. P., New algorithms and techniques for computing with geometrically continuous spline curves of arbitrary degree, Math. Modelling and Num. Analysis **26** (1992), 149–176.

Michael G. Wagner, Helmut Pottmann
Institute of Geometry
Technical University of Vienna
Vienna, AUSTRIA
wagner@ egmvs2.una.ac.at, pottmann@ egmvs2.una.ac.at